Springer Collected Works in Mathematics

For further volumes:
http://www.springer.com/series/11104

Ivan Matveevič Vinogradov
14. 9. 1891–20. 3. 1983

Ivan Matveevič Vinogradov

Selected Works

Editors

L. D. Faddeev · R. V. Gamkrelidze · A. A. Karatsuba
K. K. Mardzhanishvili · E. F. Mishchenko

Prepared by the Steklov Mathematical Institute
of the Academy of Sciences of the USSR
on the Occasion of his Ninetieth Birthday

Reprint of the 1985 Edition

Springer

Author
Ivan Matveevič Vinogradov (1891 – 1983)
Steklov Mathematical Institute
Moscow
Russia

Editors
L. D. Faddeev
R. V. Gamkrelidze
A. A. Karatsuba
K. K. Mardzhanishvili (1903 – 1981)
E. F. Mishchenko (1922 – 2010)
Steklov Mathematical Institute
Moscow and St. Petersburg
Russia

Translated by P. S. V. Naidu
Editor of the Translation: Yu. A. Bakhturin

ISSN 2194-9875
ISBN 978-3-642-55380-6 (Softcover)
978-3-540-12788-8 (Hardcover)
DOI 10.1007/978-3-642-15086-9
Springer Heidelberg New York Dordrecht London

Library of Congress Control Number: 2012954381

Printed on acid-free paper

Springer is part of Springer Science+Business Media (www.springer.com)

Глава 7.

Суммы по простым числам.

Метод, примененный в главе 4 к разысканию верхней границы модуля тригонометрической суммы T_m, распространенной на все целые числа интервала $0 < x \leq P$, пригоден и для довольно широкого класса сумм, распространенных на некоторую часть этих целых чисел. Прекрасным примером может служить сумма

$$T'_m = \sum_{p \leq P} e^{2\pi i m f(p)}$$

сходная с суммою T_m, но распространенная лишь на простые числа.

В этой главе выводится верхняя граница для модуля суммы T'_m. В отношении точности эта граница принципиально мало отличается от той, которая дана в главе 4 для модуля суммы T_m.

Обозначения. При выводе ближайших лемм мы будем пользоваться следующими обозначениями.

Предполагаем, что n — постоянное, $n \geq 5$,

$$\tau = 2[n^2(2\ln n + \ln\ln n + 4)], \quad \rho = \frac{1}{20n^2(2\ln n + \ln\ln n + 4)}.$$

Координаты точки $(\alpha_n, \ldots, \alpha_1)$ n-мерного пространства мы будем представлять в виде

$$\alpha_s = \frac{a_s}{q_s} + z_s; \quad (a_s, q_s) = 1, \quad 0 < q_s,$$

где для q_s и для $|z_s|$ будут указываться те, или иные верхние границы. Например, хорошо известно, что при $\tau_s \geq 1$ число α_s можно представить в указанном виде с условиями

$$q_s \leq \tau_s, \quad |z_s| < \frac{1}{q_s \tau_s}.$$

Символом Q будем обозначать общее наименьшее кратное чисел $q_n, \ldots, q_1$, а символом Q_0 — общее наименьшее кратное чисел $q_n, \ldots, q_2$. Число z_s будем представлять в виде $z_s = \delta_s P^{-s}$, причем примем обозначение $\delta_0 = \max(|\delta_n|, \ldots, |\delta_1|)$.

Будем считать, что P_0 превосходит 1 и выбрано настолько большим, что при $P \geq P_0$ выполняются все дальнейшие неравенства, содержащие P.

Буквою m будем обозначать целое положительное число, для которого будет указываться та, или иная верхняя граница.

Мы будем рассматривать сумму вида

$$S = \sum_y \sum_u \sum_v \sum_x e^{2\pi i m f(y_1 u v x_1)},$$

где каждое из переменных y_1, u, v, x_1 пробегает свою возрастающую последовательность положительных чисел, взаимно простых с Q, причем v пробегает значения взаимно простые с каждым значением x_1, исключаются слагаемые $(u, v) > 1$ и суммирование распространяется лишь на область

$$C' P^{0{,}245} < y_1 u \leq C'' P^{0{,}5}; \quad y_1 u v x_1 \leq P.$$

Очевидно, сумму S можно представить в виде

$$S = \sum \mu(t) S^{(t)}, \quad S^{(t)} = \sum_{y_1} \sum_{u_1} \sum_{v_1} \sum_{x_1} e^{2\pi i m f(t^2 y_1 u_1 v_1 x_1)} \qquad (1)$$

где t пробегает числа, делящие одновременно по меньшей мере одно значение u и по меньшей мере одно значение v. А в сумме $S^{(t)}$ u_1 и v_1 пробегают частные от деления на t значений u и v кратных t, причем суммирование распространяется на область

$$C' P^{0{,}245} < t y_1 u_1 \leq C'' P^{0{,}5}, \quad t^2 y_1 u_1 v_1 x_1 \leq P.$$

Далее, полагая $y_1 u_1 = y$, $v_1 x_1 = x$ и обозначая число решений уравнения $y_1 u_1 = y$ символом $\psi(y)$, получим

$$S^{(t)} = \sum_y \sum_x \psi(y) e^{2\pi i m f(t^2 y x)}$$

где y и x независимо друг от друга пробегают некоторые возрастающие последовательности натуральных чисел, взаимно простых с Q, причем суммирование распространяется на область

$$C' P^{0{,}245} t^{-1} < y \leq C'' P^{0{,}5} t^{-1}. \qquad (2)$$

В частности нам встретятся и вырожденные суммы S, где u принимает единственное значение $u = 1$. Тогда t принимает единственное значение $t = 1$ и равенство (1) обращается в такое

$$S = S^{(1)}.$$

From the handwritten manuscript of the book:
The Method of Trigonometric Sums in Number Theory
Second edition, Moscow, Nauka 1980

Preface

Springer-Verlag has invited me to bring out my *Selected Works*. Being aware that Springer-Verlag enjoys high esteem in the scientific world as a reputed publisher, I have willingly accepted the offer.

Immediately, I was faced with two problems. The first was that of acquainting the reader with the important stages in my scientific activities. For this purpose, I have included in the *Selected Works* certain of my early works that have greatly influenced my later studies. For the same reason, I have also included in the book those works that contain the first, crude versions of the proofs for many of my basic theorems.

The second problem was that of giving the reader the best possible opportunity to familiarize himself with the most important results and to learn to use my method. For this reason I have included the later improved versions of the proofs for my basic results, as well as the monographs *The Method of Trigonometric Sums in Number Theory* (Second Edition) and *Special Variants of the Method of Trigonometric Sums.*

I. M. Vinogradov

Editors' Note. The English edition of Vinogradov's *Selected Works* includes – compared to the Russian edition – only the main articles selected by the author. In addition to this selection, the English translation of the two above mentioned books are included in this selection.

Contents

Section I

Section II

Section III

Section I

A New Method of Deriving Asymptotic Expressions for Arithmetic Functions

Izvestiya Rossiiskoi Akademy Nauk, Ser. 6, **16** (1917) 1347–1378.

In searching for asymptotic expressions of arithmetic functions, the main difficulty lies in estimating an upper bound for the error. This situation is explained by the imperfection of the methods used, and therefore a need arises for improving the latter. In his eminent paper Voronoi [1] has developed a new method which can be applied to several asymptotic problems and which gives an error bound far smaller than that obtained by the Dirichlet classical method. Thus, as applied to the sum

$$\sum_{x=1}^{x \leqq a} \left[\frac{a}{x}\right]$$

Voronoi gives an upper error bound of the order of $\sqrt[3]{a} \ln a$. Applying the Voronoi method to the alternating series

$$\left[\frac{a}{1}\right] - \left[\frac{a}{3}\right] + \left[\frac{a}{5}\right] - \cdots$$

Sierpinski [2] derived an asymptotic expression with an error of the order of $\sqrt[3]{a}$, in place of the error of the order of $\sqrt{a}$ known till then. Despite its ingenuity and profundity, the Voronoi method is rather complicated which hinders its practical applications. In this paper we shall develop a new method that is far simpler than the Voronoi technique but gives almost the same upper bound for the error. Thus, for the Voronoi case the upper error bound obtained in our method is $\sqrt[3]{a} (\ln a)^5$, while for the Sierpinski case it is of the order of $\sqrt[3]{a} (\ln a)^{2/3}$.

Since a large majority of the problems on asymptotic expressions for arithmetic functions reduces to looking for asymptotic expressions for a sum of the type

$$\sum_{x>Q}^{x \leqq R} [f(x)],$$

which can easily be expressed in terms of the sum:

$$S = \sum_{x>Q}^{x \leqq R} \{f(x)\},$$

---

1 Voronoi, G.: Sur un problème du calcul des fonctions asymptotiques. J. reine angew. Math. **126** (1903) 241–282.

2 Sierpinski, W.: On a problem on asymptotic functions. [In Polish]. Prace mat.-fiz. **17** (1906) 77–118.

where

$$\{f(x)\} = f(x) - [f(x)],$$

we are motivated to find an asymptotic expression for the sum S under as general assumptions as possible on the function $f(x)$ and the numbers Q and R. Here we have derived two fundamental formulae of a quite general nature. Making certain special assumptions about the function $f(x)$ and the numbers Q and R, we can obtain several particular corollaries from these fundamental formulae. We shall confine ourselves to the most interesting examples. Amongst these examples, special mention should be made of the asymptotic expression for the sum

$$h(-1) + h(-2) + \ldots + h(-m) = \frac{4\pi}{21 \sum_{k=1}^{\infty} \frac{1}{k^3}} m^{3/2} - \frac{2}{\pi^2} m + O(m^{5/6} (\ln m)^{2/3})$$

where $h(-n)$ denotes the number of classes of proper primitive forms of discriminant $(-n)$. This expression has also been given by Gauss [1] but in a different form without the error bound. It is demonstrated in full, as far as we know, for the first time in this paper.

§1. Basic Lemmas

Our method is based on the following three lemmas which we shall, for the sake of convenience in further presentation, state in one section.

Lemma 1. If a is some real number, then for any $\tau > 1$, the system of inequalities

$$-\frac{1}{\tau} < aX - Y < \frac{1}{\tau}$$

$$0 < X \leqq \tau$$

is satisfied by coprime integers X and Y.

This is a well known theorem and its proof can be found, say, in Dirichlet's book [2].

Lemma 2. Let

$$X_1, X_2, \ldots, X_n \tag{1}$$

be a finite sequence whose terms are positive integers not exceeding a finite number N. Then, except for the last less than N terms, all the remaining terms can, without disturbing the order, be divided into subsequences satisfying the following condition: the number m of terms in each subsequence

$$X_{\alpha+1}, X_{\alpha+2}, \ldots, X_{\alpha+m}$$

is equal to the largest number in it.

---

1 Gauss: Disquisitiones arithmeticae, art. 302.

2 Dirichlet, L.: Vorlesungen über Zahlentheorie, p. 373.

Proof. If $n < N$, the lemma is self-evident.
In the contrary case, consider a subsequence consisting of the first X_1 terms of (1):

$$X_1, X_2, \ldots, X_{X_1}. \tag{2}$$

Let $X^{(2)}$ be the largest term in this subsequence. Then only two cases are possible: either $X^{(2)} = X_1$, or $X^{(2)} > X_1$. If $X^{(2)} = X_1$, the subsequence (2) satisfies the conditions of the lemma, and the process of dividing the first subsequence is concluded. If, however, $X^{(2)} > X_1$, consider a subsequence consisting of the first $X^{(2)}$ terms in (1):

$$X_1, X_2, \ldots, X_{X^{(2)}}. \tag{3}$$

Let $X^{(3)}$ denote the largest term in this subsequence. Then only two cases are possible: either $X^{(3)} = X^{(2)}$ or $X^{(3)} > X^{(2)}$. If $X^{(3)} = X^{(2)}$, the subsequence (3) satisfies the conditions of the lemma and the process of dividing the first subsequence is concluded. If, however, $X^{(3)} > X^{(2)}$, consider a subsequence consisting of the first $X^{(3)}$ terms of (1):

$$X_1, X_2, \ldots, X_{X^{(3)}}$$

and repeat the argument. Thus, we consider newer and newer subsequences. The numbers of terms in these subsequences form an increasing series:

$$X_1 < X^{(2)} < X^{(3)} < \ldots,$$

whose terms are positive integers not exceeding the finite number N. This series cannot have more than N terms. Consequently, after considering not more than N subsequences, we necessarily arrive at a subsequence satisfying the conditions of the lemma.

If after selecting the first subsequence, the sequence (1) contains less than N terms, the lemma is proved.

In the contrary case, we can select a second, third, ..., etc. subsequence satisfying the conditions of the lemma until there remain less than N terms, which will certainly happen after selecting a finite number of subsequences, because the number of terms in (1) is finite. Thus, we have demonstrated the truth of the lemma.

Note. Let

$$X_{\alpha+1}, X_{\alpha+2}, \ldots, X_{\alpha+m}$$

be one of the subsequences stated in the lemma. Then, from the conditions:

$$X_i \leqq m; \qquad i = \alpha+1, \alpha+2, \ldots, \alpha+m$$

it follows that

$$1 \leqq \sum_{i=\alpha+1}^{i=\alpha+m} \frac{1}{X_i}.$$

Writing similar inequalities for all other subsequences and then summing their left-hand and right-hand sides separately, we find that the number of subsequences does not exceed the sum

$$\sum \frac{1}{X},$$

taken over all the terms of the sequence (1).

Lemma 3. Let $\{a\}$ denote the fractional part of a number a, i.e. the difference

$$a - [a].$$

If X is a positive integer and Y is coprime to X, M an integer and $P(z)$ is a function of z such that the difference between its maximum and minimum for z from the sequence

$$z = M + 1,\ M + 2,\ \ldots,\ M + X$$

does not exceed a given number C, then

$$-C - \frac{1}{2} \leqq \sum_{z=M+1}^{z=M+X} \left\{ \frac{Yz + P(z)}{X} \right\} - \frac{1}{2} X \leqq C + \frac{1}{2}.$$

Proof. For the sake of brevity, let S denote the sum

$$\sum_{z=M+1}^{z=M+X} \left\{ \frac{Yz + P(z)}{X} \right\}$$

and consider two cases:

Case 1°. $C + \frac{1}{2} \geqq \frac{1}{2} X$.

Since the sum S contains in all X terms, each of which is $\geqq 0$ and < 1, we have

$$0 \leqq S < X.$$

Hence we easily obtain

$$-C - \tfrac{1}{2} \leqq S - \tfrac{1}{2} X < C + \tfrac{1}{2}.$$

Case 2°. $C + \frac{1}{2} < \frac{1}{2} X$.

Let P denote the minimum of $P(z)$ for z from the sequence

$$z = M + 1,\ M + 2,\ \ldots,\ M + X.$$

Now consider a function $\Phi(z)$ defined as

$$P(z) = P + \Phi(z),$$

which, for all $z = M + 1, \ldots, M + X$, obviously satisfies the condition:

$$0 \leqq \Phi(z) \leqq C.$$

Further, introducing the notation

$$K = [P]; \quad \varepsilon = \{P\},$$

we can represent the sum S as

$$S = \sum_{z=M+1}^{z=M+X} \left\{ \frac{Yz + K + \varepsilon + \Phi(z)}{X} \right\}.$$

Since the value of $\{a\}$ does not change when an integer is added to or subtracted from a, we can replace each number

$$Yz + K$$

by its least residue u modulo X. Since X and Y are coprime, u will run through the values

$$0, 1, 2, \ldots, X - 1$$

in some order when z runs through the values

$$z = M+1,\ M+2,\ \ldots,\ M+X,$$

and the sum S, after the substitution mentioned above, takes the form:

$$S = \sum_{u=0}^{u=X-1} \left\{ \frac{u+\varepsilon+\psi(u)}{X} \right\},$$

where for all $u = 0, 1, 2, \ldots, (X-1)$, the function $\psi(u)$ satisfies the condition:

$$0 \leqq \psi(u) \leqq C. \tag{4}$$

Now divide all the terms of (4) into two groups: put all the terms satisfying the condition

$$0 \leqq u < X - C - \varepsilon,$$

into the first group, and all the remaining terms satisfying the condition

$$X - C - \varepsilon \leqq u < X,$$

into the second group. It is easy to verify that for the terms of the first group the following inequalities

$$0 \leqq \frac{u+\varepsilon+\psi(u)}{X} < 1,$$

are satisfied, while for the terms of the second group we have

$$0 < \frac{u+\varepsilon+\psi(u)}{X} < 2.$$

Hence, putting

$$\left\{ \frac{u+\varepsilon+\psi(u)}{X} \right\} = \frac{u+\varepsilon+\psi(u)}{X} + \eta(u),$$

for the terms of the first group, we always have

$$\eta(u) = 0,$$

while for the terms of the second group, we find that

$$\eta(u) = 0 \quad \text{or} \quad -1.$$

But the second group contains not more than $C + \varepsilon$ terms. Hence, on collecting the terms of the sum S that correspond to both the groups, we obtain

$$\sum_{u=0}^{u=X-1} \frac{u+\varepsilon+\psi(u)}{X} - C - \varepsilon \leqq S \leqq \sum_{u=0}^{u=X-1} \frac{u+\varepsilon+\psi(u)}{X}.$$

These inequalities imply

$$\tfrac{1}{2}X - \tfrac{1}{2} - C \leqq S \leqq \tfrac{1}{2}X - \tfrac{1}{2} + \varepsilon + C,$$

which, along with the obvious inequality $-\frac{1}{2} + \varepsilon < \frac{1}{2}$, gives the final result:

$$-C - \tfrac{1}{2} \leqq S - \tfrac{1}{2}X \leqq C + \tfrac{1}{2}.$$

§2. First Fundamental Formula

Let the function $f(x)$ and two numbers Q, R be dependent on a parameter A which may take any value greater than a given positive number A_0. Furthermore, assume that as A increases to infinity, the difference $(R-Q)$ also increases to infinity. Under these conditions we can look for an asymptotic expression for the sum

$$S = \sum_{x>Q}^{x \leqq R} \{f(x)\}$$

as A increases to infinity.

Using the lemmas proved in the previous Section, we shall solve a particular case of this problem to which certain other more general cases can be reduced. That is, for any $A > A_0$, in the interval

$$Q \leqq x \leqq R$$

let the function $f(x)$ have a second derivative $f''(x)$ of the same sign and of magnitude within the limits

$$\frac{1}{kA} \quad \text{and} \quad \frac{1}{A},$$

where $k \geqq 1$ is a constant. Then, if

$$\lim_{A=\infty} \frac{(A \ln A)^{2/3}}{R-Q} = 0,$$

an asymptotic expression can be found for the sum S by means of the

First Fundamental Formula. Let k, A, Q, and R be numbers such that

$$k \geqq 1, \quad A > 29, \quad Q < R,$$

and for every x in the interval

$$Q \leqq x \leqq R$$

let the function $f(x)$ have a second derivative of the same sign and of magnitude within the limits:

$$\frac{1}{kA} \quad \text{and} \quad \frac{1}{A}.$$

Then

$$\sum_{x>Q}^{x<R} \{f(x)\} = \tfrac{1}{2}(R-Q) + G,$$

where G is numerically less than

$$2k\left(\frac{R-Q}{A} + 1\right)(A \ln A)^{2/3}.$$

Proof. 1°. Let both l and $(l+1)$ be contained in the interval (Q, R). Then

$$f'(l+1) - f'(l) = f''(l+\delta); \quad 0 < \delta < 1.$$

Hence, it is easily seen that for the sequence

$$f'([Q]+1), \quad f'([Q]+2), \ldots, f'([R]) \tag{1}$$

the difference between two neighbouring terms is numerically $\geqq \frac{1}{kA}$ and $\leqq \frac{1}{A}$. Hence, if K denotes the least number in (1), the greatest in any case will be $\leqq K + \frac{R-Q}{A}$. Moreover, obviously the number of terms in (1) whose values lie within the limits U and $V > U$ will not be greater than

$$kA(V-U)+1.$$

2°. Take a number τ such that $4 < \tau < \sqrt{A}$. According to Lemma 1 (§1), we can always find one or several pairs of coprime integers X and Y satisfying the inequalities:

$$-\frac{1}{\tau} < Xf'(x) - Y < \frac{1}{\tau},$$
$$0 < X \leqq \tau.$$

From amongst these pairs, choose some pair and denote its components by $X(x)$ and $Y(x)$. Now, putting

$$x = [Q]+1,\ [Q]+2, \ldots, [R], \tag{2}$$

in succession, we obtain a sequence

$$X([Q]+1),\ X([Q]+2), \ldots, X([R]) \tag{3}$$

of positive numbers, each of which is not greater than τ. By virtue of Lemma 2 (§1), the terms of this sequence, except for the last less than τ terms, can be divided into subsequences having the properties stated in the Lemma.

Let one such subsequence be

$$X(\alpha_s+1),\ X(\alpha_s+2), \ldots, X(\alpha_s+n_s).$$

In this sequence there may be terms equal to n_s. Let $X(x_s)$ be one such term. Then

$$-\frac{1}{\tau} < n_s f'(x_s) - Y(x_s) < \frac{1}{\tau},$$
$$0 < n_s \leqq \tau.$$

The first inequality can be expressed as

$$f'(x_s) = \frac{Y(x_s)}{n_s} + \frac{\theta}{\tau n_s}; \qquad -1 < \theta < 1. \tag{4}$$

Now consider the sum

$$\Omega_s = \sum_{x \geqq \alpha_s+1}^{x \leqq \alpha_s+n_s} \{f(x)\}.$$

Substituting $x_s + z$ for x, and using the Taylor expansion, we can easily reduce this sum to

$$\Omega_s = \sum_{z=\alpha_s-x_s+1}^{z=\alpha_s-x_s+n_s} \left\{ f(x_s) + zf'(x_s) + \frac{z^2}{2} f''[x_s + z\varrho(z)] \right\}; \quad -1 < \varrho(z) < 1.$$

Thereafter, by virtue of (4), conditions on the second derivative and the obvious inequalities

$$-n_s < z < n_s,$$

we obtain

$$\Omega_s = \sum_{z=\alpha_s - x_s + 1}^{z=\alpha_s - x_s + n_s} \left\{ \frac{n_s f(x_s) + zY(x_s) + \frac{\theta}{\tau} z + \sigma(z) \frac{n_s^3}{2A}}{n_s} \right\},$$

where for all

$$z = \alpha_s - x_s + 1, \ldots, \alpha_s - x_s + n_s$$

$\sigma(z)$ has the same sign and is numerically less than 1. But, in this form, the sum Ω_s reduces to the form stated in Lemma 3 (§1) if we put

$$M = \alpha_s - x_s; \quad X = n_s; \quad Y = Y(x_s),$$

$$P(z) = n_s f(x_s) + \frac{\theta}{\tau} z + \sigma(z) \frac{n_s^3}{2A}; \quad C = \frac{n_s}{\tau} + \frac{n_s^3}{2A}.$$

Applying this lemma, we find

$$-\frac{n_s}{\tau} - \frac{n_s^3}{2A} - \frac{1}{2} \leqq \Omega_s - \frac{1}{2} n_s \leqq \frac{n_s}{\tau} + \frac{n_s^3}{2A} + \frac{1}{2}.$$

Now writing similar inequalities for each subsequence and the obvious inequalities

$$-\tfrac{1}{2} \leqq \{f(x)\} - \tfrac{1}{2} < \tfrac{1}{2}$$

for each x of the sequence (2) such that $X(x)$ is not contained in any of the subsequences, from all these inequalities, after substituting $R - Q + \theta$ $(|\theta| < 1)$ for $[R] - [Q]$, we obtain

$$-H < \sum_{x>Q}^{x \leqq R} \{f(x)\} - \tfrac{1}{2}(R - Q) < H,$$

where

$$H = \sum_s \left(\frac{n_s}{\tau} + \frac{n_s^3}{2A} + \tfrac{1}{2} \right) + \tfrac{1}{2}\tau + \tfrac{1}{2},$$

and the summation $\sum_s$ is taken over all subsequences. This sum is divided into two subsums

$$\sum_s \left(\frac{n_s}{\tau} + \frac{n_s^3}{2A} \right) \quad \text{and} \quad \sum_s \tfrac{1}{2},$$

of which the first is less than

$$\left(\frac{1}{\tau} + \frac{\tau^2}{2A} \right) \sum_s n_s \leqq \left(\frac{1}{\tau} + \frac{\tau^2}{2A} \right) (R - Q + 1),$$

while the second can be expressed as

$$\tfrac{1}{2} T,$$

where T stands for the total number of subsequences. Since $\tau > 4$ and $< \sqrt{A}$, we obtain

$$H < \left(\frac{1}{\tau} + \frac{\tau^2}{2A}\right)(R - Q) + \tau + \tfrac{1}{2}T. \tag{5}$$

3°. To estimate the upper bound for T, we shall use the Note to Lemma 2 (§1), according to which

$$T < \sum \frac{1}{X}, \tag{6}$$

where X runs through all the terms of the sequence (3). To prove this, let us find how many times a fixed value of X occurs in the sequence (3).

With each given X only those values of Y can be associated which satisfy the inequality

$$XK - \frac{1}{\tau} < Y < X\left(K + \frac{R-Q}{A}\right) + \frac{1}{\tau},$$

which can be easily deduced from Section 1°. The number of such values does not exceed

$$X\frac{R-Q}{A} + \frac{3}{2}.$$

With each given pair of X and Y only those terms of the sequence (1) can be associated which satisfy the inequality

$$\frac{Y - \frac{1}{\tau}}{X} < f'(x) < \frac{Y + \frac{1}{\tau}}{X}.$$

The total number of such pairs is, by 1°, not greater than

$$2k\frac{A}{\tau X} + 1 < 3k\frac{A}{\tau X}.$$

Consequently, not more than

$$\left(X\frac{R-Q}{A} + \frac{3}{2}\right)3k\frac{A}{\tau X}$$

terms of the sequence (1) can be associated with a given X, therefore in (3) the same value of X occurs not more than

$$k\left(3\frac{R-Q}{\tau} + \frac{9}{2}\frac{A}{\tau X}\right)$$

times. From this result and (6), we find that

$$T < k\sum_{X>0}^{X \leqq \tau}\left(3\frac{R-Q}{\tau}\cdot\frac{1}{X} + \frac{9}{2}\cdot\frac{A}{\tau}\cdot\frac{1}{X^2}\right) < k\left(\frac{3}{2}\frac{R-Q}{\tau}\ln A + 3\frac{R-Q}{\tau} + \frac{15}{2}\frac{A}{\tau}\right).$$

Then (5) gives

$$H < (R-Q)\left(\frac{5k}{2\tau} + \frac{\tau^2}{2A} + \frac{3}{4}k\frac{\ln A}{\tau}\right) + \frac{15}{4}k\frac{A}{\tau} + \tau.$$

Now put

$$\tau = \sqrt[3]{A \ln A}.$$

Then the conditions

$$4 < \tau < \sqrt{A}$$

are satisfied and at the same time

$$H < 2k\left(\frac{R-Q}{A} + 1\right)(A \ln A)^{2/3}.$$

This inequality demonstrates the truth of the first fundamental formula.

§3. Corollaries of the First Fundamental Formula

By means of the equality

$$[f(x)] = f(x) - \{f(x)\}$$

and the first fundamental formula, we easily find that

$$\sum_{x>Q}^{x \leqq R} [f(x)] = \sum_{x>Q}^{x \leqq R} f(x) - \tfrac{1}{2}(R-Q) - G,$$

where G is numerically less than

$$2k\left(\frac{R-Q}{A} + 1\right)(A \ln A)^{2/3}.$$

This formula can be further simplified. For this purpose, we shall make use of the formula derived by N.Ya. Sonin, a proof of which can be found in his paper[1].

Let us use the notation:

$$\varrho(x) = [x] - x + \tfrac{1}{2},$$

$$\sigma(x) = \int_0^x \varrho(x)\, dx.$$

By the Sonin formula, the sum $\sum_{x>a}^{x \leqq b} f(x)$ taken over all integral x which are $>a$ and $\leqq b$ is

$$\sum_{x>a}^{x \leqq b} f(x) = \int_0^b f(x)\, dx + \varrho(b) f(b) - \varrho(a) f(a) - \sigma(b) f'(b)$$
$$+ \sigma(a) f'(a) + \int_0^b \sigma(x) f''(x)\, dx.$$

---

1 N.Ya. Sonin: On a definite integral containing the numerical function $[x]$. Izv. Varsh. Univers., 1885.

Here it should be noted that always

$$|\varrho(x)| \leqq \tfrac{1}{2} \quad \text{and} \quad |\sigma(x)| \leqq \tfrac{1}{8}.$$

Let μ denote the maximum of $f'(x)$ in the interval

$$Q \leqq x \leqq R.$$

Then, from Sonin's formula, we find

$$\sum_{x>Q}^{x\leqq R} f(x) = \int_Q^R f(x)\,dx + \varrho(R)\,f(R) - \varrho(Q)\,f(Q) + \frac{\theta}{2}\mu; \qquad -1 < \theta < 1.$$

Hence we obtain

$$\sum_{x>Q}^{x\leqq R} [f(x)] = \int_Q^R f(x)\,dx + \varrho(R)\,f(R) - \varrho(Q)\,f(Q) - \tfrac{1}{2}(R-Q) + H, \tag{1}$$

where

$$[H] < \tfrac{1}{2}\mu + 2k\left(\frac{R-Q}{A} + 1\right)(A\ln A)^{2/3}. \tag{2}$$

Geometrically, equality (1) can be interpreted as follows. Let Ω denote the domain bounded by the axis OX and the lines $x = Oq = Q$ and $x = Or = R$ and an arc nm, whose equation is $y = f(x)$. The axis OX and the line $x = Q$ are not included in the domain (Fig. 1). Let the arc nm be above the axis OX. Then, if T denotes the number of integral points in the domain Ω and S the area of this domain, by (1) we find that

$$T = S + \varrho(R)\,f(R) - \varrho(Q)\,f(Q) - \tfrac{1}{2}(R-Q) + H, \tag{3}$$

where the inequality (2) holds for H. An expression similar to (3) is also true even when the arc nm is not included in the domain (consider a function $f(x) - \varepsilon$, where ε is arbitrarily small and positive).

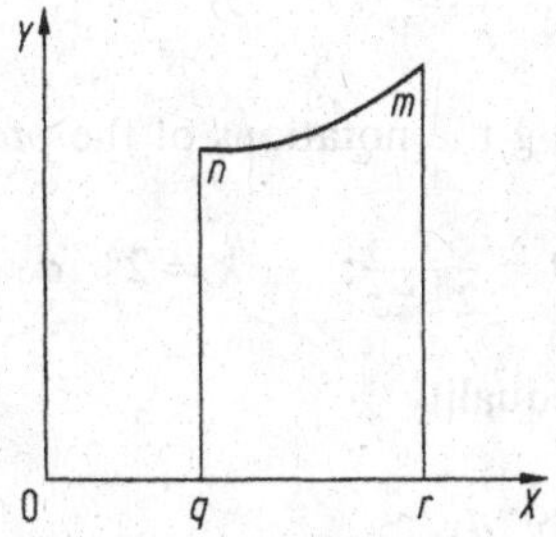

Fig. 1

As an example, assume that the arc nm belongs to the parabola:

$$y = \frac{n + x^2}{a}; \qquad n > 0.$$

Moreover, for the sake of brevity, take Q and R such that

$$-a \leqq Q < R \leqq a.$$

In this case we can put

$$A=\frac{a}{2}; \qquad k=1; \qquad \mu=2$$

and from (3) we get

$$T=\frac{1}{3a}(R^3-Q^3)+\varrho(R)\frac{n+R^2}{a}-\varrho(Q)\frac{n+Q^2}{a}+\left(\frac{n}{a}-\frac{1}{2}\right)(R-Q)+H, \tag{4}$$

where for $\frac{a}{2}>29$, $|H|$ is not greater than

$$1+10\left(\frac{a}{2}\ln\frac{a}{2}\right)^{2/3}<8(a\ln a)^{2/3}.$$

§4. Generalization of the Previous Results

The results obtained in the previous Section can be presented in a different form and generalized. Assume that on the arc nm the ratio of the maximum value, r, of the radius of curvature ϱ to its minimum value is less than a given number σ. From this condition it immediately follows that $f''(x)$ has the same sign, so that $f'(x)$ either increases or decreases on the arc from n to m. For the sake of definiteness, suppose that $f'(x)$ increases; the case where $f'(x)$ decreases is treated in a similar manner and leads to the same results.

1°. Let $f'(x)\leqq 1$ at all points of the arc nm and let it not vanish. Then from

$$y''=\frac{(1+y'^2)^{3/2}}{\varrho}$$

it is seen that $f''(x)$ satisfies the inequality

$$\frac{1}{r}<f''(x)<\frac{2^{3/2}\sigma}{r}.$$

Consequently, retaining the notations of the previous section, we can put

$$\mu=1; \qquad A=\frac{r}{2^{3/2}\sigma}; \qquad k=2^{3/2}\sigma.$$

Moreover, from the equality

$$y_1'-y_0'=\int_Q^R y''\,dx,$$

in which y_0' and y_1' denote the initial (at the point n) and the final (at the point m) values of y', we easily obtain

$$R-Q<2r.$$

With due regard for these statements, from (3) of the previous section, we find

$$T=S+\varrho(R)f(R)-\varrho(Q)f(Q)-\tfrac{1}{2}(R-Q)+H, \tag{1}$$

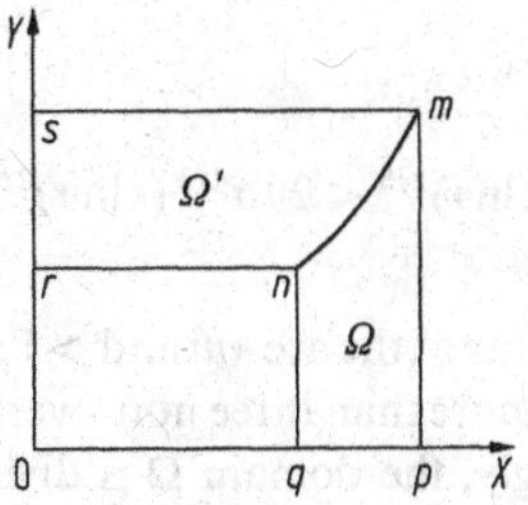

Fig. 2

where, for $\frac{r}{2^{3/2}\sigma} > 29$, we obtain

$$[H] < \tfrac{1}{2} + 2^{5/2}\sigma(2^{5/2}\sigma+1)\left(\frac{r}{2^{3/2}\sigma}\ln\frac{r}{2^{3/2}\sigma}\right)^{2/3} < 19\,\sigma^{4/3}(r\ln r)^{2/3}.$$

2°. Assume that $f'(x) \geqq 1$ everywhere on the arc nm. Let Ω' denote the domain bounded by the axis OY, lines $y=f(Q)$, $y=f(R)$ and the arc nm. The line $y=f(Q)$, the arc nm and the axis OY are not included in the domain (Fig. 2). Let T' denote the total number of integral points within the domain and let S' denote the area of this domain. Interchanging the axes OX and OY, and arguing as in Section 1° (taking $Or=h$ and $Os=l$), we find

$$T' = S' + \varrho(l)\,R - \varrho(h)\,Q - \tfrac{1}{2}(l-h) + H'',$$

where

$$|H''| < 19\sigma^{4/3}(r\ln r)^{2/3}.$$

The total number of integral points in the rectangular domain $Opms$, excepting the points on the sides Op and Os, can be represented, on one hand, as the sum

$$T + T' + [Q]\,[f(Q)],$$

and, on the other hand, as the product

$$[R]\,[f(R)].$$

Comparing both these expressions, we find

$$T = [R]\,[f(R)] - [Q]\,[f(Q)] - R\left([f(R)] - f(R) + \tfrac{1}{2}\right) + Q\left([f(Q)] - f(Q) + \tfrac{1}{2}\right) + \tfrac{1}{2}(f(R) - f(Q)) - S' - H''.$$

But, obviously, we have

$$S' = Rf(R) - Qf(Q) - S$$

and, consequently,

$$T = S + \varrho(R)\,f(R) - \varrho(Q)\,f(Q) - \tfrac{1}{2}(R-Q) + H''', \tag{2}$$

where, for $\frac{r}{2^{3/2}\sigma} > 29$,

$$|H'''| < 1 + 19\sigma^{4/3}(r\ln r)^{2/3} < 20\sigma^{4/3}(r\ln r)^{2/3}.$$

3°. Let $f'(x) \leqq 1$ at certain points of the arc nm and >1 at the other points. The arc nm can always be divided into not more than three non-overlapping parts, in each of which $f'(x)$ is $\leqq 1$ or $\geqq 1$. Accordingly, the domain Ω is divided into not more than three subdomains covering the whole domain Ω without any common points. Now we can apply either Eq. (1) or Eq. (2) to each of these subdomains. On adding the results thus obtained for all the subdomains, we finally get

$$T = S + \varrho(R)f(R) - \varrho(Q)f(Q) - \tfrac{1}{2}(R - Q) + \Gamma, \tag{3}$$

where

$$|\Gamma| < 60\sigma^{4/3}(r\ln r)^{2/3}.$$

§5. Extension to the Case of a Closed Contour

We shall now extend our results to a closed contour. Let a domain Ω of area S be bounded by a contour C on which the ratio, r, of maximum radius of curvature to its minimum does not exceed a given number σ. Let T denote the total number of integral points inside the domain Ω (Fig. 3). We can always assume that the domain Ω lies in the first quadrant; otherwise, we could shift the origin to a suitable integral point, as a result of which the total number of integral points in Ω would not change. By drawing the extreme tangents nq and mp parallel to OY, we obtain two domains, viz., the domain Ω' whose boundaries are the lines nq, mp, qp and the arc nsm, and the domain Ω'' whose boundaries are the lines nq, mp, qp and the arc nrm. Eq. (3) can be applied to each of these domains. Subtracting the results thus obtained, and denoting by γ the number of integral points on the arc nsm, we obtain

$$T + \gamma = S + K(r\ln)r^{2/3},$$

where the upper bound $|K|$ does not depend on r.

To eliminate γ, choose an arbitrary point z inside the contour C as the center of homothety and construct a curve C' lying inside C, so close to C that there are no

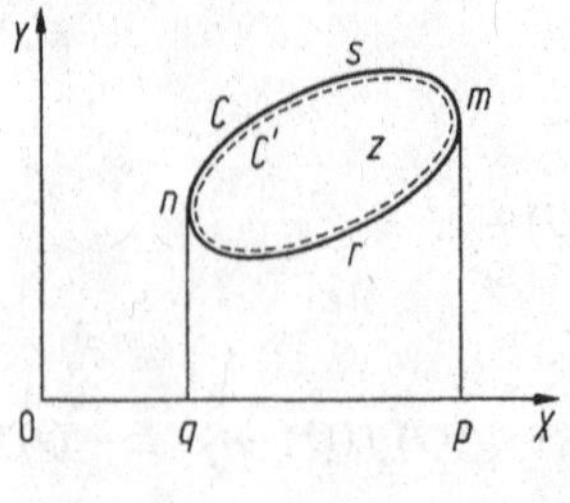

Fig. 3

integral points on it. For the curve C', denoting by r_1, r_1' and σ_1 the same quantities which for the curve C are denoted by r, r' and σ, we find

$$\sigma_1' = \sigma; \qquad r_1 = tr; \qquad r_1' = tr'; \qquad 0 < t < 1.$$

The number of integral points inside C' will also be equal to T if t is sufficiently close to 1. Hence we obtain

$$T = S + K'(r \ln r)^{2/3},$$

where the upper bound $|K'|$ does not depend on r. At the same time, we find that the number of integral points on the arc nsm (as well as on the arc nrm) does not exceed a quantity of the order of $(r \ln r)^{2/3}$. Thus, denoting by T' the number of integral points inside or on the boundary of Ω, we obtain

$$T' = S + K''(r \ln r)^{2/3},$$

where the upper bound $|K''|$ is finite and not dependent on r.

As an example, consider a domain defined by

$$ax^2 + 2bxy + cy^2 \leqq M,$$

where $ax^2 + 2bxy + cy^2$ is a positive quadratic form of discriminant $b^2 - ac = -D$. The curve C is an ellipse similar to the ellipse $ax^2 + 2bxy + cy^2 = 1$ and its center of similarity is at the origin, the ratio of similarity being $\sqrt{M}$. If ϱ is the maximum radius of curvature of the ellipse $ax^2 + 2bxy + cy^2 = 1$, then $r = \varrho\sqrt{M}$; δ is finite and can be taken to be the ratio of the maximum radius of curvature of the ellipse $ax^2 + 2bxy + cy^2 = 1$ to its minimum radius. Since the area of the domain Ω is $\pi M/\sqrt{D}$, we have

$$T' = \frac{\pi M}{\sqrt{D}} + KM^{1/3}(\ln M)^{2/3},$$

where the upper bound $|K|$ is a finite quantity not dependent on M. This equation can be interpreted as follows. If $F(m)$ denotes the number of representations of m by the form (a, b, c), then

$$\sum_{m=1}^{M} F(m) = \frac{\pi M}{\sqrt{D}} + KM^{1/3}(\ln M)^{2/3}.$$

For the particular case of the form $x^2 + y^2$, a slightly more exact result:

$$\sum_{m=1}^{M} F(m) = \pi M + KM^{1/3}$$

has been obtained by Sierpinski with the help of the Voronoi method in the paper already cited.

§6. Modification of the First Fundamental Formula

Using the notation

$$A(x) = \frac{1}{|f''(x)|}$$

we shall show that

$$(R-Q)\sqrt[3]{\frac{(\ln A)^2}{A}} = \lambda \sqrt[3]{k} \int_Q^R \sqrt[3]{\frac{(\ln A(x))^2}{A(x)}}\, dx, \tag{1}$$

where λ satisfies the inequality

$$0 < \lambda < 1.$$

Indeed, a well-known theorem in integral calculus shows that we can write

$$J = \int_Q^R \sqrt[3]{\frac{(\ln A(x))^2}{A(x)}}\, dx = (R-Q)\sqrt[3]{\frac{(\ln A(\xi))^2}{A(\xi)}}; \qquad Q < \xi < R.$$

Therefore the ratio of

$$H = (R-Q)\sqrt[3]{\frac{(\ln A)^2}{A}}$$

to the integral J can be represented as

$$\frac{H}{J} = \sqrt[3]{\left(\frac{\ln A}{\ln A(\xi)}\right)^2 \frac{A(\xi)}{A}}.$$

Hence, since $A \leqq A(\xi)$ is not greater than $\frac{A(\xi)}{k}$, we find that

$$\frac{H}{J} < \sqrt[3]{k}.$$

This proves equality (1).

By (1), we can write the first fundamental formula as

$$\sum_{x>Q}^{x \leqq R} \{f(x)\} = \tfrac{1}{2}(R-Q) + G, \tag{2}$$

where

$$|G| < 2k^{4/3}\left(1 + \frac{A}{R-Q}\right) \int_Q^R \sqrt[3]{\frac{(\ln A(x))^2}{A(x)}}\, dx.$$

§7. Second Fundamental Formula

Fundamental Formula II. For each x in the interval

$$Q \leqq x \leqq R$$

let the function $f(x)$ have a second derivative of the same sign and assume that it does not vanish. Furthermore, suppose the function $A(x)$ defined by the equality

$$A(x) = \frac{1}{|f''(x)|}$$

is >29 in this interval and the ratio of its maximum to its minimum is $\geqq 2$. Assume that $A'(x)$ is of the same sign in the interval (Q, R) and that it is not greater than a given number σ. Then,

$$\sum_{x>Q}^{x \leqq R} \{f(x)\} = \tfrac{1}{2}(R-Q) + T,$$

where T is less than

$$13(1+\sigma) \int_Q^R \sqrt[3]{\frac{(\ln A(x))^2}{A(x)}}\, dx.$$

Proof. We shall restrict ourselves to the case where $A'(x) \geqq 0$ in the interval $Q \leqq x \leqq R$, because the case where $A'(x) \leqq 0$ is treated in exactly the same way and leads to the same results. Since $A'(x) \geqq 0$, as x increases from Q to R, $A(x)$ will be a non-decreasing function; its least value is $A(Q)$ and greatest value is $A(R)$. By our supposititon

$$\frac{A(R)}{A(Q)} \geqq 2.$$

Therefore, we can always find such an integer v that

$$2 \leqq \sqrt[v]{\frac{A(R)}{A(Q)}} < 4.$$

Then we can write

$$\frac{A(R)}{A(Q)} = k^v, \qquad \text{where } 2 \leqq k < 4.$$

Now having determined $Q_1, Q_2, \ldots, Q_{v-1}$ from

$$A(Q_i) = k^i A(Q); \qquad i = 1, 2, \ldots, v-1,$$

it is easy to verify that

$$Q < Q_1 < Q_2 < \ldots < Q_{v-1} < R.$$

Therefore the sum

$$\sum_{x>Q}^{x \leqq R} \{f(x)\}$$

can be represented as

$$\sum_{x>Q}^{x\leqq R}\{f(x)\} = \sum_{x>Q}^{x\leqq Q_1}\{f(x)\} + \sum_{x>Q_1}^{x\leqq Q_2}\{f(x)\} + \ldots + \sum_{x>Q_{\nu-1}}^{x\leqq Q_\nu}\{f(x)\}, \tag{1}$$

where, for the sake of uniformity, we have put $R = Q_\nu$. For any $i = 0, 1, 2, \ldots, \nu-1$, from

$$A(Q_{i+1}) - A(Q_i) = (Q_{i+1} - Q_i)A'(Q_i'); \qquad Q_i < Q_i' < Q_{i+1},$$
$$A(Q_{i+1}) - A(Q_i) = (k-1)A(Q_i)$$

we obtain

$$A(Q_i)Q_{i+1} - Q_i \leqq \frac{\sigma}{k-1} \leqq \sigma.$$

Moreover, it is easy to verify that in each interval (Q_i, Q_{i+1}) $(i = 0, 1, 2, \ldots, \nu-1)$ the second derivative $f''(x)$ satisfies the inequality:

$$\frac{1}{kA(Q_i)} \leqq f''(x) \leqq \frac{1}{A(Q_i)}.$$

Using these facts, and then applying the first fundamental formula to each sum on the right-hand side of (1), we easily obtain:

$$\sum_{x>Q}^{x\leqq R}\{f(x)\} = \tfrac{1}{2}(R-Q) + T,$$

where T is numerically less than

$$13(1+\sigma)\int_Q^R \sqrt[3]{\frac{(\ln A(x))^2}{A(x)}}\,dx.$$

§8. Application of the Second Fundamental Formula to Particular Cases

By way of example, take

$$f(x) = \frac{a}{x}.$$

For this function $f''(x) = -\frac{2a}{x^3}$, and therefore

$$A(x) = \frac{x^3}{2a}; \qquad A'(x) = \frac{3}{2}\cdot\frac{x^2}{a}.$$

It is easy to verify that if we take

$$Q = \sqrt[3]{58a} \quad \text{and} \quad R = \sqrt{a},$$

then for all $a > 8^2 \cdot 58^2$, the conditions stipulated in the previous section are satisfied. Hence, $\sigma = 3/2$, and T is less than

$$13\left(1+\frac{3}{2}\right)\int\limits_{\sqrt[3]{58a}}^{\sqrt{a}} \frac{\sqrt[3]{2a}\left(\ln\frac{x^3}{2a}\right)^{2/3}}{x}\,dx < 13\sqrt[3]{a(\ln a)^5}\,.$$

Consequently, on applying the second fundamental formula to this case, we obtain

$$\sum_{x>\sqrt[3]{58a}}^{x\leqq\sqrt{a}} \left\{\frac{a}{x}\right\} = \tfrac{1}{2}(\sqrt{a}-\sqrt[3]{58a}) + \lambda\cdot 13\sqrt[3]{a(\ln a)^5}\,; \qquad -1<\lambda<1.$$

On the other hand,

$$\sum_{x=1}^{x\leqq\sqrt[3]{58a}} \left\{\frac{a}{x}\right\} = \tfrac{1}{2}\sqrt[3]{58a} + \frac{\lambda'}{2}\sqrt[3]{58a}; \qquad -1\leqq\lambda'<1.$$

Adding these two equalities, we arrive at

$$\sum_{x=1}^{x\leqq\sqrt{a}} \left\{\frac{a}{x}\right\} = \tfrac{1}{2}\sqrt{a} - \lambda''\cdot 15\sqrt[3]{a(\ln a)^5}\,.$$

Hence, since

$$\left[\frac{a}{x}\right] = \frac{a}{x} - \left\{\frac{a}{x}\right\},$$

we readily find that

$$\sum_{x=1}^{x\leqq\sqrt{a}} \left[\frac{a}{x}\right] = \sum_{x=1}^{x\leqq\sqrt{a}} \frac{a}{x} - \tfrac{1}{2}\sqrt{a} + \lambda''\cdot 15\sqrt[3]{a(\ln a)^5}\,.$$

Substituting it into the formula

$$\sum_{x=1}^{x\leqq a} \left[\frac{a}{x}\right] = 2\sum_{x=1}^{a\leqq\sqrt{a}} \left[\frac{a}{x}\right] - [\sqrt{a}]^2,$$

we obtain

$$\sum_{x=1}^{x\leqq a} \left[\frac{a}{x}\right] = 2\sum_{x=1}^{x\leqq\sqrt{a}} \frac{a}{x} - \sqrt{a} - [\sqrt{a}]^2 + \lambda''\cdot 30\sqrt[3]{a(\ln a)^5}. \tag{1}$$

For further transformation, it is more convenient to use the Sonin formula which we have already applied in § 3. In the particular case where

$$\lim_{a=\infty} f'(x) = 0$$

and $f''(x)$ does not change its sign as x increases from a to ∞, from the Sonin formula we obtain

$$\sum_{x>a}^{x\leqq b} f(x) = C + \int\limits_a^b f(x)\,dx + \varrho(b)f(b) + \frac{\theta}{4}f'(b), \tag{2}$$

where C is not dependent on b and

$$-1 < \theta < 1.$$

Applying formula (2) to the sum

$$a + \sum_{x>1}^{x \leq \sqrt{a}} \frac{a}{x},$$

and after some simplifications, we find

$$\sum_{x=1}^{x \leq \sqrt{a}} \frac{a}{x} = aE + \frac{a}{2} \ln a + \sqrt{a}\,[\sqrt{a}] - a + \tfrac{1}{2}\sqrt{a} + \frac{\theta}{4},$$

where

$$-1 < \theta < 1$$

and E stands for the Euler constant. Substituting the last formula into Eq. (1), and since

$$[\sqrt{a}]^2 = 2\sqrt{a}\,[\sqrt{a}] - a + \theta'; \quad 0 \leqq \theta' < 1$$

we finally obtain the asymptotic expression:

$$\sum_{x=1}^{x \leqq a} \left[\frac{a}{x}\right] = a(\ln a + 2E - 1) + \varrho \cdot 32 \sqrt[3]{a(\ln a)^5}.$$

This is precisely the example which Voronoi considered in his paper, where the reminder term was found to be of the order of $\sqrt[3]{a} \ln a$, which is slightly more exact than the result obtained by our method.

§ 9. Proof of the Gauss Formula for the Mean of the Number of Classes of Quadratic Forms

In conclusion we shall apply our method to determine the mean of the number of classes of proper primitive forms. If $h(-\Delta)$ denotes the number of proper primitive classes of discriminant $-\Delta$, the problem reduces to finding an asymptotic expression for the sum:

$$\sum_{\Delta=1}^{\Delta=m} h(-\Delta).$$

Using the results published in Lipschitz's memoirs[1], which Bachman[2] reproduces in Chapter 13 of his book, we shall start with the equality

$$\sum_{\Delta=1}^{\Delta=m} h(-\Delta) = \sum \mu(k)\, F\left(\frac{m}{k^2}\right); \quad k = 1, 3, 5, \ldots,$$

1 Lipschitz, R.: Über die asymptotischen Gesetze von gewissen Gattungen zahlentheoretischer Funktionen. Monatsber. Berlin Acad. (1865) 174.

2 Bachman, H.: Análytische Zahlentheorie. (1894).

where $\mu(k)$ denotes the Moebius function: $F(n) = 0$ for $n < 1$, and for $n > 1$, it is defined by the following equation

$$F(n) = \psi(n) + \sigma(n) - \chi(n) - \tau(n),$$

in which $\psi(n)$, $\sigma(n)$, $\chi(n)$ and $\tau(n)$ stand for the number of systems of integral values of the variables x, y, z satisfying the inequalities:

$$\left.\begin{array}{l} 0 < xz - y^2 \leqq n \\ 0 < x < z \\ -x/2 < y \leqq x/2 \end{array}\right\} \quad \text{for} \quad \psi(n),$$

$$\left.\begin{array}{l} 0 < xz - y^2 \leqq n \\ x = z \\ 0 \leqq y < x/2 \end{array}\right\} \quad \text{for} \quad \sigma(n),$$

$$\left.\begin{array}{l} 0 < 4xz - y^2 \leqq n \\ 0 < x < z \\ -x < y \leqq x \end{array}\right\} \quad \text{for} \quad \chi(n),$$

$$\left.\begin{array}{l} 0 < 4xz - y^2 \leqq n \\ x = z \\ 0 \leqq y < x \end{array}\right\} \quad \text{for} \quad \tau(n).$$

First consider the sum

$$\psi(n) + \sigma(n).$$

Now we find how many systems of integral values of y and z satisfy the inequalities:

$$\begin{array}{l} 0 < xz - y^2 \leqq n, \\ z > x, \\ -\tfrac{1}{2}x < y \leqq \tfrac{1}{2}x, \end{array}$$

for a given integral $x \geqq 1$; in other words how many integral points lie in a domain Ω of y and z defined by

$$\begin{array}{l} 0 < xz - y^2 \leqq n, \\ z > x, \\ -\tfrac{1}{2}x < y \leqq \tfrac{1}{2}x. \end{array}$$

It is not difficult to verify that integral points may exist inside this domain if and only if $x < \sqrt{\tfrac{4}{3}n}$. If this condition is satisfied, the domain has then different shapes, depending on whether $x \leqq \sqrt{n}$ or $x > \sqrt{n}$.

1°. If $x \leqq \sqrt{n}$, the domain Ω is bounded by the lines $y = -x/2$, $y = x/2$, $z = x$ and the parabola $z = (n + y^2)/x$. The points on $y = -x/2$ and $z = x$ are not included in the domain (Fig. 4). The results of § 3 can be applied to this domain. Since the area of the domain is

$$n - \tfrac{11}{12}x^2.$$

The number, T, of integral points belonging to this domain is

$$T = n - \tfrac{11}{12}x^2 + \left(\frac{n + x^2/4}{x} - x\right)\left[\varrho\left(\frac{x}{2}\right) - \varrho\left(-\frac{x}{2}\right)\right] - \frac{x}{2} + L(x\ln x)^{2/3},$$

where the upper bound $|L|$ does not depend on x, and, for $x \geqq 2$, on n either. It can be easily verified that for any integral x, always

$$\varrho\left(\frac{x}{2}\right) - \varrho\left(-\frac{x}{2}\right) = 0.$$

Consequently, if $x \leqq \sqrt{n}$, we obtain a simple expression:

$$T = n - \tfrac{11}{12}x^2 - x/2 + L(x\ln x)^{2/3}.$$

2°. If $x > \sqrt{n}$, the parabola $z = (n + y^2)/x$ intersects the line $z = x$ at $y = \pm\sqrt{x^2 - n}$ and the domain Ω is divided into two separate subdomains Ω' and Ω'' (Fig. 5). The domain Ω' is bounded by the lines $y = x/2$, $z = x$ and the parabola $z = (n + y^2)/x$. The line $z = x$ is not included in the domain.

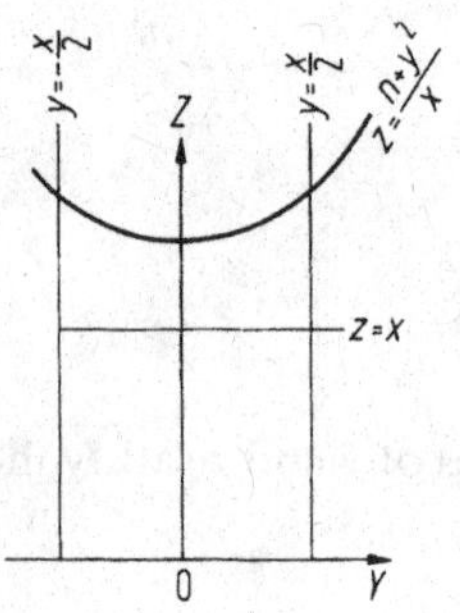

Fig. 4

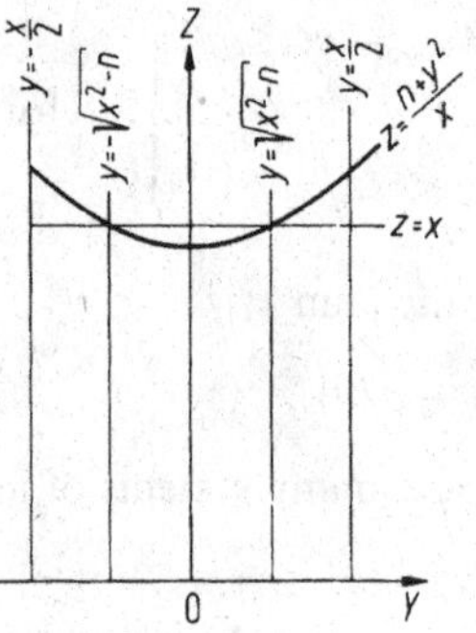

Fig. 5

The domain Ω'' is bounded by the lines $y = -x/2$, $z = x$ and the parabola $z = (n + y^2)/x$. The straight line parts of the contour are not included in the domain.

We shall now apply the results of § 3 to the subdomain Ω'. Since its area is

$$\frac{n}{2} - \tfrac{11}{24}x^2 + \frac{2}{3x}(x^2 - n)^{3/2}.$$

Eq. (4) of § 3 gives an expression for the number T' of integral points inside the domain Ω':

$$T' = \frac{n}{2} - \tfrac{11}{24}x^2 + \frac{2}{3x}(x^2 - n)^{3/2} + \varrho\left(\frac{x}{2}\right)\left(\frac{n + x^2/4}{x} - x\right)$$
$$- \tfrac{1}{2}\left(\tfrac{1}{2}x - \sqrt{x^2 - n}\right) + L'(x\ln x)^{2/3},$$

where the upper bound $|L'|$ does not depend on either n or x. The number T'' of integral points inside the domain Ω'' is (by symmetry) equal to T' if x is odd, and if x is even it is less than T' by the number of the integral points on the line $y = x/2$ for which $z > x$ and $z \leqq (n + y^2)/x$. It is equal to

$$\frac{n + x^2/4}{x} - x - \delta; \qquad 0 \leqq \delta < 1.$$

By virtue of this fact and since $\varrho(x/2) = 0$ for odd x and $\varrho(x/2) = 1/2$ for even x, we find that the number T of integral points belonging to the domain Ω is

$$T = n - \tfrac{11}{12}x^2 + \frac{4}{3x}(x^2 - n)^{3/2} - \tfrac{1}{2}x + \sqrt{x^2 - n} + L(x\ln x)^{2/3} + \delta,$$

which is valid even for $x > \sqrt{n}$.

Summing up the results of 1° and 2°, we easily obtain an expression for $\psi(n)$:

$$\psi(n) = \sum_{x>0}^{x \leqq \sqrt{\frac{4}{3}}} (n - \tfrac{11}{12}x^2) + \tfrac{4}{3} \sum_{x>\sqrt{n}}^{x \leqq \sqrt{\frac{4}{3}n}} \frac{(x^2 - n)^{3/2}}{x} - \tfrac{1}{2} \sum_{x>0}^{x \leqq \sqrt{\frac{4}{3}n}} x$$

$$+ \sum_{x>\sqrt{n}}^{x \leqq \sqrt{\frac{4}{3}n}} \sqrt{x^2 - n} + \sum_{x>0}^{x \geqq \sqrt{\frac{4}{3}n}} L(x\ln x)^{2/3} + \sum_{x>\sqrt{n}}^{x \leqq \sqrt{\frac{4}{3}n}} \delta.$$

On the other hand, for $x \leqq \sqrt{n}$ the number of integral y satisfying the inequalities

$$0 < x^2 - y^2 \leqq n,$$
$$0 \leqq y < \tfrac{1}{2}x,$$

is equal to $[x/2] + 1$; and for $x > \sqrt{n}$ but $< \sqrt{\frac{4}{3}n}$, this number can be expressed as

$$[x/2] - [\sqrt{x^2 - n}] + h,$$

where $|h| \leqq 1$. Consequently,

$$\sigma(n) = \sum_{x>0}^{x \leqq \sqrt{\frac{4}{3}n}} \left[\frac{x}{2}\right] + \sum_{x>0}^{x \leqq \sqrt{n}} 1 - \sum_{x>\sqrt{n}}^{x \leqq \sqrt{\frac{4}{3}}} [\sqrt{x^2 - n}] + \sum_{x>0}^{x \leqq \sqrt{\frac{4}{3}n}} h.$$

By virtue of the expression for $\psi(n)$, we obtain

$$\psi(n) + \sigma(n) = \sum_{x>0}^{x \leqq \sqrt{\frac{4}{3}n}} (n - \tfrac{11}{12}x^2) + \tfrac{4}{3} \sum_{x>\sqrt{n}}^{x \leqq \sqrt{\frac{4}{3}n}} \frac{(x^2 - n)^{3/2}}{x}$$

$$+ \sum_{x>0}^{x \leqq \sqrt{\frac{4}{3}n}} L(x\ln x)^{2/3} + K\sqrt{n},$$

where K is bounded for any n.

We shall calculate the sums in this expression by the Sonin summation formula which we have already used elsewhere. It can now be rewritten as

$$\sum_{x>a}^{x \leqq b} f(x) = \int_a^b f(x)\,dx + \varrho(b) f(b) - \varrho(a) f(a) + \frac{\theta}{2}\mu; \qquad -1 < \theta < 1,$$

if $f''(x)$ does not change its sign between a and b, and μ is the upper bound of $f'(x)$ in (a, b). Since $f(x) = n - \frac{11}{12}x^2$, we find

$$\sum_{y>0}^{x \leqq \sqrt{\frac{4}{3}n}} \left(n - \tfrac{11}{12}x^2\right) = \frac{32}{27\sqrt{3}}\, n^{3/2} - \frac{n}{2} - \frac{2}{9}\, n\varrho \sqrt{\tfrac{4}{3}n} + \theta \frac{11}{6\sqrt{3}} \sqrt{n}.$$

Putting $f(x) = \dfrac{(x^2 - n)^{3/2}}{x}$, we get

$$f'(x) = (x^2 - n)^{1/2}\left(2 + \frac{n}{x^2}\right),$$

$$f''(x) = (x^2 - n)^{-1/2}\left(2x + \frac{2n^2}{x^3} - \frac{n}{x}\right),$$

Hence, it is seen that $f''(x) > 0$ for $\sqrt{n} < x \leqq \sqrt{\frac{4}{3}n}$, and under these conditions $0 < f'(x) < \dfrac{11}{4\sqrt{3}} \sqrt{n}$. Therefore

$$\sum_{x > \sqrt{n}}^{x \leqq \sqrt{\frac{4}{3}n}} \frac{(x^2 - n)^{3/2}}{x} = \int_{\sqrt{n}}^{\sqrt{\frac{4}{3}n}} \frac{(x^2 - n)^{3/2}}{x}\, dx + \frac{n}{6}\, \varrho \sqrt{\tfrac{4}{3}n} + \theta' \frac{11}{8\sqrt{3}} \sqrt{n}.$$

But since

$$\int_{\sqrt{n}}^{\sqrt{\frac{4}{3}n}} \frac{(x^2 - n)^{3/2}}{x}\, dx = n^{3/2}\left(\frac{\pi}{6} - \frac{8}{9\sqrt{3}}\right),$$

we finally obtain

$$\frac{4}{3} \sum_{x > \sqrt{n}}^{x \leqq \sqrt{\frac{4}{3}n}} \frac{(x^2 - n)^{3/2}}{x} = \frac{2\pi}{9}\, n^{3/2} - \frac{32}{27\sqrt{3}}\, n^{3/2} + \frac{2}{9}\, n\varrho \sqrt{\tfrac{4}{3}n}$$

$$+ \theta' \frac{11}{6\sqrt{3}} \sqrt{n}.$$

From these formulae it is easy to derive

$$\sum_{x>0}^{x \leqq \sqrt{\frac{4}{3}n}} \left(n - \tfrac{11}{12}x^2\right) + \tfrac{4}{3} \sum_{x > \sqrt{n}}^{x \leqq \sqrt{\frac{4}{3}n}} \frac{(x^2 - n)^{3/2}}{n} = \frac{2\pi}{9}\, n^{3/2} - \frac{n}{2} + K' \sqrt{n},$$

where K' is bounded for all $n \geqq 2$. Now the sum $\psi(n) + \sigma(n)$ can be represented as

$$\psi(n) + \sigma(n) = \frac{2\pi}{9}\, n^{3/2} - \frac{n}{2} + Tn^{5/6} (\ln n)^{2/3},$$

where the upper bound of $|T|$ does not depend on n for $n \geqq 2$. Similarly, we find

$$\chi(n) + \tau(n) = \frac{\pi}{18} n^{3/2} - \frac{n}{4} + \Gamma' n^{5/6} \ln n)^{2/3}$$

and $\quad F(n) = \pi/6\, n^{3/2} - n/4 + T'' n^{5/6} (\ln n)^{2/3}$,

where T' and T'' are finite for all $n \geqq 2$. Substituting the expression for $F(n)$ into the sum

$$S = \sum \mu(k)\, F\left(\frac{m}{k^2}\right),$$

taken over all odd $k \leqq \sqrt{m}$, we obtain

$$S = \frac{\pi}{6} m^{3/3} \sum \frac{\mu(k)}{k^3} - \frac{m}{4} \sum \frac{\mu(k)}{k^2} + m^{5/6} \sum \mu(k)\, T'' \frac{\left(\ln \frac{m}{k^2}\right)^{2/3}}{k^{5/3}},$$

where all the sums are taken over odd $k \leqq \sqrt{m}$. Now, putting

$$e = 1 + \frac{1}{2^3} + \frac{1}{3^3} + \cdots,$$

we get

$$\sum \frac{\mu(k)}{k^3} = \frac{8}{7e} + \frac{L}{m}, \qquad \sum \frac{\mu(k)}{k^2} = \frac{8}{\pi^2} + \frac{L'}{\sqrt{m}},$$

$$\left| \sum \mu(k)\, T'' \frac{\left(\ln \frac{m}{k^2}\right)^{2/3}}{k^{5/3}} \right| < L'' (\ln m)^{2/3},$$

where L, L' and L'' are finite for any $m \geqq 2$. Consequently,

$$S = \frac{4\pi}{21e} m^{3/2} - \frac{2}{\pi^2} m + \Gamma m^{5/6} (\ln m)^{2/3},$$

where Γ is finite and

$$\sum_{\Delta=1}^{\Delta=m} h(-\Delta) = \frac{4\pi}{21e} m^{3/2} - \frac{2}{\pi^2} m + \Gamma m^{5/6} (\ln m)^{2/3}.$$

This is the expression that we intended to derive. It is, of course, given by Gauss in a different form with the error bound in his paper cited elsewhere. So far, if we are not mistaken, the Gauss formula has not been demonstrated to the fullest extent, because Lipschitz and Mertens[1] give only the principal term $\frac{4\pi}{21e} m^{3/2}$, while the term $-\frac{2}{\pi^2} m$, mentioned by Gauss, cannot be obtained by their technique which is not adequate for this purpose.

---

[1] Mertens, F.: Über einige asymptotische Gesetze der Zahlentheorie. J. reine angew. Math. 77 (1874) 289–338.

On the Mean Value of the Number of Classes of Proper Primitive Forms of Negative Discriminant

Soobchsheniya Khar'kovskogo matematicheskogo obshestva (Transactions of the Kharkov Mathematical Society), **16** no. 1–2 (1918) 10–38.

Gauss gives without proof the formula for the mean of the number of classes of proper primitive forms of negative discriminant, mentioning that it has been obtained as a result of strenuous theoretical study (*per disquisitionem theoreticam satis difficilem*) [1].

In this paper we intend to derive the Gauss formula with an upper bound for the error from quite elementary considerations. In a paper, which will appear shortly, we have treated this problem in an entirely different way, so that we could estimate an upper bound for the error of the order of $m^{5/6} \ln m$, whereas in the present paper the upper error bound is found to be of the order of $m^{3/4} (\ln m)^2$.

§ 1. Derivation of an Asymptotic Expression for the Generalized Gauss Sum

Let n and λ be numbers satisfying the conditions:

$$n \geqq 8, \tag{1}$$

$$0 \leqq \lambda < 1, \tag{2}$$

ξ an integer satisfying the condition

$$0 < \xi \leqq \frac{n}{4} - \lambda, \tag{3}$$

and η, T be positive numbers such that η decreases unboundedly, while T increases unboundedly.

In the complex plane $z = x + iy$, take a domain Ω bounded by the contour $A\alpha\beta BC\gamma\delta DA$ (Fig. 1) consisting of the lines:

$$x = 0; \qquad x = \xi; \qquad y = T, \; y = -T$$

and semicircles of radius η described about the points 0 and ξ. With the help of these semicircles the points 0 and ξ are excluded from the domain. In the domain (inside it and on its contour) the function

$$\frac{e^{\frac{2\pi i}{n}(\lambda + k)^2}}{e^{2\pi i s} - 1}$$

[1] Gauss: Disquisitiones arithmeticae, art. 302.

has no other singularities than its poles at $z=k$ where the residues are $e^{\frac{2\pi i}{n}(\lambda+k)^2}/2\pi i$; k runs through all integers greater than zero and less than ξ. Hence, from the Cauchy contour integration theorem we find that the sum

$$\sum_{k>0}^{k<\xi} e^{\frac{2\pi i}{n}(\lambda+z)^2}$$

can be expressed as the sum of the integrals

$$\int_{A\alpha}+\int_{\alpha\beta}+\int_{\beta B}+\int_{BC}+\int_{C\gamma}+\int_{\gamma\delta}+\int_{\delta D}+\int_{DA}.$$

But it is easy to show that the integrals $\int_{BC}$ and $\int_{DA}$, as T tends to infinity, vanish and that

$$\lim_{\eta=0}\left(\int_{\alpha\beta}+\int_{\gamma\delta}\right)=-\frac{1}{2}e^{\frac{2\pi i}{n}\lambda^2}-\frac{1}{2}e^{\frac{2\pi i}{n}(\lambda+\xi)^2}.$$

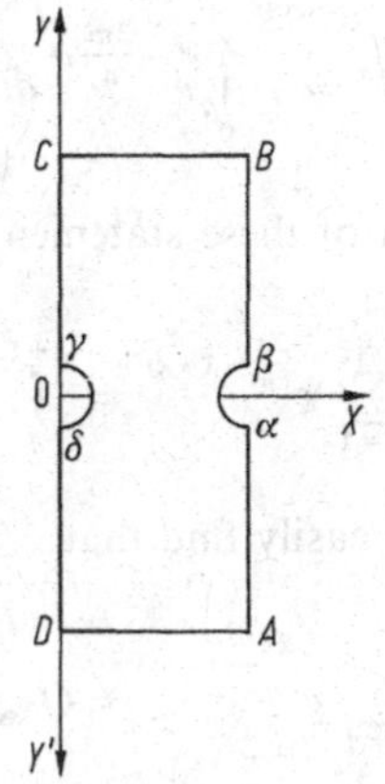

Fig. 1

Therefore we can write

$$\frac{1}{2}e^{\frac{2\pi i}{n}\lambda^2}+\frac{1}{2}e^{\frac{2\pi i}{n}(\lambda+\xi)^2}+\sum_{k>0}^{k<\xi} e^{\frac{2\pi i}{n}(\lambda+k)^2}$$

$$=\lim_{\eta=0,\,T=\infty}\left(-i\int_\eta^T\frac{e^{\frac{2\pi i}{n}(\lambda+yi)^2}}{e^{-2\pi y}-1}dy-i\int_\eta^T\frac{e^{\frac{2\pi i}{n}(\lambda-yi)^2}}{e^{2\pi y}-1}dy\right)$$

$$+\lim_{\eta=0,\,T=\infty}\left(i\int_\eta^T\frac{e^{\frac{2\pi i}{n}(\lambda+\xi+yi)^2}}{e^{-2\pi y}-1}dy+i\int_\eta^T\frac{e^{\frac{2\pi i}{n}(\lambda+\xi-yi)^2}}{e^{2\pi y}-1}dy\right). \tag{4}$$

Denote the first term on the right-hand side of (4) by S' and the second term by S''. After some simple transformations, we obtain

$$S'=i\int_0^\infty e^{\frac{2\pi i}{n}(\lambda+yi)^2}dy+ie^{\frac{2\pi i}{n}\lambda^2}\int_0^\infty\frac{e^{-\frac{4\pi\lambda}{n}x}-e^{\frac{4\pi\lambda}{n}x}}{e^{2\pi x}-1}e^{-\frac{2\pi i}{n}x^2}dy. \tag{5}$$

But we can easily derive

$$i\int_0^\infty e^{\frac{2\pi i}{n}(\lambda+xi)^2}dy=\int_0^\infty e^{-\frac{2\pi i}{n}t^2}dt-\int_0^\lambda e^{\frac{2\pi i}{n}t^2}dt.$$

The second term on the right-hand side of (5), by conditions (1) and (2), is in absolute value not greater than

$$\int_0^\infty \frac{e^{\pi y} - e^{-\pi y}}{e^{2\pi y} - 1}\, dy = \frac{1}{\pi}.$$

Consequently,

$$S' = \int_0^\infty e^{-\frac{2\pi i}{n}t^2}\, dt - \int_0^\lambda e^{\frac{2\pi i}{n}t^2} + \frac{\theta'}{\pi}; \qquad |\theta'| \leqq 1.$$

Arguing in a similar manner and using condition (3), we find

$$S'' = -\int_0^\infty e^{-\frac{2\pi i}{n}t^2}\, dt + \int_0^{\lambda+\xi} e^{\frac{2\pi i}{n}t^2}\, dt + \frac{\theta''}{\pi}; \qquad |\theta''| \leqq 1.$$

As a result of these statements, we can express equality (4) as

$$\sum_{k=1}^{k=\xi} e^{\frac{2\pi i}{n}(\lambda+k)^2} = \int_\lambda^{\lambda+\xi} e^{\frac{2\pi i}{n}t^2}\, dt + 2\theta'''; \qquad |\theta'''| < 1.$$

Hence, we easily find that

$$\sum_{k=1}^{k=\xi} e^{\frac{2\pi i}{n}(\lambda+k)^2} = O(\sqrt{n}). \tag{6}$$

Assume that

$$\xi = \left[\frac{n}{4} - \lambda\right].$$

In this case the absolute value of the difference $\xi - [n/4]$ is not greater than 1. Now it is easy to verify that each of the integrals

$$\int_0^\lambda e^{\frac{2\pi i}{n}t^2}\, dt; \qquad \int_{\lambda+\xi}^\infty e^{\frac{2\pi i}{n}t^2}\, dt$$

is in absolute value not greater than 1. Therefore, since

$$\int_0^\infty e^{\frac{2\pi i}{n}t^2}\, dt = \frac{1+i}{4}\sqrt{n},$$

we have

$$\sum_{k>0}^{k<\frac{n}{4}} e^{\frac{2\pi i}{n}(\lambda+k)^2} = \frac{1+i}{4}\sqrt{n} + 5\theta; \qquad |\theta| < 1. \tag{7}$$

§ 2. Determination of the Order Which is not Exceeded by the Order of the Sums $\sum_{x>P}^{x \leqq Q} \sin 2\pi\left(\frac{n}{x} - Ax\right)$ and $\sum_{x>P}^{x \leqq Q} \cos 2\pi\left(\frac{n}{x} - Ax\right)$

1. Let n denote a number satisfying the condition

$$n \geqq 16, \tag{1}$$

R and A be any numbers, and λ, P, Q be numbers such that

$$0 \leqq \lambda \leqq \tfrac{3}{4}, \quad \sqrt[3]{n} \leqq \lambda + P \leqq \lambda + Q \leqq \sqrt{\frac{n}{2}}. \tag{2}$$

Defining x_s as

$$\lambda + x_s = \sqrt{\frac{n}{s-\lambda}}, \tag{3}$$

and then with the help of μ and v found from the conditions

$$x_{\mu+1} < P \leqq x_\mu, \qquad x_v < Q \leqq x_{v-1},$$

we form the series

$$x_{\mu+1} < x_\mu < x_{\mu-1} < \ldots < x_{v+1} < x_v < x_{v-1}.$$

Decompose the sum

$$S = \sum_{x>P}^{x=Q} \sin 2\pi\left(R + \frac{n}{\lambda+x} - Ax\right)$$

as follows:

$$S = \sum_{y>P}^{y \leqq y_\mu} + \sum_{y>y_\mu}^{y \leqq y_{\mu-1}} + \sum_{y>y_{\mu-1}}^{y \leqq y_{\mu-2}} + \ldots + \sum_{y>y_{v+1}}^{y \leqq y_v} + \sum_{y>y_v}^{y \leqq Q}. \tag{4}$$

Now we shall determine the order which is not exceeded by the order of one of these sums. In order to examine simultaneously all the cases, consider the sum

$$H = \sum_{x>a}^{x \leqq b} \sin 2\pi\left(R + \frac{n}{\lambda+x} - Ax\right),$$

where a and b are integers such that

$$x_s \leqq a \leqq b \leqq x_{s-1},$$

and s stands for one of the numbers

$$\mu+1,\ \mu-1,\ \ldots,\ v+1,\ v.$$

Since the function sx takes integral values for all integral x, we can write

$$H = \sum_{x>a}^{x \leqq b} \sin 2\pi\left(R + \frac{n}{\lambda+x} - Ax + sx\right) = \sum_{x>a}^{x \leqq b} \sin 2\pi\left(R + \frac{n}{\lambda+x} + \frac{nx}{(\lambda+x_s)^2}\right).$$

Hence, after some obvious transformations, we obtain

$$|H| < \left| \sum_{x>a}^{x\leqq b} \sin 2\pi \frac{n(x-x_s)^2}{(\lambda+x_s)^2(\lambda+x)} \right| + \left| \sum_{x>a}^{x\leqq b} \cos 2\pi \frac{n(x-x_s)^2}{(\lambda+x_s)^2(\lambda+x)} \right|. \tag{5}$$

2. Taking z to be a complex variable

$$z = x + iy$$

we shall express the function

$$\Phi(z) = \frac{n(z-x_s)^2}{(\lambda+x_s)^2(\lambda+z)}$$

as follows

$$\Phi(z) = \varphi(x, y) + i\psi(x, y),$$

where

$$\varphi(x, y) = \frac{n}{(\lambda+x_s)^2} \frac{(x-x_s)^2(\lambda+x) + (-2x_s+x-\lambda)y^2}{(\lambda+x)^2+y^2},$$

$$\psi(x, y) = \frac{n}{(\lambda+x_s)^2} \frac{[(\lambda+x)^2-(\lambda+x_s)^2]y+y^3}{(\lambda+x)^2+y^2}.$$

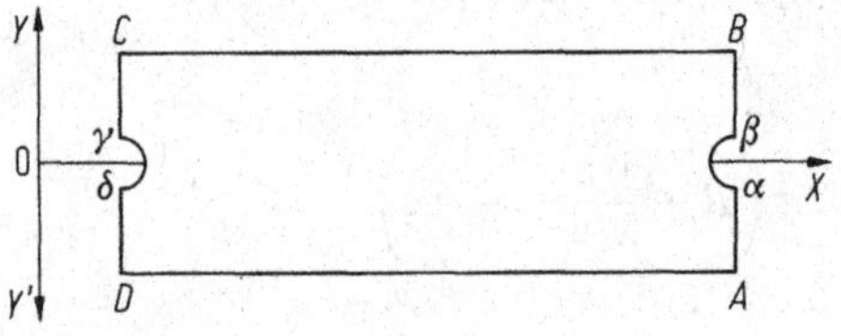

Fig. 2

Except for the poles at $z = k$, where k runs through all integers $>a$ and $<b$, the function $e^{2\pi i\Phi(z)}/(e^{2\pi iz}-1)$ has no other singularities in the domain (inside and on the contour) bounded by the contour $A\alpha\beta BC\gamma\delta DA$ (Fig. 2) consisting of the lines $x = a$, $x = b$, $y = T$, $y = -T$ and semicircles of radius η described about the points a and b, by which these points are excluded from the domain. Applying the Cauchy contour integration theorem and arguing as in § 1, we obtain the general formula

$$\sum_{y>a}^{y<b} e^{2\pi i \frac{n(y-y_s)^2}{(\lambda+y_s)^2(\lambda+y)}} = -\lim_{\eta=0} i \int_{\eta}^{T} \left(\frac{e^{-2\pi\psi(a,x)}}{e^{-2\pi x}-1} + \frac{e^{2\pi\psi(a,x)}}{e^{2\pi x}-1} \right) e^{2\pi i\varphi(a,y)}\, dy$$

$$+ \lim_{\eta=0} i \int_{\eta}^{T} \left(\frac{e^{-2\pi\psi(b,y)}}{e^{-2\pi y}-1} + \frac{e^{2\pi\psi(b,y)}}{e^{2\pi y}-1} \right) e^{2\pi i\varphi(b,y)}\, dy$$

$$+ \int_{a}^{b} \left(\frac{-e^{-2\pi\psi(x,T)}}{e^{2\pi ix-2\pi T}-1} + \frac{e^{2\pi\psi(x,T)}}{e^{2\pi ix+2\pi T}-1} \right) e^{2\pi i\varphi(x,T)}\, dx + \theta;$$

$$|\theta| \leqq 1.$$

3. Take T to be equal to

$$T = \sqrt{x_{s-1} - x_s}. \tag{7}$$

Then, for

$$x_s \leqq x \leqq x_{s-1}; \qquad 0 \leqq y \leqq T,$$

we can write

$$\psi(x, y) = \frac{n}{(\lambda + x_s)^2} \frac{[(\lambda + x)^2 - (\lambda + x_s)^2]\, y}{(\lambda + x)^2 + y^2} + \frac{\theta'}{2}; \qquad |\theta'| \leqq 1. \tag{8}$$

Indeed, by (3), (1) and (2), we find

$$\frac{1}{6} < \frac{\sqrt{n}}{2(s - A)^{3/2}} < x_{s-1} - x_s < \frac{\sqrt{n}}{(s - A)^{3/2}}$$

and, consequently,

$$\frac{1}{3} < \frac{1}{2} \sqrt{\frac{(\lambda + x_s)^3}{n}} < T < \sqrt{\frac{(\lambda + x_s)^3}{n}}. \tag{9}$$

With the help of these inequalities, it is not difficult to conclude that the second term on the right-hand side of

$$\psi(x, y) = \frac{n}{(\lambda + x_s)^2} \frac{[(\lambda + x)^2 - (\lambda + x_s)^2]\, y}{(\lambda + x)^2 + y^2} + \frac{n y^3}{(\lambda + x_s)^2 [(\lambda + x)^2 + y^2]}$$

is not greater than $\frac{1}{n^{1/4}} \leqq \frac{1}{2}$, thus proving the equality (8). Using (3), from (8) we easily obtain

$$\psi(x, y) \leqq y + \tfrac{1}{2}. \tag{10}$$

Turning our attention to the first term in (6), we find that its absolute value is not greater than

$$\int_0^T \left| e^{-2\pi\psi(a, y)} + \frac{e^{2\pi\psi(a, y)} - e^{-2\pi\psi(a, y)}}{e^{2\pi y} - 1} \right| dy.$$

Since $\psi(a, y) \geqq 0$, in the interval $(0, T)$, we have

$$\int_0^T e^{-2\pi\psi(a, y)}\, dy < T.$$

Now decomposing the integral

$$\int_0^T \frac{e^{2\pi\psi(a, y)} - e^{-2\pi\psi(a, y)}}{e^{2\pi y} - 1}\, dy$$

as the sum

$$\int_0^T = \int_0^{\frac{\ln 2}{2\pi}} + \int_{\frac{\ln 2}{2\pi}}^T$$

and since $2\psi(a, y) \leqq 3$ for $y \leqq 1$, we find that the first integral in the above sum is not greater than

$$\int_0^{\frac{\ln 2}{2\pi}} \frac{e^{6\pi x} - 1}{e^2 \pi y - 1}\, dy < 1,$$

while the second integral is not greater than

$$2 \int_0^T e^{2\pi\psi(a, y) - 2\pi y}\, dy,$$

which, by (10), is less than

$$2e^{\pi} T.$$

Thus, the absolute value of the first term on the right-hand side of (6) is less than

$$(2e^{\pi} + 1)\, T + 1.$$

Similar arguments are also applicable to the second term, and thus we find that it is less than

$$(2e^{\pi} + 1)\, T + 1.$$

4. Finally, consider the third term on the right-hand side of (6). By (9), it is less in absolute value than

$$\frac{1}{1 - e^{-\frac{2\pi}{3}}} \left(\int_a^b e^{-2\pi\psi(x, T)}\, dx + \int_a^b e^{-2\pi T + 2\pi\psi(x, T)}\, dx \right). \tag{11}$$

But, from (8) it follows that

$$-\psi(x, y) < -\frac{x - x_s}{4T} + \frac{1}{2}$$

therefore the first integral is less than

$$e^{\pi} \int_{x_s}^{x_{s-1}} e^{-\frac{(x - x_s)\pi}{2T}}\, dx < \frac{2e^{\pi}}{\pi} T.$$

The second integral is, by (8), less than

$$e^{\pi} \int_{x_s}^{x_{s-1}} e^{2\pi T \left(\frac{n}{(\lambda + x_s)^2} - \frac{n}{(\lambda + x)^2} - 1 \right)}\, dx. \tag{12}$$

But the derivative of the function

$$F(x) = \frac{n}{(\lambda + x_s)^2} - \frac{n}{(\lambda + x)^2} - 1$$

is equal to $2n/(\lambda + x)^3$. As x increases from x_s to x_{s-1}, it decreases and its least value is $2n/(\lambda + x_{s-1})^3$, while the function $F(x)$ increases and vanishes at $x = x_{s-1}$. Therefore in the interval (x_s, x_{s-1}) we can take

$$F(x) < -\frac{2n(x_{s-1}-x)}{(\lambda+x_{s-1})^3}.$$

Thus, expression (12) is less than

$$e^{\pi}\int_{x_s}^{x_{s-1}} e^{-\frac{4\pi Tn(x_{s-1}-x)}{(\lambda+x_{s-1})^3}}\,dx < \frac{e^{\pi}(\lambda+x_{s-1})^3}{4\pi Tn} < e^{\pi}T,$$

and expression (11) is less than

$$\frac{1}{1-e^{2\pi/3}}\left(\frac{2e^{\pi}}{\pi}+e^{\pi}\right)T < 2e^{\pi}T.$$

From this result and the statement proved in §3, from (6) we find that

$$\left|\sum_{x>a}^{x\leqq b} e^{2\pi i\frac{n(x-x_s)^2}{(\lambda+x_s)^2(\lambda+x)}}\right| < (6e^{\pi}+2)T+4 < k\sqrt{\frac{(\lambda+x_s)^3}{n}},$$

where k is a constant not dependent on a, b, n and λ. Hence, from (5) we get

$$|H| < 2k\sqrt{\frac{(\lambda+x_s)^3}{n}}$$

and from (4)

$$|S| < 2k\sum_{s=\mu+1}^{s=\nu}\sqrt{\frac{(\lambda+x_s)^3}{n}} = 2k\sum_{s=\nu}^{s=\mu+1}\frac{n^{1/4}}{(s-A)^{3/4}} < L\sqrt{\frac{n}{P}}, \tag{13}$$

where L is a constant not dependent on the numbers in the sum S.

5. In the sum

$$S' = \sum_{x>P}^{x\leqq Q}\sin 2\pi\left(\frac{n}{x}-Ax\right)$$

let P, Q and n be constrained by the inequality:

$$2\sqrt[3]{n} \leqq P \leqq Q \leqq \sqrt{2n}. \tag{14}$$

We can write

$$|S'| < \sum_{\varepsilon=0}^{\varepsilon=3}\left|\sum_{x>\frac{1}{4}P}^{x\leqq\frac{1}{4}Q}\sin 2\pi\left(\frac{n/4}{x+\varepsilon/4}-4Ax-A\varepsilon\right)\right| + 8.$$

But under the conditions (14) we have

$$\sqrt[3]{\frac{n}{4}} \leqq \frac{1}{4}P \leqq \frac{1}{4}Q \leqq \sqrt{\frac{n}{8}},$$

Hence the result derived in the last paragraph can be applied to each of the above sums, if $n/4$ is substituted for n. If $n/4$ is $\geqq 16$, we find

$$|S'| < 4L\sqrt{\frac{n}{P}} + 8.$$

In a similar way, under the conditions (14) we can show that

$$\left|\sum_{x>P}^{x\leqq Q} \cos 2\pi\left(\frac{n}{x} - Ax\right)\right| < 4L\sqrt{\frac{n}{P}} + 8,$$

and then formulate the results of this paragraph as the following.

Theorem. If n is a given number $\geqq 64$, A is some other given number, and P, Q satisfy the inequalities

$$2\sqrt[3]{n} \leqq P \leqq Q \leqq \sqrt{2n},$$

then

$$\left|\sum_{x>P}^{x\leqq Q} \sin 2\pi\left(\frac{n}{x} - Ax\right)\right| < N\sqrt{\frac{n}{P}},$$

$$\left|\sum_{x>P}^{x\leqq Q} \cos 2\pi\left(\frac{n}{x} - Ax\right)\right| < N\sqrt{\frac{n}{P}},$$

where N is a constant not dependent on n, A, P and Q.

§3. The Formula of N. Ya. Sonin

Henceforth we will frequently use Sonin's formula. Its proof is given in his paper[1]. According to this formula, on assuming

$$\varrho(x) = [x] - x + \tfrac{1}{2}; \qquad \sigma(x) = \int_0^y \varrho(x)\,dx,$$

we find that

$$\sum_{x>a}^{x\leqq b} f(x) = \int_a^b f(x)\,dx + \varrho(b)\,f(b) - \varrho(a)\,f(a) - \sigma(b)\,f'(b)$$
$$+ \sigma(a)\,f'(a) + \int_a^b \sigma(x)\,f''(x)\,dx.$$

Let μ denote the greatest value of $|f'(x)|$ in (a, b). Since

$$|\varrho(x)| \leqq \tfrac{1}{2}; \qquad |\sigma(x)| \leqq \tfrac{1}{8},$$

[1] Sonin, N. Ya.: On a definite integral containing the numerical function [x]. Bull. univ. Varsovie (1885) [in Russian].

we obtain

$$\sum_{x>a}^{x\leqq b} f(x) = \int_a^b f(x)\,dx + \varrho(b)\,f(b) - \varrho(a)\,f(a) + \frac{\theta}{2}\mu; \qquad |\theta| \leqq 1. \tag{1}$$

If the limit of $f'(\xi)$ is equal to 0 and $f''(\xi)$ does not change its sign in the interval (a, ∞), we obtain

$$\sum_{x>a}^{x\leqq b} f(x) = C + \int_a^b f(x)\,dx + \varrho(b)\,f(b) + \frac{\theta}{4} f'(b), \tag{2}$$

where $|\theta| \leqq 1$ and C is a constant not dependent on b.

§4. An Auxiliary Theorem from the Theory of Fourier Series

Let $\{\alpha\}$ denote the fractional part of a number α, i.e. the difference $\alpha - [\alpha]$.

Using the well known expansion in the theory of Fourier series

$$x = 2\left(\frac{\sin x}{1} - \frac{\sin 2x}{2} + \frac{\sin 3x}{3} - \dots\right),$$

which is valid for

$$-\pi < x < \pi \dots, \tag{1}$$

and since for non-integral α,

$$x = 2\pi\{\alpha\} - \pi$$

always satisfies the conditions (1), we obtain

$$\{\alpha\} = \frac{1}{2} - \frac{1}{\pi}\left(\frac{\sin 2\pi\alpha}{1} + \frac{\sin 4\pi\alpha}{2} + \frac{\sin 6\pi\alpha}{3} + \dots\right),$$

because

$$\sin k(2\pi\{\alpha\} - \pi) = (-1)^k \sin 2\pi k\alpha.$$

Applying the Abel transformation to the series

$$\frac{\sin 2\pi\alpha}{1} + \frac{\sin 4\pi\alpha}{2} + \frac{\sin 6\pi\alpha}{3} + \dots$$

and since, for any integers k and l,

$$|\sin 2\pi k\alpha + \sin 2\pi(k+1)\alpha + \sin 2\pi(k+2)\alpha + \dots + \sin 2\pi(k+l)\alpha| < \frac{1}{|\sin \pi\alpha|}$$

is always true, we find

$$\{\alpha\} = \frac{1}{2} - \frac{1}{\pi}\left(\frac{\sin 2\pi\alpha}{1} + \frac{\sin 4\pi\alpha}{2} + \frac{\sin 6\pi\alpha}{3} + \dots + \frac{\sin 2\pi N\alpha}{N}\right) + \frac{\theta}{\pi(N+1)\sin \pi\alpha}, \tag{2}$$

where $|\theta| \leqq 1$. If α is an integer, we have the obvious equality

$$\{\alpha\} = \frac{1}{2} - \frac{1}{\pi}\left(\frac{\sin 2\pi\alpha}{1} + \frac{\sin 4\pi\alpha}{2} + \frac{\sin 6\pi\alpha}{3} + \ldots + \frac{\sin 2\pi N\alpha}{N}\right) - \frac{1}{2}. \tag{3}$$

§5. The Order of the Function $\tau(a)$

Expressing an integer a in the form

$$a = p_1^{\alpha_1} p_2^{\alpha_2} p_3^{\alpha_3} \ldots p_k^{\alpha_k},$$

where $p_1, p_2, p_3, \ldots, p_k$ are different prime divisors of a, we obtain an expression for $\tau(a)$ the total number of all divisors of a:

$$\tau(a) = (\alpha_1 + 1)(\alpha_2 + 1) \ldots (\alpha_k + 1).$$

Denoting by ε some positive number $\leqq 1$, we find

$$\frac{\tau(a)}{a^\varepsilon} < \frac{\alpha_1 + 1}{2^{\varepsilon\alpha_1}} \cdot \frac{\alpha_2 + 1}{3^{\varepsilon\alpha_2}} \cdot \frac{\alpha_3 + 1}{4^{\varepsilon\alpha_3}} \cdots \frac{\alpha_k + 1}{(k+1)^{\varepsilon\alpha_k}}.$$

Separate the product on the right-hand side of this inequality into two products, of which the first consists of factors of the type $(\alpha_i + 1)/(i+1)^{\varepsilon\alpha_i}$, where $(i+1) < e^{1/\varepsilon}$, or is equal to 1, if there are no such factors. The second product consists of all the remaining factors or is equal to 1, if there are no such factors. There are less than $e^{1/\varepsilon}$ factors in the first product, each of which is not greater than the greatest value of $2x/2^{\varepsilon y}$ (for $x \geqq 1$) equal to $2/^\varepsilon e \ln 2$. In the second product, if it is not equal to 1, each factor $(\alpha_i + 1)/(i+1)^{\varepsilon\alpha_i} \leqq \dfrac{\alpha_i + 1}{e^{\alpha_i}}$, and consequently, it is not greater than the greatest value of $2x/e^y$ (for $x \geqq 1$), equal to $2/e < 1$. Hence, it follows that $\tau(a) < Ma^\varepsilon$, where $M = (2/\varepsilon e \ln 2)^{e^{1/\varepsilon}}$ is finite for any positive ε.

§6. Asymptotic Expression for the Sum $\sum_{y=1}^{y=n} \left[\frac{n}{x}\right]$

Let n and x be integers. If n is not divisible by x, then using formula (2) of §4, and taking $N = \left[\dfrac{x}{2\sqrt[3]{n}}\right]$, we obtain

$$\left\{\frac{n}{x}\right\} - \frac{1}{2} = -\frac{1}{\pi}\left(\frac{\sin 2\pi\frac{n}{x}}{1} + \frac{\sin 4\pi\frac{n}{x}}{2} + \ldots + \frac{\sin 2\pi\left[\frac{x}{2\sqrt[3]{n}}\right]\frac{n}{x}}{\left[\frac{x}{2\sqrt[3]{n}}\right]}\right) + \frac{\theta}{\pi\frac{x}{2\sqrt[3]{n}}\sin\pi\frac{n}{x}}; \quad |\theta| \leqq 1. \tag{1}$$

Denoting by $R(x)$ the least residue of n modulo x, and using the inequality:

$$\sin \pi \frac{|R(x)|}{x} \geqq \frac{2|R(x)|}{x},$$

(which can be easily derived), we can express the remainder of (1) in the form:

$$\theta' \frac{\sqrt[3]{n}}{|\pi R(x)|}; \qquad |\theta'| \leqq 1.$$

With the help of formula (3) of §4, we can in general write

$$\left\{\frac{n}{x}\right\} - \frac{1}{2} = -\frac{1}{\pi}\left(\frac{\sin 2\pi \frac{n}{x}}{1} + \frac{\sin 4\pi \frac{n}{x}}{2} + \ldots + \frac{\sin 2\pi \left[\frac{x}{2\sqrt[3]{n}}\right]\frac{n}{x}}{\left[\frac{x}{2\sqrt[3]{n}}\right]}\right) + \vartheta(x),$$

where $\vartheta(x) = -\frac{1}{2}$ or $|\vartheta(x)| \leqq \sqrt[3]{n}/\pi\,|R(x)|$, depending on whether $n \equiv 0 \pmod{x}$ or not. Hence, we find

$$\sum_{x>2\sqrt[3]{n}}^{x\leqq\sqrt{n}} \left(\left\{\frac{n}{x}\right\} - \frac{1}{2}\right) = -\frac{1}{\pi} \sum_{x>2\sqrt[3]{n}}^{x\leqq\sqrt{n}} \left(\frac{\sin 2\pi \frac{n}{x}}{1} + \frac{\sin 4\pi \frac{n}{x}}{2} + \ldots + \frac{\sin 2\pi \left[\frac{x}{2\sqrt[3]{n}}\right]\frac{n}{x}}{\left[\frac{x}{2\sqrt[3]{n}}\right]}\right)$$

$$+ \theta'' \sum_{x>2\sqrt[3]{n}}^{x\leqq\sqrt{n}} \vartheta(x); \qquad |\theta''| \leqq 1. \tag{2}$$

But from the equality

$$S = \sum_{x\geqq 2\sqrt[3]{n}}^{x\leqq\sqrt{n}} \left(\frac{\sin 2\pi \frac{n}{x}}{1} + \frac{\sin 4\pi \frac{n}{x}}{2} + \ldots + \frac{\sin 2\pi \left[\frac{x}{2\sqrt[3]{n}}\right]\frac{n}{x}}{\left[\frac{x}{2\sqrt[3]{n}}\right]}\right)$$

$$= \sum_{t=1}^{t\leqq\frac{1}{2}n^{1/6}} \frac{1}{t} \sum_{x\geqq 2t\sqrt[3]{n}}^{x\leqq\sqrt{n}} \sin 2\pi \frac{tn}{x},$$

and, by the theorem proved in §2, it follows that the sum S is of order

$$\sum_{t=1}^{t\leqq\frac{1}{2}n^{1/6}} \frac{1}{t} \sqrt{\frac{tn}{2t\sqrt[3]{n}}} = O(\sqrt[3]{n}\ln n).$$

Now turning our attention to the second term on the right-hand side of (2), we find that the number of x, for which $n \equiv 0 \pmod{x}$, is not greater than $\tau(n)$. In general, the number of x, for which $n \equiv \pm k \pmod{x}$, where $k < n$ is a given positive integer, will not be greater than $\tau(n-k) + \tau(n+k)$. Since the greatest value of the sum

$$\tau(n-k) + \tau(n+k); \qquad k = 0, 1, 2, \ldots, [\sqrt{n}],$$

according to §5, is less than $M'n^{\varepsilon'}$, where $\varepsilon' > 0$ and M' is a constant, we find

$$\sum_{x \geqq 2\sqrt[3]{n}}^{x \leqq \sqrt{n}} \vartheta(x) = O\left[n^{\frac{1}{3}+\varepsilon'}\left(1+\frac{1}{1}+\frac{1}{2}+\frac{1}{3}+\ldots+\frac{1}{[\frac{1}{2}\sqrt{n}]}\right)\right]$$

$$= O\,(n^{\frac{1}{3}+\varepsilon'}\ln n)\,.$$

Therefore, using (2), and since

$$\sum_{x=1}^{x<2\sqrt[3]{n}} \left(\left\{\frac{n}{x}\right\}-\frac{1}{2}\right) = O\,(n^{1/3}),$$

we can write

$$\sum_{x>0}^{x \leqq \sqrt{n}} \left\{\frac{n}{x}\right\} = \frac{\sqrt{n}}{2} + O\,(n^{\frac{1}{3}+\varepsilon}), \tag{3}$$

where ε is an arbitrarily small positive constant. Applying formula (2) of §3 to the sum

$$n + \sum_{x>1}^{x \leqq \sqrt{n}} \frac{n}{x}$$

we obtain

$$\sum_{x>0}^{x \leqq \sqrt{n}} \frac{n}{x} = nE + \frac{n}{2}\ln n + \sqrt{n}\,[\sqrt{n}] - n + \frac{1}{2}\sqrt{n} + \frac{\theta}{4}, \tag{4}$$

where E stands for the Euler constant and $|\theta| \leqq 1$. From (3) and (4) it follows that

$$\sum_{x>0}^{x \leqq \sqrt{n}} \left[\frac{n}{x}\right] = nE + \frac{n}{2}\ln n + \sqrt{n}\,[\sqrt{n}] - n + O\,(n^{\frac{1}{3}+\varepsilon}).$$

Substituting this expression into the well-known formula

$$\sum_{y=1}^{y=n} \left[\frac{n}{x}\right] = 2\sum_{x=1}^{x \leqq \sqrt{n}} \left[\frac{n}{x}\right] - [\sqrt{n}]^2,$$

and after certain simplifications we obtain the asymptotic equality:

$$\sum_{y=1}^{y=n} \left[\frac{n}{x}\right] = n\,(\ln n + 2E - 1) + O\,(n^{\frac{1}{3}+\varepsilon}).$$

§7. Auxiliary Propositions Needed in the Derivation of the Gauss Formula. Asymptotic Expression of the Sum $\sum_{x=1}^{x=\alpha}\left\{\frac{m+x^2}{a}\right\}$

1. Let m, a and α denote certain given integers, where a and α are positive. Then the number of solutions of the congruence

$$x^2+m\equiv l \pmod a$$

is not greater than $4\delta^{0.5}\tau(a/\delta)$, where $\delta=(l-m,a)$. Let

$$\mu=[\alpha/a]+1\,.$$

Then in the series of absolutely least residues of the numbers

$$m+1^2,\ m+2^2,\ m+3^2,\ \ldots,\ m+a^2 \tag{1}$$

there will be not more than $4\mu\delta^{0.5}\tau(a/\delta)$ residues of absolute value equal to l, if l satisfies the condition:

$$-a/2<l\leqq a/2\,.$$

Hence, applying formula (3) of §4 to the fractional part $\{m+x^2/a\}$, if it is zero, or applying formula (2) otherwise, we obtain

$$\left\{\frac{m+x^2}{a}\right\}=\frac{1}{2}-\frac{1}{\pi}\left(\frac{\sin 2\pi\frac{m+x^2}{a}}{1}+\ldots+\frac{\sin 2\pi N\frac{m+x^2}{a}}{N}\right)+\psi(x),$$

where $|\psi(x)|=\frac{1}{2}$ when $l=0$, and $\psi(x)<\dfrac{1}{\pi(N+1)\left|\sin\pi\frac{l}{a}\right|}<\dfrac{a}{(N+1)\,|l|}$ in the contrary case. If $N\leqq a$, we also have

$$|\psi(x)|<\tfrac{1}{2}+\frac{1}{\pi}(\ln N+1)<\frac{1}{\pi}(\ln a+1)\,.$$

Therefore

$$\sum_{y=1}^{y=\alpha}\left\{\frac{m+x^2}{a}\right\}=\frac{\alpha}{2}-\frac{1}{\pi}\sum_{t=1}^{N}\frac{1}{t}\sum_{y=1}^{y=\alpha}\sin 2\pi t\,\frac{m+x^2}{a}+R,$$

$$|R|<2\sum_{\delta/a}4\mu\delta^{0.5}\tau(a/\delta)\,((1/\pi)\ln a+1)+2a\sum_{\delta/a}\frac{4\mu\delta^{0.5}\tau(a/\delta)}{N+1}\sum_{z=1}^{z\leqq\frac{a}{2\delta}}\frac{1}{\delta z}$$

$$=O\left(r\,\frac{\mu a\ln a}{N}\right), \tag{2}$$

$$r=\sum_{\delta/a}\frac{\tau(a/\delta)}{\sqrt{\delta}}\,.$$

2. The sum

$$S_t = \sum_{x=1}^{x=\alpha} \sin 2\pi t \frac{m + x^2}{a}$$

is the coefficient of i in the expansion of the sum

$$\Phi_t = \sum_{x=1}^{x=\alpha} e^{2\pi i \frac{m+x^2}{4n}},$$

where

$$n = \frac{a}{4t}. \tag{3}$$

Expand the sum Ω_t as follows

$$\Omega_t = \sum_{y>0}^{y\leqq n} + \sum_{y>n}^{y\leqq 2n} + \sum_{y>2n}^{y\leqq 3n} + \ldots + \sum_{x>\left[\frac{\alpha}{n}\right]n}^{x=\alpha} \tag{4}$$

From the relationships

$$\sum_{x>2sn}^{x\leqq(2s+1)n} e^{2\pi i \frac{m+x^2}{4n}} = e^{2\pi i \left(\frac{m}{4n} - ns^2\right)} \sum_{k>0}^{k<n} e^{\frac{2\pi i}{4n}(-\{2sn\}+k)^2} + O(1),$$

$$\sum_{x>(2s-1)n}^{x\leqq 2sn} e^{2\pi i \frac{m+x^2}{4n}} = e^{2\pi i \left(\frac{m}{4n} - ns^2\right)} \sum_{k>0}^{k<n} e^{\frac{2\pi i}{4n}(\{2sn\}+k)^2} + O(1),$$

which can be easily derived, and with the help of (7) of §1, we get

$$\sum_{x>2sn}^{x\leqq(2s+1)n} e^{2\pi i \frac{m+x^2}{4n}} = \frac{1+i}{2}\sqrt{n}\, e^{2\pi i \left(\frac{m}{4n} - ns^2\right)} + O(1),$$

$$\sum_{x>(2s-1)n}^{x\leqq 2sn} e^{2\pi i \frac{m+x^2}{4n}} = \frac{1+i}{2}\sqrt{n}\, e^{2\pi i \left(\frac{m}{4n} - ns^2\right)} + O(1).$$

Finally, applying similar transformations to the last sum in (4), we obtain, by Equation (6) of §1:

$$\sum_{x>\left[\frac{\alpha}{n}\right]n}^{x=\alpha} e^{2\pi i \frac{m+x^2}{4n}} = O(\sqrt{n}).$$

Hence, from (4) we get

$$\Omega_t = (1+i)\sqrt{n} \sum_{s=1}^{s\leqq\frac{\alpha}{2n}} e^{2\pi i \left(\frac{m}{4n} - ns^2\right)} \times O\left(\frac{\alpha}{n} + \sqrt{n}\right).$$

Therefore, from what has been said in the beginning of this paragraph, and using (3), we find that

$$S_t = \sqrt{\frac{a}{4t}} \sum_{s=1}^{s\leqq\frac{2\alpha}{a}t} \left[\sin 2\pi\left(\frac{tm}{a} - \frac{as^2}{4t}\right) + \cos 2\pi\left(\frac{tm}{a} - \frac{as^2}{4t}\right)\right] + O\left(\frac{\alpha}{a}t + \sqrt{\frac{a}{t}}\right). \tag{5}$$

Now consider a number $M \geqq 1$, to be defined later. Let

$$N = [M].$$

Then, using (5), write (2) as follows:

$$\sum_{y=1}^{y=\alpha} \left\{\frac{m+x^2}{a}\right\} = \frac{a}{2} - \frac{1}{2\pi} \sum_{t>0}^{t \leqq M} \frac{\sqrt{a}}{t^{3/2}} \sum_{s>0}^{s \leqq \frac{2\alpha}{a} t} \left[\sin 2\pi \left(\frac{tm}{a} - \frac{as^2}{4t}\right)\right.$$
$$\left. + \cos 2\pi \left(\frac{tm}{a} - \frac{as^2}{4t}\right)\right] + O\left(\frac{M\alpha}{a} + \sqrt{a}\right) + \left(\frac{\alpha}{a} + 1\right) \frac{r a \ln a}{M}. \tag{6}$$

3. Putting $\alpha \leqq a$, we can write (6) as

$$\sum_{x=1}^{x=\alpha} \left\{\frac{m+x^2}{a}\right\} = \frac{\alpha}{2} + O\left(\sqrt{a} \sum_{t=0}^{t \leqq M} \frac{1}{\sqrt{t}} + M + \frac{r a \ln a}{M}\right).$$

Hence, assuming

$$M = r^{2/3} a^{1/3} (\ln a)^{2/3},$$

we find, from § 5, that

$$\sum_{x=0}^{x=\alpha} \left\{\frac{m+x^2}{a}\right\} = \frac{\alpha}{2} + O(a^{2/3} \sqrt[3]{r \ln a}) = \frac{\alpha}{2} + O(a^{2/3+\varepsilon}).$$

4. As a particular case, take $\alpha = 2a$. Then, in (6), the sum for a given t can be written as follows:

$$S_t' = \sum_{s>0}^{s=4t} \left[\sin 2\pi \left(\frac{tm}{a} - \frac{as^2}{4t}\right) + \cos 2\pi \left(\frac{tm}{a} - \frac{as^2}{4t}\right)\right].$$

Let d be the greatest common divisor of a and $4t$. With the help of the integers

$$\sigma = \frac{a}{d}; \qquad \tau = \frac{4t}{d}$$

rewrite the sum S_t' as

$$S_t' = \sum_{s>0}^{s=\tau d} \left[\sin 2\pi \left(\frac{tm}{a} - \frac{\sigma s^2}{\tau}\right) + \cos 2\pi \left(\frac{tm}{a} - \frac{\alpha s^2}{\tau}\right)\right].$$

Representing the sum S_t' as

$$S_t' = \sum_{s>0}^{s=\tau} + \sum_{s>\tau}^{s=2\tau} + \ldots + \sum_{s>\tau d-\tau}^{s=\tau d},$$

we find that each of the d sums in this expansion can be reduced to the sum

$$\sum_{s>0}^{s=\tau} \left[\sin 2\pi \left(\frac{tm}{a} - \frac{\sigma s^2}{\tau}\right) + \cos 2\pi \left(\frac{tm}{a} - \frac{\alpha s^2}{\tau}\right)\right],$$

which, since σ, τ are coprime, is of order $\sqrt{\tau}$. Consequently, S_t' is of order

$$d\sqrt{\tau} = O(\sqrt{td}).$$

Let us denote by $\delta(t)$ the greatest common divisor of a and t. Then $d \leqq 4\delta(t)$, and consequently

$$S_t' = O(\sqrt{t\delta(t)}).$$

Using this result, from (6) we get

$$\sum_{x=1}^{x=2a} \left\{\frac{m+x^2}{a}\right\} = a + O\left(\sum_{t>0}^{t\leqq M} \frac{\sqrt{a\delta(t)}}{t} + M + \sqrt{a} + \frac{r a \ln a}{M}\right). \tag{7}$$

Now let σ run through all different divisors of a. Then

$$\sum_{t>0}^{t\leqq M} \frac{\sqrt{a\delta(t)}}{t} = O\left(\sum_{\sigma} \sqrt{a\sigma} \sum_{t>0}^{t\leqq \frac{M}{\sigma}} \frac{1}{\sigma t}\right) = O\left(\sqrt{a}\ln M \sum_{\sigma} \frac{1}{\sqrt{\sigma}}\right),$$

and since

$$\sum_{\alpha} \frac{1}{\sqrt{\alpha}} = O\left(\frac{1}{\left(1-\frac{1}{\sqrt{p_1}}\right)\left(1-\frac{1}{\sqrt{p_2}}\right)\dots\left(1-\frac{1}{\sqrt{p_k}}\right)}\right) = O(2^k),$$

where $p_1, p_2, \dots, p_k$ are different prime divisors of a, we can write

$$\sum_{t>0}^{t\leqq M} \frac{\sqrt{a\delta(t)}}{t} = O(2^k\sqrt{a}\ln M).$$

Therefore, taking $M = \sqrt{a}$, from (7) we find

$$\sum_{x=1}^{x=2a} \left\{\frac{m+x^2}{a}\right\} = a + O(r\sqrt{a}\ln a).$$

Hence we obtain the asymptotic equalities:

$$\sum_{x=1}^{x=a} \left\{\frac{m+x^2}{a}\right\} = \frac{a}{2} + O(r\sqrt{a}\ln a) \tag{8}$$

and

$$\sum_{x=1}^{x\leqq \frac{a}{2}} \left\{\frac{m+x^2}{a}\right\} = \frac{a}{4} + O(r\sqrt{a}\ln a). \tag{9}$$

5. In conclusion we shall prove the following

Lemma. If $P \geqq 0$, $M \geqq P$ and $\Delta > 0$ are given numbers and if, for any i satisfying the condition $0 < i \leqq M$, the inequality

$$\left|\sum_{x=P+i}^{x=P+M} f(x)\right| \leqq \Delta,$$

is satisfied, then

$$\left|\sum_{x=P+i}^{x=P+M} \sqrt{x}\, f(x)\right| \leqq \Delta \sqrt{M+P}.$$

Proof. Putting

$$\beta_i = \sum_{x=P+i}^{x=P+M} f(x)$$

and since $|\beta_i| \leqq \Delta$, we find that

$$\begin{aligned}\left|\sum_{x=P+i}^{x=P+M} \sqrt{x}\, f(x)\right| &= |\beta_1 \sqrt{P+1} + \beta_2(\sqrt{P+2} - \sqrt{P+1}) \\ &\quad + \beta_3(\sqrt{P+3} - \sqrt{P+2}) + \ldots + \beta_M(\sqrt{P+M} - \sqrt{P+M-1})| \\ &\leqq \Delta\, |\sqrt{P+1} + (\sqrt{P+2} - \sqrt{P+1}) + (\sqrt{P+3} - \sqrt{P+2}) \\ &\quad + \ldots + (\sqrt{P+M} - \sqrt{P+M-1})| = \Delta \sqrt{P+M},\end{aligned}$$

which proves the lemma.

Using this lemma and the theorem demonstrated in §2, we find

$$\sum_{x>P}^{x=Q} \sqrt{x} \sin 2\pi \left(\frac{n}{x} - Ax\right) = O\left(\sqrt{n\frac{Q}{P}}\right),$$

$$\sum_{x>P}^{x\leqq Q} \sqrt{x} \cos 2\pi \left(\frac{n}{x} - Ax\right) = O\left(\sqrt{n\frac{Q}{P}}\right).$$

§8. Proof of the Gauss Formula

We shall now determine the mean of the number of classes of proper positive quadratic forms. If $h(-\Delta)$ denotes the number of proper primitive classes of discriminant $-\Delta$, then our problem reduces to finding the asymptotic expansion for the sum

$$\sum_{\Delta=1}^{\Delta\leqq m} h(-\Delta).$$

By the definition given on page 24, we have

$$\sum_{\Delta=1}^{\Delta\leqq m} h(-\Delta) = \sum \mu(k)\, F\left(\frac{m}{k^2}\right); \qquad k = 1, 3, 5, \ldots, \tag{1}$$

where $\mu(k)$ denotes the Moebius function and $F(n)$ is a function which is equal to zero when $n < 1$, and for $n > 1$ is defined by the expression:

$$F(n) = \psi(n) + \sigma(n) - \kappa(n) - \eta(n), \tag{2}$$

where $\psi(n)$, $\sigma(n)$, $\chi(n)$ and $\eta(n)$ denote the numbers of systems of integral values of x, y and z satisfying the inequalities

$$\left.\begin{aligned} xz - y^2 &\leqq n \\ 0 < x &< z \\ -\tfrac{1}{2}x < y &\leqq \tfrac{1}{2}x \end{aligned}\right\} \quad \text{for} \quad \psi(n), \tag{3}$$

$$\left.\begin{aligned} xz - y^2 &\leqq n \\ x &= z \\ 0 \leqq y &< \frac{x}{2} \end{aligned}\right\} \quad \text{for} \quad \sigma(n), \tag{4}$$

$$\left.\begin{aligned} 4xz - y^2 &\leqq n \\ 0 < x &< z \\ -x < y &\leqq x \end{aligned}\right\} \quad \text{for} \quad \chi(n),$$

$$\left.\begin{aligned} 4xz - y^2 &\leqq n \\ x &= z \\ 0 \leqq y &< x \end{aligned}\right\} \quad \text{for} \quad \eta(n).$$

Putting $z = x + t$, we find that $\psi(n)$, $\sigma(n)$, $\eta(n)$ and $\chi(n)$ can also be regarded as the numbers of systems of integral values of x, y and t satisfying the inequalities:

$$\left.\begin{aligned} tx + x^2 - y^2 &\leqq n \\ x > 0; \qquad t &> 0 \\ -\tfrac{1}{2}x < y &\leqq \tfrac{1}{2}x \end{aligned}\right\} \quad \text{for} \quad \psi(n),$$

$$\left.\begin{aligned} x^2 - y^2 &\leqq n \\ t &= 0 \\ 0 \leqq y &< \frac{x}{2} \end{aligned}\right\} \quad \text{for} \quad \sigma(n),$$

$$\left.\begin{aligned} 4tx + 4x^2 - y^2 &\leqq n \\ x > 0; \qquad t &> 0 \\ -x < y &\leqq x \end{aligned}\right\} \quad \text{for} \quad \chi(n),$$

$$\left.\begin{aligned} 4x^2 - y^2 &\leqq n \\ t &= 0 \\ 0 \leqq y &< x \end{aligned}\right\} \quad \text{for} \quad \eta(n).$$

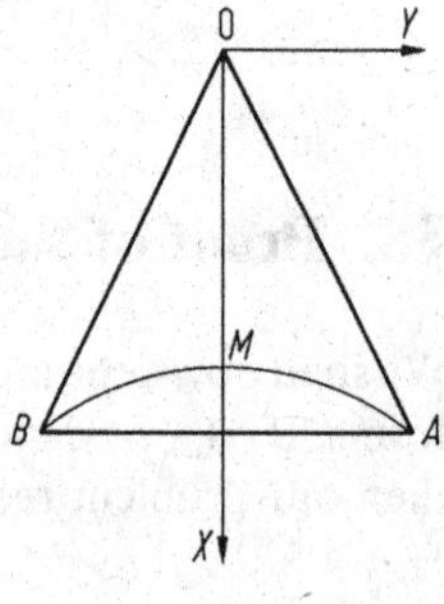

Fig. 3

First we shall determine the function $\psi(n)$. Construct a domain Ω bounded by the lines OA and OB (Fig. 3) defined by

$$y = \frac{x}{2}; \qquad y = -\frac{x}{2}$$

and an arc AMB of the curve $x^2 - y^2 = n$. The line $y = -x/2$ is not included in the domain.

It is easy to find that the system x_1, y_1 and t_1 of integers can satisfy (3) only if the point (x_1, y_1) lies in the domain Ω. On the other hand, it is not difficult to verify that to the point (x_1, y_1) in Ω, there correspond $\left[\frac{n+y_1^2}{x} - x\right]$ different values of t, which along with x_1 and y_1, satisfy the inequality. Hence, it follows that

$$\psi(n) = \sum_{\Omega}\left[\frac{n+y^2}{x} - x\right] = \sum_{\Omega}\left(\frac{n+y^2}{x} - x\right) - \sum_{\Omega}\left\{\frac{n+y^2}{x}\right\}, \tag{5}$$

where the summation $\sum_{\Omega}$ is taken over all pairs of x, y belonging to the domain Ω. Now we readily find that $x = \sqrt{4n/3}$ at A and B, and $x = \sqrt{n}$ at M. Let Ω' denote the domain bounded by the lines OA, OB and AB, where OB does not belong to this domain. Let Ω'' denote the domain bounded by the line AB and the arc AMB, while the arc AMB does not belong to this domain. Then equality (5) can be written as

$$\begin{aligned}\psi(n) &= \sum_{\Omega'}\left(\frac{n+y^2}{x} - x\right) - \sum_{\Omega''}\left(\frac{n+y^2}{x} - x\right)\\ &\quad - \sum_{\Omega'}\left\{\frac{n+y^2}{x}\right\} + \sum_{\Omega''}\left\{\frac{n+y^2}{x}\right\}.\end{aligned} \tag{6}$$

1. Consider the sum

$$S_3 = \sum_{\Omega'}\left\{\frac{n+y^2}{x}\right\}.$$

It can be rewritten as

$$S_3 = \sum_{x=1}^{x<\sqrt{\frac{4}{3}n}} \sum_{y>-x/2}^{y\leqq x/2} \left\{\frac{n+y^2}{x}\right\}.$$

Applying formula (9) of §7, we find

$$S_3 = \sum_{x=1}^{x\leqq\sqrt{\frac{4}{3}n}} \frac{x}{2} + O\left(\sum_{x=1}^{x\leqq\sqrt{\frac{4}{3}n}} r(x)\sqrt{x}\ln x\right); \qquad r(x) = \sum_{\delta/x} \frac{\tau(x/\delta)}{\sqrt{\delta}}.$$

But the sum

$$\sum_{x=1}^{x\leqq\sqrt{\frac{4}{3}n}} r(x)\sqrt{x}\ln x$$

is evidently of the order of

$$n^{1/4}\ln n \sum_{x=1}^{x\leqq\sqrt{\frac{4}{3}n}} r(x).$$

But, since we know that

$$\sum_{x=1}^{x\leqq i} \tau(x) = O(i \ln i),$$

we easily find that

$$\sum_{x=1}^{x\leqq\sqrt{\frac{4}{3}n}} r(x) = O(n^{3/4}\ln n)$$

and, consequently, we can finally write

$$S_3 = \sum_{x=1}^{x\leqq\sqrt{\frac{4}{3}n}} \frac{x}{2} + O(n^{3/4}\ln n)^2). \tag{7}$$

2. Now consider the sum

$$S_4 = \sum_{\Omega''} \left\{\frac{n+y^2}{x}\right\}.$$

It can be represented in the form

$$S_4 = \sum_{x>\sqrt{n}}^{x\leqq\sqrt{\frac{4}{3}n}} \sum_{y>-\sqrt{x^2-n}}^{y<\sqrt{x^2-n}} \left\{\frac{n+y^2}{x}\right\}.$$

Applying (6) of §7 to the sum

$$\sum_{y>-\sqrt{x^2-n}}^{y<\sqrt{x^2-n}} \left\{\frac{n+y^2}{x}\right\}$$

and putting $M=\sqrt{x}$, we find that

$$S_4 = \sum_{x>\sqrt{n}}^{x\leqq\sqrt{\frac{4}{3}n}} \sqrt{x^2-n} - \frac{1}{\pi} \sum_{x>\sqrt{n}}^{x\leqq\sqrt{\frac{4}{3}n}} \sum_{t=1}^{t\leqq\sqrt{x}} \sum_{s>0}^{s\leqq 2\frac{\sqrt{x^2-n}}{x}t} \frac{\sqrt{x}}{t^{3/2}} \left[\sin 2\pi\left(\frac{tn}{x}-\frac{xs^2}{4t}\right)\right.$$

$$\left. + \cos 2\pi\left(\frac{tn}{x}-\frac{xs^2}{4t}\right)\right] + O\left(\sum_{x>\sqrt{n}}^{x\leqq\sqrt{\frac{4}{3}n}} r(x)\sqrt{x}\ln x\right),$$

where $r(x)$ has the same meaning as in §1.

Now consider a domain Ω''' of the variables x, t, and s bounded by the surfaces $x=\sqrt{n}$, $x=\sqrt{\frac{4}{3}n}$, $t=0$, $t=\sqrt{x}$, $s=0$ and $s=2\frac{\sqrt{x^2-n}}{x}t$. The points on the surfaces $x=\sqrt{n}$, $t=0$ and $s=0$ (Fig. 4) are excluded from the domain. Then the second term in the expression for S_4, if the factor $-1/\pi$ is omitted, can be expressed as the sum

$$H = \sum_{\Omega'''} \frac{\sqrt{x}}{t^{3/2}} \left[\sin 2\pi\left(\frac{tn}{x}-\frac{xs^2}{4t}\right) + \cos 2\pi\left(\frac{tn}{x}-\frac{xs^2}{4t}\right)\right].$$

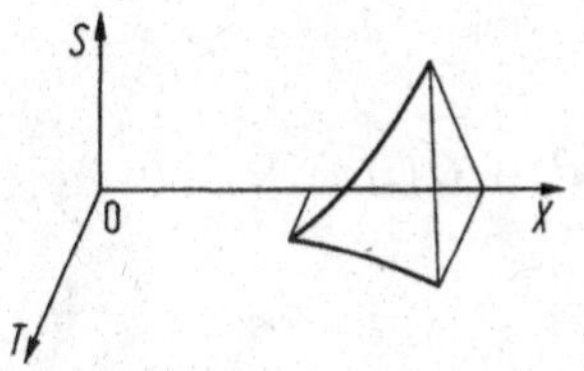

Fig. 4

Now draw a line in this domain that corresponds to a given pair of s and t. On this line x varies from a certain value $\alpha \geqq \sqrt{n}$ to $\sqrt{\frac{4}{3}n}$. Consequently, the part of the sum H that corresponds to this line can be expressed as

$$\frac{1}{t^{3/2}} \sum_{x>\alpha}^{x \leqq \sqrt{\frac{4}{3}n}} \left[\sin 2\pi\left(\frac{tn}{x} - \frac{xs^2}{4t}\right) + \cos 2\pi\left(\frac{tn}{x} - \frac{xs^2}{4t}\right)\right] \sqrt{x},$$

which, by Paragraph 5 of §7, is of order $\frac{\sqrt{tn}}{t^{3/2}} = \frac{\sqrt{n}}{t}$. Therefore we can write

$$H = O\left(\sum_{\Omega^{\mathrm{IV}}} \frac{\sqrt{n}}{t}\right),$$

where the summation $\sum_{\Omega^{\mathrm{IV}}}$ is taken over all pairs of t, s in Ω^{IV} bounded by the lines $s = 0$, $t = \sqrt[4]{\frac{4}{3}n}$, $s = t$. Therefore we can write:

$$H = O\left(\sum_{t>0}^{t \leqq \sqrt[4]{\frac{4}{3}n}} \sqrt{n}\right) = O(n^{3/4}).$$

Finally, repeating the arguments used at the end of Paragraph 1, we find

$$\sum_{x>n}^{x \leqq \sqrt{\frac{4}{3}n}} r(x) \sqrt{x} \ln x = O(n^{3/4} (\ln n)^2).$$

Thus, we infer that

$$S_4 = \sum_{x>\sqrt{n}}^{x \leqq \sqrt{\frac{4}{3}n}} \sqrt{x^2 - n} + O(n^{3/4} (\ln n)^2). \tag{8}$$

3. Consider the sum

$$S_1 = \sum_{\Omega'} \left(\frac{n + y^2}{x} - x\right).$$

Rewrite it as

$$S_1 = \sum_{x>0}^{x \leqq \sqrt{\frac{4}{3}n}} \sum_{y>-x/2}^{y \leqq x/2} \left(\frac{n + y^2}{x} - x\right).$$

Applying formula (1) of §3, since

$$\varrho(x/2) - \varrho(-x/2) = 0,$$

we obtain

$$S_1 = \sum_{x>0}^{x \leqq \sqrt{\frac{4}{3}n}} (n - \tfrac{11}{12}x^2) + O(\sqrt{n}). \tag{9}$$

4. Finally, consider the sum

$$S_2 = \sum_{\Omega''} \left(\frac{n+y^2}{x} - x\right),$$

Express it as

$$S_2 = \sum_{x>\sqrt{n}}^{x \leqq \sqrt{\frac{4}{3}n}} \sum_{y>-\sqrt{x^2-n}}^{y \leqq \sqrt{x^2-n}} \left(\frac{n+y^2}{x} - x\right).$$

Applying formula (1) of §3, we obtain

$$S_2 = -\sum_{x>\sqrt{n}}^{x \leqq \sqrt{\frac{4}{3}n}} \frac{4}{3x}(x^2-n)^{3/2} + O(\sqrt{n}). \tag{10}$$

By virtue of (7)–(10) and (6), we find that

$$\psi(n) = \sum_{x>0}^{x \leqq \sqrt{\frac{4}{3}n}} (n - \tfrac{11}{12}x^2) + \tfrac{4}{3} \sum_{x>\sqrt{n}}^{x \leqq \sqrt{\frac{4}{3}n}} (x^2-n)^{3/2}$$

$$- \sum_{x>0}^{x \leqq \sqrt{\frac{4}{3}n}} \frac{x}{2} + \sum_{x>\sqrt{n}}^{x \leqq \sqrt{\frac{4}{3}n}} \sqrt{x^2-n} + O(n^{3/4}(\ln n)^2). \tag{11}$$

5. From (4) it follows that $\sigma(n)$ can be regarded as the number of systems of integral values of x and y satisfying the inequalities:

$$x^2 - y^2 \leqq n; \quad 0 \leqq y < x/2.$$

Hence we easily obtain

$$\sigma(n) = \sum_{x>0}^{x \leqq \sqrt{\frac{4}{3}n}} \frac{x}{2} - \sum_{x>\sqrt{n}}^{x \leqq \sqrt{\frac{4}{3}n}} \sqrt{x^2-n} + O(\sqrt{n}).$$

Now from (11) we find

$$\psi(n) + \sigma(n) = \sum_{x>0}^{x \leqq \sqrt{\frac{4}{3}n}} (n - \tfrac{11}{12}x^2) + \tfrac{4}{3} \sum_{x>\sqrt{n}}^{x \leqq \sqrt{\frac{4}{3}n}} \frac{(x^2-n)^{3/2}}{x} + O(n^{3/4}(\ln n)^2). \tag{12}$$

We calculate the sums in this expression by formula (1) of §3. Putting

$$f(x) = n - \tfrac{11}{12}x^2,$$

we find

$$\sum_{y>0}^{x \leqq \sqrt{\frac{4}{3}n}} (n - \tfrac{11}{12}x^2) = \frac{32}{27\sqrt{3}} n^{3/2} - \frac{n}{2} - \tfrac{2}{9} n \varrho (\sqrt{\tfrac{4}{3}n}) + O(\sqrt{n}). \tag{13}$$

Taking

$$f(x) = \frac{(x^2 - n)^{3/2}}{x},$$

we find

$$f'(x) = (x^2 - n)^{1/2} \left(2 + \frac{n}{x^2}\right),$$

$$f''(x) = (x^2 - n)^{-1/2} \left(2x + \frac{2n^2}{x^3} - \frac{n}{x}\right),$$

Hence, $f'(x) > 0$ when $\sqrt{n} < x \leqq \sqrt{\frac{4}{3}n}$ and so $f'(x) < \frac{11}{4\sqrt{3}} \sqrt{n}$ under these conditions. Therefore we have

$$\sum_{x > \sqrt{n}}^{x \leqq \sqrt{\frac{4}{3}n}} \frac{(x^2 - n)^{3/2}}{x} = \int_{\sqrt{n}}^{\sqrt{\frac{4}{3}n}} \frac{(x^2 - n)^{3/2}}{x} dx + \frac{n}{6} \varrho (\sqrt{\tfrac{4}{3}n}) + O(\sqrt{n}). \tag{14}$$

From (13) and (14), and since

$$\int_{\sqrt{n}}^{\sqrt{\frac{4}{3}n}} \frac{(x^2 - n)^{3/2}}{x} dx = n^{3/2} \left(\frac{\pi}{6} - \frac{8}{9\sqrt{3}}\right),$$

we easily obtain, by (12), the following

$$\psi(n) + \sigma(n) = \frac{2\pi}{9} n^{3/2} - \frac{n}{2} + O(n^{3/4} (\ln n)^2).$$

Similarly, we find

$$\kappa(n) + \eta(n) = \frac{\pi}{18} n^{3/2} - \frac{n}{4} + O(n^{3/4} (\ln n)^2)$$

and then, by (2), we find

$$F(n) = \frac{\pi}{6} n^{3/2} - \frac{n}{4} + O(n^{3/4} (\ln n)^2).$$

Substituting this expression into (1), we obtain

$$\sum_{\Delta=1}^{\Delta \leqq m} h(-\Delta) = \frac{\pi}{6} m^{3/2} \sum \frac{\mu(k)}{k^3} - \frac{m}{4} \sum \frac{\mu(k)}{k^2}$$

$$+ O\left(m^{3/4} \sum \frac{\left(\ln \frac{m}{k}\right)^2}{k^{3/2}}\right); \qquad k = 1, 3, 5, \ldots \leqq \sqrt{m}.$$

Hence, putting

$$e = 1 + \frac{1}{2^3} + \frac{1}{3^3} + \dots,$$

we obtain

$$\sum \frac{\mu(k)}{k^3} = \frac{8}{7e} + O\left(\frac{1}{m}\right),$$

$$\sum \frac{\mu(k)}{k^2} = \frac{8}{\pi^2} + O\left(\frac{1}{\sqrt{m}}\right).$$

Thus, without any difficulty we find that

$$\sum_{\Delta=1}^{\Delta \leqq m} h(-\Delta) = \frac{4\pi}{21e} m^{3/2} - \frac{2}{\pi^2} m + O(m^{3/4} (\ln m)^2).$$

On the Distribution of Power Residues and Non-Residues

Zhurnal Fiz. Mat. Obshchestva pri Permskom Universitet (Journal of the Physico-Mathematical Society, Perm University), **1** (1918) 94–98.

The aim of this paper is to find the asymptotic laws governing the distribution of power residues and non-residues with respect to an indefinitely increasing modulus and to derive similar laws for numbers belonging to a given index for a given modulus.

Of the two methods which we have established, we shall only illustrate the application of the first method which is more effective in the case of a simple modulus. We hope to present the second method in detail in one of the papers to be published in near future. As regards the results derived in this paper, we attribute special significance to the law of distribution of primitive roots of a prime in an arithmetic progression whose common difference is not divisible by this prime.

§ 1. Let n be a prime greater than 3, g be one of its primitive root, ind z denote the index modulo n with respect to the base g of an integer not divisible by n, e and h be integers satisfying the conditions

$$eh = n - 1; \quad e > 1; \quad h \geqq 1,$$

and u an integer not divisible by e. Let

$$\varrho = \cos\frac{2\pi}{e} + i\sin\frac{2\pi}{e}, \qquad r = \cos\frac{2\pi}{n} + i\sin\frac{2\pi}{n},$$

$$(\varrho^u, r) = \sum_{z=1}^{n-1} \varrho^{u \operatorname{ind} z} \cdot r^z.$$

Then, according to Bachman[1], we find

$$(\varrho^u, r)(\varrho^{-u}, r) = (-1)^{uh} n.$$

Hence, we conclude that the absolute value of at least one of the numbers (ϱ^u, r) and (ϱ^{-u}, r) does not exceed $\sqrt{n}$.

Let $|(\varrho^{u\varphi(u)}, r)| \geqq \sqrt{n}$, where $\varphi(u)$ is $+1$ or -1. Let τ be an integer and if it is divisible by n, assume that $\varrho^{\delta \operatorname{ind}\tau} = 0$, where δ is some integer. It is easy to derive the following expansion

$$\varrho^{-u\varphi(u)\operatorname{ind}\tau}(\varrho^{u\varphi(u)}, r) = \sum_{z=1}^{z=n-1} \varrho^{u\varphi(u)\operatorname{ind} z} r^{z\tau}.$$

[1] Bachman, H.: Die Lehre von der Kreistheilung und ihre Beziehungen zur Zahlentheorie. Ch. 8, eq. 29 (1872).

§ **2.** Let a be an integer not divisible by n, c an integer and v be an integer such that

$$0 < v < n.$$

Applying the expansion derived in the previous section, we obtain

$$(\varrho^{u\varphi(u)}, r) \sum_{x=0}^{x=v-1} \varrho^{-u\varphi(u)\operatorname{ind}(ax+c)} = \sum_{z=1}^{z=n-1} \varrho^{u\varphi(u)\operatorname{ind} z}\, r^{zc} \cdot \frac{r^{azv}-1}{r^{az}-1}.$$

Hence, observing that

$$\left|\sum_{x=0}^{x=v-1} \varrho^{-u\varphi(u)\operatorname{ind}(ax+c)}\right| = \left|\sum_{x=0}^{x=v-1} \varrho^{u\operatorname{ind}(ax+c)}\right|.$$

we easily find

$$\left|\sum_{x=0}^{x=v-1} \varrho^{u\operatorname{ind}(ax+c)}\right| \leq \frac{1}{\sqrt{n}} \sum_{z=1}^{z=n-1} \frac{1}{\left|\sin\dfrac{\pi z a}{n}\right|} = \frac{2}{\sqrt{n}} \sum_{z=1}^{z=\frac{n-1}{2}} \frac{1}{\sin\dfrac{\pi z}{n}}.$$

But we know that

$$\sin\omega > \frac{2}{\pi}\,\omega,$$

when

$$0 < \omega < \frac{\pi}{2}.$$

Hence we easily obtain

$$\frac{2}{\sqrt{n}} \sum_{z=1}^{z=\frac{n-1}{2}} \frac{1}{\sin\dfrac{\pi z}{n}} < \sqrt{n} \sum_{z=1}^{z=\frac{n-1}{2}} \frac{1}{z}.$$

Applying the Euler expansion, we find

$$\sum_{z=1}^{z=\frac{n-1}{2}} \frac{1}{z} = E + \ln(n+1) - \ln 2 - \frac{1}{n+1} - \frac{\lambda}{3(n+1)^2}; \qquad 0 < \lambda < 1,$$

where E is the Euler constant. Hence, it follows that

$$\sum_{z=1}^{z=\frac{n-1}{2}} \frac{1}{z} < \ln n.$$

Using this inequality, we obtain

$$\left|\sum_{x=0}^{x=v-1} \varrho^{u\operatorname{ind}(ax+c)}\right| < \sqrt{n}\ln n.$$

§ **3.** Let m be an integer not divisible by n. Applying the inequality derived in the previous Section, we find

$$\left|\sum_{u=1}^{u=e-1} \sum_{x=0}^{x=v-1} \varrho^{u\operatorname{ind}(ax+c)} \varrho^{-u\operatorname{ind} m}\right| < (e-1)\sqrt{n}\ln n.$$

Hence, if q denotes the number of terms in the arithmetic progression

$$c, a+c, 2a+c, \ldots, va-a+c,$$

which are not divisible by n and are congruent modulo n to the product of the e-th power of an integer and m, and since the above A. P. does not contain more than one number divisible by n, we obtain

$$|q(e-1)-(v-q)| < (e-1)\sqrt{n}\ln n + 1.$$

Consequently, we can write

$$q = \frac{v}{e} + \eta\sqrt{n}\ln n; \quad |\eta| < 1,$$

which evidently remains true even when $e=1$ and $e=(n-1)$.

§ 4. Let e be one of the divisors of $(n-1)$, s one of the divisors of e, $F(s)$ the number of residues of index $\dfrac{(n-1)}{s}$ modulo n contained in the arithmetic progression:

$$c, a+c, 2a+c, \ldots, va-a+c$$

not divisible by n, and $f(s)$ the number of terms in this A. P. belonging to index s modulo n.

Obviously, we have

$$F(s) = \sum f(\sigma),$$

where σ runs through all the divisors of s. Hence, applying the rule of inversion of arithmetic functions, we find

$$f(e) = \sum \mu(s)\, F\left(\frac{e}{s}\right),$$

where s runs through all the divisors of e.

Applying the last equality of the previous Section, and denoting by k the number of different prime divisors of e, we find

$$f(e) = \sum \mu(s) \frac{ve}{(n-1)s} + \varepsilon \cdot 2^k \sqrt{n}\ln n$$

$$= \frac{v}{n-1}\varphi(e) + \varepsilon \cdot 2^k \sqrt{n}\ln n; \quad |\varepsilon| < 1.$$

Summing up these results, we come to the following conclusion. Let n be an integer greater than 3, e a divisor of $(n-1)$, k be the number of different prime divisors of e, a an integer not divisible by n, c an integer, m an integer not divisible by n, and v be an integer such that

$$0 < v < n.$$

Under these conditions:

1°. The number q of terms in the arithmetic progression

$$c, a+c, 2a+c, \ldots, va-a+c,$$

which are not divisible by n and congruent modulo n to the product of the e-th power of an integer and m, is

$$q = \frac{v}{e} + \eta\sqrt{n}\ln n; \quad |\eta| < 1.$$

2°. The number p of terms in this arithmetic progression which belong to exponent e modulo n is

$$p = \frac{v}{n-1}\,\varphi(e) + \varepsilon \cdot 2^k\sqrt{n}\ln n; \quad |\varepsilon| < 1.$$

In particular, putting $e = (n-1)$, we find that the number t of terms in the arithmetic progression

$$c, a + c, 2a + c, \ldots, va - a + c,$$

which are primitive roots of n, is equal to

$$t = v\,\frac{\varphi(n-1)}{n-1} + \varepsilon \cdot 2^k\sqrt{n}\ln n,$$

where k is the number of different prime divisors of $(n-1)$.

This formula shows that the least primitive root of n does not exceed

$$2^k\,\frac{n-1}{\varphi(n-1)}\sqrt{n}\ln n \leqq 2^{2k}\sqrt{n}\ln n.$$

Hence: To find a primitive root of a number n, it is sufficient to test the numbers:

$$2, 3, 4, \ldots, [2^{2k}\sqrt{n}\ln n].$$

An Elementary Proof of a General Theorem in Analytic Number Theory

Izvestiya Akademii Nauk SSSR, Ser. 6, **19** no. 16–17 (1925) 785–796.

In the present paper I shall present a new method for solving the problems of distribution of power residues and non-residues.

In my previous papers devoted to this question (the first of which was submitted in 1915 to the Physico-mathematical Faculty, Petrograd University) I solved this problem by different methods, simple but essentially dependent on the use of infinitesimals.

In the present paper I shall show that the basic theorem on the distribution of numbers which are congruent to numbers of the type $Ax^n \pmod p$ and are terms of the arithmetic progression $ax+b$; $x=0, 1, \ldots, (h-1)$ $(0<h<p)$ can be proved with the help of quite simple considerations, using the concept of incommensurable number solely for the sake of simplifying the expressions of remainder terms.

For the sake of brevity, I shall confine myself only to the derivation of the basic theorem, because all important corollaries (say, for instance, the laws of distribution of numbers belonging to a given exponent, laws of distribution of primitive roots, the upper bound

$$p^{1/2k}; \quad k = e^{\frac{n-1}{n}}$$

of the least residue of index n modulo p) can be derived from this theorem as has been done in my previous papers[1].

Lemma I. If p is a prime > 2, α is coprime to p and k is a positive integer, then there exist coprime integers x and y satisfying the conditions:

$$ax \equiv y \pmod p; \quad 0 < x \leqq k; \quad 0 < |y| < \frac{p}{k}.$$

Proof. Consider a system of congruences

$$\alpha r \equiv \beta_r \pmod p; \quad r = 1, 2, \ldots, k,$$

the right-hand sides of which are the least positive residues of the left-hand sides.

---

[1] On an asymptotic equality in the theory of quadratic forms, Zhur fiz.-mat. otd. Perm inst. (1918, no. 1). On the distribution of power residues and non-residues, *ibid.* = This collection, pp. 53–56. On the distribution of square residues and non-residues, *ibid* (1919, no. 11). On the mean value of the number of classes of proper primitive forms of negative discriminant. Soobsh. Kharkov mat. ob.-va = This collection, pp. 28–52.

Rearranging these congruences in such a way that the right-hand sides increase, and then adding the obvious congruence $\alpha.0 \equiv p \pmod p$, we obtain the system

$$\begin{aligned} \alpha\gamma_1 &\equiv \lambda_1 \pmod p, \\ \alpha\gamma_2 &\equiv \lambda_2 \pmod p, \\ &\dots\dots\dots\dots \\ \alpha\gamma_k &\equiv \gamma_k \pmod p, \\ \alpha \cdot 0 &\equiv p \pmod p), \end{aligned}$$

From this system, by subtraction we obtain

$$\begin{aligned} \alpha\gamma_1 &\equiv \lambda_1 \pmod p \\ \alpha(\gamma_2 - \gamma_1) &\equiv \lambda_2 - \lambda_1 \pmod p, \\ \alpha(\gamma_3 - \gamma_2) &\equiv \lambda_3 - \lambda_2 \pmod p, \\ &\dots\dots\dots\dots \\ \alpha(-\gamma_k) &\equiv p - \lambda_k \pmod p. \end{aligned}$$

Among $\lambda_1, \lambda_2 - \lambda_1, \lambda_3 - \lambda_2, \dots, p - \lambda_k$, certainly there exists at least one number $\leqq p(k+1)^{-1}$, because the number of these quantities is $(k+1)$, each of which is > 0, and their sum is p. Therefore, in the second system of congruences there exists at least one congruence of the type

$$\alpha x_1 \equiv y_1 \pmod p; \quad 0 < |x_1| \leqq k; \quad 0 < y_1 \leqq p(k+1)^{-1}.$$

Hence, since x_1 and y_1 can always be divided by their g.c.d., we find that the lemma is true.

Lemma II. Let k be any number $\geqq 1$, q be a positive integer $\leqq k$, c an integer, m a positive integer $\leqq kq^{-1}$, B an arbitrary number, and A a number of the type

$$A = \frac{t}{q} + \frac{\theta}{kq},$$

where t is an integer coprime to q and $|\theta| < 1$. Then

$$S = \sum_{x=c}^{c+mq-1} \{Ax + B\} = \tfrac{1}{2} mq + \tfrac{1}{2} \varrho (m+1); \quad |\varrho| < 1^{*},$$

where $\{z\}$ stands for the fractional part of z.

Proof. First take $q > 1$. We have

$$Ax + B = \frac{tx}{q} + \frac{\theta x}{kq} + B = \frac{tx + f(x)}{q}; \quad f(x) = Bq + \frac{\theta x}{k}.$$

The values of the function $f(x)\,(x = 0, 1, \dots, mq-1)$ form an arithmetic progression. Consider only the case where $\theta \geqq 0$. The case where $\theta < 0$ can be treated

---

* This lemma in a slightly different form has been proved in my paper: A new method of deriving asymptotic expressions for arithmetic functions. Izv. Ross. akad. nauk (ser. 6) **11** (1917) 1347–1378 = This collection, pp. 3–27.

in a similar manner. Let $n = [f(c)]$. Then two cases are possible:
a) All the values of $f(x)$ are less than $(n+1)$ or
b) at least one of these values is $\geqq (n+1)$.

First consider the case (a). Resolving the sum S as

$$S = \sum_{x=c}^{c+q-1} + \sum_{x=c+q}^{c+2q-1} + \ldots + \sum_{x=c+mq-q}^{c+mq-1}, \tag{1}$$

consider one of the sums in this expansion

$$I_s = \sum_{x=c+sq}^{c+sq+q-1} \left\{ \frac{tx+n+\lambda(x)}{q} \right\}; \quad 0 \leqq \lambda(x) < 1.$$

Substituting the least positive residue $r \bmod q$ for the numbers $tx+n$ (which can always be done as $\{z\}$ does not change when an integer is added to or subtracted from z) and putting $\lambda(x) = \nu(z)$, we obtain

$$I_s = \sum_{r=0}^{q-1} \left\{ \frac{r+\nu(r)}{q} \right\}; \quad 0 \leqq \nu(r) < 1.$$

Obviously,

$$\left\{ \frac{r+\nu(r)}{q} \right\} = \frac{r+\nu(r)}{q}; \quad r = 0, 1, \ldots, q-1.$$

Therefore

$$I_s = \frac{1}{2}q - \frac{1}{2} + \frac{1}{q}\sum_{r=0}^{q-1} \nu(r) = \frac{1}{2}q - \frac{1}{2} + \theta'; \quad 0 \leqq \theta' < 1,$$

Hence

$$I_s = \tfrac{1}{2}q + \tfrac{1}{2}\varrho_s; \quad |\varrho_s| \leqq 1. \tag{2}$$

At the same time, we have

$$S = \tfrac{1}{2}mq + \tfrac{1}{2}m\varrho; \quad |\varrho| \leqq 1,$$

which proves the lemma for the case (a).

Now we shall take up case (b). Let σ be the greatest integer satisfying the condition: $f(c+\sigma q) < n+1$. Then, for the sums I_s in (1), for which $s \leqq \sigma$, put

$$f(x) = n + \lambda(x),$$

and for the sums for which $s > \sigma$, take

$$f(x) = n + 1 + \lambda(x).$$

Under these conditions, for any I_s, $s \gtrless \delta$, we have $0 \leqq \lambda(x) < 1$. Therefore, for all $s \gtrless \sigma$, the equality (2) holds valid as before. This equality holds true also when $s = \sigma$, if

$$\lambda(c + \sigma q + q - 1) < 1.$$

It now remains to consider the sum I_σ under the following conditions $\lambda(c+\sigma q) < 1 \leqq \lambda(c+\sigma q+q-1) < 2$. We have

$$\frac{1}{2} \leqq \frac{1}{q} \sum_{x=c+\sigma q}^{c+\sigma q+q-1} \lambda(x) < \frac{3}{2}.$$

As in the case (a), after reducing the sum I_σ to the form:

$$I_\sigma = \sum_{r=0}^{q-1} \left\{\frac{r+\nu(r)}{q}\right\},$$

we can write

$$\left\{\frac{r+\nu(r)}{q}\right\} = \frac{r+\nu(r)}{q}$$

only when $r = 0, 1, \ldots,$ (because here it is possible that $1 \leqq \nu(\tau) \leqq 2$), while when $r = q-1$ this equality has to be replaced by the following equality:

$$\left\{\frac{r+\nu(r)}{q}\right\} - \frac{r+\nu(r)}{q} = \delta,$$

where δ can be either 0 or 1. Thus, we have

$$I_\sigma = \tfrac{1}{2} q - \tfrac{1}{2} + \delta + \frac{1}{q} \sum_{x=c+\sigma q}^{c+\sigma q+q-1} \lambda(x) = \tfrac{1}{2} q + \varrho_\sigma; \quad |\varrho_\sigma| < 1.$$

On substituting this expression for I_σ and the expression (2) for I_s, $s \gtrless \sigma$, into (1), we find that the lemma is true.

2) Now put $q = 1$. Then, evidently

$$-\tfrac{1}{2} m \leqq S - \tfrac{1}{2} m q \leqq \tfrac{1}{2} m.$$

Thus, the proof of the lemma is complete.

Lemma III. Let p be a prime > 2, α an integer not divisible by p, h a positive integer $< p$, and β_α be any integer which varies with α. Furthermore, let

$$S_\alpha = \sum_{x=0}^{h-1} \left\{\frac{\alpha x + \beta_\alpha}{p}\right\}; \qquad L_\alpha = S_\alpha - \tfrac{1}{2} h.$$

Then the sum $\sum |L_\alpha|$ taken over all the numbers of the sequence

$$1, 2, \ldots, p-1,$$

is less than

$$T = \sum_{x=1}^{h} \sum_{y=1}^{[p x^{-1}]} \left(\frac{p}{xy} + 1\right),$$

where for each x, summation with respect to y is taken over only those y which are coprime to x.

Proof. Take any sum S_α. In Lemma I put $k = h$. Then we find x_0 and y_0, coprime such that

$$\alpha x_0 \equiv y_0 \pmod{p}; \quad 0 < x_0 \leqq h; \quad 0 < |y_0| < \frac{p}{h}.$$

Hence $\alpha x_0 = y_0 + t_0 p$, where t_0 is an integer. At the same time we obtain

$$\frac{\alpha}{p} = \frac{t_0}{x_0} + \frac{y_0}{x_0 p} = \frac{t_0}{x_0} + \frac{\theta_0}{x_0 h}; \qquad |\theta_0| < 1.$$

Putting $m = [h x_0^{-1}]$; $\quad h_1 = h - m x_0$, we find

$$S_\alpha = \sum_{x=0}^{m x_0 - 1} \left\{\frac{\alpha x + \beta_\alpha}{p}\right\} + S'_\alpha; \quad S'_\alpha = \sum_{x = m x_0}^{m x_0 + h_1 - 1} \left\{\frac{\alpha x + \beta_\alpha}{p}\right\}; \qquad 0 \leqq h_1 < h.$$

Hence applying Lemma II, we find

$$S_\alpha = \tfrac{1}{2} m x_0 + \tfrac{1}{2} \varrho (m+1) + S'_\alpha$$

$$= \tfrac{1}{2}(h - h_1) + \tfrac{1}{2} \varrho_0 \left(\frac{h}{x_0} + 1\right) + S'_\alpha; \qquad |\varrho_0| < 1.$$

Putting $k_1 = h_1$, and treating the sum S'_α in the same way as we did with the sum S_α, we obtain

$$S'_\alpha = \tfrac{1}{2}(h_1 - h_2) + \tfrac{1}{2} \varrho_1 \left(\frac{h_1}{x_1} + 1\right) + S''_\alpha; \qquad |\varrho_1| < 1; \quad 0 \leqq h_2 < h_1,$$

where the sum S''_α will contain h_2 terms. Similarly, we find

$$S''_\alpha = \tfrac{1}{2}(h_2 - h_3) + \tfrac{1}{2} \varrho_2 \left(\frac{h_2}{x_2} + 1\right) + S'''_\alpha; \qquad |\varrho_2| < 1; \quad 0 \leqq h_3 < h_2,$$

and so on till we arrive at some $h_{n+1} = 0$. Then

$$S_\alpha = \tfrac{1}{2} h + \tfrac{1}{2} \sigma \left[\left(\frac{h}{x_0} + 1\right) + \left(\frac{h_1}{x_1} + 1\right) + \ldots + \left(\frac{h_n}{x_n} + 1\right)\right]; \qquad |\sigma| < 1.$$

The Lemma will be proved, if we can demonstrate that

$$\Omega = \tfrac{1}{2} \sum_\alpha \left[\left(\frac{h}{x_0} + 1\right) + \left(\frac{h_1}{x_1} + 1\right) + \ldots + \left(\frac{h_n}{x_n} + 1\right)\right] < T,$$

where the summation is taken over all the terms of (3) and the number n, as well as h_1, $h_2, \ldots, h_n, x_0, x_1, \ldots, x_n$, depends on the choice of α.

In order to estimate an upper bound for the sum Ω, we shall first calculate an upper bound for the sum of those terms $k/x + 1$, which correspond to the same value of x. To a given x correspond only those values of α that satisfy the congruence $\alpha x \equiv y \pmod{p}$ where y is an integer coprime to x satisfying the condition $|y| < pk^{-1}$ and consequently $|y| < px^{-1}$. Therefore, for a given x, y may take only those values from

$$\pm 1, \quad \pm 2, \ldots, \pm [px^{-1}],$$

which are coprime to x. Accordingly, to each such y, we find an α. To each system of numbers x, y and α, there corresponds a k satisfying the conditions $|y| < pk^{-1}$ or $k < p|y|^{-1}$, i.e. $kx^{-1} < p(x|y|)^{-1}$. Therefore the sum of all terms in Ω that correspond to a given x will be less than $\sum_{y=1}^{[px^{-1}]} \left(\frac{p}{xy} + 1\right)$, where summation is taken over only those y's which are coprime to x. Hence, Lemma III follows directly.

Lemma IV. Let p be a prime > 2, α an integer not divisible by p, β_α an integer that varies with α; h and γ be integers such that $0 < h < p$ and $0 < \gamma < p$. Let R_α denote the least positive residue of the numbers

$$\alpha x + \beta_\alpha; \quad x = 0, 1; \ldots, h-1,$$

that are less than γ, and let

$$R_\alpha = h\gamma p^{-1} + H_\alpha.$$

Then, summing over all $\alpha = 1, 2, \ldots, (p-1)$, we find

$$\sum |H_\alpha| < 2T.$$

Proof. By Lemma III, on putting

$$S'_\alpha = \sum_{x=0}^{h-1} \left\{\frac{\alpha x + \beta - \gamma}{p}\right\} = \frac{1}{2}h + L'_\alpha; \quad S_\alpha = \sum_{x=0}^{h-1} \left\{\frac{\alpha x + \beta}{p}\right\} = \frac{1}{2}h + L_\alpha,$$

we obtain

$$\sum |L'_\alpha| < T; \quad \sum |L_\alpha| < T; \quad \sum |S'_\alpha - S_\alpha| < 2T.$$

But we easily find that

$$1) \text{ if } \left\{\frac{\alpha x + \beta}{p}\right\} < \frac{\gamma}{p}, \quad \text{then} \quad \left\{\frac{\alpha x + \beta - \gamma}{p}\right\} = \left\{\frac{\alpha x + \beta}{p}\right\} + 1 - \frac{\gamma}{p},$$

$$2) \text{ if } \left\{\frac{\alpha x + \beta}{p}\right\} \geqq \frac{\gamma}{p}, \quad \text{then} \quad \left\{\frac{\alpha x + \beta - \gamma}{p}\right\} = \left\{\frac{\alpha x + \beta}{p}\right\} - \frac{\gamma}{p}.$$

Therefore

$$S'_\alpha - S_\alpha = R_\alpha - \frac{h\gamma}{p} = H_\alpha,$$

which proves the lemma, because

$$\sum |S'_\alpha - S_\alpha| < 2T.$$

Theorem. Let p be a prime > 2, e one of the divisors of $(p-1)$, a an integer not divisible by p, and b any integer. After dividing the numbers $1, 2, \ldots, (p-1)$ into e classes $(0, 1, \ldots, (e-1))$ such that the i-th class contains all the numbers whose indices[1] are $\equiv i \pmod e$, the number of elements in any class that occur in the arithmetic progression $ax + b$; $(x = 0, 1, \ldots, h-1)$; $(0 < h < p)$ can be represented in the following form:

$$\frac{h}{e} + \Delta; \quad \Delta^2 < 2T.$$

Proof. Let $(p-1)e^{-1} = f$ and consider fh numbers of the type:

$$\alpha(ax + b), \tag{4}$$

where α runs through all numbers of the i-th class, while x runs, independently of α,

---

[1] with respect to the chosen primitive root.

through all numbers $0, 1, \ldots, (h-1)$. Accordingly, for each number in (4) we find one and only one u satisfying the conditions

$$au + b \equiv \alpha(ax+b) \pmod{p}; \qquad 0 \leqq u < p.$$

This number u, on introducing an a' satisfying the congruence $aa' \equiv 1 \pmod{p}$, can be determined from the following conditions:

$$u \equiv \alpha x + \beta_\alpha \pmod{p}; \qquad \beta_\alpha = \alpha b a' - b a'; \qquad 0 \leqq u < p. \tag{5}$$

Let D be the number of figures less than h found in this manner. The idea of the proof lies in calculating the number D in two different ways.

1) If α is constant, then β_α will also remain unchanged. Therefore, by virtue of the congruence (5), the number of u's less than h that correspond to all the terms of the sequence

$$\alpha(ax+b); \qquad x = 0, 1, \ldots, h-1,$$

according to Lemma IV, can be represented as

$$\frac{h^2}{p} + H_\alpha.$$

After summing over all $\alpha = 1, 2, \ldots, (p-1)$ not necessarily belonging only to the i-th class, we obtain

$$\sum |H_\alpha| < 2T. \tag{6}$$

But, since the number of α belonging to the i-th class is f, we have

$$D = f\frac{h^2}{p} + \sum_i |H_\alpha|,$$

where $\sum_i$ is the sum taken over all the numbers in the i-th class.

2) Now assume that the sequence

$$ax + b; \qquad x = 0, 1, \ldots, h-1 \tag{7}$$

contains c_0 numbers from the 0-th class, c_1 numbers from the 1-st class, ..., and c_{e-1} numbers from the $(e-1)$-th class. We shall also use the symbol c_s even for $s \geqq e$, to denote the total number of figures in a class whose index is the least positive residue of s (mod e). On multiplying one of the numbers in the j-th class of the sequence (7) by all the numbers in the i-th class, and then replacing the products by $au + b$ $(0 \leqq u < p)$ which are congruent to them, we obtain f numbers of the type $au + b$ evidently belonging to the $(i+j)$-th class. Among these numbers, there will be c_{i+j} numbers for which $u < h$. Therefore, since j can take only the values $0, 1, \ldots\ldots, (e-1)$, we find

$$D \equiv c_0 c_i + c_{i+1} + \ldots + c_{e-1} c_{i+e-1}.$$

Comparing this value of D with the value found earlier, we obtain

$$c_0 c_i + c_1 c_{i+1} + \ldots + c_{e-1} c_{i+e-1} = f\frac{h^2}{p} + \sum_i H_\alpha.$$

In particular, for $i = 0$, we have

$$c_0^2 + c_1^2 + \ldots + c_{e-1}^2 = f\frac{h^2}{p} + \sum_0 H_\alpha.$$

Subtracting the previous result from this equality, by virtue of (6), we obtain

$$\tfrac{1}{2}\left[(c_0 - c_i)^2 + (c_1 - c_{i+1})^2 + \ldots + (c_{e-1} - c_{i+e-1})^2\right] < 2T,$$

Hence for any r

$$(c_r - c_{r+i})^2 + (c_{r-i} - c_r)^2 < 4T.$$

Adding this equality for $i = 1, 2, \ldots, (e-1)$ to the identity $(c_r - c_r)^2 + (c_r - c_r)^2 = 0$, we obtain:

$$2\sum_{s=0}^{e-1} (c_r - c_s)^2 < 4T(e-1).$$

Hence, putting $(c_0 + c_1 + \ldots + c_{e-1})e^{-1} = g$; $\quad c_i = g + \delta_i$,

$$\sum_{s=0}^{e-1} (\delta_r - \delta_s)^2 < 2T(e-1); \quad e\delta_r^2 + \sum_{s=0}^{e-1} \delta_s^2 < 2T(e-1);$$

$$(e+1)\delta_r^2 < 2T(e-1); \quad \delta_r^2 < 2T\frac{e-1}{e+1}.$$

But $c_0 + c_1 + \ldots + c_{e-1}$ is equal to h or $(h-1)$ (because one of the numbers in (7) may be divisible by p). Therefore, $g = (h - \sigma)e^{-1}$, where $\sigma = 0$ or 1. At the same time, we have

$$\left(c_r - \frac{h}{e} + \frac{\sigma}{e}\right)^2 < 2T\frac{e-1}{e+1}; \quad \Delta_r^2 < 2T\frac{e-1}{e+1} + 2\frac{|\Delta_r|}{e} < 2 \text{ (*)},$$

which proves the theorem.

Note 1. By roughly estimating the upper bound of $2T$, it is not difficult to establish that

$$2T < p(\ln p)^2,$$

which also gives the same order for the remainder term as the analytic methods.

Note 2. Obviously, the result I have derived includes the basic theorem stated at the beginning as a particular case.

(*) Δ_r denotes the remainder term in the r-th class, i.e. $\Delta_z = C_z - \frac{h}{e}$.

On the Distribution of Indices

Doklady Akademii Nauk SSSR, Ser. A (1926) 73–76.

I have derived the asymptotic laws governing the distribution of power residues and non-residues and also the laws of distribution of numbers belonging to a given exponent modulo a given prime, in particular, the number of primitive roots[1].

In the present note I shall show that the same method also permits us to find the general asymptotic laws describing the distribution of indices of numbers belonging to an arithmetic progression whose common difference is a number coprime to the modulus. For the sake of brevity, I shall confine myself only to the case of a prime modulus p, though the method is applicable to the general case.

Lemma 1. Let c be an integer $\geqq 2$ and u, v be numbers which run, independently of each other, through the system of integers

$$u = u_1, u_2, \ldots, u_n; \qquad v = v_1, v_2, \ldots, v_s.$$

If m denotes any integer not divisible by c, then

$$\left|\sum_u e^{2\pi i \frac{mu}{c}}\right| \leqq \Delta; \qquad \sum_{m=1}^{c-1} \left|\sum_v e^{2\pi i \frac{mv}{c}}\right| \leqq D.$$

The number T of solutions of the congruence

$$u \equiv v \pmod{c}$$

can be represented in the form

$$T = \frac{ns}{c} + \theta \frac{\Delta D}{c}; \qquad |\theta| \leqq 1.$$

Proof. Consider the sum

$$S = \sum_u \sum_v \sum_{m=0}^{c-1} e^{2\pi i \frac{m(u-v)}{c}}.$$

1°. On summing with respect to m for given u and v, we find that, for the case $u \equiv v \pmod{c}$ the sum is equal to c and otherwise to zero.

Therefore

$$S = cT.$$

---

[1] On the distribution of power residues and non-residues. Zhur. Perm fiz.-mat. ob.-va 1 (1918) 94–98 = This collection, pp. 53–56.

2°. On summing with respect to u and v, for $m=0$, we obtain the sum ns. The result of summation over other values of u, v, m, by the conditions of the lemma, will not be greater in magnitude than ΔD. Therefore we have

$$S = ns + \theta\Delta D; \qquad |\theta| \leqq 1.$$

Comparing both these results, we find that

$$cT = ns + \theta\Delta D.$$

Hence the validity of the lemma follows.

Lemma 2. Let p be a prime greater than 2, c a divisor of $(p-1)$ satisfying the condition $1 < c \leqq (p-1)$, a an integer not divisible by p and b be any integer. Then for any integer m not divisible by c and for any integer p_1 satisfying the condition $0 < p_1 < p$, we have(*)

$$\left|\sum_{x=0}^{p_1-1} e^{2\pi i m \frac{\operatorname{ind}(ax+b)}{c}}\right| < \sqrt{p}\ln p,$$

where summation is taken over only those values of x for which $(ax+b)$ are not divisible by p.

The proof of this lemma can be found in my paper quoted above.

Lemma 3. Let c be an integer $\geqq 3$, h and f divisors of c satisfying the condition $hf = c$, f_1 an integer satisfying the condition $0 < f_1 < f$, α an integer coprime to f and ε any real number. Then

$$S = \sum_{m=1}^{c-1}\left|\sum_{y=0}^{f_1-1} e^{2\pi i m\left(\frac{ay}{f}+\varepsilon\right)}\right| < c\ln c.$$

Proof. For the case where $c = f$, this lemma can be demonstrated by repeating the arguments that were applied to a similar sum in my paper cited above. Therefore we shall only consider a case where $f < c$, i.e. $h > 1$. When $f < 3$, the lemma is self-evident. Therefore, assume that $f \geqq 3$.

If we disregard the terms corresponding to those values of m that are divisible by f, whose sum is obviously equal to $(h-1)f_1$, the remainder of the sum S can be divided into h sums, each of which can be represented as

$$\sum_{m=1}^{f-1}\left|\sum_{y=0}^{f_1-1} e^{2\pi i m\left(\frac{ay}{f}+\varepsilon\right)}\right|.$$

Consequently, according to the case examined above, this sum will be less than $f\ln f$.

Therefore the whole sum S is less than

$$(h-1)f_1 + hf\ln f < c - f + c\ln\frac{c}{h} = c\ln c + c - f - c\ln h.$$

But for $h > 2$, we have

$$c - f - c\ln h < c - c\ln 3 < 0,$$

(*) The index is taken with respect to the chosen primitive root.

and when $h = 2$, we find

$$c - f - c \ln h = c - \frac{c}{2} - c \ln 2 < 0 .$$

Thus the lemma has been proved.

Theorem 1. Let p be a prime $\geqq 5$, c a divisor of $(p-1)$ exceeding 2, f and h divisors of c satisfying the condition $hf = c$, a an integer not divisible by p, α an integer coprime to f, b and β any integers and finally, p_1 and f_1 integers satisfying the condition

$$0 < p_1 \leqq p; \qquad 0 < f_1 \leqq f .$$

Then, if T denotes the number of solutions of the congruence:

$$\operatorname{ind}(ax + b) \equiv h\alpha y + \beta \pmod{c}, \tag{1}$$

where x and y run, independently of each other, through the systems of values

$$x = 0, 1, 2, \ldots, p_1 - 1; \qquad y = 0, 1, 2, \ldots, f_1 - 1$$

(omitting those values of x for which $(ax + b)$ is divisible by p), we find

$$T = \frac{p_1 f_1}{c} + \theta \sqrt{p} \ln p \ln c; \qquad |\theta| < 1 .$$

Proof. In Lemma 1 put

$$n = p_1, \qquad s = f_1, \qquad u_i = \operatorname{ind}[a(i-1) + b], \qquad v_i = h\alpha(i-1) + \beta .$$

By Lemma 2, the number Δ can be taken to be equal to $\sqrt{p} \ln p$, and according to Lemma 3, D can be taken to be $c \ln c$ ($\varepsilon = \beta c^{-1}$). Applying Lemma 1 under these conditions, we find that the theorem has been proved. Now put $h = 1$, $c = (p-1)$, then from Theorem 1 we obtain the following

Corollary. The number T of terms in the sequence

$$\operatorname{ind}(ax + b); \qquad x = 0, 1, \ldots, p_1 - 1 ,$$

which are congruent modulo $(p-1)$ to the terms of the sequence

$$\alpha y + \beta; \qquad y = 0, 1, \ldots, f_1 - 1; \qquad 0 < f_1 \leqq p - 1 ,$$

can be expressed as

$$T = \frac{p_1 f_1}{p-1} + \theta \sqrt{p} (\ln p)^2 .$$

A different interpretation of these results. Let g denote a primitive root and $c = (p-1)$. Then the congruence (1) is equivalent to

$$g^{h\alpha y + \beta} \equiv ax + b \pmod{p} .$$

Put $\beta = h\gamma$ and $g^h = A$. Then A will be a number belonging to the exponent f modulo p (by A we mean any number belonging to the exponent f since the choice of the primitive root is arbitrary). From Theorem 1 follows:

Theorem 2. Let p be a prime $\geqq 5$, f and h divisors of $(p-1)$ such that $fh=(p-1)$, α an integer coprime to f and γ an integer. Furthermore, let a be an integer not divisible by p, b any integer, p_1 and f_1 integers satisfying the conditions $0<p_1\leqq p$, $0<f_1\leqq f$, and A any integer belonging to the exponent f modulo p. Then the numbers T of terms in the sequence

$$A^{\alpha y+\gamma}; \qquad y=0,1,\ldots,f_1-1,$$

that are congruent modulo p to the terms of the sequence

$$ax+b; \qquad x=0,1,\ldots,p_1-1,$$

is

$$T=\frac{p_1 f_1}{p-1}+\theta\sqrt{p}\,(\ln p)^2; \qquad |\theta|<1. \tag{2}$$

Put $A=g$, $\alpha=1$, $\gamma=0$, $a=1$ and $b=0$. Then, from Theorem 2 we obtain the following

Corollary. For any $0<f_1\leqq(p-1)$, the number of those terms in the sequence

$$g^0, g^1, g^2, \ldots, g^{f_1-1},$$

whose least positive residue modulo p is less than p_1 is given by (2).

Note. From Theorem 1 we can easily deduce that

$$\sum_{x=0}^{p_1-1}\operatorname{ind}(ax+b)=\frac{(p-1)p_1}{2}+\theta_1 2p\sqrt{p}\,(\ln p)^2; \qquad |\theta_1|<1,$$

and from Theorem 2 that

$$\sum_{y=0}^{f_1-1}\left\{\frac{A^{\alpha y+\gamma}+n}{p}\right\}=\frac{f_1}{2}+\theta_2 2\sqrt{p}\,(\ln p)^2; \qquad |\theta_2|<1,$$

where $\{z\}$ stands for the fractional part of z, i.e. the difference $z-[z]$.

On a Bound for the Least n^{th} Power Non-Residue

Izvestiya Akademii Nauk SSSR, Ser. 6, **20** (1926) 47–58.

1. I have shown[1] that the least quadratic non-residue of a prime modulus p is less than

$$p^{\frac{1}{2\sqrt{e}}} (\ln p)^2$$

for all sufficiently large p. Using this method, we can prove a more general theorem.

Theorem 1. If p is a prime and n a divisor of $(p-1)$ distinct from 1, the least n^{th} power non-residue modulo p is less than

$$p^{\frac{1}{2k}} (\ln p)^2, \qquad k = e^{\frac{n-1}{n}},$$

for all sufficiently large p.

This bound can be improved still further, if the method is slightly modified. The larger the value of the exponent n, the greater the reduction in the bound. Thus, for example, we can prove the following theorems.

Theorem 2. If p is a prime and n a divisor of $(p-1)$ greater than 20, the least n^{th} power non-residue modulo p is less than

$$p^{1/6},$$

for all sufficiently large p.

Theorem 3. If p is a prime and n a divisor of $(p-1)$ greater than 204, the least n^{th} power non-residue modulo p is less than

$$p^{1/8},$$

for all sufficiently large p.

Finally, a more general theorem can be established.

Theorem 4. If p is a prime and n a divisor of $(p-1)$ greater than m^m, where m is an integer > 8, the least n^{th} power non-residue p is less than $p^{1/m}$ for all sufficiently large p.

We now proceed to prove these theorems.

2. Let

$$P = p^{1/2} (\ln p)^2; \qquad T = p^{1/2k} (\ln p)^2; \qquad k = e^{\frac{n-1}{n}}.$$

[1] On the distribution of power residues and non-residues. Zhur. Perm fiz.-mat. ob.-va **1** (1918) 94–98 = This collection, pp. 53–56.

We shall show that the least n^{th} power non-residue modulo p is less than T for all sufficiently large p.

Indeed, assume the contrary. Then the nth power non-residues modulo p can be only those terms in the sequence

$$1, 2, \ldots, [P], \tag{1}$$

which are divisible by primes greater than T and less than P. But the number of such numbers is obviously not more than

$$\sum_{q<T}^{q>P} \left[\frac{P}{q}\right],$$

where q runs only through primes. By the law of distribution of primes, this expression can be written as

$$P \ln \frac{\ln P}{\ln T} + O\left(\frac{P}{\ln p}\right) = P\left(\frac{n-1}{n} + \ln \frac{1 + \dfrac{4 \ln \ln p}{\ln p}}{1 + \dfrac{4k \ln \ln p}{\ln p}}\right) + O\left(\frac{P}{\ln p}\right)$$

$$= P\left(\frac{n-1}{n} + (4-4k)\frac{\ln \ln p}{\ln p} + O\left(\frac{1}{\ln p}\right)\right).$$

On the other hand, according to my previous paper[1], the number of nth power residues modulo p contained in the sequence (1) is

$$\frac{[P]}{n} - \Delta; \qquad |\Delta| < \sqrt{p} \ln p,$$

Hence the total number of non-residues in that sequence can be represented in the form

$$P\frac{n-1}{n} - \varrho; \qquad |\varrho| < \sqrt{p} \ln p + 1.$$

Therefore, we should have

$$P\frac{n-1}{n} + \varrho \leqq P\left(\frac{n-1}{n} + (4-4k)\frac{\ln \ln p}{\ln p} + O\left(\frac{1}{\ln p}\right)\right),$$

which leads to an inequality of the type

$$(4-4k) \ln \ln p \geqq O(1).$$

For sufficiently large p, this inequality is impossible because $k > 1$. Thus, Theorem 1 has been proved.

3. Take

$$P = p^{1/2} (\ln p)^2; \qquad T = p^{1/6}$$

[1] On the distribution of power residues and non-residues. Zhur. Perm fiz.-mat. ob.-va **1** (1918) 94–98 = This collection, pp. 53–56.

and assume that there are no n^{th} power non-residues modulo p that are less than T. Then, only those numbers which are divisible by primes greater than T and less than P can be the non-residues less than P. The number of such numbers is obviously less than

$$\sum_{q>T}^{q<T_1}\left[\frac{P}{q}\right]+\sum_{q>T_1}^{q<P}\left[\frac{P}{q}\right]-\sum_{q>T_1}^{q<\sqrt{P}}\sum_{q_1>q}^{q_1<P/q}\left[\frac{P}{qq_1}\right],$$

where

$$T_1=p^{1/6}\ln p.$$

But, according to the law of distribution of primes, the first sum can be represented as

$$P\ln\frac{\ln T_1}{\ln T}+O\left(\frac{P}{\ln p}\right)=O\left(\frac{P\ln\ln p}{\ln p}\right),$$

the second sum as

$$P\frac{\ln P}{\ln T_1}+O\left(\frac{P}{\ln p}\right)=P\ln 3+O\left(\frac{P\ln\ln p}{\ln p}\right)=P\cdot 1.098612\ldots+O\left(\frac{P\ln\ln p}{\ln p}\right),$$

and the third sum as

$$P\sum_{q>T_1}^{q<\sqrt{P}}\frac{1}{q}\ln\frac{\ln P/q}{\ln q}+O\left(\frac{P}{\ln p}\right),$$

which is

$$P\sum_{q>p^{1/6}}^{q<p^{1/4}}\frac{1}{q}\ln\frac{\ln\dfrac{p^{1/2}}{q}}{\ln q}+O\left(\frac{P\ln\ln p}{\ln q}\right)=P\sum_{q>p^{1/6}}^{q<p^{1/4}}\frac{1}{q}\ln\frac{1-\dfrac{2\ln q}{\ln p}}{2\dfrac{\ln q}{\ln p}}+O\left(\frac{P\ln\ln p}{\ln p}\right).$$

Applying the law of distribution of primes, we can easily transform the last expression as follows:

$$P\int_{p^{1/6}}^{p^{1/4}}\ln\frac{1-\dfrac{2\ln z}{\ln p}}{2\dfrac{\ln z}{\ln p}}\cdot\frac{dz}{z\ln z}+O\left(\frac{P\ln\ln p}{\ln p}\right)$$

$$=P\int_{1/3}^{1/2}\ln\frac{1-u}{u}\frac{du}{u}+O\left(\frac{P\ln\ln p}{\ln p}\right)=P\cdot 0.147\ldots+O\left(\frac{P\ln\ln p}{\ln p}\right).$$

Thus, for sufficiently large p, the number of nth power non-residues modulo p is less than

$$P(1.0987-0.147)=P\cdot 0.9517.$$

On the other hand, the number of nth power non-residues modulo p that are contained in the sequence

$$1, 2, \ldots, [P],$$

as we have demonstrated in the proof of Theorem 1, is equal to

$$P\left(1-\frac{1}{n}\right)+O\left(\frac{P}{\ln p}\right).$$

Thus, for sufficiently large p, we have

$$P\left(1-\frac{1}{n}\right)+O\left(\frac{P}{\ln p}\right)<P\cdot 0.9517,$$

Hence it follows that for sufficiently large p, we have

$$1-\frac{1}{n}<0.952;\qquad \frac{1}{n}>0.048;\qquad n<\frac{1}{0.048}=20.8\ldots$$

The fact that this inequality is impossible for all integers $n>20$ demonstrates the truth of Theorem 2.

4. Take

$$P=p^{1/2}(\ln p)^2;\qquad T=p^{1/8}$$

and assume that there are no n^{th} power non-residues modulo p that are less than T. Then only those numbers which are divisible by primes greater than T and less than P can be the non-residues less than P. In the same way as in the proof of Theorem 2, we easily find that the upper bound of the number of such non-residues can be represented as

$$\sum_{q>p^{1/8}}^{q<p^{1/2}}\frac{P}{q}-\sum_{q>p^{1/8}}^{q<p^{1/4}}\ \sum_{q_1>q}^{q_1<\frac{p^{1/2}}{q}}\frac{P}{qq_1}+\sum_{q>p^{1/8}}^{q<p^{1/6}}\ \sum_{q_1>q}^{q_1<\frac{p^{1/4}}{\sqrt{q}}}\ \sum_{q_2>q_1}^{q_2<\frac{p^{1/2}}{qq_1}}\frac{P}{qq_1q_2}+O\left(\frac{P\ln\ln p}{\ln p}\right).\tag{2}$$

a) The first sum can be expressed as

$$P\ln 4+O\left(\frac{P}{\ln p}\right)=P\cdot 1.386294\ldots+O\left(\frac{P}{\ln p}\right).$$

b) Arguing in a similar manner as in the proof of Theorem 2, we can express the second sum, which is a double sum, in the form:

$$-P\int_{1/4}^{1/2}\ln u\frac{du}{u}-P\int_{1/4}^{1/2}\left(1+\frac{u}{2}+\frac{u^2}{3}+\frac{u^3}{4}+\ldots\right)du+O\left(\frac{P}{\ln p}\right)$$
$$=P\cdot 0.40609\ldots+O\left(\frac{P}{\ln p}\right).$$

c) It now remains to calculate the third, triple sum. We find

$$\sum_{q_2>q_1}^{q_2<\frac{p^{1/2}}{qq_1}}\frac{P}{qq_1q_2}=\frac{P}{qq_1}\ln\frac{\frac{1}{2}\ln p-\ln q-\ln q_1}{\ln q_1}+O\left(\frac{P}{qq_1\ln p}\right).$$

Hence we easily obtain

$$\sum_{\substack{q_1 < \frac{p^{1/4}}{\sqrt{q}} \\ q_1 > q}} \sum_{\substack{q_2 < \frac{p^{1/2}}{qq_1} \\ q_2 > q_1}} \frac{P}{qq_1q_2} = \frac{P}{q} \int_q^{\frac{p^{1/4}}{\sqrt{q}}} \frac{dy}{y \ln y} \ln \frac{\frac{1}{2}\ln p - \ln q - \ln y}{\ln y} + O\left(\frac{P}{q \ln p}\right)$$

$$= \frac{P}{q} \int_v^{\frac{1}{4}-\frac{1}{2}v} \frac{dz}{z} \ln \frac{\frac{1}{2} - v - z}{z} + O\left(\frac{P}{q \ln p}\right); \qquad v = \frac{\ln q}{\ln p}.$$

At the same time, the triple sum can be represented as

$$P \int_{1/8}^{1/6} \frac{dv}{v} \int_v^{\frac{1}{4}-\frac{1}{2}v} \frac{dz}{z}$$

$$\cdot \left[\ln\left(\tfrac{1}{2} - v\right) - \ln z - \frac{z}{\frac{1}{2} - v} - \frac{z^2}{2\left(\frac{1}{2} - v\right)^2} - \frac{z^3}{3\left(\frac{1}{2} - v\right)^3} - \ldots\right] + O\left(\frac{P \ln \ln p}{\ln p}\right)$$

$$= P \int_{1/8}^{1/6} \ln \frac{\frac{1}{2}\left(\frac{1}{2} - v\right)}{v} \cdot \ln \sqrt{\frac{2\left(\frac{1}{2} - v\right)}{v}} \cdot \frac{dv}{v}$$

$$- P \int_{1/8}^{1/6} \left[\frac{1}{2} + \frac{1}{4 \cdot 4} + \frac{1}{8 \cdot 9} + \frac{1}{16 \cdot 16} + \ldots\right] \frac{dv}{v}$$

$$+ P \int_{1/8}^{1/6} \left[\frac{v}{\frac{1}{2} - v} + \frac{1}{4}\left(\frac{v}{\frac{1}{2} - v}\right)^2 + \frac{1}{9}\left(\frac{v}{\frac{1}{2} - v}\right)^3 + \ldots\right] \frac{dv}{v} + O\left(\frac{P \ln \ln p}{\ln p}\right),$$

which, after making the substitution $u = \frac{\frac{1}{2} - v}{v}$ in the first integral and $u = \frac{v}{\frac{1}{2} - v}$ in the third integral, is transformed into

$$P \int_2^3 \ln \frac{u}{2} \cdot \ln \sqrt{2u} \, \frac{du}{1 + u} - P\left[\frac{1}{2} + \frac{1}{4 \cdot 4} + \frac{1}{8 \cdot 9} + \ldots\right] \ln \tfrac{4}{3}$$

$$+ P \int_{1/3}^{1/2} \left(1 + \tfrac{1}{4}u + \tfrac{1}{9}u^2 + \ldots\right) \frac{du}{1 + u} + O\left(\frac{P \ln \ln p}{\ln p}\right).$$

This expression, when calculated correct to the fifth decimal place, is equal to

$$P\,[0.04933\ldots - 0.16750\ldots + 0.13305\ldots] + O\left(\frac{P \ln \ln p}{\ln p}\right),$$

which, for sufficiently large p, is less than

$$P \cdot 0.01490\,.$$

Comparing this result with those proved in (a) and (c), we find that the expression (2), for sufficiently large p, is less than

$$P(1.38630 - 0.40609 + 0.01490) < P\left(1 - \tfrac{1}{205}\right).$$

Hence, arguing in a similar manner as in the proof of Theorem 2, we find that the least n^{th} power non-residue modulo p for $n > 204$, is less than

$$p^{1/8}$$

for all sufficiently large p. Thus Theorem 3 has been proved.

5. Before proceeding to prove Theorem 4, we shall establish the following lemma.

Lemma. If k is a positive number which can increase unboundedly and s is an integer $\geqq 2$, the number of integers less than t_s, where t_s is any number satisfying the condition

$$k^s < t_s < k^{s+\frac{1}{s+2}},$$

which are not divisible by primes greater than k, is greater than

$$\frac{t_s}{s!\,(s+2)^s}$$

for sufficiently large k.

Proof. Put

$$\varepsilon = \frac{1}{s+2}.$$

1°. Take any number t_1 satisfying the condition

$$k < t_1 \leqq k^{2-2\varepsilon}.$$

We shall find the lower bound T_1 of numbers which are $\leqq t_1$ and are divisible by at least one prime greater than $k^{1-\varepsilon}$ and $\leqq k$. If q is one such prime, then the total number of those numbers which are $\leqq t_1$ and are divisible by q is equal to

$$\left[\frac{t_1}{q}\right].$$

Consequently,

$$T_1 = \sum_{q>k^{1-\varepsilon}}^{q\leqq k} \left[\frac{t_1}{q}\right],$$

where q runs only through primes. By virtue of the law of distribution of primes, this number can be represented as

$$T_1 = t_1 \ln \frac{\ln k}{(1-\varepsilon)\ln k} + O\left(\frac{t}{\ln k}\right).$$

But this expression is greater than

$$t_1 \ln \frac{1}{1-\varepsilon} + O\left(\frac{t_1}{\ln k}\right),$$

which, for sufficiently large k, is greater than εt_1. Thus, for sufficiently large k, we have

$$T_1 > \varepsilon t_1.$$

2°. Take any number t_2 satisfying the condition

$$k^2 < t_2 \leqq k^{3-3\varepsilon}.$$

We shall find the lower bound T_2 of numbers which are $\leqq t_2$ and are divisible by the product of any two primes greater than $k^{1-\varepsilon}$ and $\leqq k$. Here the products in which the factors are permuted are to be considered different.

Let q be a prime greater than $k^{1-\varepsilon}$ and $\leqq k$. The numbers which are not greater than t_2 and are divisible by q are

$$q,\ 2q, \ldots, \left[\frac{t_2}{q}\right] q.$$

Therefore, we have to find how many numbers in the sequence

$$1,\ 2, \ldots, \left[\frac{t_2}{q}\right]$$

are divisible by primes greater than $k^{1-\varepsilon}$ and $\leqq k$. Since

$$k = k^{2-1} < \frac{t_2}{q} < k^{3-3\varepsilon-1+\varepsilon} = k^{2-2\varepsilon},$$

according to 1°, for sufficiently large k, this number is greater than

$$\varepsilon \frac{t_2}{q}.$$

Hence, as in 1°, we find that

$$T_2 > \varepsilon^2 t_2$$

for all sufficiently large k.

3°. Similarly, taking any number t_3 satisfying the condition

$$k^3 < t_3 \leqq k^{4-4\varepsilon},$$

and denoting by T_3 the total number of numbers $\leqq t_3$ which are divisible by the product of any three primes greater than $k^{1-\varepsilon}$ and $\leqq k$ (taking the products in which the factors are permuted to be different) we obtain

$$T_3 \geqq \varepsilon^3 t_3.$$

4°. Proceeding in this way, finally, if t_s is any number satisfying the condition

$$k^s < t_s \leqq k^{s+1-(s+1)\varepsilon},$$

and T_s denotes the total number of numbers which are $\leqq t_s$ and are divisible by the product of s primes greater than $k^{1-\varepsilon}$ and $\leqq k$ (taking the products in which the factors are permuted to be different), for sufficiently large k, we find that

$$T_s \geqq \varepsilon^s t_s = \frac{t_s}{(s+2)^s}.$$

Since

$$T = \frac{T_s}{s!},$$

the proof of the lemma is complete.

6. *Proof of Theorem 4.* We have already shown that if n is a divisor of $(p-1)$, the number R of nth power residues modulo p that are less than $\sqrt{p}(\ln p)^2$ can be represented as

$$R = \frac{\sqrt{p}}{n}(\ln p)^2 + O(\sqrt{p}\ln p). \tag{3}$$

Taking $k = p^{1/m}$, $s = m/2$ for even m, and $k = p^{1/(m+1)}$, $s = (m+1)/2$ for odd m, according to the lemma proved above, we find that the total number of numbers less than $\sqrt{p}(\ln p)^2$, which are divisible by only those primes which are less than $p^{1/m}$, for sufficiently large p, is greater than

$$\frac{\sqrt{p}(\ln p)^2}{s!(s+2)^s}.$$

Assuming that there is no n^{th} power non-residue modulo p among the numbers less than $p^{1/m}$, we obtain

$$R > \frac{\sqrt{p}(\ln p)^2}{s!(s+2)^s}.$$

Comparing this inequality with the inequality (3), we find

$$\frac{1}{n} + O\left(\frac{1}{\ln p}\right) > \frac{1}{s!(s+2)^s},$$

Hence

$$n < s!(s+2)^s + \delta_p,$$

where δ_p tends to zero as p tends to infinity. Now applying the Stirling formula, since $s \leqq (m+1)/2$, after some simple calculations, we find that

$$s!(s+2)^s < m^m.$$

Hence it follows that for sufficiently large p,

$$n < m^m.$$

Consequently, for $n > m^m$ there should exist at least one non-residue less than $p^{1/m}$. Thus, Theorem 4 has been proved.

Note. Evidently, the limit $n > m^m$ which we have derived is quite rough. Thus, even for $m = 8$ we obtain $n > 16777216$ in place of $n > 204$ which we derived earlier.

7. We know that to determine the primitive root with respect to a prime modulus p, it is sufficient to find various prime divisors

$$2, q_1, q_2, \ldots, q_r$$

of $(p-1)$ and then to find one non-residue $v_0, v_1, \ldots, v_r$ of power $2, q_1, q_2, \ldots, q_r$ modulo p. Then the primitive root can easily be found from the numbers $v_0, v_1, \ldots, v_r$. From the theorem proved above, we readily infer that:

1°. If p is sufficiently large, then all the numbers $v_0, v_1, \ldots, v_r$ are found in the sequence

$$1, 2, \ldots, \left[p^{\frac{1}{2\sqrt{e}}}(\ln p)^2\right]. \tag{4}$$

If, in addition, p is not of the form $(8N+1)$, and all the numbers $q_1, q_2, \ldots, q_r$ are sufficiently large, we can take a sequence shorter than the above sequence. The greater the lower bound of the numbers $q_1, q_2, \ldots, q_r$, the shorter the sequence. Thus, for instance, if all the numbers $q_1, q_2, \ldots, q_r$ are greater than 20, we can take the sequence:

$$1, 2, \ldots, [p^{1/6}], \tag{5}$$

while, if all the numbers $q_1, q_2, \ldots, q_r$ are greater than 204, we can take the sequence

$$1,2, \ldots, [p^{1/8}]. \tag{6}$$

Finally, applying Theorem 4, we infer that

2°. If p is a sufficiently large prime which cannot be represented in the form $(8N+1)$, and all odd prime divisors of $(p-1)$ are greater than m^m, then all the numbers $v_0, v_1, \ldots, v_r$ can be found among the terms of the sequence

$$-1, 1, 2, \ldots, [p^{1/m}]. \tag{7}$$

These results can be formulated in a slightly different form as follows.

1. If p is a sufficiently large prime, then the complete system of residues modulo p can be obtained by multiplying the powers of the terms in the sequence (4).

If, in addition, p cannot be represented in the form $(8N+1)$, and all odd prime divisors of $(p-1)$ are greater than 20, the complete system of residues modulo p can be obtained by multiplying the terms in the sequence (5).

If p cannot be represented in the form $(8N+1)$, and all odd prime divisors of $(p-1)$ are greater than 204, the complete system of residues modulo p can be obtained by multiplying the terms in the sequence (6).

2. If p cannot be represented in the form $(8N+1)$, and all odd prime divisors of $(p-1)$ are greater than m^m, where m is an integer >8, the complete system of residues modulo p can be obtained by multiplying the powers of the terms in the sequence (7).

On the Distribution of Fractional Parts of Values of a Function of One Variable

Zhurnal Leningradskogo fiz. matem. obshestva (Journal of the Leningrad Physico-Mathematical Society), 1 no. 1 (1926) 56–65.

In this paper I shall solve the problems related to the distribution of fractional parts of the values of a function $f(x)$ satisfying certain conditions. The asymptotic equalities derived give some new information on the distribution of fractional parts and the estimates found for the remainder terms seem to be the best possible.

Using a similar method, we can also obtain analogous results for the distribution of fractional parts of the values of a function of several variables. But I shall present these results in some other communication.

1°. Let Δ and Δ_1 be two such positive numbers that $\Delta + \Delta_1 = 1$. Consider a periodic function $\varphi(x)$ of period 1 defined by the expressions:

1. $\varphi(x) = \dfrac{x}{\Delta}$ in the interval $(0, \Delta)$

and

2. $\varphi(x) = \dfrac{1-x}{\Delta_1}$ in the interval $(\Delta, 1)$.

This function can be expanded as the following Fourier series:

$$\varphi(x) = \frac{1}{2} + \sum_{m=1}^{\infty} (A_m \cos 2\pi mx + B_m \sin 2\pi mx), \tag{1}$$

whose coefficients satisfy the inequalities

$$\begin{aligned} |A_m| &< \frac{1}{\pi^2 m^2}\left(\frac{1}{\Delta} + \frac{1}{\Delta_1}\right); \qquad |A_m| < \frac{2}{\pi m}, \\ |B_m| &< \frac{1}{\pi^2 m^2}\left(\frac{1}{\Delta} + \frac{1}{\Delta_1}\right); \qquad |B_m| < \frac{2}{\pi m}. \end{aligned} \tag{2}$$

2°. Besides the Fourier expansion of $\varphi(x)$, we shall also make use of the following lemma.

Lemma. Let A, U, q and r denote quantities satisfying the conditions

$$U \geqq A \geqq k; \qquad A \geqq 2; \qquad 0 < r - q \leqq U,$$

where k is a constant $\geqq 1$. Furthermore, let $f(x)$ be a function whose second derivative in the interval $q \leqq x \leqq r$ satisfies the conditions

$$A^{-1} \leqq f''(x) \leqq kA^{-1}.$$

Then

$$\left|\sum_{x>q}^{x\leqq r} e^{2\pi i f(x)}\right| < 10k\,\frac{U}{\sqrt{A}}\,{}^{(*)}.$$

3°. Assuming that the conditions of the above lemma are satisfied, take an $m \geqq 1$. The second derivative of the function $mf(x)$ in the interval $q \leqq x \leqq r$ will then satisfy the conditions:

$$mA^{-1} \leqq mf''(x) \leqq kmA^{-1}.$$

Therefore it is not difficult to verify (for $A \geqq km$ and $A \geqq 2m$) that

$$\left|\sum_{x>q}^{x\leqq r} e^{2\pi i m f(x)}\right| < 10k\,\frac{\sqrt{m}\,U}{\sqrt{A}} \tag{3}$$

(This inequality is obviously true even when $A < km$ or $A < 2m$).

4°. Assuming that the conditions of Lemma 2° are satisfied, we shall take that r and q are integers and that $(r-q) = n$. Consider the sum

$$S = \sum_{x>q}^{x\leqq r} \varphi(f(x)).$$

Applying the expansion (1), we can express this sum as

$$S = \frac{n}{2} + \sum_{x>q}^{x\leqq r}\sum_{m=1}^{\infty} (A_m \cos 2\pi m f(x) + B_m \sin 2\pi m f(x)),$$

which, by virtue of (2) and (3), can be represented in the form:

$$S = \frac{n}{2} + O\left(\frac{U}{\sqrt{A}}\left(\frac{1}{\sqrt{\Delta}} + \frac{1}{\sqrt{\Delta_1}}\right)\right). \tag{4}$$

5°. For the sake of greater clarity, we shall give a geometrical interpretation of these results. On arranging the fractions(**)

$$\{f(x)\}; \qquad x = q+1;\ q+2, \ldots, r$$

in ascending order, we obtain a sequence

$$\beta_0, \beta_1, \beta_2, \ldots, \beta_{n-1}.$$

We shall represent each product $y_i = n_{\beta_i}$ in the form of a rectangle (Fig. 1) bounded by the lines $x = i-1$, $x = i$, $y = 0$ and $y = y_i$.

(*) The proof of this lemma is given in my dissertation "On the number of integral points in two and three dimensional domains" submittes in May 1920 to the Faculty of Physics and Mathematics, Petrograd University. It is an extension of the method used in my paper: On the mean value of the number of classes of proper primitive forms of negative discriminant. Soobsh. Kharkov mat. ob-va. **16** (1918) 10–38 = This collection, pp. 28–52.

(**) The symbol $\{z\}$ stands for the fractional part of a number z, i.e. the difference $z - [z]$.

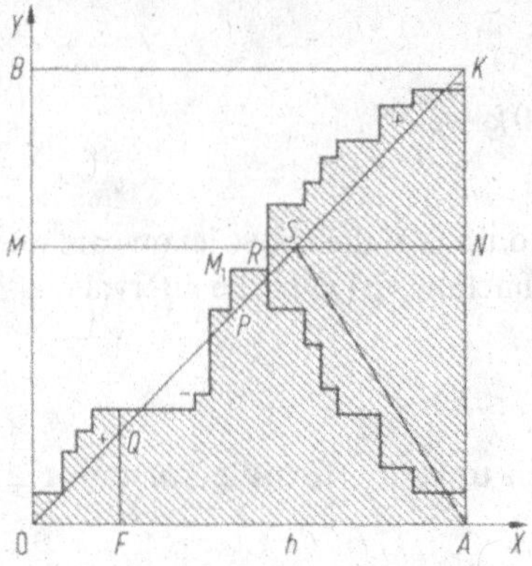

Fig. 1

Then, the sum

$$\sum_{i=0}^{n-1} y_i$$

is represented by the shaded area.

In this figure $OA = OB = n$ and OK is the diagonal of the square of sides OA and OB.

Let C denote the difference between the shaded area and the area of the triangle $\Delta\, OKA$. Thus, C represents the area bounded by the broken line ORK and the diagonal OK. The area lying above OK is taken positive while the area lying below OK, negative. Now let h be any number between 0 and n, and MN be a line defined by the equation $y = h$. Then, putting $\Delta = h/n$, we obtain $\Delta_1 = n - h/n$ and, by (4), we have

$$\sum_{i=0}^{n-1} \varphi(\beta_i) = \frac{n}{2} + O(H); \qquad H = \frac{U\sqrt{n}}{\sqrt{A}}\left(\frac{1}{\sqrt{h}} + \frac{1}{\sqrt{n-h}}\right). \tag{5}$$

But for $y_i \leqq h$, we have

$$\varphi(\beta_i) = \frac{n\beta_i}{h} = \frac{y_i}{h},$$

and for $y_i \geqq h$:

$$\varphi(\beta_i) = \frac{n - n\beta_i}{n-h} = 1 - \frac{y_i - h}{n-h}.$$

In the same way as we represented the number y_i, now represent $h\varphi(\beta_i) = z_i$. Thus, we obtain another broken line ORA whose form can be easily found, since

$$z_i = y_i \quad \text{for} \quad y_i \leqq h \quad \text{and} \quad z_i = h - \frac{(y_i - h)\,h}{n-h} \quad \text{for} \quad y_i \geqq h.$$

From (5) it follows that

$$\sum_{i=0}^{n-1} z_i = \frac{hn}{2} + O(hH).$$

The left-hand side of this expression represents the area of the figure ORA bounded above by the broken line ORA and below by the axis OX. On the other hand, the area of the triangle OSA is equal to $nh/2$. Hence, the difference T between these areas (ORA and OSA) is a quantity of the order of hH. Let α denote that part of the area C which lies below MN and β the part above MN. Now the difference T can be expressed as

$$T = \alpha - \beta \frac{h}{n-h}.$$

Thus we have

$$\alpha - \beta \frac{h}{n-h} = O(hH).$$

Hence, since $\alpha + \beta = C$, we find

$$\alpha = \frac{Ch}{n} + O\left(\frac{h(n-h)}{n} H\right). \tag{6}$$

6°. For the sake of definiteness, assume that $C \geqq 0$ and take

$$h = \frac{U^{2/3} n^{1/3}}{A^{1/3}}.$$

(It suffices to assume that $U < n\sqrt{A}$.) Hence, we obtain

$$\alpha = \frac{CU^{2/3}}{A^{1/3} n^{2/3}} + O\left(\frac{U^{4/3} n^{2/3}}{A^{2/3}}\right).$$

On the other hand, from Fig. 1, it is obvious that

$$\alpha \leqq \tfrac{1}{2} h^2 = O\left(\frac{U^{4/3} n^{2/3}}{A^{2/3}}\right).$$

Thus, we have

$$C = O(L); \qquad L = \frac{U^{2/3} n^{4/3}}{A^{1/3}}.$$

We would have arrived at the same result, had we assumed that $C < 0$.

7°. Furthermore, it is not difficult to show that the whole broken line ORK lies in a strip between two lines parallel to OK, the difference between the initial ordinates of which being a quantity of the order of Ln^{-1}. For this purpose, on the broken line ORK let M_1 be the point with ordinate y_1 most remote from OK and let $PM_1 = W$. Now, in place of the function $f(x)$, take the function $f_1(x) = f(x) - y_1 n^{-1}$. For this function $f_1(x)$, in place of C, we have to take the number $C_1 = C - Wn$. Since $f_1(x)$ satisfies the conditions of 2°, we have $C_1 = O(L)$, i.e.

$$C - Wn = O(L); \qquad Wn = O(L); \qquad W = O\left(\frac{L}{n}\right),$$

which proves our statement.

8°. Now let n_1 be any integer $\leqq n$. Consider the sum

$$\Omega = \sum_{i=0}^{n_1-1} y_i .$$

Let n_1 on Fig. 1 be represented by the segment OF. Take $h = n_1$. By (6), the corresponding value of α can be represented as

$$\alpha_1 = \frac{Cn_1}{n} + O\left(U\sqrt{\frac{n_1(n-n_1)}{A}}\right).$$

On the other hand, by the theorem proved in 7°, it is clear that on adding α_1 to the area of the triangle OFQ, $n_1^2/2$, we obtain a number which differs from Ω only by a quantity of the order $(Ln^{-1})^2 = U^{4/3} n^{2/3} A^{-2/3}$. Hence we can write

$$\Omega = \frac{n_1^2}{2} + \frac{Cn_1}{n} + O\left(U\sqrt{\frac{n_1(n-n_1)}{A}} + \frac{U^{4/3} n^{2/3}}{A^{2/3}}\right).$$

These results can be formulated as the following (noting that $y_i = n\beta_i$).

Theorem. Let A, U be any numbers, q and r integers satisfying the conditions $A \geqq 2$; $U \geqq A \geqq k$; $0 < n = r - q \leqq U$, where k is a constant $\geqq 1$. Furthermore, let the function $f(x)$ have a second derivative in the interval $q \leqq x \leqq r$ satisfying the conditions

$$A^{-1} \leqq f''(x) \leqq kA^{-1}.$$

On arranging the fractions

$$\{f(x)\}; \qquad x = q+1,\ q+2, \ldots, r$$

in ascending order of magnitude

$$\beta_0 \leqq \beta_1 \leqq \beta_2 \leqq \ldots \leqq \beta_{n-1},$$

for any integral positive $n_1 \leqq n$, we find

$$\sum_{i=0}^{n_1-1} \beta_i = \frac{n_1^2}{2n} + Rn_1 + O\left(\frac{U}{\sqrt{A}}\sqrt{\frac{n_1(n-n_1)}{n^2}} + \sqrt{\frac{U^4}{nA^2}}\right),$$

where R is of the order

$$U^{2/3} A^{-1/3} n^{-2/3},$$

independent of n_1.

Corollary. If the ratios U/A and U/n are bounded, then

$$\sum_{i=0}^{n_1-1} \beta_i = \frac{n_1^2}{2n} + Rn_1 + O(\sqrt{A}); \qquad R = O(A^{-1/3}). \tag{7}$$

The order $\sqrt{A}$ of the remainder term in this expansion cannot be reduced any further, and is therefore the best possible estimate.

9°. **Appendix.** An interesting example, where the number R can be expressed in a more definite manner, is provided by the functions:

$$\text{a) } f(x)=\frac{x^2}{p};\qquad \text{b) } f(x)=\frac{x^2}{p}+\frac{1}{4},$$

where p is taken (for the sake of simplicity) to be a prime $\geqq 3$.

a) First consider the function $f(x)=x^2/p$, putting $q=0$ and $r=p$. Then we can take $A=p/2$, $k=1$. The number R will obviously be equal to

$$R=\frac{1}{p}\sum_{x=1}^{p}\left(\left\{\frac{x^2}{p}\right\}-\frac{1}{2}\right),$$

which can be represented in the form of a trigonometrical series:

$$R=-\frac{1}{\pi p}\sum_{x=1}^{p}\sum_{m=1}^{\infty}\frac{\sin 2\pi m\dfrac{x^2}{p}}{m}.$$

Using the expansion of the Gauss sums and denoting by $h(-p)$ the number of classes of proper primitive quadratic forms of negative discriminant $-p$, we find that: if p is of the form $(4N+1)$, $R=0$, and if p is of the form $(4N+3)$, then

$$R=-\frac{h(-p)}{2p\left(1-\frac{1}{2}(2|p)\right)}.$$

The formula (7) takes the form

$$\sum_{i=0}^{n_1-1}\beta_i=\frac{n_1^2}{2p}+Rn_1+O(\sqrt{p}).$$

b) Now put $f(x)=x^2/p+\frac{1}{4}$; $q=0$; $r=p$.
Then: if p is of the form $(4N+1)$,

$$R=-\frac{h(-p)}{2p}-\frac{1}{4p},$$

and if p is of the form $(4N+3)$, then

$$R=\frac{h(-p)\,(1-(2|p))}{4p\left(1-\frac{1}{2}(2|p)\right)}-\frac{1}{4p}.$$

10°. Let p be a prime $\geqq 3$ and c be a divisor of $(p-1)$ exceeding 1. From cyclotomy theory, we know that[(*)]

$$\left|\sum_{x=1}^{p}\sin 2\pi\frac{mx^c}{p}\right|<c\sqrt{p}.$$

(*) This inequality can easily be derived from

$$\left|\sum_{z=1}^{p-1}\varrho^{u\,\mathrm{ind}\,z}\,r^z\right|=\sqrt{p};\quad \varrho=e^{\frac{2\pi i}{c}};\qquad r=e^{\frac{2\pi i}{p}a};\qquad (a,m)=1,$$

which holds for u not divisible by c, by summing over all $u=0,1,\ldots,(c-1)$.

Now take

$$\sum_{x=1}^{p} \sin 2\pi \frac{mx^c}{p} = \alpha_m c \sqrt{p}.$$

Then we obtain $|\alpha_m| < 1$.

Now let $f(x) = ax^c/p$, where a is an integer not divisible by p. Put $q = 0$ and $r = p$. Then we have

$$R = -\frac{c}{\pi\sqrt{p}} \sum_{m=1}^{\infty} \frac{\alpha_m}{m}.$$

We have to estimate the remainder term ϱ in

$$\sum_{i=0}^{n_1-1} \beta_i = \frac{n_1^2}{2p} + Rn_1 + \varrho$$

in a special manner because the conditions of Theorem 8° are not satisfied in this case. In place of (4), now we obtain

$$S = \frac{p}{2} + O\left(c\sqrt{p}\ln\frac{1}{\Delta\Delta_1}\right).$$

Now in (5) we can take H to be equal to

$$H = c\sqrt{p}\ln\frac{p^2}{h(p-h)}.$$

By (6), the remainder term is of the order of

$$c\frac{(p-h)h}{\sqrt{p}}\ln\frac{p^2}{h(p-h)},$$

L can be taken to be equal to $cp\sqrt{p}\ln p$. Then W proves to be a quantity of the order of $\sqrt{p}\ln p$ and we find (it is sufficient to assume that $c < \sqrt{p}$) that

$$\varrho = O\left(c\sqrt{p}\ln\frac{p^2}{n_1(p-n_1)}\right).$$

In particular, if the ratios p/n_1 and $p/(p-n_1)$ are bounded, then

$$\varrho = O(c\sqrt{p}).$$

On the Distribution of Fractional Parts of Values of a Function of Two Variables

Izvestiya Leningradskogo Politekhnicheskogo Instituta (Annals of the Leningrad Polytechnical Institute), **30** (1927) 31–52.

In paper[1] I considered only the case of a function of one variable. In the present paper I shall solve problems concerning the distribution of fractional parts of the values of a function of two variables.

I shall use the method first developed by me in paper[2] which in a more refined form constituted the subject of my doctorate dissertation "On the number of integral points in two and three dimensional domains" submitted in May 1920 at the Faculty of Physics and Mathematics, Petrograd University[3]. After appropriately supplementing the statements in the dissertation, I have derived a quite general theorem on the distribution of fractional parts of a function of two variables from which my previous results follow as particular cases.

1°. First we shall derive the asymptotic approximations to the sum

$$\sum_{M>q}^{M\leqq r} e^{2\pi i f(M)},$$

where summation is taken over all integral values of M within the limits specified, and $f(x)$ satisfies certain conditions. The derivation is based on the following two identities:

$$\sum_{M=Q}^{M=R} \int_{-1/2}^{1/2} \frac{\sin(2m+1)\pi x}{\sin \pi x} e^{2\pi i F(M+x)}\, dx = \sum_{n=-m}^{n=m} \int_{Q-1/2}^{R+1/2} e^{2\pi i(-nx+F(x))}\, dx, \tag{1}$$

$$\int_{-1/2}^{1/2} \frac{\sin(2m+1)\pi x}{\sin \pi x} e^{2\pi i F(M+x)}\, dx$$

$$= e^{2\pi i F(M)} + e^{2\pi i F(M)} \int_{-1/2}^{1/2} \frac{\sin(2m+1)\pi x}{\sin \pi x} (e^{2\pi i \varphi(x)} - 1)\, dx, \tag{2}$$

where Q, R and m are such integers that $Q < R$, $m > 0$; $F(x) = tx + f(x)$, t is an integer and $\varphi(x) = F(M+x) - F(M)$.

---

[1] On the distribution of fractional parts of values of a function of one variable. Zhur. Leningrad fiz.-mat. ob-va. **1** (1926) 56–65 = This collection, pp. 78–84.

[2] On the mean value of the number of classes of proper primitive forms of negative discriminant. Soobsh. Kharkov mat. ob-va. **16** (1918) 10–38 = This collection, pp. 28–52.

[3] The main results were presented by me in Autumn 1916 at the meeting of the Mathematical Association, Petrograd University.

The first identity can easily be verified by summing the right-hand side over n and then dividing the interval $(Q-\frac{1}{2}, R+\frac{1}{2})$ into intervals of length 1. The second identity is established by virtue of the following:

$$\int_{-1/2}^{1/2} \frac{\sin(2m+1)\pi x}{\sin \pi x}\, dx = 1\,.$$

Now let k be a constant $\geqq 2$, $A \geqq k$, $U \geqq A$, $m = |4U^3+1|$, $\mu=(2m+1)^{1/3}$, $R-Q \leqq U-1$, and let in the interval $Q-\frac{1}{2} \leqq x \leqq R+\frac{1}{2}$ $f(x)$ have a second derivative satisfying the conditions:

$$A^{-1} \leqq f''(x) \leqq kA^{-1}\,.$$

It is readily seen that $\mu \geqq 2U$ and the integer t can be so chosen that in the interval $Q-\frac{1}{2} \leqq x \leqq R+\frac{1}{2}$ we have

$$-\frac{\mu}{2} \leqq F'(x) \leqq \frac{\mu}{2},$$

which we shall always assume in future.

2°. On integrating by parts, we find

$$\int_{-1/2}^{1/2} \frac{e^{2\pi i \varphi(x)}-1}{\sin \pi x} \sin(2m+1)\pi x\, dx$$

$$= \int_{-1/2}^{1/2} \frac{\cos(2m+1)\pi x}{\pi\mu^3}\, \frac{\sin \pi x \cdot 2\pi i \varphi'(x)\, e^{2\pi i \varphi(x)} - (e^{2\pi i \varphi(x)}-1)\, \pi \cos \pi x}{\sin^2 \pi x}\, dx\,.$$

Now applying Maclaurin's theorem to the second factor in the integrand with a remainder in place of the first derivative, and since $|\sin \pi x| \geqq 2x$ for $|x| \leqq \frac{1}{2}$, we find that the absolute value of the integrand of the last integral, and therefore of the integral itself, is less than $2\mu^{-1}$. Hence, we can rewrite (2) as

$$\int_{-1/2}^{1/2} \frac{\sin(2m+1)\pi x}{\sin \pi x}\, e^{2\pi i F(M+x)}\, dx = e^{2\pi i F(M)} + \varrho U^{-1}; \qquad |\varrho| < 1\,,$$

and (1) takes the form

$$\sum_{M=Q}^{M=R} e^{2\pi i F(M)} = \sum_{n=-m}^{n=m} \int_{Q-1/2}^{R+1/2} e^{2\pi i(-nx+F(x))}\, dx + \lambda; \qquad |\lambda| < 1\,. \tag{3}$$

3°. We shall now estimate the order of the integral

$$I = \int_\alpha^\beta e^{2\pi i \eta(x)}\, dx\,,$$

where $\alpha < \beta$, $\eta'(\alpha) \geqq 0$ and $A^{-1} \leqq \eta''(x) \leqq kA^{-1}$ for $\alpha \leqq x \leqq \beta$.

On transforming this integral by means of the substitution $\eta(x) = z$, and then separating the real and imaginary parts, we find

$$I = \int_{\eta(\alpha)}^{\eta(\beta)} \frac{\cos 2\pi z}{\eta'(x)}\, dx + i \int_{\eta(\alpha)}^{\eta(\beta)} \frac{\sin 2\pi z}{\eta'(x)}\, dz\,,$$

where $\eta'(x)$ should be considered a function of z. By virtue of the geometrical meanings of these integrals, and since $[\eta'(x)]^{-1}$ is a decreasing function of x, and consequently of z, we readily find that for $\eta'(\alpha) > 0$,

$$|I| < \frac{2}{\pi\eta'(\alpha)}.$$

Now, since for $\beta - \alpha \geqq \sqrt{A}$, $\eta(x)$ increases at least by 1/2 as x increases from α to $\alpha + \sqrt{A}$, we obtain a different inequality which holds valid even when $\eta'(\alpha) = 0$:

$$|I| < 2\sqrt{A}.$$

Now, in place of $\eta'(\alpha) > 0$, assume that $\eta'(\beta) \leqq 0$, other conditions being the same as before. Putting $x = -\xi$, reduce the integral I to the form considered above. Hence, we find that in this case

$$|I| < \frac{2}{\pi|\eta'(\beta)|}; \qquad |I| < 2\sqrt{A}.$$

If

$$F'(Q - \tfrac{1}{2}) < n \leqq F'(R + \tfrac{1}{2}),$$

the integral

$$V_n = \int_{Q-1/2}^{R+1/2} e^{2\pi i(-nx + F(x))}\, dx$$

in the sum (3) can be expanded as

$$V_n = \int_{Q-1/2}^{x_n} + \int_{x_n}^{R+1/2}; \qquad F'(x_n) = n.$$

Putting $\eta_n(x) = -nx + F(x)$, from what has been said above it follows that $|V_n| < 4\sqrt{A}$ and

$$\sum_{n > F'(Q-1/2)}^{n \leqq F'(R+1/2)} V_n = O(UA^{-1/2}).$$

Furthermore, since $\eta'_n(x) - \eta'_{n+1}(x) = 1$, we easily find that the absolute value of the sum $\sum V_n$ taken over the remaining values of n is less than

$$8\sqrt{A} + 3\sum_{s=1}^{s=m} \frac{1}{s} = O(\sqrt{A} + \ln U).$$

Thus, we can now rewrite (3) as

$$\sum_{M=Q}^{M=R} e^{2\pi i F(M)} = O\left(\frac{U}{\sqrt{A}}\right).$$

4°. Before we derive a more exact expression, let us find an asymptotic expression for the integral

$$I = \int_0^v e^{2\pi i \eta(x)}\, dx,$$

where $0 < v < U, \eta(0) = \eta'(0) = 0; A^{-1} \leqq \eta''(x) \leqq kA^{-1}$ and $|\eta'''(x)| < l(AU)^{-1}$ for $0 \leqq x \leqq v$, where l is a constant. Furthermore, we shall assume that the number of different real roots of the equation

$$(\eta'(x))^6 - 8\eta''(0)(\eta(x))^3(\eta''(x))^2 = 0,$$

in the interval $0 \leqq x \leqq v$ does not exceed a constant h or that $\eta'''(x) = 0$ everywhere in the interval. Putting $z = \eta(x)$ we obtain

$$I = \int_0^{z_1} \frac{e^{2\pi i z}}{\eta'(x)} \cdot dz; \qquad z_1 = \eta(v),$$

where $\eta(x)$ should be regarded a function of z. Put $[\eta''(0)]^{-1} = A_1$, and compare the integral thus found with the integral

$$J = \int_0^{z_1} e^{2\pi i z} \sqrt{\frac{z_1}{2z}}\, dz.$$

In case $\eta'''(x) = 0$ everywhere in the interval, evidently $I = J$. In the first case, consider the difference

$$I - J = \int_0^{z_1} e^{2\pi i z}\, \Psi(z)\, dz; \qquad \Psi(z) = \frac{1}{\eta'(x)} - \sqrt{\frac{A_1}{2z}}.$$

Since

$$\Psi'(z) = \frac{\sqrt{A_1}\,(\eta'(x))^3 - \sqrt{8}\, z^{3/2}\, \eta''(x)}{\sqrt{8}\, z^{3/2}\, (\eta'(x))^3},$$

the whole interval $0 \leqq z \leqq z_1$ can be divided into not more than $2h+2$ subintervals, in each of which the function $\Psi(z)$ is monotonic and is of constant sign. Denoting by L the greatest value of $|\Psi(z)|$ in the interval $0 \leqq z \leqq z_1$, we find

$$|I - J| < \frac{2L}{\pi}(2h+2).$$

But

$$|\Psi(z)| < \left| \frac{2\eta(x) - A_1(\eta'(x))^2}{2\eta(x)\,\eta'(x)} \right|,$$

and since

$$\eta(x) = \frac{x^2}{2A_1} + \frac{\theta l x^3}{6AU}; \quad \eta'(x) = \frac{x}{A_1} + \frac{\theta_1 l x^2}{2AU}; \quad |\theta| < 1; \quad |\theta_1| < 1,$$

$$\eta'(x) > \frac{x}{A}; \quad 2\eta(x) > \frac{x^2}{A},$$

we find

$$|\Psi(z)| < \frac{lA}{U}\left(\frac{4}{3} + \frac{l}{4}\right)$$

and, consequently,

$$|I-J|<l\left(2+\frac{l}{3}\right)(h+1)AU^{-1}.$$

Furthermore,

$$\left|\int_{z_1}^{\infty} e^{2\pi iz}\sqrt{\frac{A_1}{2z}}\,dz\right|<T;\qquad T\leqq 2\sqrt{A};\qquad T\leqq\frac{1}{2}\sqrt{\frac{A_1}{z_1}},$$

$$\int_0^{\infty} e^{2\pi iz}\sqrt{\frac{A_1}{2z}}\,dz=\frac{1+i}{2}\sqrt{\frac{A_1}{2}}.$$

Therefore

$$I=\frac{1+i}{2}\sqrt{\frac{A_1}{2}}+\varrho(NAU^{-1}+T);\qquad |\varrho|<1;\quad N-\text{const.}$$

Now substitute the interval $v_1\leqq x\leqq 0$ for the interval $0\leqq x\leqq v$, the conditions being the same as before. Putting $x=-\xi$, we obtain the integral

$$I=\int_0^{v_1} e^{2\pi i\eta(-\xi)}\,d\xi,$$

which satisfies all the conditions of the case under consideration. Consequently, the integral I will have the same asymptotic expression as before and $z_1=\eta(v_1)$.

5°. We shall now proceed to find an asymptotic expression for the integral V_n in (3) under the condition

$$F'(Q-\tfrac{1}{2})<n\leqq F'(R+\tfrac{1}{2})$$

and the additional conditions:

1. When $Q-\frac{1}{2}\leqq x\leqq R+\frac{1}{2}$, assume that

$$|f'''(x)|\leqq l(AU)^{-1},$$

where l is a constant.

2. For any x_1 between $Q-\frac{1}{2}$ and $R+\frac{1}{2}$, either the number of different real roots of the equation

$$(F'(x)-F'(x_1))^6-8f''(x_1)(F(x)-F(x_1)-F'(x_1)(x-x_1))^3(f''(x))^2=0, \tag{4}$$

in the same interval does not exceed a constant h or $f'''(x)=0$ everywhere in the interval.

Putting $x=x_n+z$, where $F'(x_n)=n$, we obtain

$$V_n=\int_{Q-\frac{1}{2}-x_n}^{R+\frac{1}{2}-x_n} e^{2\pi i(-nx_n+F(x_n)+\eta_n(z))}\,dz,$$

where $\eta_n(z)=F(x_n+z)-F(x_n)-nz$ satisfies the conditions specified in 4°.

Therefore, on decomposing V_n into two subintervals $(Q-\frac{1}{2}-x_n, 0)$ and $(0, R+\frac{1}{2}-x_n)$, we obtain

$$V_n = \frac{1+i}{\sqrt{2}} \frac{e^{2\pi i(-nx_n+F(x_n))}}{\sqrt{F''(x_n)}} + \varrho_n\left(2N\frac{A}{U}+T_n\right); \qquad |\varrho_n|<1,$$

where T_n is the least of 2α, $\alpha+\beta$, $\alpha+\gamma$, $\beta+\gamma$, where

$$\alpha = 2\sqrt{A}; \quad \beta = \frac{1}{2}\sqrt{\frac{A}{\eta(Q-\frac{1}{2}-x_n)}}; \quad \gamma = \frac{1}{2}\sqrt{\frac{A}{\eta(R+\frac{1}{2}-x_n)}}.$$

Therefore, since the number of all values of n which are $> F'(Q-\frac{1}{2})$ and $\leqq F'(R+\frac{1}{2})$ is not greater than $UkA^{-1}+1$, from (3) we obtain

$$\sum_{M=Q}^{M=R} e^{2\pi i f(M)} = \frac{1+i}{\sqrt{2}} \sum_{n>F'(Q-\frac{1}{2})}^{n\leqq F'(R+\frac{1}{2})} \frac{e^{2\pi i(-nx_n+F(x_n))}}{\sqrt{f''(x_n)}} + P, \tag{5}$$

where

$$|P| < 2NAU^{-1}(kUA^{-1}+1) + \sum_{n=-m}^{n=m} T_n,$$

and T_n has the same meaning as before if $F'(Q-\frac{1}{2}) < n \leqq F'(R+\frac{1}{2})$ and $T_n = |V_n|$ otherwise.

a) First we shall estimate the sum

$$\sum_{n>F'(Q-\frac{1}{2})}^{n\leqq F'(R+\frac{1}{2})} T_n. \tag{6}$$

From the equality $F'(x_{n+1}) - F'(x_n) = 1$, it follows that $x_{n+1} - x_n \leqq Ak^{-1}$. Now it is not difficult to verify that

$$\eta_n(x) \geqq \frac{(x-x_n)^2}{2A}.$$

in general. Consequently, in estimating T_n we can substitute

$$\beta' = \frac{A}{\sqrt{2}(x_n - Q + \frac{1}{2})}; \qquad \gamma' = \frac{A}{\sqrt{2}(R+\frac{1}{2}-x_n)}.$$

for β and γ, respectively. Let x_{n_1} be the least x_n exceeding $Q-\frac{1}{2}$. Then

$$F'(x_{n_1}) = n_1; \qquad F'(x_{n_1}) - F'(Q-\tfrac{1}{2}) < (x_{n_1} - Q + \tfrac{1}{2})kA^{-1},$$

i.e. denoting by (u) the difference between u and an integer nearest to u, we obtain

$$(F'(Q-\tfrac{1}{2})) < \frac{(x_{n_1} - Q + \frac{1}{2})k}{A}.$$

Therefore, if $n = n_1$, in place of β' we can take

$$\beta'' = \frac{k}{\sqrt{2}\,(F'(Q-\frac{1}{2}))}.$$

Similarly, if $n=n_2$, where n_2 is the nearest integer less than $R+\frac{1}{2}$, in place of γ' we can take

$$\gamma''=\frac{k}{\sqrt{2}\,(F'(R+\frac{1}{2}))}.$$

Putting $T_n=\beta'+\gamma'$ in the remaining cases, we find that the absolute value of the sum (6) is less than

$$T+k\sqrt{2}\sum_{s=1}^{s=U}\frac{1}{s}=O(T+\ln U),$$

where T is a quantity of order the smallest of

$$\sqrt{A},\quad \frac{1}{(F'(Q-\frac{1}{2}))},\quad \frac{1}{(F'(R+\frac{1}{2}))}.$$

b) For the sum

$$\sum{}'T_n=\sum{}'|V_n|,$$

taken over all the remaining values of n, applying similar arguments as in 3°, we find that it is a quantity of the order of $T+\ln U$. Thus, the remainder term P in (5) will be of the order of

$$P=O(T+\ln U).$$

It is now quite clear that the right-hand side of (5) does not change if $f(x_n)$, $f'(Q-\frac{1}{2})$ and $f'(R+\frac{1}{2})$ are substituted for $F(x_n)$, $F'(Q-\frac{1}{2})$ and $F'(R+\frac{1}{2})$, respectively, x_n being defined as $f'(x_n)=n$ (n is substituted for $n-t$). Moreover, in place of Q and R, we can take any q and r satisfying only the conditions $2\leqq r-q\leqq U$, and defining then $Q-\frac{1}{2}$ and $R+\frac{1}{2}$ to be numbers of the type $\lambda+\frac{1}{2}$ (λ an integer), nearest to q and r, one of which is $\geqq q$ and the other $\leqq r$. On replacing the limits $Q, R, f'(Q-\frac{1}{2}), f'(R+\frac{1}{2})$ by $q, r, f'(q), f'(r)$, we change the left-hand and right-hand sides of (5) by a quantity of the order of T. Since the conditions of 1° and the additional conditions of 5° are satisfied in the interval $q\leqq x\leqq r$, we obtain

$$\sum_{M>q}^{M<r} e^{2\pi i f(M)}=\frac{1+i}{\sqrt{2}}\sum_{n>f'(q)}^{n<f'(r)}\frac{e^{2\pi i(-nx_n+f(x_n))}}{\sqrt{f''(x_n)}}+O(T+\ln U);\quad f'(x_n)=n.$$

Finally, of greatest interest are the special cases where $f(x)$ is an algebraic function defined by an equation whose degree is a constant v. With the help of Eq. (4), we easily find that h can be taken to be a constant dependent on v. Therefore the results derived can be formulated as the following

Lemma 1. a) Let A, U, q, r satisfy the conditions $U\geqq A\geqq k$; $0<r-q\leqq U$, where k is a constant $\geqq 2$. Let $f(x)$ be a function, which has a second derivative in the interval $q\leqq x\leqq r$ that satisfies the conditions $A^{-1}\leqq f''(x)\leqq kA^{-1}$. Then

$$\sum_{x>q}^{x<r} e^{2\pi i f(x)}=O\left(\frac{U}{\sqrt{A}}\right).$$

b) If, in addition, $f(x)$ is an algebraic function defined by an equation whose power is a constant ν, and in the interval $q \leqq x \leqq r$ its third derivative satisfies the condition

$$|f'''(x)| \leqq l(AU)^{-1},$$

where l is a constant, then

$$\sum_{x>q}^{x<r} e^{2\pi i f(x)} = \frac{1+i}{\sqrt{2}} \sum_{n>f'(q)}^{n<f'(r)} \frac{e^{2\pi i(-nx_n+f(x_n))}}{\sqrt{f''(x_n)}} + O(T+\ln U),$$

where T is the smallest of the numbers

$$\sqrt{A}, \quad \frac{1}{(f'(Q))}, \quad \frac{2}{(f'(R))}.$$

6°. **Lemma 2.** If the conditions (a) of Lemma 1 are satisfied and $\tau(x)$ denotes a function which is monotonic and of constant sign in the interval $q \leqq x \leqq r$, then

$$\sum_{x>q}^{x<r} \tau(x) e^{2\pi i f(x)} = O(MUA^{-\frac{1}{2}}),$$

where M is the greatest value of $|\tau(x)|$ in the interval $q \leqq x \leqq r$.

To prove this lemma, consider all integers σ that are $>q$ and are $\leqq r$. According to Lemma 1 we can write

$$\sum_{x>q}^{x\leqq\sigma} e^{2\pi i f(x)} = L_\sigma UA^{-\frac{1}{2}},$$

where $|L_\sigma|$ is less than a certain constant. Hence, applying the Abel transformation, we find that our lemma has been demonstrated.

7°. **Lemma 3.** In the domain Ω of the variables x and y bounded by a closed contour, let the function $z=f(x,y)$ be defined by an algebraic equation whose degree is a constant number n. Furthermore, let there be four parameters A, B, U and W which may increase indefinitely with the changing domain Ω and the function $f(x,y)$. Moreover, let the four constants r, s, l and p satisfy the following conditions:

$$A \geqq B \geqq 1; \quad U \geqq 2; \quad W \geqq 2; \quad U \geqq A; \quad W \geqq B,$$
$$1 \leqq r; \quad 1 \leqq s; \quad 0 \leqq l; \quad 0 \leqq p < 1.$$

In the domain Ω impose the following restrictions on the function $z=f(x,y)$:

$$A^{-1} \leqq \frac{\partial^2 z}{\partial x^2} \leqq rA^{-1}; \qquad B^{-1} \leqq \frac{\partial^2 z}{\partial y^2} \leqq sB^{-1},$$

$$\left(\frac{\partial^2 z}{\partial x \partial y}\right)^2 \leqq p^2 \frac{\partial^2 z}{\partial x^2} \cdot \frac{\partial^2 z}{\partial y^2}; \qquad \left|\frac{\partial^3 z}{\partial x^3}\right| \leqq l(AU)^{-1}.$$

Moreover, assume that the differences between the maximum and minimum values of the abscissae and ordinates of the points in the domain Ω are not greater than U and W, respectively.

The following restrictions are imposed on the contour of the domain:

a) The number of points of intersection of the contour with each line $y = \text{const.}$ does not exceed 2.

b) The number of points of intersection of the contour with each curve $\partial z/\partial x = \text{const.}$ does not exceed 2.

c) For any integral $m = 1, 2, \ldots, [A]$, the following inequality

$$\sum T_m < K \frac{UWm \ln A}{\sqrt{AB}}$$

is satisfied, where K is a constant independent of m; summation is taken over all points of intersection of $y = $ (an integer) with the contour and T_m is the least of the numbers

$$\sqrt{\frac{A}{m}}; \qquad \frac{1}{\left(m\frac{\partial z}{\partial x}\right)}.$$

1. When these conditions are satisfied, and summation is taken over all points of the domain, we obtain, for $1 \leqq m \leqq B$

$$S_m = \sum \sum e^{2\pi i m f(x, y)} = O\left(\frac{UW \ln A}{\sqrt{AB}}\right).$$

If the validity of condition (b) is not known, then

2. for $1 \leqq m \leqq B$

$$S_m = O\left(\frac{UWm}{\sqrt{AB}} + W\sqrt{\frac{A}{m}}\right);$$

3. for $B \leqq m \leqq A$

$$S_m = O\left(\frac{UW\sqrt{m}}{\sqrt{A}}\right).$$

Proof. We shall first demonstrate (1). For $m \leqq B$, all the conditions of the lemma are satisfied, if instead of A, B and $f(x, y)$, we take $A_1 = Am^{-1}$, $B_1 = Bm^{-1}$ and $\Psi(x, y) = mf(x, y)$, respectively. First consider a part

$$S_0 = \sum e^{2\pi i \Psi(x, y_0)}$$

of the double sum corresponding to a given $y = y_0$. Putting $\Psi'_x(x, y) = u$ we obtain $x = \varphi(u, y_0)$. Applying the second formula of Lemma 1 to the sum S_0 we get

$$S_0 = \frac{1+i}{\sqrt{2}} \sum_{u > \Psi'_x(x_1, y_0)}^{u \leqq \Psi'_x(x_2, y_0)} \frac{e^{2\pi i(\Psi(x, y_0) - uz)}}{\sqrt{\Psi''_{xx}(x, y_0)}} + O(T + \ln u); \qquad x = \varphi(u, y_0),$$

where x_1 and x_2 are the abscissae of the points of intersection of the line $y = y_0$ with the contour of the domain; T is the smallest of the numbers

$$\sqrt{\frac{A}{m}}; \qquad \frac{1}{(\Psi'_x(x_1, y_0))}; \qquad \frac{1}{(\Psi'_x(x_2, y_0))}.$$

On summing the above equality over all possible integral y_0, we get

$$S=\sum\sum \frac{1+i}{\sqrt{2}} \frac{e^{2\pi i(\Psi(x,y)-ux)}}{\sqrt{\Psi''_{xx}(x,y)}}+O\left(\frac{UWm\ln U}{\sqrt{AB}}\right), \tag{7}$$

where the summation on the right is taken over all possible pairs of integral u and y. But defining x as an implicit function of y by the equation

$$\Psi'_x(x,y)=u, \tag{8}$$

we find that its y-derivative is a ratio whose denominator $\Psi''_{xx}(x,y)$ does not vanish. Consequently, the curve (8) intersects any line $y=$ const. at not more than one point. Therefore no integral value of y corresponding to a given u is repeated or omitted when y runs from its smallest value $y=y_1$ to its largest value $y=y_2$.

Hence, denoting by S_1 that part of the sum (7) which corresponds to a given $u=u_1$, we get

$$S_1=\sum_{y=y_1}^{y=y_2} \frac{1+i}{\sqrt{2}} \frac{e^{2\pi i\theta(y)}}{\sqrt{\Psi''_{xx}(x,y)}}; \qquad \theta(y)=\Psi(x,y)-u_1 x; \qquad x=\varphi(u_1,y).$$

But, because of the restrictions imposed on $z=f(x,y)$, the whole interval $y_1 \leqq y \leqq y_2$ can be divided into a finite number of subintervals (the upper bound for this number depends only on n), in each of which the function

$$\Psi''_{xx}(x,y); \qquad x=\varphi(u_1,y)$$

is monotonic with respect to y.

Let $\eta_1<y<\eta_2$ be one such interval and S_2 be the part of the sum S_1 corresponding to this interval. Thus,

$$\frac{d^2\theta}{dy^2}=\frac{\dfrac{\partial^2\Psi}{\partial y^2}\dfrac{\partial^2\Psi}{\partial x^2}-\left(\dfrac{\partial^2\Psi}{\partial x\partial y}\right)^2}{\dfrac{\partial^2\Psi}{\partial x^2}}.$$

Hence

$$\frac{1-p^2}{2B_1} \leqq \frac{d^2\theta}{dy^2} \leqq \frac{2s}{B_1}.$$

Applying Lemma 2 we find $M=O(\sqrt{A_1})$. Thus, we get

$$S_2=O(W\sqrt{A_1/B_1})$$

and consequently, the sum S_1 is also of the same order.

If $W \leqq \sqrt{B_1/A_1}\,U$, then let u_1 be the least and u_2 be the greatest possible value of u. Then

$$u_2-u_1=\Psi''_{xx}(\xi',\eta')(x''-x')+\Psi''_{x,y}(\xi',\eta')(y''-y'),$$

where (ξ', η') is a certain point in the domain. Consequently,

$$u_2 - u_1 \leqq rUA_1^{-1} + pW(A_1 B_1)^{-1/2} \leqq (r+p)\, UA_1^{-1}.$$

Comparing this result with the estimate derived for S_1, we find that the double sum in (7) is of the order

$$\frac{UW}{\sqrt{A_1 B_1}} = \frac{UWm}{\sqrt{AB}}.$$

If, however, $W > \sqrt{B_1/A_1}\, U$, then all the y's in the double sum can be distributed among $O\,(W\sqrt{B_1/A_1}\, U^{-1})$ subintervals of length $\leqq \sqrt{B_1/A_1}\, U$. Accordingly, the sum is divided into $O\,(W\sqrt{B_1 A_1}\, U^{-1})$ partial sums, each of which satisfies the conditions of the previous case, and consequently, is of order

$$\frac{U}{\sqrt{A_1 B_1}} \sqrt{\frac{B_1}{A_1}}\, U = \frac{U^2}{A_1}.$$

Therefore the double sum is of order

$$W\sqrt{\frac{A_1}{B_1}}\, U^{-1} \frac{U^2}{A_1} = \frac{UW}{\sqrt{A_1 B_1}} = \frac{UWm}{\sqrt{AB}}.$$

Thus, statement (1) of the lemma has been proved.

Now we shall demonstrate the validity of (2). If condition (c) is discarded, then, in place of T, we can take $\sqrt{A_1}$. Hence we get

$$S = O\left(\frac{UWm}{\sqrt{AB}} + W\sqrt{\frac{A}{m}}\right).$$

Finally, we shall prove (3). Here the condition imposed on $\dfrac{\partial^2 \Psi}{\partial x^2}$ is satisfied. Therefore, we shall directly apply the first formula of Lemma 1 to estimate S_0. Thus, we get

$$S_0 = O\left(\frac{U\sqrt{m}}{\sqrt{A}}\right); \qquad S = O\left(\frac{UW\sqrt{m}}{\sqrt{A}}\right).$$

Hence the lemma has been proved.

8°. Now we shall use the method developed in my paper "On the distribution of fractional parts of the value of a function of one variable". Let Δ and Δ_1 be such positive numbers that $\Delta + \Delta_1 = 1$. Consider a periodic function $\varphi(x)$ of period 1, defined by

$$\varphi(x) = x \qquad \text{in} \quad 0 \leqq x \leqq \Delta,$$

$$\varphi(x) = \frac{\Delta}{\Delta_1}(1 - x) \quad \text{in} \quad \Delta \leqq x \leqq 1.$$

This function can be expanded as a trigonometrical series:

$$\varphi(x) = \frac{\Delta}{2} + \sum_{m=1}^{\infty} (A_m \cos 2\pi m x + B_m \sin 2\pi m x), \tag{9}$$

whose coefficients satisfy the conditions:

$$|A_m| < \frac{1}{\pi^2 m^2}\left(1 + \frac{\Delta}{\Delta_1}\right); \qquad |A_m| < \frac{2\Delta}{\pi m},$$

$$|B_m| < \frac{1}{\pi^2 m^2}\left(1 + \frac{\Delta}{\Delta_1}\right); \qquad |B_m| < \frac{2\Delta}{\pi m}.$$

Assuming that all the conditions of Lemma 3 are satisfied, consider the sum $S = \sum\sum \varphi(f(x, y))$,

taken over the domain Ω. Let $\Delta \leqq \Delta_1$. Then the coefficients A_m and B_m in (9) will be of the order of Δm^{-1} if $m \leqq \Delta^{-1}$ and of the order of m^{-2}, if $m > \Delta^{-1}$.
Let $\Delta^{-1} \leqq B$. Applying (9) to the sum S, and denoting by n the number of integral points in Ω, by virtue of Lemma 3, we obtain

$$S = \frac{\Delta}{2} n + O\left(\sum_{m=1}^{m \leqq \Delta^{-1}} \frac{\Delta UW \ln U}{\sqrt{AB}} + \sum_{m > \Delta^{-1}}^{m \leqq B} \frac{UW \ln U}{m\sqrt{AB}}\right.$$

$$\left. + \sum_{m > B}^{m \leqq A} \frac{UW}{m\sqrt{m}\sqrt{A}} + \sum_{m > A}^{m = \infty} \frac{UW}{m^2}\right) = \frac{\Delta}{2} n + O\left(\frac{UW(\ln U)^2}{\sqrt{AB}}\right).$$

The same result is obtained even if $B < \Delta^{-1}$.

Arguing in a similar manner for the case where $\Delta > \Delta_1$, we obtain a remainder $\Delta\Delta_1^{-1}$ times greater in order of magnitude. Hence we can write

$$S = \frac{\Delta}{2} n + O\left(\left(1 + \frac{\Delta}{\Delta_1}\right) H\right); \qquad H = \frac{UW(\ln U)^2}{\sqrt{AB}}.$$

Arranging the fractions $\{f(x, y)\}$ in an ascending order, we obtain the sequence:

$$y_1, y_2, \ldots, y_n.$$

We shall represent each y_i (see the figure) by a rectangle bounded by the lines $x = i - 1$, $x = i$, $y = 0$ and $y = y_i$ (for the sake of convenience, the scale of the OY axis is taken to be n times the scale of the OX axis). The sum

$$\sum_{i=1}^{i=n} y_i$$

is shown by the shaded area. Let C denote the difference between the shaded area and the area OAK. This difference is represented by the area bounded by the broken line ORK and the straight line OK; the area above the line is taken positive, while the area below the line, negative.

Let MN be a line defined by the equation $y = \Delta$ and let α be that part of the area C which lies below MN and β that part which lies above MN. According to the expression found above, we have

$$\sum_{i=1}^{i=n} \varphi(y_i) = \frac{\Delta}{2} n + O\left(\left(1 + \frac{\Delta}{\Delta_1}\right) H\right). \tag{10}$$

We shall now represent $z_i = \varphi(y_i)$ in the same manner as y_i was represented before. Thus, we obtain another broken line OSA whose form is defined by the equations:

$$z_i = y_i \qquad \text{for} \quad y_i \leqq \Delta,$$

$$z_i = \frac{\Delta}{\Delta_1}(1 - y_i) \quad \text{for} \quad y_i > \Delta.$$

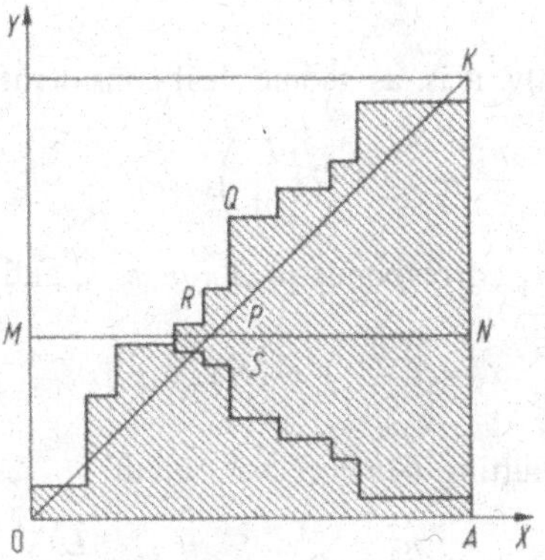

Expression (10) shows that

$$\sum z_i - \frac{\Delta}{2} n = O\left(\left(1 + \frac{\Delta_1}{\Delta}\right) H\right).$$

The left-hand side of this equality, as is not difficult to verify, is equal to $\alpha - \beta \Delta \Delta_1^{-1}$. Consequently,

$$\alpha - \beta \frac{\Delta}{\Delta_1} = O\left(\left(1 + \frac{\Delta}{\Delta_1}\right) H\right).$$

On the other hand, since $\alpha + \beta = C$, we find

$$\begin{aligned} \alpha &= C\Delta + O(H), \\ \beta &= C\Delta_1 + O(H). \end{aligned} \tag{11}$$

For the sake of definiteness, assume that $C \geqq 0$, then we obtain (whatever Δ)

$$\alpha \leqq \frac{\Delta^2 n}{2}; \qquad C\Delta + O(H) \leqq \frac{\Delta^2 n}{2}; \qquad C = O\left(\frac{H}{\Delta} + \Delta n\right).$$

Taking $\Delta = \sqrt{Hn^{-1}}$, we find

$$C = O(\sqrt{Hn}) = O(L), \qquad L = \frac{UW \ln U}{(AB)^{1/4}}.$$

The same result is obtained also for the case where $C < 0$.

It is now not difficult to verify that the whole broken line ORK lies in a strip between the lines $y=(1/n)x+b_1$ and $y=(1/n)x+b_2$, where $b_2-b_1=O(Ln^{-1})$. Indeed, let $Q(\xi,\eta)$ be the point on the broken line ORK that is most remote from the line OK. Consider the function $f_1(x,y)=f(x,y)-N$, where $N=PQ$ is the difference between the ordinates of the points Q and P on the line OK with the same abscissa. For this function we obtain another number C_1 of the same order as C. Evidently,

$$C-C_1=Nn.$$

Thus, $Nn=O(L)$, hence our statement has been proved.

Now let n_1 be a certain positive integer. Consider the sum

$$S_1=\sum_{i=1}^{i=n_1} y_i.$$

Evidently, it is, as geometrical considerations show, equal to

$$\frac{n_1^2}{2n}+\alpha_1+O\left(\frac{L^2}{n}\right),$$

where α_1 corresponds to $\Delta=n_1n^{-1}$, and consequently, according to (11):

$$\alpha_1=C\frac{n_1}{n}+O(H).$$

Thus, putting $C/n=R$, we obtain

$$S_1=\frac{n_1^2}{2n}+Rn_1+O\left(H+\frac{L^2}{n}\right),$$

where R does not depend on n_1, and

$$R=O\left(\frac{L}{n}\right).$$

Thus we obtain the following theorem.

Theorem. Let all the condition of Lemma 3 be satisfied. Let n denote the number of integral points in the domain Ω and arrange the fractional parts $\{f(x,y)\}$ corresponding to all integral points in Ω in ascending order:

$$y_1\leqq y_2\leqq\ldots\leqq y_n.$$

Then, for any positive integral $n_1\leqq n$

$$\sum_{i=1}^{i=n_1} y_i=\frac{n_1^2}{2n}+Rn_1+O\left(\frac{U^2W^2\ln U)^2}{n\sqrt{AB}}\right),$$

where R is a quantity of order

$$\frac{UW\ln U}{n(AB)^{1/4}},$$

independent of n_1.

9°. By slightly complicating the arguments, we can obtain a lower order

$$O\left(\frac{U^2W^2}{n\sqrt{AB}}\right)$$

for the remainder term (modified conditions being imposed on the contour) which cannot be improved any further.

For the order of the constant R, by complicating the method, we can obtain the form

$$\frac{UW}{n(AB)^\sigma},$$

where σ is slightly greater than $\frac{1}{4}$. But such a reduction requires considerable complications in the method and additional restrictions to be imposed on the function. Here we shall confine ourselves to demonstrating how the factor $\ln U$ can be eliminated from the expression of the order.

For this purpose, assume that $\Delta = (AB)^{-1/4}$ and consider two periodic functions $\varphi(x)$ and $\psi(x)$ defined as follows. Let

$$\Lambda = \frac{1}{\Delta\sqrt{2}};$$

then

1. the function $\varphi(x)$ is equal to

$$\begin{array}{rll} \dfrac{x}{1-2\Delta} & \text{for} & 0 \leqq x \leqq 1-2\Delta, \\ 1-\Lambda^2(x-1+2\Delta)^2 & \text{for} & 1-2\Delta \leqq x \leqq 1-\Delta, \\ \Lambda^2(1-x)^2 & \text{for} & 1-\Delta \leqq x \leqq 1. \end{array}$$

2. the function $\psi(x)$ is equal to

$$\begin{array}{rll} 1-\Lambda^2x^2 & \text{for} & 0 \leqq x \leqq \Delta, \\ \Lambda^2(x-2\Delta)^2 & \text{for} & \Delta \leqq x \leqq 2\Delta, \\ 0 & \text{for} & 2\Delta \leqq x \leqq 1-4\Delta, \\ \Lambda^2(x-1+4\Delta)^2 & \text{for} & 1-4\Delta \leqq x \leqq 1-3\Delta, \\ 1-\Lambda^2(x-1+2\Delta)^2 & \text{for} & 1-3\Delta \leqq x \leqq 1-2\Delta, \\ 1 & \text{for} & 1-2\Delta \leqq x \leqq 1. \end{array}$$

These functions can be expanded as trigonometric series:

$$\varphi(x) = \frac{1}{2} + \sum_{m=1}^{\infty} (A_m \cos 2\pi mx + B_m \sin 2\pi mx),$$

$$\Psi(x) = 4\Delta + \sum_{m=1}^{\infty} (A'_m \cos 2\pi mx + B'_m \sin 2\pi mx),$$

where the coefficients depend only on m and Δ, and are of the order of

$$1/m \quad \text{for} \quad m \leqq 1/\Delta,$$
$$1/\Delta^2 m^3 \quad \text{for} \quad m > 1/\Delta.$$

Taking the sum over all integral points in the domain and assuming that all conditions, except condition (b), of Lemma 3 are satisfied, we obtain, after some simple calculations,

$$\sum\sum \varphi(f(x,y)) = \frac{1}{2}n + O\left(\frac{UW}{(AB)^{1/4}}\right),$$

$$\sum\sum \Psi(f(x,y)) = O\left(\frac{UW}{(AB)^{1/4}}\right),$$

if we suppose that AB^{-3} is bounded.

The second equality shows that the number of fractional parts $\{f(x,y)\}$ within the limits $1-2\Delta$ and 1 is a quantity of the order of

$$N = \frac{UW}{(AB)^{1/4}}.$$

Furthermore, since for $0 \leqq \{f(x,y)\} \leqq 1-2\Delta$, the difference $\varphi[f(x,y)] - \{f(x,y)\}$ is a quantity of the order of Δ, we obviously get

$$\sum\sum \varphi(f(x,y)) - \sum\sum \{f(x,y)\} = O(N).$$

Hence it follows that

$$\sum\sum \varphi(f(x,y)) - \tfrac{1}{2}n = O(N).$$

Thus, assuming that AB^{-3} is bounded and that all the conditions except (c) are satisfied, we obtain

$$R = O\left(\frac{UW}{(AB)^{1/4}}\right).$$

On Waring's Theorem

Izvestiya Akademii Nauk SSSR, Otd. Fiz.-Mat. Nauk, no. 4 (1928) 393–400.

Although quite a large number of papers are devoted to Waring's theorem, yet we believe that the method proposed in this paper is not devoid of interest. So far the most refined results in this field have been derived in the works of Hardy and Littlewood, Nevertheless, their method, even after the simplifications made by, say, Landau[1], is quite complicated.

In the present paper, using essentially the same techniques as Hardy and Littlewood's but in a different way and introducing certain additional considerations, we shall derive the same result with incomparable brevity and simplicity. This fact is rather important not only for the reader, but also because it simplifies the application of the method which we shall use in our work in future.

We shall use the following notations:

a) $D = D(a, \ldots, k)$ denotes that D depends only on $a, \ldots, k$; $D < D(a, \ldots, k)$ means that $|D|$ is less than a number that depends only on $a, \ldots, k$;
b) ε, ε_1 stand for arbitrary positive numbers;
c) n is an integer >2; r is an integer $\geqq r_0 = (n-2)2^{n-1} + 5$; $\sigma = 2^{-n+1}$; $\sigma_1 = 2^{-n}$ (so that $\sigma r \geqq 9/4$; $\sigma(r-1) \geqq 2$);
d) $C, C_1, \ldots$ depend only on r and n; the absolute values of $c, c_1, \ldots$ do not exceed the numbers which depend only on r and n.

Lemma 1. Let P be an integer >0, z a real number,

$$z = a/q + y; \quad (a, q) = 1; \quad 0 < q \leqq P^n; \quad |y| < q^{-2}; \quad S = \sum_{x=1} e^{2\pi i z x^n}.$$

Then

$$|S| < DP^{1+\varepsilon}(\max(q^{-1}, P^{-1}, qP^{-n}))^{\sigma}; \qquad D = D(n, \varepsilon).^2$$

Lemma 2. Let P be an integer >0, and z a real number,

$$z = a/q + y; \quad (a, q) = 1; \quad 0 < q \leqq P; \quad S = \sum_{x=1}^{P} e^{2\pi i z x^n},$$

---

1 Landau, E.: Vorlesungen über Zahlentheorie, vol. 1 (Leipzig, 1927).

2 An analytic proof of the theorem on the distribution of the fractional parts of an integral polynomial. Izv. akad. nauk SSSR (ser. mat.) (1927, no. 7–8) 567–578: Lemma 3, a special case.

$$S_{a,q}=\sum_{\alpha=0}^{q-1} e^{2\pi i\frac{a}{q}\alpha^n};\qquad I=\int_0^P e^{2\pi i y z^n}\,dz;\qquad S_1=\max\left(|S|,|q^{-1}S_{a,q}I|\right),$$

then

$$|S^r-q^{-r}S_{a,q}^r I^r|<CS_1^{r-1}q\max\left(|y|P^n,1\right).$$

Proof. Putting

$$x=qv+\alpha;\qquad \omega=[(P-\alpha)q^{-1}],$$

we find

$$S=\sum_{\alpha=1}^{q} e^{2\pi i\frac{a}{q}\alpha^n}\Omega_\alpha;\qquad \Omega_a=\sum_{0\leqq v\leqq\omega} e^{2\pi i(qv+\alpha)^n y}=q^{-1}I+L(|y|P^n+1);$$

$$L<L(n).$$

Consequently,

$$S=q^{-1}S_{a,q}I+L_1 q(|y|P^n+1);\qquad |L_1|<L_1(n).$$

Hence the lemma has been demonstrated.

Lemma 3. The number R of the solutions to the Diophantine equation

$$h_1^n+h_2^n=m;\qquad h_1\geqq 0,\ h_2\geqq 0$$

satisfies the inequality[1]:

$$R<D_1 m^{\varepsilon_1};\qquad D_1=D_1(n,\varepsilon_1).$$

Lemma 4. The number of solutions in positive integral $x_1,\dots,x_r$ of the inequality

$$x_1^n+\dots+x_r^n\leqq V$$

is given by

$$\frac{n}{r}KV^{r/n}+c_1 V^{\frac{r-1}{n}};\qquad K=\frac{(\Gamma(1+1/n))^r}{\Gamma(r/n)}.$$

Lemma 5. If $N_1\leqq N_2$ are integers and q is an integer >1, then[2]

$$\sum_{a=1}^{q-1}\left|\sum_{N=N_1}^{N_2} e^{-2\pi i\frac{a}{q}N}\right|<q\ln q.$$

Number of representations. Let N_0 be an integer >0; $P=[N_0^{1/n}]+1$; N an integer $=N_0(1-\theta P^{-0.5})$ $(0\leqq\theta\leqq 1)$; I_N be the number of solutions in positive integral $x_1,\dots,x_r$ of the equation:

$$x_1^n+\dots+x_r^n=N.$$

---

1 Landau, E. Op. cit. Satz 262.

2 We believe Lemmas 4 and 5 are well known.

Then,

$$I_N = \int_0^1 S^r e^{-2\pi i z N}\, dz; \qquad S = \sum_{z=1}^{P} e^{2\pi i z x^n}.$$

Representing z as

$$z = a/q + y; \quad (a, q) = 1; \quad 0 < q \leqq P^{n-1}; \quad |y| < q^{-1} P^{-n+1},$$

divide the interval $0 \leqq z \leqq 1$ into subintervals of the following four classes:

I. $1 \leqq q \leqq P^{0.6}$; y takes all values from $-P^{-n+0.12}$ to $+P^{-n+0.12}$ (*);

II. $P \leqq q$;

III. $P^{0.6} \leqq q \leqq P$; $|y| \leqq P^{-n}$;

IV. The remaining part of the intervals corresponding to the case $1 \leqq q \leqq P$.

Now express the integral I_N as the sum of four terms:

$$I_N = H_1 + H_2 + H_3 + H_4.$$

In the estimations we shall take $\varepsilon r = 0.02$ unless assumed otherwise.

Class I. Applying Lemmas 1 and 2, we find

$$|S^r - q^{-r} S_{a,q} I^r| < C(DP^{1+\varepsilon} q^{-\sigma})^{r-1} qP^{0.12} < C_2 P^{r-0.86} q^{-1}.$$

Then we obtain

$$S^r e^{-2\pi i z N} = S^r e^{-2\pi i \left(\frac{a}{q} N + N_0 y - \theta y N_0 P^{-0.5}\right)}$$

$$= S^r e^{-2\pi i \left(\frac{a}{q} N + N_0 y\right)} + c_3 P^{r-0.36} q^{-2.25}.$$

Hence it is easily found that

$$\left| S^r e^{-2\pi i z N} - q^{-r} e^{-2\pi i \left(\frac{a}{q} N + N_0 y\right)} S^r_{a,q} I^r \right| < C_4 P^r (P^{-0.86} q^{-1} + P^{-0.36} q^{-2.25}).$$

Integrating this inequality from $-P^{-n+0.12}$ to $+P^{-n+0.12}$, and since for a given q we have $\leqq q$ intervals, we obtain

$$|H_1 - T\Delta_N| < C_5 P^{r-n-\sigma_1}; \qquad T = T(r, n, N_0),$$

$$\Delta_N = \sum_{q=1}^{P^{0.6}} A_q; \qquad A_q = q^{-r} \sum S^r_{a,q} e^{-2\pi i \frac{a}{q} N},$$

where a runs through those of $1, \ldots, q$ that are coprime to q.

Class II. On assuming $S_2 = \max |S|$ over the subintervals of Class II, we obtain

$$|H_2| < S_2^{r-4} \int_0^1 |S|^4\, dz.$$

(*) Evidently, the subintervals in Class I do not overlap.

But, by Lemma 1, putting $\varepsilon r = 1/2\sigma_1$, we obtain

$$S_2^{r-4} < C_6 P^{(1-\sigma)(r-4)+\frac{1}{2}\sigma_1}.$$

On the other hand, the integral

$$\int_0^1 |S|^4\,dz = \int_0^1 \sum_{x=1}^{P}\sum_{y=1}^{P}\sum_{u=1}^{P}\sum_{v=1}^{P} e^{2\pi i z(x^n+y^n-u^n-v^n)}\,dz$$

is equal to the number of solutions in positive integral x, y, u, v of the equation

$$x^n + y^n = u^n + v^n.$$

Therefore, by Lemma 3, it is less than

$$P^2(D_1(2P^n)^{\varepsilon_1})^2 < D_2 P^{2+\varepsilon_2}; \qquad D_2 = D_2(n, \varepsilon_2).$$

Hence, putting $\varepsilon_2 = \frac{1}{2}\sigma_1$, we find

$$|H_2| < C_7 P^{r-n-\sigma_1}.$$

Class III. Applying Lemma 1, we obtain

$$|S|^r < C_8 P^{r+0.02} q^{-2.25}.$$

The sum of the lengths of the subintervals corresponding to a given q is less than $2qP^{-n}$. Consequently,

$$|H_3| < 2C_8 P^{r-n+0.02-0.15} < 9C_8 P^{r-n-\sigma_1}.$$

Class IV. Assuming $1 \leqq l \leqq 1/2P$, consider the subintervals which are parts of the intervals corresponding to a given q, where z can be expressed as

$$z = \frac{a_1}{q_1} + y_1; \quad (a_1, q_1) = 1; \quad |y_1| \leqq \frac{l}{q_1 P^n}; \quad q_1 \leqq \frac{P^n}{l}; \quad q_1 < q, \tag{1}$$

But it cannot be expressed in this form when $l_1 = 2l$ is substituted for l.

Since $q_1 < P$ is not possible, and $P \leqq q_1 \leqq P^{n-1}$ is contained in Class II, we have $q_1 > P^{n-1}$. Therefore, by Lemma 1, we have

$$|S|^r < C_9 P^{r+0.02} l^{-2.25}.$$

Obviously, we also have

$$z = \frac{a}{q} + y; \qquad |y| \leqq \frac{2l}{qP^n}.$$

Therefore the sum of the lengths of the intervals is $\leqq 4lP^{-n}$. Hence the absolute value of H_4 corresponding to these intervals is

$$|H_2| < S_2^{r-4} \int_0 |S|^4\,dz.$$

Taking $E = qP^{0.12}$, when $1 \leqq q \leqq P^{0.6}$;

and $E = q$, when $q > P^{0.6}$

and for each given q, assigning the following values of l

$$l=\frac{P}{2^k},\ldots,\frac{P}{2^2},\frac{P}{2};\qquad \frac{P}{2^k}\leqq E<\frac{P}{2^{k-1}},$$

we exhaust all the subintervals of Class IV. At the same time

$$|H_4|<C_{11}\left(\sum_{q=1}^{P^{0.6}}P^{r-n+0.02-0.15}q^{-1.25}+\sum_{q>P^{0.6}}^{P}P^{r-n+0.02}q^{-1.25}\right)<C_{12}P^{r-n-\sigma_1}.$$

Collecting the results obtained for all the four classes, we get

$$|I_N-T\Delta_N|<C_{13}P^{r-n-\sigma_1}. \tag{2}$$

Calculation of T. Summing (2) for N within the limits

$$N_0-N_0P^{-0.5}<N\leqq N_0,$$

we obtain

$$\left|\sum I_N-T\sum\Delta_N\right|<C_{14}P^{r-0.5-\sigma_1}. \tag{3}$$

a) $\sum I_N$ is the number of solutions in positive integral $x_1,\ldots,x_r$ of the inequality

$$N_0(1-P^{-0.5})<x_1^n+\ldots+x_r^n\leqq N_0.$$

Therefore, by Lemma 4, we have

$$\sum I_N=\frac{n}{r}KN_0^{r/n}(1-(1-P^{-0.5})^{r/n})+c_{15}N_0^{r-1/n}$$
$$=KN_0^{r/n}P^{-0.5}+c_{16}N_0^{r-1/n}.$$

b) Now we shall calculate $\sum\Delta_N$. According to Lemma 1, we have

$$|S_{a,q}|^r<C_{17}q^{r-2.23};\qquad |A_q|<C_{17}q^{-1.23}. \tag{4}$$

Hence, applying Lemma 5, we find that for $q>1$

$$\left|\sum_N A_q\right|\leqq q^{-r}\sum_a|S_{a,q}|^r\left|\sum_N e^{-2\pi i\frac{a}{q}N}\right|<C_{17}q^{-2.23}q\ln q<C_{18}q^{-1.22}.$$

When $q=1$, we have $A_1=1$, therefore

$$\sum_N A_1=N_0P^{-0.5}+\theta_1;\qquad |\theta_1|<1.$$

Hence

$$\sum\Delta_N=N_0P^{-0.5}+c_{19}.$$

By the statements proved in (a) and (b), inequality (3) gives

$$KN_0^{r/n}P^{-0.5}=T(N_0P^{-0.5}+c_{19})+c_{20}N_0^{r/n}P^{-0.5-\sigma_1}.$$

Hence

$$T = KN_0^{r/n-1} + c_{21} N^{r/n-1-\eta}; \qquad \eta = \frac{1}{n}\sigma_1 .$$

On the other hand, by (4), we find

$$\sum_{q>P^{0.6}}^{\infty} |A_q| < C_{17} \sum_{q>P^{0.6}}^{\infty} q^{-1.23} < C_{22} P^{-\sigma_1}; \qquad \sum_{q=1}^{\infty} |A_q| < C_{23} .$$

Therefore, putting

$$\mathfrak{S} = \sum_{q=1}^{\infty} A_q ,$$

from (2), we obtain

$$|I_N - KN_0^{r/n-1}\,\mathfrak{S}| < C_{24} N_0^{r/n-1-\eta} .$$

Hence, for $N = N_0$, we obtain

$$|I_{N_0} - KN_0^{r/n-1}\,\mathfrak{S}| < C_{25} N_0^{r/n-1-\eta} .$$

All the corollaries derived by Hardy and Littlewood follow from this inequality.

A New Estimate for $G(n)$ in Waring's Problem

Doklady Akademii Nauk SSSR, 5 no. 5–6 (1934) 249–253.

In this paper I shall give a new estimate for the well-known function $G(n)$, namely

$$G(n) < 6n \ln n + 10n,$$

which is a quantity of the order of $n \ln n$.

This order, because of the inequality $G(n) > n$, cannot be improved any further.

1. If B is positive, then $A \ll B$ or $A = O(B)$ denotes that the ratio of $|A|$ to B does not exceed a constant dependent only on n. Consider the integral

$$I_N = \int_0^1 S_\alpha^2 V_\alpha \; T_\alpha^{4n-2} e^{-2\pi i \alpha N} d\alpha,$$

where n is an integer > 3 and

$$S_\alpha = \sum_{u_1} e^{2\pi i \alpha u_1}; \quad V_\alpha = \sum_{y=1}^{Y} \sum_{u_2} e^{2\pi i \alpha y^n u_2}; \quad T_\alpha = \sum_{x=1}^{3P} e^{2\pi i \alpha x^n}.$$

Here, denoting by N_0 a sufficiently large integer, we shall take

$$\nu = \frac{1}{n}; \quad P = [\tfrac{1}{3} N_0^\nu + 1]; \quad Y = \left[\left(\frac{\sqrt{P}}{2}\right)^\nu\right]; \quad N_0 - N_0 P^{-1/4} < N \leqq N_0.$$

We shall take u_1 and u_2 as follows. Let

$$P_2 = [n^\nu P^{1-\nu}]; \quad R_1 = [P^{1-(\nu/2)}]; \quad R_2 = [n^\nu R_1^{1-\nu}]; \quad k = [2n \ln n + n \ln 6].$$

Now let us form the numbers

$$v_1 = 1^n, 2^n, \ldots, P_2^n.$$

Insert the numbers of the types $v_1' + w^n$ between two consecutive values v_1' and v_1'' of v_1, where w runs through all positive integers satisfying the condition $v_1' + w^n < v_1''$. Taking the complement of these numbers in the sequence of numbers v_1, we obtain a new sequence of numbers v_2.

Repeat this process with the numbers v_2. Thus we obtain a new sequence of numbers v_3. In this way we form the numbers v_4, v_5 and finally the numbers $\xi_1 = v_k$. Let u_1 take the values

$$u_1 = \xi_1 + x_1^n,$$

where ξ_1 runs through the values specified above, while x_1 runs through the following values

$$x_1 = P,\ P+1, \ldots, 2P-1.$$

It is readily seen that the numbers u_1 are positive and do not overlap each other. The greatest of them is less than $2^n P^n$ and the total number X_1 of these numbers satisfies the condition

$$X_1 \gg P^{n-\sigma}; \qquad \sigma = n(1-\nu)^k.$$

Now let u_2 take the values

$$u_2 = \xi_2 + x_2^n,$$

where ξ_2 and x_2 run through the values which are similar to the values specified for ξ_1 and x_1, but here R_1 and R_2 have to be substituted for P and P_2, respectively. The numbers u_2 are also positive and do not overlap each other. The greatest of them is less than $2^n R_1^n$ and the total number X_2 of these numbers satisfies the condition

$$X_2 \gg R_1^{n-\sigma}.$$

Taking

$$\tau = 2n \cdot 3^{n-1} \cdot P^{n-1/2},$$

replace the interval $0 \leqq \alpha \leqq 1$ by the following interval:

$$-\tau^{-1} \leqq \alpha \leqq 1 - \tau^{-1}.$$

Divide the latter interval into subintervals of two classes satisfying the conditions:

Class I: $\alpha = \dfrac{a}{q} + z; \quad (a,q) = 1; \quad 1 \leqq q \leqq \sqrt{P}; \quad -\tau^{-1} \leqq z \leqq \tau^{-1}.$

Class II: All other remaining subintervals.

Accordingly, the integral I_N is decomposed into a sum of two terms

$$I_N = H_1 + H_2.$$

2. If α belongs to a subinterval of Class II, then

$$\sqrt{P} < q \ll P^{n-1/2},$$

$$V_2 \ll \sqrt{X_2 \sum_{y=1}^{Y} \sum_{y_1=1}^{Y} \left| \sum_{x=1}^{2^n R_1^n} e^{2\pi i \alpha (y^n - y_1^n) x} \right|} \ll X_2 Y \sqrt{\frac{R_1^n}{X_2 Y}\left(1 + \frac{\tau^{1+\varepsilon}}{R_1^n Y}\right)}$$

$$\ll X_2 Y P^{\left(1-\frac{\nu}{2}\right)\frac{\sigma}{2} - \frac{\nu}{4}}.$$

Hence, we find

$$H_2 \ll P^{4n-2} X_2 Y P^{\left(1-\frac{\nu}{2}\right)\frac{\sigma}{2} - \frac{\nu}{4}} \int_0^1 \sum_{u_1} \sum_{u_1'} e^{2\pi i \alpha (u_1 - u_1')}\, d\alpha$$

$$\ll P^{4n-2} X_2 Y P^{\left(1-\frac{\nu}{2}\right)\frac{\sigma}{2} - \frac{\nu}{4}} \frac{X_1^2}{P^n} P^{\sigma} \ll X_1^2 Y_2 Y P^{3n-2-\delta},$$

where δ is a positive number dependent only on n.

3. The term H_1 is calculated by means of a procedure similar to those which I applied in my previous papers dealing with Waring's problem. Thus, we obtain

$$H_1 = \sum_{y=1}^{Y} F_y \sum_{q=1}^{\leq\sqrt{P}} \sum_{a} E_y + O(X_1^2 X_2 Y P^{3n-2-\delta}),$$

where a runs through $0, 1, \ldots, (q-1)$ coprime to q. Here F_y depends only on y, n and N_0. And we have

$$E_y = e^{-2\pi i \frac{a}{q} N} \left\{ \sum_{\xi_1} e^{2\pi i \frac{a}{q} \xi_1} \right\}^2 \sum_{u_2} e^{2\pi i \frac{a}{q} y^n u_2} \left(\frac{1}{q} \sum_{r=0}^{q-1} e^{2\pi i \frac{a}{q} r^n} \right)^{4n}.$$

Here

$$F_y = D_y \{1 + O(P^{-\delta})\}; \qquad \sum_{q=1}^{\leq\sqrt{P}} \sum_{a} E_y = J_y \{1 + O(P^{-\delta})\},$$

where D_y and J_y are real and

$$D_y \gg P^{3n}; \qquad J_y \gg X_1^2 X_2 P^{-2}.$$

4. The results derived in 2° and 3° give

$$I_{N_0} \gg X_1^2 X_2 Y P^{3n-2} \{1 + O(P^{-\sigma})\} \gg X_1^2 X_2 Y P^{3n-2},$$

which shows that $I_{N_0} > 0$. But I_{N_0} is the number of representations of N_0 in the form:

$$N_0 = u_1 + u_1' + y^n u_2 + m_1^n + m_2^n + \ldots + m_{4n-2}^n.$$

The right-hand side of this equality is the sum of

$$\leq 3(k+1) + 4n - 2 < 6n \ln n + 10n$$

terms of the type t^n. Hence we obtain the following

Theorem. Let n be an integer >3,

$$s = [6n \ln n + 10n].$$

Then there exists such a constant c, dependent only on n, that any integer $N \geq c$ can be represented in the form

$$N = t_1^n + t_2^n + \ldots + t_s^n,$$

where $t_1, t_2, \ldots, t_s$ are integers ≥ 0.

On the Upper Bound for $G(n)$ in Waring's Problem

Izvestiya Akademii Nauk SSSR, Otd. Mat. Estestv. Nauk, no. 10 (1934) 1455–1469.

Let $G(n)$ denote the well known function expressing the least integer under the following condition: there exists an $N' > 0$, dependent only on n, such that every integer $N \geqq N'$ can be represented in the form of a sum of s terms of the type t^n, where t is an integer greater than zero, but not every integer $N \geqq N'$ can be represented in the form of a sum of $(s-1)$ terms of the same type.

In my previous papers[1] I have developed a new method for solving this problem by the analytic number theory; moreover, for the function $G(n)$ I have obtained the following estimates:

$$G(n) = O((n \ln n)^2); \qquad G(n) = O(n^2).$$

Here, by slightly modifying the proof, I have obtained

$$G(n) = O(n \ln n),$$

or to be more precise

$$G(n) < 6n \ln n + (\ln 216 + 4)\, n.$$

This result is remarkable in that it gives almost the final answer to the question about the order of $G(n)$. Indeed, from the well known inequality $G(n) > n$, it follows that the order of $G(n)$ cannot be less than $O(n)$.

A brief account of my new result has already been reported in one of my recent papers[2]. Here we shall present it in detail.

§1. Notations and Lemmas

We shall adopt the following notations:

$$n \text{ – an integer} > 3; \qquad \nu = \frac{1}{n}; \qquad \delta = \frac{\nu^2}{40};$$

$$k = [2n \ln n + n \ln 6]; \qquad \sigma = n(1-\nu)^k,$$

---

[1] A new solution of Waring's problem; Dokl. akad. nauk SSSR (1934, vol. 2, no. 4) 337–341. On some new problems in the theory of numbers *Ibid.* (1934, vol. 3, no. 1) 1–6; Sur quelques nouveaux résultats en la théorie analytique des nombres; Comptes rendues Acad. Sci. Paris **199**, no. 3 (1934) 174–5. [In French].

[2] A new estimate for $G(n)$ in Waring's problem; Dokl. akad. nauk SSSR (1934, vol. 4, no. 5–6) 249–253 = This collection, pp. 107–109.

c', c, c_0, c_1, ... are real numbers dependent only on n; θ', θ, θ_1 are certain numbers whose moduli do not exceed 1, $A \ll B$, $B \gg A$ or $A = O(B)$ for positive B denotes that $|A| \leqq c'B$. If z is a real number, then (z) denotes the distance from z to the nearest integer.

$$N_0 \text{ an integer} > 0; \quad P = [3^{-1} N_0^{\nu} + 1]; \quad P_1 = [n^{\nu} P^{1-\nu}],$$
$$R = [P^{1-\nu/2}]; \quad R_1 = [n^{\nu} R^{1-\nu}],$$
$$Y = [P^{\nu-\nu^2/2}]; \quad N - \text{integer}; \quad N_0 - N_0 P^{-1/4} < N \leqq N_0;$$
$$\tau = 2n 3^{n-1} P^{n-1/2},$$

$N_0 > c_0$; c_0 is so chosen (which is possible) that

$$N_0 \geqq 64 \cdot 3^{n-4} \cdot P^n; \quad Y^n \leqq 2^{-1}\sqrt{P} \text{ and } \tau \leqq P^n. \tag{1}$$

Moreover, for an integer $q > 0$ and $(a, q) = 1$, let

$$D_{aq} = \frac{1}{q} \sum_{r=0}^{q-1} e^{2\pi i \frac{a}{q} r^n}; \quad A_q(N) = \sum_a D_{aq}^{4n} e^{-2\pi i \frac{a}{q} N},$$

where a in the last sum runs through a reduced system of residues modulo q.

The lemmas, on which my proof is based, are well known. In particular, they have already been used in my previous papers devoted to Waring's theorem. Therefore here I shall state them without proof.

Lemma 1. We have

$$D_{aq} \ll q^{-\nu}.$$

Lemma 2. We have

$$\sum_{q=1}^{\sqrt{P}} A_q(N) = b(N) + O(P^{-1}),$$

where $b(N) > c_1$ and $c_1 > 0$.

Lemma 3. In the interval $g \leqq t \leqq h$ let

$$0 \leqq f'(t) \leqq \tfrac{1}{2}; \quad f''(t) > 0.$$

Then

$$\sum_{t>g}^{h} e^{\pm 2\pi i f(t)} = \int_g^h e^{\pm 2\pi i f(t)}\, dt + 5\theta'.$$

if the same sign $+$ or $-$ is taken on both sides.

Lemma 4. Let $x_1, x_2, \ldots, x_{4n-2}$ run independently of each other through positive values. Then the number of solutions of the inequality

$$x_1^n + x_2^n + \ldots + x_{4n-2}^n \leqq V$$

is given by the expression:

$$c_2 V^{4-2\nu} + O(V^{4-3\nu}); \quad c_2 > 0.$$

Lemma 5. If G and H are integers such that $G < H, \lambda$ is a real noninteger, then

$$\left|\sum_{x>G}^{H} e^{2\pi i\lambda x}\right| < \frac{1}{2(\lambda)}.$$

§2. The Number u

Form the numbers

$$s_1 = 1^n, 2^n, \ldots, P_1^n.$$

The greatest difference between two consecutive values of s_1 is

$$\leqq P_1^n - (P_1 - 1)^n < nP_1^{n-1}.$$

We shall insert numbers of the type

$$s_1 + w^n,$$

between each pair of two consecutive values s_1' and s_1'' of s_1, where w runs through all integral values satisfying the condition

$$0 < w^n < s_1'' - s_1'.$$

On adding these numbers to the sequence of numbers s_1, we obtain a new sequence of numbers s_2. The numbers s_2 do not overlap each other. By the method used to fill the interval between s_1' and s_1'', it follows that the greatest difference between two consecutive values of s_2 satisfying the condition

$$s_1' \leqq s_2 \leqq s_1'',$$

is equal to

$$\leqq \{(s_1'' - s_1')^{\nu}\}^n - \{(s_1'' - s_1')^{\nu} - 1\}^n < n(nP_1^{n-1})^{\nu(n-1)} = n^{1+(1-\nu)} P_1^{(n-1)(1-\nu)}.$$

Consequently, the difference between any pair of two consecutive values of s_2 does not also exceed the above value. Repeat the same procedure with the numbers s_2 as with the numbers s_1. Thus we obtain new numbers s_3, then s_4, s_5 ... and finally $\xi = s_{k-1}$.

The numbers ξ do not overlap each other, and satisfy the condition

$$1 \leqq \xi \leqq P_1^n.$$

The difference between each two consecutive values of ξ will be

$$< n^{1+(1-\nu)+(1-\nu)^2+\ldots+(1-\nu)^{k-2}} P_1^{(n-1)(1-\nu)^{k-2}} < n^n P_1^{n(1-\nu)^{k-1}}.$$

The total number of all values of ξ is

$$> n^{-n} P_1^{n-n(1-\nu)^{k-1}} \gg P^{n-1-\sigma}.$$

Let u run through the values

$$\xi + v^n,$$

where ξ independently runs through the values formed above, and v runs through the values

$$P, P+1, \ldots, 2P-1.$$

It is readily seen that the numbers u do not overlap each other and satisfy the condition

$$P^n < u < 2^n P^n.$$

The total number of u's, X, satisfies the condition

$$X \gg P^{n-\sigma}.$$

Exactly in the same way as the numbers ξ, v and u were formed with the help of P and P_1, we shall form new numbers ξ_1, v_1 and u_1 with the help of the numbers R and R_1. Let X_1 denote the total number of values of u_1.

§3. The Main Integral

Taking

$$T_\alpha = \sum_{x=1}^{3P} e^{2\pi i \alpha x^n}; \qquad S_\alpha = \sum_u e^{2\pi i \alpha u}; \qquad S_{1\alpha} = \sum_{y=1}^{Y} \sum_{u_1} e^{2\pi i \alpha y^n u_1},$$

we shall consider the integral

$$I_N = \int_0^1 T_\alpha^{4n-2} S_\alpha^2 S_{1\alpha} e^{-2\pi i \alpha N} \, d\alpha,$$

where the range of integration can be taken as $-\tau^{-1} < \alpha \leqq 1 - \tau^{-1}$. Any α in the last interval can be represented in the form

$$\alpha = \frac{a}{q} + z; \qquad (a, q) = 1; \qquad 1 \leqq q \leqq \tau; \qquad |z| \leqq \frac{1}{q\tau}. \tag{2}$$

We shall divide this range into subranges of two classes. Class I contains the subranges in which

$$\alpha = \frac{a}{q} + z; \qquad (a, q) = 1; \qquad 1 \leqq q \leqq \sqrt{P}; \qquad -\tau^{-1} \leqq z \leqq \tau^{-1}.$$

Class II contains all other remaining subranges.

It is readily seen that the subranges belonging to Class I do not overlap. Moreover, they include all values of α which can be represented in the form (2) with $q \leqq \sqrt{P}$. Therefore every α belonging to the subranges of Class II can be represented in the form (2) with $q > \sqrt{P}$.

In accordance with this division of the integration range into two classes, the integral I_N can be expressed as the sum of two terms

$$I_N = H_1 + H_2.$$

§ 4. Estimate of H_2

If α belongs to one of the subranges of Class II, then

$$\sqrt{P} < q \leqq \tau \ll P^{n-1/2}; \qquad Y < \left(\frac{q}{2}\right)^{\nu}.$$

And we find

$$S_{1\alpha} \ll \sqrt{X_1 \sum_{x=1}^{2^n R^n} \left| \sum_{y=1}^{Y} e^{2\pi i \alpha y^n z} \right|^2} = \sqrt{X_1 \sum_{y=1}^{Y} \sum_{y_1=1}^{Y} \left| \sum_{x=1}^{2^n R^n} e^{2\pi i \alpha (y^n - y_1^n) x} \right|} \ll \sqrt{X_1 \Omega},$$

where

$$\Omega = \sum_{y=1}^{Y} \sum_{y_1=1}^{Y} \min(R^n, (\alpha t)^{-1}); \qquad t = y^n - y_1^n.$$

We have $|t| < q/2$. Furthermore, the number of solutions of the equation:

$$y^n - y_1^n = 0$$

is equal to Y. Therefore, that part of the sum Ω which corresponds to $t = 0$ is equal to

$$YR^n.$$

For t distinct from zero, the number of solutions of the equation

$$y^n - y_1^n = t$$

is $\ll q^{\delta}$. Moreover, since $(a, q) = 1$ and $|t| < q/2$, the least positive residue r of at modulo q is different from zero, and different values of r correspond to different values of t.

The absolute values s of the least residue of at modulo q which correspond to non-zero t are also different from zero. Moreover, s may take the same value $\ll q^{\delta}$ times. Since

$$(\alpha t)^{-1} = \left(\frac{at}{q} + \frac{\theta t}{q\tau}\right)^{-1} = \left(\frac{s}{q} + \frac{\theta_1 q}{2q\tau}\right)^{-1} \gg \frac{q}{s},$$

we find that part of the sum Ω which corresponds to the values of t distinct from zero is

$$\ll \sum_{s=1}^{q} \frac{q^{1+\delta}}{s} \ll q^{1+2\delta}.$$

Collecting all these results, we obtain

$$S_{1\alpha} \ll \sqrt{X_1 (YR^n + q^{1+2\delta})} \ll X_1 Y \sqrt{R^n X_1^{-1} \cdot Y^{-1}},$$

and by §2, we have

$$\ll X_1 Y \sqrt{R^{\sigma} Y^{-1}} \ll X_1 Y P^{\left(1-\frac{\nu}{2}\right)\frac{\sigma}{2} - \frac{\nu}{4} + \frac{\nu 2}{4}}.$$

Hence

$$H_2 \ll P^{4n-2} X_1 Y P^{\left(1-\frac{\nu}{2}\right)\frac{\sigma}{2} - \frac{\nu}{4} + \frac{\nu 2}{4}} \int_0^1 \sum_{u} \sum_{u'} e^{2\pi i \alpha (u-u')}\, d\alpha.$$

But the integral here is equal to

$$X \ll X^2 P^{-n} P^{\sigma}$$

therefore

$$H_2 \ll P^{3n-2} X^2 X_1 Y P^{\frac{3}{2}\sigma - \frac{\nu\sigma}{4} - \frac{\nu}{4} + \frac{\nu^2}{4}}.$$

Now we have

$$\frac{\nu}{4} - \frac{\nu^2}{4} - \frac{3}{2}\sigma + \frac{\nu\sigma}{4} = \frac{\nu}{4}(1 - \nu - 6n\sigma + \sigma),$$

$$\ln(6n\sigma - \sigma) < \ln 6 + 2\ln n - k\left(\nu + \frac{\nu^2}{2}\right) - \frac{\nu}{6}$$

$$\leq \ln 6 + 2\ln n - \frac{2n\ln n + n\ln 6 - 1}{n} - \frac{2n\ln n + n\ln 6 - 1}{2n^2} - \frac{\nu}{6}$$

$$< -1.32\nu,$$

$$6n\sigma - \sigma < 1 - 1{,}1\,\nu; \qquad \frac{\nu}{4}(1 - \nu - 6n\sigma + \sigma) > \delta.$$

Therefore

$$H_2 \ll P^{3n-2} X^2 X_1 Y P^{-\delta}.$$

§5. Subintervals of Class I. The Sum T and Its Homologues

Before calculating the value of H_1, we shall find an asymptotic expression for the sum T_α under the condition that α belongs to a subinterval of Class I. Make the substitution

$$x = qt + r; \qquad r = 0, 1, \ldots, q-1,$$

where for each given r the number t runs through the values which satisfy the inequalities

$$-rq^{-1} < t \leq (3P - r)q^{-1}. \tag{3}$$

The expression for T_α takes the form

$$T_\alpha = \sum_{r=0}^{q-1} \sum_t e^{2\pi i\left(\frac{a}{q} + z\right)(qt+r)^n} = \sum_{r=0}^{q-1} e^{2\pi i \frac{a}{q} r^n} K_r,$$

$$K_r = \sum_t e^{2\pi i z q^n (t + rq^{-1})^n},$$

where t takes the values in the interval (3). But in this interval the function

$$f(t) = |z|\, q^n (t + rq^{-1})^n$$

satisfies the conditions of Lemma 3. Therefore

$$K_r = \int_{-rq^{-1}}^{(3P-r)q^{-1}} e^{2\pi i z q^n (t+rq^{-1})^n} dt + 5\theta_2 = \frac{\varrho(z)}{q} + 5\theta_2;$$

$$\varrho(z) = \int_0^{3P} e^{2\pi i z x^n} dx.$$

Hence the expression for T_a takes the form

$$T_\alpha = \varrho(z) D_{aq} + O(q). \tag{4}$$

It is not difficult to verify that always

$$\varrho(z) \ll P.$$

If, however,

$$|z| \geqq P^{-n},$$

it is more convenient to use another easily derivable inequality:

$$\varrho(z) \ll |z|^{-\nu}.$$

Therefore, putting

$$Z = P \text{ when } |z| \leqq P^{-n} \quad \text{and} \quad Z = |z|^{-\nu} \text{ when } |z| \geqq P^{-n}$$

we always have

$$\varrho(z) \ll Z.$$

Hence, by virtue of Lemma I, the principal term on the right-hand side of (4) is

$$\ll Z q^{-\nu}, \text{ i.e. } \gg P^{1-\frac{\nu}{2}} \cdot q^{-\nu} \geqq q.$$

From (4) we easily find that

$$T_\alpha^{4n-2} = \{\varrho(z)\}^{4n-2} D_{aq}^{4n-2} + O(Z^{4n-3} q^{-3+3\nu}), \tag{5}$$

where the principal term is

$$\ll Z^{4n-2} q^{-4+2\nu}, \text{ i.e. } \gg Z^{4n-3} q^{-3+3\nu}.$$

In the same way as we treated T_α, consider the sum

$$V_\alpha = \sum_{v=P}^{2P-1} e^{2\pi i \alpha v^n},$$

where

$$V_\alpha = \varrho_0(z) D_{aq} + O(q); \qquad \varrho_0(z) = \int_P^{2P} e^{2\pi i z x^n} dx, \tag{6}$$

and the principal term on the right-hand side is

$$\ll P, \text{ i.e. } \gg q.$$

Finally, consider the sum

$$V_{y\alpha} = \sum_{v_1=R}^{2R-1} e^{2\pi i \alpha y^n v_1^n}; \qquad 1 \leqq y \leqq Y.$$

We obtain

$$V_{y\alpha} = \varrho_y(z) \frac{1}{q} \sum_{r=0}^{q-1} e^{2\pi i \frac{a}{q} y^n r^n} + O(q); \qquad \varrho_y(z) = \int_R^{2R} e^{2\pi i z y^n x^n}\, dx.$$

But, we have

$$\sum_{v_1=R}^{2R-1} e^{2\pi i \frac{a}{q} y^n v_1^n} = \frac{R}{q} \sum_{r=0}^{q-1} e^{2\pi i \frac{a}{q} y^n r^n} + O(q).$$

Hence

$$\frac{1}{q} \sum_{r=0}^{q-1} e^{2\pi i \frac{a}{q} y^n r^n} = \frac{1}{R} \sum_{v_1=R}^{2R-1} e^{2\pi i \frac{a}{q} y^n v_1^n} + O\left(\frac{q}{R}\right).$$

Therefore

$$V_{y\alpha} = \frac{\varrho_y(z)}{R} \sum_{v_1=R}^{2R-1} e^{2\pi i \frac{a}{q} y^n v_1^n} + O(q), \tag{7}$$

where the principal term is

$$\ll R, \quad \text{i.e.} \quad \gg q.$$

§6. Estimation of the Sums S_α and $S_{1\alpha}$

We find that

$$U_\alpha = \sum_\xi e^{2\pi i \alpha \xi} = \sum_\xi e^{2\pi i \frac{a}{q} \xi} + O(|z| X P^{n-2}),$$

where the principal term is

$$\ll X P^{-1}, \quad \text{i.e.} \quad \gg |z| X P^{n-2}$$

Hence, from (6) we find

$$S_\alpha = V_\alpha U_\alpha = \varrho_0(z) D_{aq} \sum e^{2\pi i \frac{a}{q} \xi} + O(X |z| P^{n-1} + X P^{-1} q),$$

where the principal term is

$$\ll X, \quad \text{i.e.} \quad \gg X |z| P^{n-1} + X P^{-1} q.$$

Hence we obtain

$$S_\alpha^2 = \left(\varrho_0(z) D_{aq} \sum_\xi e^{2\pi i \frac{a}{q} \xi} \right)^2 + O(X^2 |z| P^{n-1} + X^2 P^{-1} q), \tag{8}$$

where the principal term is

$$\ll X^2, \quad \text{i.e.} \quad \gg X^2|z|P^{n-1}+X^2P^{-1}q.$$

Moreover, we find

$$U_{y\alpha}=\sum_{\xi_1} e^{2\pi i\alpha y^n\xi_1}=\sum_{\xi_1} e^{2\pi i\frac{a}{q}y^n\xi_1}+O(W), \tag{9}$$

where

$$W=X_1R^{-1}|z|Y^nR^{n-1}\ll X_1R^{-1}|z|P^{\frac{1-\nu}{2}+\left(1-\frac{\nu}{2}\right)(n-1)}\ll X_1R^{-1}|z|P^{n-1}.$$

The principal term in (9) is

$$\ll X_1R^{-1}, \quad \text{i.e.} \quad \gg X_1R^{-1}|z|P^{n-1}.$$

Hence, from (7) and (9), we find

$$\begin{aligned} S_{1\alpha}=\sum_{y=1}^{Y}V_{y\alpha}U_{y\alpha}&=\sum_{y=1}^{Y}\frac{\varrho_y(z)}{R}\sum_{u_1}e^{2\pi i\frac{a}{q}y^nu_1}\\ &+O(X_1Y|z|P^{n-1}+X_1YqR^{-1}), \end{aligned} \tag{10}$$

where the principal term is

$$\ll X_1Y, \quad \text{i.e.} \quad \gg X_1Y|z|P^{n-1}+X_1YqR^{-1}.$$

§7. Calculation of H_1

Since

$$N_0-N_0P^{-1/4}<N\leqq N_0,$$

we have

$$e^{-2\pi i\alpha N}=e^{-2\pi i\frac{a}{q}N+2\pi izN_0}+O(|z|P^{-1/4}).$$

Hence, from (5), (8) and (10) we get

$$\begin{aligned} T_\alpha^{4n-2}S_\alpha^2S_{1\alpha}e^{-2\pi i\alpha N}&=\left(\{\varrho(z)\}^{4n-2}D_{aq}^{4n-2}+O(Z^{4n-3}q^{-3+3\nu})\right)\\ &\times\left(\left\{\varrho_0(z)D_{aq}\sum_{\xi}e^{2\pi i\frac{a}{q}\xi}\right\}^2+O(X^2|z|P^{n-1}+X^2P^{-1}q)\right)\\ &\times\left(\sum_{y=1}^{Y}\frac{\varrho_y(z)}{R}\sum_{u_1}e^{2\pi i\frac{a}{q}y^nu_1}+O(X_1Y|z|P^{n-1}+X_1YqR^{-1})\right)\\ &\times\left(e^{-2\pi i\frac{a}{q}N-2\pi izN_0}+O(|z|P^{n-1/4})\right), \end{aligned}$$

where the principal terms of the factors in large brackets are

$$\ll Z^{4n-2}q^{-4+2\nu}; \qquad \ll X^2; \qquad \ll X_1Y; \qquad \ll 1,$$

respectively. According to §5, we have

$$Z=P, \quad \text{if} \quad |z| \leqq P^{-n};$$
$$Z=|z|^{-\nu}, \quad \text{if} \quad |z| \geqq P^{-n}.$$

Hence we find

$$T_\alpha^{4n-2} S_\alpha^2 S_{1\alpha} e^{-2\pi i\alpha N} = \sum_{y=1}^{Y} E_y F_y + L, \tag{11}$$

where

$$E_y = D_{aq}^{4n} \left(\sum_{\xi} e^{2\pi i \frac{a}{q}\xi} \right)^2 \sum_{u_1} e^{2\pi i \frac{a}{q} y^n u_1} e^{-2\pi i \frac{a}{q} N}$$

and F_y depends only on z, y, n and N_0. Now

$$\begin{aligned} L &\ll Z^{4n-2} q^{-4+2\nu} X^2 X_1 Y (Z^{-1} q^{1+\nu} + |z| P^{n-1} + P^{-1} q + |z| P^{n-1} \\ &\quad + qP^{-1+\nu/2} + |z| P^{n-1/4}) \\ &\ll X^2 X_1 Y Z^{4n-2} P^{-1/4} q^{-3} (|z| P^n + 1). \end{aligned}$$

On multiplying (11) by dz, and then integrating from $z=-\tau^{-1}$ to $z=+\tau^{-1}$, we find that the term H_1 corresponding to an interval with given a and q is equal to

$$\sum_{y=1}^{Y} E_y Q_y + L_1,$$

where Q_y depends only on y, n and N_0. Furthermore,

$$\begin{aligned} L_1 &\ll X^2 X_1 Y P^{-1/4} q^{-3} \left\{ \int_0^{P^{-n}} P^{4n-2}\, dz + \int_{P^{-n}}^{\tau^{-1}} P^n z^{-3+2\nu}\, dz \right\} \\ &\ll X^2 X_1 Y P^{3n-2-1/4} q^{-3}. \end{aligned}$$

In view of this fact, and by making a run through the values $0, 1, \ldots, (q-1)$ coprime to q, and q through all the values which satisfy the inequality

$$0 < q \leqq \sqrt{P},$$

we obtain

$$H_1 = \sum_{y=1}^{Y} Q_y \sum_{q=1}^{\sqrt{P}} \sum_{a} E_y + O(X^2 X_1 Y P^{3n-2-1/4}).$$

§8. General Expression for I_N

Collecting the results established in §7 and 4, we obtain

$$I_N = \sum_{y=1}^{Y} Q_y B_y + O(X^2 X_1 Y P^{3n-2-\delta}); \quad B_y = \sum_{q=1}^{\sqrt{P}} \sum_{a} E_y,$$

where

$$E_y = D_{aq}^{4n} \left(\sum_{\xi} e^{2\pi i \frac{a}{q}\xi} \right)^2 \sum_{u_1} e^{2\pi i \frac{a}{q} y^n u_1} e^{-2\pi i \frac{a}{q} N}.$$

§9. Estimation of Q_y

Let I_{yN} denote that part of the integral I_N which corresponds to a given y. In the same way as we did with I_N, we shall represent I_{yN} in the form of a sum of two terms:

$$I_{yN} = H_{1y} + H_{2y}. \tag{12}$$

Now to H_{1y} we shall apply the same arguments which we applied to H_1. Evidently, we obtain (only the summation over y is omitted)

$$H_{1y} = Q_y B_y + O(X^2 X_1 P^{3n-2-\delta}).$$

Sum (12) over all integral N satisfying the inequalities

$$N_0 - N_0 P^{-1/4} < N \leqq N_0.$$

The result of summing the left-hand side is evidently equal to the number of solutions of the system of inequalities:

$$N_1 - N_0 P^{-1/4} < M \leqq N_1, \tag{13}$$

where

$$N_1 = N_0 - u - u' - y^n u_1; \qquad M = x_1^n + x_2^n + \ldots + x_{4n-2}^n.$$

The values which $u, u', u_1, x_1, x_2, \ldots, x_{4n-2}$ take are evident from a consideration of I_{yN}. By virtue of (1), we have

$$N_0 \geqq \tfrac{64}{81}(\tfrac{3}{2})^n \cdot 2^n P^n \geqq 4 \cdot 2^n P^n.$$

Moreover,

$$u + u' + y^n u_1 < 3 \cdot 2^n P^n; \qquad N_1 > 2^n P^n.$$

Hence, from Lemma (4), we find that for a given N_1, the number of solutions of the inequality

$$M \leqq N_1$$

is equal to

$$c_2 N_1^{4-2\nu} + O(P^{4n-3}),$$

and the number of solutions of the inequality

$$M \leqq N_1 - N_0 P^{-1/4}$$

is equal to

$$c_2 (N_1 - N_0 P^{-1/4})^{4-2\nu} + O(P^{4n-3}).$$

Therefore, for a given N_1, the number of solutions of the system (13) is

$$c_2 \{N_1^{4-2\nu} - (N_1 - N_0 P^{-1/4})^{4-2\nu}\} + O(P^{4n-3}).$$

For all N_0 exceeding a certain $c_3 \geqq c_0$, this expression is

$$\gg P^{3n-2} N_0 P^{-1/4}.$$

Hence it follows that the total number of solutions of the system (13) is

$$\gg X^2 X_1 P^{3n-2} N_0 P^{-1/4}.$$

We shall now sum the right-hand side of (12). First we sum H_{1y}, i.e. that part which corresponds to $q = 1$. It is equal to

$$Q_y X^2 X_1 P^{-2}$$

and on summing it over all values of N mentioned above, we get

$$Q_y X^2 X_1 P^{-2} \{N_0 P^{-1/4} + O(1)\}.$$

We now sum that part of H_{1y} which corresponds to any $q > 1$. According to Lemma 5, we have

$$\left| \sum_N e^{-2\pi i \frac{a}{q} N} \right| < \frac{1}{2(a/q)}.$$

But, since

$$E_y \ll X^2 X_1 P^{-2} q^{-4}; \qquad \sum_a \frac{1}{2(a/q)} < q \ln q,$$

the result of its summation gives

$$\ll |Q_y| X^2 X_1 P^{-2} q^{-2}.$$

Hence it is not difficult to verify that the result of summation of H_{1y} is

$$Q_y X^2 X_1 P^{-2} \{N_0 P^{-1/4} + O(1)\} + O(X^2 X_1 P^{3n-2-\delta} N_0 P^{-1/4}).$$

Now in the summation of H_{2y}, we find that in the subintervals of Class II, we have

$$(\alpha) \geqq \tau^{-1} \gg P^{-n+1/2}.$$

Therefore

$$\sum_N e^{-2\pi i \alpha N} \ll \frac{1}{(\alpha)} \ll P^{n-1/2}.$$

Hence we obtain

$$\sum_N H_{2y} \ll P^{4n-2} X_1 P^{n-1/2} \int_0^1 \sum_u \sum_{u'} e^{2\pi i \alpha (u-u')} \, d\alpha$$
$$\ll P^{4n-2} X X_1 P^{n-1/2} \ll P^{4n-2} X^2 X_1 P^{\sigma-1/2}.$$

Since

$$\ln \sigma < \ln n - kv < -\ln n - \ln 6 + v < -\ln 8; \quad \sigma - \tfrac{1}{2} < \tfrac{1}{8} - \tfrac{1}{2} = -\tfrac{3}{8},$$

the latter is

$$\ll X^2 X_1 P^{4n-2} P^{-1/8-1/4} \ll X^2 X_1 P^{3n-2-\delta} \cdot N_0 P^{-1/4}.$$

Collecting all these results, we find that the summation of the right-hand side of (12) gives

$$Q_y X^2 X_1 P^{-2}\{N_0 P^{-1/4}+O(1)\}+O(X^2 X_1 P^{3n-2-\delta}\cdot N_0 P^{-1/4}).$$

Remembering that the result of summation of the left-hand side is a real number

$$\gg X^2 X_1 P^{3n-2} N_0 P^{-1/4},$$

we find that

$$Q_y = C_y\{1+O(P^{-\delta})\},$$

where C_y is real and

$$C_y \gg P^{3n}.$$

§ 10. Estimation of B_y

From the expression derived for E_y in §8, it follows that

$$E_y=\sum_{N_2} e^{-2\pi i \frac{a}{q} N_2} D_{aq}^{4n},$$

where

$$N_2 = N-\xi-\xi'-y^n u_1 .$$

Therefore

$$B_y=\sum_{N_2}\sum_{q=1}^{\sqrt{P}} A_q(N_2),$$

which, by Lemma 2, is equal to

$$\sum_{N_2} b(N_2)+O(X^2 X_1 P^{-3}).$$

The principal term here is a real number which

$$\gg X^2 X_1 P^{-2}.$$

Consequently,

$$B_y = J_y\cdot\{1+O(P^{-1})\},$$

where J_y is real and

$$J_y \gg X^2 X_1 P^{-2}.$$

§ 11. Derivation of Waring's Theorem

From the formulae of §8, after what has been proved in §9 and 10, it follows that (now we take $N=N_0$)

$$I_{N_0} \gg \sum_{y=1}^{Y} C_y J_y\{1+O(P^{-\delta})\}+O(X^2 X_1 Y P^{3n-2-\delta}),$$

where $C_y J_y$ is real and

$$\sum_{y=1}^{Y} C_y J_y \gg X^2 X_1 P^{3n-2} Y.$$

Hence, it is evident that for all $N_0 \geqq c$, where c is sufficiently large and $\geqq c_3$, we have

$$I_{N_0} \gg X^2 X_1 P^{3n-2} Y.$$

Consequently $I_{N_0} > 0$. But I_{N_0} is the number of representations of N_0 in the form

$$N_0 = u + u' + y^n u_1 + x_1^n + x_2^n + \ldots + x_{4n-2}^n, \tag{14}$$

where all notations have the same meaning as before. That means N_0 can be represented in the form (14) at least in one way. Consequently it is the sum of

$$3k + 4n - 2 = 3\,[2n \ln n + n \ln 6] + 4n - 2 < 6n \ln n + (\ln 216 + 4)\, n$$

terms of the type t^n, where $t \geqq 0$. Thus we have proved the following

Theorem. Let $n > 3$ be an integer and

$$s = [n(6 \ln n + \ln 216 + 4)].$$

Then there exists a positive constant c dependent only on n such that any integer

$$N \geqq c$$

can be represented in the form

$$N = t_1^n + t_2^n + \ldots + t_s^n,$$

where $t_1, t_2, \ldots, t_s$ are integers $\geqq 0$.

New Estimates for Weyl Sums

Doklady Akademii Nauk SSSR, **8** no. 5 (1935) 195–198.

§1. We shall adopt the following notations: n is an integral constant $\geqq 20$; (α) denotes the difference between α and its nearest integer; $\{a\}$ stands for the fractional part of a; $A \ll B$ or $B \gg A$ is equivalent to $A = O(B)$; $a_0, a_1, \ldots, a_n$ are real numbers; Q and P are integers;

$$f(x) = a_0 x^n + \ldots + a_n; \qquad S = \sum_{x=Q+1}^{Q+P} e^{2\pi i f(x)};$$

$$\nu = \frac{1}{n}; \qquad a_0 = \frac{a}{q} + \theta q^{-2+\nu^2}; \qquad (a, q) = 1; \qquad q > 0; \qquad |\theta| < 1;$$

$$p = q^\nu; \qquad \sigma = (1-\nu)^{k-1}; \qquad 0 < P \ll P_1; \qquad \varepsilon = \frac{\delta n^2}{2}.$$

In our arguments we shall assume that q is sufficiently large. We shall base all our proofs on the following two main assumptions:

1. $k = [2n \lg n + 2n]; \qquad P_1 = q^{2\nu - 2\nu^2 + \nu^3}; \qquad Y = [q^{\nu - 2\nu^2}]; \qquad \delta = \dfrac{\nu^3}{3k}.$

2. $k = [3n \lg n + 3n]; \qquad P_1 = q^{\frac{2}{2n-1}}; \qquad Y = [q^{\frac{\nu 2}{2}}]; \qquad \delta = \dfrac{\nu^4}{5k}.$

Furthermore, let p_k denote a positive integer:

$$q_s = \left[2n^{\frac{1}{1-\nu}} p_s^{\frac{\nu}{1-\nu}} + 1\right]; \qquad p_{s-1} = q_s p_s. \qquad (s = 2, \ldots, k)$$

The number p_k can be so chosen that

$$p \ll p_1 \ll p.$$

§2. Assuming that

$$A_s = \left| \sum_{u_s=0}^{p_s-1} e^{2\pi i f(x + \xi_2 p_2 y + \ldots + \xi_s p_s y + u_s y)} \right|^{2n(k+1-s)},$$

we obtain

$$S=\frac{1}{p_1 Y}\sum_x\sum_{y=1}^{Y}\sum_{u_1=0}^{p_1-1} e^{2\pi i f(x+u_1 y)}+O(pY)$$

$$\ll \frac{1}{pY}(PY)^{1-\frac{1}{2nk}}\left(\sum_x\sum_y A_1\right)^{\frac{1}{2nk}}+pY. \qquad (1)$$

$$A_1\ll\left|\sum_{u_1} e^{2\pi i f(x+u_1 y)}\right|^{2n} q_2^{2n(k-1)-1}\sum_{\xi_2=0}^{q-1} A_2,$$

$$\cdots\cdots\cdots\cdots\cdots\cdots\cdots\cdots\cdots$$

$$A_{k-1}\ll\left|\sum_{u_{k-1}} e^{2\pi i f(x+\xi_2 p_2 y+\ldots+\xi_{k-1}p_{k-1}y+u_{k-1}y)}\right|^{2n} q_k^{2n-1}\sum_{\xi_k=0}^{q_k^{-1}} A_k.$$

Hence we find that A_1 can be represented as the sum of

$$(q_2^{k-1}q_3^{k-2}\ldots q_k)^{2n}\ll p^{2nk}(p_1\ldots p_k)^{-2n} \qquad (2)$$

terms of the type $|B|^2$, where after changing the summation indices, each B can be expressed as (x_0 is not dependent on t):

$$B=\prod_{t=1}^{k} S_t^n; \qquad S_t=\sum_{v_t=-r_t}^{p_t-r_t-1} e^{2\pi i f(x_0+v_t y)}. \qquad (0<r_t<p_t)$$

Divide each sum S_t into $\ll p^{\varepsilon}$ sums of the type:

$$\sum_{v_{ts}} e^{2\pi i f(x_0+v_{ts}y)}$$

the length of summation range being $\leqq p_t p^{-\varepsilon}$.

Then S_t^n can be divided into $\ll p^{\varepsilon n}$ sums, and B into $\ll p^{\varepsilon nh}$ sums C of the type:

$$C=\prod_{t=1}^{k} e^{2\pi i[f(x_0+v_{t1})+\ldots+f(x_0+v_{tn})]}.$$

At the same time, we have

$$|B|^2\ll p^{\varepsilon nk}\sum\nolimits_1 |C|^2. \qquad (3)$$

Applying the Taylor expansion, we obtain

$$C=\sum\nolimits_2 e^{2\pi i(a_0 U_n y^n+a_1 U_{n-1}y^n+\ldots+a_n)};$$

$$U_r=\sum_{t=1}^{k} V_{tr}; \qquad V_{tr}=v_{t1}^r+\ldots+v_{tn}^r,$$

where, in the expression for C, summation is taken over all possible systems of values of

$$v_{11},\ldots,v_{1n};\ldots;v_{k,1},\ldots,v_{kn}.$$

If we discard those terms of $|C|^2$, for which at least one of the differences of the type:

$$v_{tg}-v_{th}$$

is less than $p_t p^{-\delta}$ even for one value of t, then we find that the estimate of the sum S changes by a quantity

$$\ll Pp^{-\frac{\delta}{2kn}}, \tag{4}$$

and the estimate of A_1 contains

$$\ll p^{2kn}(p_1 \dots p_k)^{-2n} p^{2\varepsilon nk}$$

terms of $|C|^2$ for which, for all $t = 1, \dots, k$ and for arbitrarily chosen $v_{t1}, \dots, v_{tn}$, we have

$$\left\| \begin{matrix} v_{t1}^{n-1}, \dots, v_{tn}^{n-1} \\ \dots\dots\dots \\ v_{t1}, \dots, v_{tn} \\ 1, \dots, 1 \end{matrix} \right\| \gg p_t^{\frac{n(n-1)}{2}} p^{-\frac{\delta n^2}{2}(1-\nu)}. \tag{5}$$

§3. Now consider some C from the remaining values. For each $r = 1, \dots, n$, take an interval of length

$$p_t^{r-1}$$

and determine the number N_1' of systems of values of $v_{t1}, \dots, v_{tn}$, for which all $V_{tr}(r = 1, \dots, n)$ are found simultaneously in the above interval. If

$$\eta_1, \dots, \eta_n; \qquad \eta_1 + s_1, \dots, \eta_n + s_n$$

are two such systems of values of $v_{t1}, \dots, v_{tn}$, then

$$\begin{gathered} (\eta_1 + s_1)^n - \eta_1^n + \dots + (\eta_n + s_n)^n - \eta_n^n = O(P_t^{n-1}) \\ \dots\dots\dots\dots\dots\dots\dots\dots \\ (\eta_1 + s_1)^2 - \eta_1^2 + \dots + (\eta_n + s_n)^2 - \eta_n^2 = O(P_t) \\ \eta_1 + s_1 - \eta_1 + \dots + (\eta_n + s_n) - \eta_n = O(1), \end{gathered}$$

Hence, by virtue of (5), we readily find

$$s_i \ll p^{\frac{\delta n^2}{2}}; \qquad N_1' \ll p^{\frac{\delta n^3}{2}}.$$

For each $r = 1, \dots, n$, take an interval of length

$$p_t^{r(1-\nu)}.$$

Then the number N_t of systems of values $v_{t1}, \dots, v_{tn}$, for which all $V_{tr}(r = 1, \dots, n)$ are found simultaneously in these new intervals, is equal to

$$N_t \ll p_t^{\frac{n-1}{2}} p^{\frac{\delta n^3}{2}} \tag{6}$$

§4. Now we shall estimate the number T of solutions of the system

$$U_r = W_r; \qquad (r = 1, \dots, n)$$

for given integral W_r $(r=1, \ldots, n)$. All those values which the sum

$$V_{t+1,r} + \ldots + V_{kr},$$

can take lie obviously in a definite interval of length $\leqq p_t^{r(1-\nu)}$. Hence, we find that all V_{1r} $(r=1, \ldots, n)$ lie in definite intervals of lengths $p_1^{r(1-\nu)}$. Thus, for each given system of numbers V_{1r}, the numbers V_{2r} lie in definite intervals of lengths $p_2^{r(1-\nu)}$, and so on. Finally, for each given set of systems of numbers $V_{1r}, \ldots, V_{k-2,r}$ $(r=1, \ldots, n)$, the numbers $V_{k-1,r}$ lie in definite intervals of lengths $p_{k-1}^{r(1-\nu)}$. Since the number of systems of values of V_{kr} $(r=1, \ldots, n)$, is

$$\ll p_k^n,$$

by (6) and all the preceeding, we obtain

$$T \ll (p_1 \ldots p_{k-1})^{\frac{n-1}{2}} p_k^n p^{\frac{\delta n^3 k}{2}}.$$

§5. We have

$$\sum_y |C|^2 = \sum_y \sum\nolimits_2 \sum\nolimits_2' e^{2\pi i [a_0(U_n - U_n')y^n + \ldots + a_{n-1}(U_1 - U_1')y]}$$

$$\ll (p_1 \ldots p_k)^n p^{-\varepsilon nk} (p_1 \ldots p_{k-1})^{\frac{n-1}{2}} p_k^n p^{\frac{\delta n^3 k}{2}} \Omega;$$

$$\Omega = \sum_{z_n \geqslant -p^n}^{\ll p^n} \ldots \sum_{z_1 \geqslant -p}^{\ll p} \left| \sum_y e^{2\pi i [a_0 z_n y^n + \ldots + a_{n-1} z_1 y]} \right|$$

$$\ll p^{\frac{n(n-1)}{2}} \sqrt{p^n \sum_y \sum_{y_1} \min p^n, \left(\frac{1}{(a_0 (y^n - y_1^n))}\right)} \ll p^{\frac{n(n+1)}{2}} Y^{\frac{1}{2}}.$$

Therefore, from (1), (2), (3), and (4), we obtain

$$S \ll Pq^{\frac{1+\nu}{4k}\sigma + \frac{\delta n}{2}} Y^{-\frac{1}{4nk}} + Pq^{-\frac{\delta \nu^2}{2k}} + Yq^{\nu}.$$

§6. Thus, from the assumptions (1) and (2) made in §1, we obtain

Theorem. Let n be an integer $\geqq 20$; $a_0, \ldots, a_n$ be real numbers; Q and P be integers $(0 < P \ll P_1)$;

$$f(x) = a_0 x^n + \ldots + a_n; \qquad S = \sum_{x=Q+1}^{Q+P} e^{2\pi i f(x)}; \qquad \nu = \frac{1}{n};$$

$$a_0 = \frac{a}{q} + \theta q^{-2+\nu^2}; \qquad (a, q) = 1; \qquad q > 0; \qquad |\theta| \leqq 1.$$

Then

$$S \leqq c P_1 q^{-\varrho},$$

where c is a constant that depends only on n, and

$$\varrho = \frac{\nu^7}{24(\lg n + 1)^2}, \quad \text{if} \quad P_1 = q^{2\nu - 2\nu^2 + \nu^3};$$

$$\varrho = \frac{\nu^8}{90(\lg n + 1)^2}, \quad \text{if} \quad P_1 = q^{\frac{2}{2n-1}}.$$

Corollary. Let Q and P be integers,

$$A \gg 1; \; P \gg A^{2\nu}$$

and the real valued function $f(x)$ satisfy the condition

$$\frac{1}{A} \ll f^{(n)}(x) \ll \frac{1}{A}; \qquad f^{(n+1)}(x) \ll \frac{1}{A^2}.$$

in the interval $Q \leqq x \leqq Q + P$, then

$$\sum_{x=Q+1}^{Q+P} e^{2\pi i f(x)} \ll PA^{-\varrho}; \qquad \varrho = \frac{\nu^7}{24(\lg n + 1)^2}.$$

Example 1. The asymptotic expression for the number of fractions

$$\{f(x)\}; \qquad x = 1, \ldots, P',$$

lying in a given interval can be obtained accurate to a quantity of the order of $P'q^{-\varrho_1}$ provided $P' \gg P_1$ (the theorem) and to a quantity of the order of $P'A^{-\varrho}$ provided $P' \gg A^{2\nu}$ (the corollary).

Example 2. The asymptotic expression derived by Hardy and Littlewood for the number of representations of an integer $N \geqq N_0(n)$ in the form

$$N = x_1^n + \ldots + x_r^n$$

holds valid for all

$$r > 183\, n^9 (\lg n + 1)^2.$$

A detailed consideration of the determinant given in §2 shows that the power of the numerator in the expression of ϱ can be reduced by 1, for example

$$\varrho = \frac{\nu^6}{24(\lg n + 1)^2}.$$

Representation of an Odd Number as the Sum of Three Primes

Doklady Akademii Nauk SSSR, **15** no. 6–7 (1937) 291–294.

Simple examples[1] illustrating the application of my method to the theory of primes were given in 1934.

In this paper I shall apply this method to estimate the sum

$$\sum_{p \leqq N} e^{2\pi i \alpha p}.$$

Using this estimate and a new theorem on the distribution of primes in an arithmetical progression[2] (the common difference increasing slowly with the increasing number of terms), I shall derive an asymptotic expression for the number of representations of an odd number $N > 0$ in the form

$$N = p_1 + p_2 + p_3 .$$

It directly follows that any sufficiently large odd number can be represented in the form of a sum of three primes. This is the complete solution of the Goldbach problem for odd numbers.

The estimates derived in this paper can be improved further.

Notations. $N > 0$ is a sufficiently large odd number; $n = \ln N$; h, h_1, h_2 are arbitrary large constants > 3; $\tau = Nn^{-3h}$; $\tau_1 = Nn^{-h}$; θ is a real number; $|\theta| \leqq 1$; $A \ll B$ or $A = O(B)$ denotes that the ratio $|A|/B$ does not exceed a certain constant; (d) stands for a sequence of divisors $d \leqq N$ of the product H of all primes $\leqq \sqrt{N}$; (d_0) denotes that part of this sequence which contains all d with an even number of prime divisors, while (d_1) denotes that part which contains all d with an odd number of prime divisors. The sequence (d) can also be divided into two subsequences (d') and (d'') as follows: the first contains the numbers d satisfying the condition that all prime divisors are less than n^{3h} and the second contains all the remaining numbers d.

Accordingly, the sequences (d_0) and (d_1) are divided into subsequences (d_0'), (d_0'') and (d_1'), (d_1''), respectively.

Lemma. Let (x) and (y) denote two increasing sequences of positive integers:

$$1 < U_0 < U_1 \leqq N_1 \leqq N; \quad m = \text{an integer} > 0;$$

---

1 On some new problems in the theory of numbers; Dokl. akad. nauk SSSR (1934, vol. 3, no. 1) 1–6. Some theorems in analytic number theory; Dokl. akad. nauk SSSR (1934, vol. 4, no. 4) 185–187.

2 Walfisz, A.: Zur additiven Zahlentheorie II; Math. Z. **40** (1935) 592–607.

$$\alpha = \frac{a}{q} + \frac{\theta}{q\tau}; \quad (a, q) = 1; \quad 0 < q \leqq \tau; \quad m = m_1\delta; \quad q = q_1\delta;$$

$$\delta = (m, q), \quad T = \sum_x \sum_y e^{2\pi i \alpha m x y},$$

where x runs through all the members of (x) satisfying the condition

$$U_0 < x \leqq U_1$$

and y, for a given x, runs through all the members of (y) satisfying the condition

$$0 < y \leqq \frac{N_1}{x}.$$

Then

$$T \ll N_1 n \sqrt{\frac{n}{U_0} + \frac{U_1}{N_1} + \frac{q_1 n}{N_1} + \frac{1}{q_1} + \frac{m_1}{\tau}}.$$

Theorem 1. Let

$$\alpha = \frac{a}{q} + \frac{\theta}{q\tau}; \qquad (a, q) = 1; \qquad n^{3h} \leqq q \leqq \tau.$$

Then

$$S = \sum_{p \leqq N} e^{2\pi i \alpha p} \ll N n^{2-h}.$$

Proof. We have

$$S = \sum_{(d)} \mu(d) S_d + O(\sqrt{N}); \qquad S_d = \sum_{m=1}^{N/d} e^{2\pi i \alpha m d}. \tag{1}$$

Hence we find

$$\begin{aligned} S &= \sum_{d > \tau_1} \mu(d) S_d + O(Nn^{-h+1}) = T_0 - T_1 + O(Nn^{-h+1}); \\ T_0 &= \sum_{(d_0)} S_d; \qquad T_1 = \sum_{(d_1)} S_d, \end{aligned} \tag{2}$$

where d runs through those values which are $> \tau_1$. We shall only estimate T_0, because T_1 can be estimated in the same manner. By changing the order of summation, we obtain

$$T_0 = \sum_m T(m); \qquad T(m) = \sum_d e^{2\pi i \alpha m d}, \tag{3}$$

where m runs through

$$m = 1, \ldots, [n^h],$$

and d, for each given m, runs through the numbers (d_0) satisfying the condition:

$$\tau_1 < d \leqq \frac{N}{m}.$$

Furthermore, we find that

$$T(m) = T''(m) + O\left(\frac{N}{m} n^{-h}\right), \tag{4}$$

where $T''(m)$ contains only those terms in the sum $T(m)$ that correspond to the values of d in the sequence (d_0''). Indeed, the part $T'(m)$ of the sum $T(m)$ which corresponds to the values of d in the sequence (d_0') does not exceed the number of those terms in the sequence (d') which are not greater than N/m. But the order of this number is far less than

$$\frac{N}{m} n^{-h}.$$

If d belongs to (d_0'') and k is the number of prime divisors of d, then

$$k < n.$$

Therefore

$$T''(m) = \sum_{k<n} T_k(m), \tag{5}$$

where $T_k(m)$ contains those terms in $T''(m)$ which contain exactly k prime divisors greater than n^{3h}. Moreover,

$$T_k(m) = \frac{1}{k} T_{k_0}(m) + O\left(\frac{N}{mk} u^{-3h}\right), \tag{6}$$

where

$$T_{k_0}(m) = \sum_u \sum_v e^{2\pi i \alpha m u v}.$$

Here u runs through the primes $\geqq n^{3h}$ belonging to (d), while v runs, for a given u, through the numbers satisfying the condition

$$\frac{\tau_1}{u} < v \leqq \frac{N}{mu}.$$

Applying the lemma to the sum T_{k_0}, we obtain

$$T_{k_0}(m) \ll N \frac{n^{(3/2)-(3h/2)}}{\sqrt{m}}.$$

Thus, from (6), (5), (4), (3) and (2) we obtain Theorem 1.

Theorem 2. The number I_N of the representations of N in the form

$$N = p_1 + p_2 + p_3$$

is given by the expression

$$I_N = RS + O(N^2 n^{-c}),$$

where c is an arbitrarily large constant >3 and

$$S=\sum_{q=1}^{\infty}\frac{\mu(q)}{q_1^3}\sum_{\substack{0\leqq a<q\\(a,q)=1}} e^{2\pi i\frac{a}{q}N};\quad R=\frac{N^2}{2n^3}(1+\lambda);\quad \lim_{N\to\infty}\lambda=0;\quad q_1=\varphi(q).$$

Proof. a) We have

$$I_N=\int_0^1 S_\alpha^3 e^{-2\pi i\alpha N}\cdot d\alpha;\qquad S_\alpha=\sum_{p\leqq N} e^{2\pi i\alpha p}.$$

Divide the range of integration into intervals of two classes as follows:
Class I contains all intervals satisfying the conditions

$$\alpha=\frac{a}{q}+z;\qquad (a,q)=1;\qquad 0<q\leqq n^{3h};\qquad -\frac{1}{\tau}\leqq z\leqq\frac{1}{\tau}.$$

Class II contains all other remaining intervals for which

$$\alpha=\frac{a}{q}+z;\qquad (a,q)=1;\qquad n^{3h}<q\leqq\tau;\qquad |z|\leqq\frac{1}{q\tau}.$$

Accordingly, we have

$$I_N=I_{N_1}+I_{N_2}. \tag{7}$$

b) Applying Theorem 1 we find that

$$I_{N_2}\ll Nn^{2-h}\int_0^1|S_\alpha|^2\,d\alpha\ll Nn^{2-h}\int_0^1\sum_{p\leqq N}\sum_{p_1\leqq N}e^{2\pi i\alpha(p-p_1)}d\alpha$$
$$\ll Nn^{2-h}\frac{N}{n}\ll N^2n^{1-h}.$$

c) It is not difficult to calculate I_{N_1}. It is calculated by the procedure that was used in Waring's problem, but here we have to use a new theorem on the distribution of primes in an arithmetical progression.

If α belongs to an interval of the first class, then we find

$$S_\alpha=\frac{\mu(q)}{q_1}V(z)+O(Nn^{-h_1});\qquad V(z)=\int_2^N\frac{e^{2\pi izx}}{\ln x}dx,$$

Hence that part of I_{N_1} which corresponds to a given fraction a/q is represented as

$$R\frac{\mu(q)}{q_1^3}e^{-2\pi i\frac{a}{q}N}+O(Nn^{-h_2});\qquad R=\int_{-1/\tau}^{1/\tau}[V(z)]^3e^{-2\pi izN}dz, \tag{8}$$

where R can be expressed in the form

$$R=\frac{N^2}{2n^3}(1+\lambda);\qquad \lim_{N\to\infty}\lambda=0.$$

Hence, without any difficulty we find that

$$I_{N_1}=RS+O(N^2n^{-h_3}),$$

and by virtue of (7) and (8) the theorem follows immediately.

Estimates of Certain Simple Trigonometric Sums with Prime Numbers

Izvestiya Akademii Nauk SSSR, Ser. Mat., **3** no. 4 (1939) 371–398.

The most important estimates reported in my recent papers[1] for trigonometric sums with primes that can be evaluated by my methods developed in the years 1937–38 are the best possible results. In some fairly important particular cases, however, the accuracy of my result can be improved further. This paper is devoted to a reconsideration of one such particular case. Here I shall estimate sums of the type

$$\sum_{N-A<p\leqq N} e^{2\pi i\frac{a}{q}p^n}; \qquad (a,q)=1. \tag{1}$$

It should be noted that for small q ($q<e^u$, $u=r^\delta$, where $r=\ln N$ and δ is a proper fraction slightly greater than 1/2) and A a quantity of an order close to N, quite exact estimates of the sum (1) follow directly from the theorems on the distribution of primes in arithmetic progressions (Siegel, Esterman and others).

My method gives estimates for (1) close to the limiting value under sufficiently general condition; the main difficulty lies in applying my method to the case where q is small. Nevertheless, as I have already shown[2], my method together with Brun's method gives a possibility of treating the case of small q with success.

At the end of the paper I shall consider a sum more general than (1) for the case of $n=1$. Thus, I shall derive a quite general theorem on the distribution of remainders for the division of primes by a given large number (distribution of primes with respect to a given modulus) with high accuracy.

Notations. The symbolic inequalities $A \ll B$ or $B \gg A$ denote that the ratio $|A|/|B|$ does not exceed a certain constant. The letter p stands for a prime, θ for a number satisfying the condition $|\theta| \leqq 1$. Introduce the symbols:

$$\{x\}=x-[x]; \qquad (x)=\min(\{x\},1-\{x\}).$$

Lemma 1 (Brun). Let ε, c_0, c be any positive constants <1 and ϱ an arbitrarily small positive constant. Let $N>2$, $r=\ln N$

$$0<q\leqq e^{r^c}; \qquad 0<q_1<e^{r^c}; \qquad (q_1,q)=1; \qquad 0\leqq l<q; \qquad (q,l)=1,$$

---

1 Some general lemmas and their application to the estimation of trigonometrical sums, Math. sbornik (n.s.) **3(45)** (1938) 435–471. Estimation of certain sums containing primes, Izv. akad. nauk SSSR (ser. mat.) **2** (1938) 399–416.

2 A new improvement of the method of estimation of trigonometrical sums with primes [in English], Dokl. akad. nauk **22**, no. 2 (1939) 59.

$$Ne^{-r^{c_0}} \ll A < N; \qquad p_0 = e^{r^{1-\varepsilon}},$$

and let D be the product of all primes $\leqq p_0$ which are not factors of qq_1.

Then the number T of numbers $\equiv 1 \pmod q$ coprime to D lying in the interval

$$N - A < qx + l \leqq N,$$

satisfies the inequality

$$T \ll \frac{A\,(qq_1)^\varrho}{r^{1-\varepsilon} q}.$$

Proof.[1] It suffices to take $N > C$, where C is a sufficiently large constant > 2. Let δ denote the number of prime divisors $p_1, \ldots, p_\delta$ of D.

$$m = 2\,[3 \ln r - 1],$$

$\Omega(d)$ denotes the number of different prime divisors of d. Then we have

$$T < \sum_{\substack{d \setminus D \\ \Omega(d) \leqq m}} \mu(d) \left(\frac{A}{qd} + \theta_d\right).$$

Hence we find

$$T < \sum_{\substack{d \setminus D \\ \Omega(d) \leqq m}} \mu(d) \frac{A}{qd} + \sum_{h=0}^{m} \binom{\sigma}{h}.$$

But

$$\sum_{d \setminus D} \mu(d) \frac{A}{qd} = \frac{A}{q} \frac{\prod_{p \leqq p_\sigma} \left(1 - \frac{1}{p}\right)}{\prod_{p \setminus qq_1} \left(1 - \frac{1}{p}\right)} \ll \frac{A\,(qq_1)^\varrho}{r^{1-\varepsilon} q}.$$

Now we have

$$\left| \sum_{\substack{d \setminus D \\ m < \Omega(d) \leqq \sigma}} \mu(d) \frac{A}{qd} \right| < \frac{A}{q} \sum_{\substack{d \setminus D \\ m < \Omega(d) \leqq \sigma}} \frac{1}{d} < \frac{A}{q} \sum_{n > m}^{\sigma} S_n,$$

where S_n denotes the n-th elementary symmetric function of

$$\frac{1}{p_1}, \ldots, \frac{1}{p_\sigma}.$$

It is readily seen that

$$\frac{A}{q} \sum_{n>m}^{\sigma} S_n < \frac{A}{q} \sum_{n>m}^{\sigma} \frac{S_1^n}{n!} < \frac{A}{q} \sum_{n>m}^{\sigma} \left(\frac{eS_1}{n}\right)^n \ll \frac{A}{q} \sum_{n>m}^{\sigma} \left(\frac{4 \ln r}{m}\right)^n \ll \frac{A}{q} \left(\frac{4}{5}\right)^{6 \ln r} \ll \frac{A}{rq}.$$

[1] Brun, V.: Videnskapsselskapets Skrifter I; Mat.-naturv. Kl. **3**. Titchmarsh, E.C.: A divisor problem, Rend. circ. mat. Palermo 54 (1930) 414–524 (Correction) *ibid.* **57** (1932) 478–9. Landau, E.: Vorlesungen über Zahlentheorie, vol. 1, II, ch. 2 (Leipzig, 1927).

Finally,

$$\sum_{h=0}^{m} \binom{\sigma}{h} < \sum_{h=0}^{m} \sigma^h < p_\sigma^m < \frac{A}{rq} \frac{rq}{A} e^{6r^{1-\varepsilon}\ln r} < \frac{A}{rq},$$

which proves our lemma.

Lemma 2. Let η be an arbitrarily small positive constant, n a positive constant, q an integer >0, and let x run through the reduced system of residues modulo q. Then for $(a, q) = 1$, the number of solutions of the congruence

$$x^n \equiv a \pmod{q}$$

is

$$\ll q^\eta .$$

Proof. Let

$$q = p_1^{\alpha_1} \dots p_k^{\alpha_k}$$

be the canonical expansion of q. The congruence

$$x^n \equiv a \pmod{q}$$

is equivalent to the system of congruences

$$x^n \equiv a \pmod{p_1^{\alpha_1}},$$

$$\dots\dots\dots\dots$$

$$x^n \equiv a \pmod{p_k^{\alpha_k}}.$$

Each of these congruences has at most n solutions, except possibly one which may have $\leqq 2n$ solutions (the case of $p_1 = 2$). Hence, from

$$n^k \ll q^\eta$$

the lemma follows directly.

Lemma 3. Let ϱ be an arbitrarily small positive constant and n a positive constant,

$$(a, q) = 1; \qquad q > 1;$$

Let $\varphi(x)$ and $\psi(y)$ denote the functions which take non-negative values for the values of x and y under consideration and let

$$\sum_{\substack{0 \leqq x < q \\ (x,q)=1}} (\varphi(x))^2 \leqq X; \qquad \sum_{\substack{0 \leqq y < q \\ (y,q)=1}} (\psi(y))^2 \leqq Y.$$

Then the sum

$$S = \sum_{\substack{0 \leqq x < q \\ (x,q)=1}} \sum_{\substack{0 \leqq y < q \\ (y,q)=1}} \varphi(x) \psi(y) e^{2\pi i \frac{a}{q} x^n y^n}.$$

obeys the inequality

$$S^2 \ll XYq^{1+\varrho}.$$

Proof. We have

$$S^2 \ll X \sum_x \sum_y \sum_{y_1} \psi(y)\,\psi(y_1)\, e^{2\pi i \frac{a}{q} x^n (y^n - y_1^n)}$$

$$\ll X q^{\frac{\varrho}{3}} \sum_{n=0}^{q-1} \sum_y \sum_{y_1} \psi(y)\,\psi(y_1)\, e^{2\pi i \frac{a}{q} n (y^n - y_1^n)} .$$

First summing over u, we obtain a sum equal to either q or 0, depending on whether $y^n \equiv y_1^n \pmod q$ or not. Hence the lemma follows.

Lemma 4. Let ϱ be an arbitrarily small positive constant, n a positive constant

$$(a, q) = 1 ; \qquad q > 1$$

and let z run through a reduced system of residues modulo q. Then the sum

$$T = \sum_z e^{2\pi i \frac{a}{q} z^n}$$

obeys the inequality

$$T \ll q^{1/2+\varrho} .$$

Proof. Evidently

$$T = \frac{1}{\Phi(q)} S ; \qquad S = \sum_{\substack{0 \leqq x < q \\ (x,q)=1}} \sum_{\substack{0 \leqq y < q \\ (y,q)=1}} e^{2\pi i \frac{a}{q} x^n y^n} ,$$

where $\Phi(q)$ is the Euler totient function. According to Lemma 3, we have

$$S \ll q^{3/2+\varrho/2} .$$

Hence, since

$$\Phi(q) \gg q^{1-\varrho/2} ,$$

Lemma 4 follows immediately.

Note. The same estimate will also hold true for a more general sum

$$T_1 = \sum_{0 \leqq z < q} \chi(z)\, e^{2\pi i \frac{a}{q} z^n} ,$$

where χ is a certain character modulo q. Indeed,

$$T_1 = \frac{1}{\Phi(q)} S_1 ; \qquad S_1 = \sum_{0 \leqq x < q} \sum_{0 \leqq y < q} \chi(x)\,\chi(y)\, e^{2\pi i \frac{a}{q} x^n y^n} .$$

The sum S_1 is estimated in a similar way to the sum S in Lemma 3. Consequently, we always have

$$T_1 \ll q^{\frac{1}{2}+\varrho} .$$

Theorem 1. Let ε be a positive constant $\leqq 1/3$, η an arbitrarily small positive constant $<\varepsilon$, $N>2$, $r=\ln N$

$$Ne^{-r^{1-2\varepsilon}} \ll A \leqq N, \quad (a,q)=1, \quad 0<q\leqq e^{r^{\varepsilon}},$$

n a positive integer,

$$S=\sum_{N-A<p\leqq N} e^{2\pi i\frac{a}{q}p^n}.$$

Then

$$S \ll \frac{A}{r^{1-4\varepsilon}q^{\frac{1}{2}-\eta}}.$$

Proof. 1°. It suffices to assume that $N\geqq N_0$, where N_0 is a sufficiently large number >2. Let

$$\eta=3\varrho; \quad p_0=e^{r^{1-\varepsilon}}$$

and D denote the product of all primes $\leqq p_0$ which do not divide q. Now let d run through all the divisors of D and P through the numbers coprime to Dq. Then

$$\sum_{N-A<P\leqq N} e^{2\pi i\frac{a}{q}P^n}=\sum_d \mu(d)S_d; \quad S_d=\sum_{\substack{\frac{N-A}{d}<m\leqq\frac{N}{d}\\(m,q)=1}} e^{2\pi i\frac{a}{q}d^n m^n}. \tag{1}$$

2°. From the left-hand side, omit all those terms where P is divisible by the square of an integer >1. The error arising from such an elimination is

$$\ll \sum_{p_0<P\leqq\sqrt{N}}\left(\frac{A}{P^2}+1\right)\ll\frac{A}{p_0}+\sqrt{N}\ll\frac{A}{r\sqrt{q}}.$$

3°. Now let us estimate the error Δ that arises if those terms which correspond to

$$d>N^{3/4}.$$

are omitted from the right-hand side. Let such a d have k prime divisors. Then, we find

$$p_0^k\geqq N^{3/4}; \quad k\geqq\tfrac{3}{4}r^{\varepsilon}.$$

But

$$\Delta\leqq\sum_{N^{3/4}<d\leqq N}\ \sum_{\frac{N-A}{d}<m\leqq\frac{N}{d}} 1\leqq\sum_{0<m\leqq N^{1/4}}\ \sum_{\frac{N-A}{m}<d\leqq\frac{N}{m}} 1.$$

Now from a well-known formula, it follows that

$$\tau\left[\frac{N-A}{m}+1\right]+\ldots+\tau\left[\frac{N}{m}\right]\ll\frac{A}{m}r.$$

Therefore

$$2^{\frac{3}{4}r^{\varepsilon}} \sum_{\frac{N-A}{m}<d\leqq\frac{N}{m}} 1 \ll \frac{A}{m} r; \qquad \sum_{\frac{N-A}{m}<d\leqq\frac{N}{m}} 1 \ll \frac{A}{m} r \cdot 2^{-\frac{3}{4}r^{\varepsilon}},$$

$$\Delta \ll Ar^2 \cdot 2^{-\frac{3}{4}r^{\varepsilon}} \ll \frac{A}{r\sqrt{q}}.$$

4°. By the results of 2° and 3°, Eq. (1) takes the form

$$\sum_{N-A<Q\leqq N} e^{2\pi i \frac{a}{q} Q^n} = \sum_{0<d\leqq N^{3/4}} \mu(d)\, S_d + O\left(\frac{A}{r\sqrt{q}}\right),$$

$$S_d = \sum_{\substack{\frac{N-A}{d}<m\leqq\frac{N}{d} \\ (m,q)=1}} e^{2\pi i \frac{a}{q} d^n m^n}. \tag{2}$$

Here Q in the left-hand side runs through only those numbers which are not divisible by the square of an integer >1 and coprime to Dq.

5°. Let us now estimate one of the sums S_d in (2). We have

$$S_d = \left[\frac{A}{qd}\right] T + O(q),$$

where T does not depend on d, and by Lemma 4,

$$T \ll q^{\frac{1}{2}+\varrho}.$$

Therefore

$$\sum_{0<d\leqq N^{3/4}} \mu(d)\, S_d = T \sum_{0<d\leqq N^{3/4}} \mu(d) \left[\frac{A}{qd}\right] + O(N^{3/4} q).$$

Arguing as in 3°, we find

$$\left| T \sum_{N^{3/4}<d\leqq\frac{A}{q}} \mu(d) \left[\frac{A}{qd}\right] \right| < |T| \sum_{N^{3/4}<d\leqq\frac{A}{q}} \; \sum_{0<m\leqq\frac{A}{qd}} 1 \ll \frac{A}{r\sqrt{q}}.$$

Consequently,

$$\sum_{0<d\leqq N^{3/4}} \mu(d)\, S_d \ll q^{\frac{1}{2}+\varrho} Z + O\left(\frac{A}{r\sqrt{q}}\right); \qquad Z = \sum_{0<d\leqq\frac{A}{q}} \mu(d) \left[\frac{A}{qd}\right].$$

Obviously, Z is the number of those numbers which are coprime to D and are not greater than A/q. Therefore, by Lemma 1 (with 1 in place of q, q in place of q_1, A/q in place of N and A, and $\ln A/q$ in place of r), we have

$$Z \ll \frac{Aq^{\varrho}}{qr^{1-\varepsilon}}.$$

Therefore Eq. (2) can be rewritten as

$$\sum_{N-A<Q\leqq N} e^{2\pi i \frac{a}{q} Q^n} \ll \frac{A}{r^{1-\varepsilon} q^{\frac{1}{2}-2\varrho}}. \tag{3}$$

6°. Consider the left-hand side of (3). Let k be the number of prime divisors of some Q. Then we have

$$p_0^k \leqq N; \qquad k \leqq r^{\varepsilon}.$$

Consequently, Eq. (3) can be rewritten as

$$\sum_{0<k\leqq r^{\varepsilon}} \Omega_k \ll \frac{A}{r^{1-\varepsilon} q^{\frac{1}{2}-2\varrho}}, \tag{4}$$

where Ω_k is the sum of the terms on the left-hand side of (3) which correspond to the values of Q having exactly k prime divisors.

7°. Now we shall estimate the value of a term Ω_k, assuming that $k>1$. First we shall estimate the sum

$$T_k = \sum_u \sum_v e^{2\pi i \frac{a}{q} u^n v^n},$$

where u runs through such primes as satisfy the condition

$$p_0 < u \leqq N p_0^{-k+1}$$

and v, for a given u, runs through Q's in the interval

$$\frac{N-A}{u} < v \leqq \frac{N}{u},$$

which are not divisible by the square of an integer >1 and have exactly $(k-1)$ prime factors $>p_0$ and $\leqq \sqrt{N}$.

We shall divide the interval

$$p_0 < u \leqq N p_0^{-k+1}$$

into $\ll r$ subintervals of the type

$$U < u \leqq U_0; \qquad 2U \leqq U_0 \leqq 4U$$

and then assuming that

$$H_0 = e^{(\ln U)^{1-\varepsilon}},$$

we shall divide each subinterval into $\ll H_0$ still smaller subintervals of the type

$$u_0 < u \leqq u'; \qquad UH_0^{-1} < u' - u_0 \leqq 2UH_0^{-1}.$$

Let

$$W = \sum_{u_0<u\leqq u'} \sum_{\frac{N-A}{u}<v\leqq \frac{N}{u}} e^{2\pi i \frac{a}{q} u^n v^n}.$$

Then we have

$$\frac{N-A}{u}=\frac{N-A}{u_0}+O\left(\frac{N}{U}H_0^{-1}\right); \qquad \frac{N}{u}=\frac{N}{u_0}+O\left(\frac{N}{U}H_0^{-1}\right).$$

Therefore the sum W will differ from the sum

$$W_1=\sum_{u_0\leqq u\leqq u'}\ \sum_{\frac{N-A}{u_0}<v\leqq\frac{N}{u_0}} e^{2\pi i\frac{a}{q}u''v''}$$

by a quantity of the order of

$$\frac{N}{U}H_0^{-1}UH_0^{-1}=NH_0^{-2}.$$

The sum of all such errors for all W into which T_k is divided will be

$$\ll N\sum_U H_0^{-1}\ll Ae^{r^{1-2\varepsilon}-r^{(1-\varepsilon)^2}}r\ll\frac{A}{rq}.$$

8°. Let l be a number in the sequence $0, 1, \ldots, (q-1)$ such that $(l, q)=1$. Applying Lemma 1, we shall estimate the number of u's of the form $(qx+l)$ in the interval

$$u_0<u\leqq u'.$$

Here we shall put $q_1=1$. In place of N, now we have to take u'. Consequently, in place of r we have to take $\ln u'$. For ε we may take the same ε as in the theorem that is being proved. For A we take $u'-u_0$ which is

$$\gg u'e^{-(\ln u')^{1-\varepsilon}}.$$

Then the conditions of the lemma are satisfied and we have

$$B\ll\frac{u'-u_0}{r^{(1-\varepsilon)^2}q^{1-\varrho}}\ll M; \qquad M=\frac{UH_0^{-1}}{r^{(1-\varepsilon)^2}q^{1-\varrho}}.$$

9°. Now, applying Lemma 1, we shall also estimate the number B_1 of numbers $v\equiv l \pmod q$ which are coprime to D and lie in the interval

$$\frac{N-A}{u_0}<v\leqq\frac{N}{u_0}.$$

Here instead of the N used in Lemma 1, we have to take N/u_0, and instead of r we have to take $\ln N/u_0$, and finally in place of A we have to take A/u_0. Evidently, then

$$r^{1-\varepsilon}\leqq\ln\frac{N}{u_0}\leqq r-r^{1-\varepsilon}.$$

Therefore we obtain

$$B_1\ll M_1; \qquad M_1=\frac{A}{Ur^{(1-\varepsilon)^2}q^{1-\varrho}}.$$

10°. We can now express W_1 in the form

$$W_1=\sum_l\sum_{l_1}e^{2\pi i\frac{a}{q}l''l_1''},$$

where l and l_1 run, independently of each other, through those numbers from 0, 1, ..., $(q-1)$ which are coprime to q. Moreover, l takes each value $\ll M$ times and l_1 takes each value $\ll M_1$ times. Therefore, according to Lemma 3, we have

$$W_1 \ll MM_1 q^{\frac{3}{4}+\varrho} \ll \frac{AH_0^{-1}}{r^{2-4\varepsilon+2\varepsilon^2} q^{\frac{1}{4}-3\varrho}}.$$

At the same time, from 7°, we find that

$$T_k \ll \frac{A}{r^{1-4\varepsilon+2\varepsilon^2} q^{\frac{1}{4}-3\varrho}}.$$

11°. Evidently, every term

$$e^{2\pi i \frac{a}{q} Q^n}$$

of the sum Ω_k is also contained in the sum T_k as well, and each term occurs exactly k times. Moreover, the sum T_k may also contain terms of the type

$$e^{2\pi i \frac{a}{q} u^{2n} v_1^n}.$$

But the number of such terms is

$$\ll \sum_{p_0 < u \leqq \sqrt{N}} \left(\frac{A}{u^2} + 1\right) \ll \frac{A}{p_0}.$$

Therefore, for $k > 1$,

$$\Omega_k = \frac{1}{k} T_k + O\left(\frac{1}{k} \frac{A}{p_0}\right) \ll \frac{A}{kr^{1-4\varepsilon+2\varepsilon^2} q^{\frac{1}{4}-3\varrho}}.$$

Hence from (4) our theorem follows immediately.

Note. Looking for asymptotic formulae in additive problems with prime numbers (I mean the modified method of Hardy and Littlewood used in my papers and the problems which have already been studied) reduces to solving two problems, viz. (1) estimation of an integral on large arcs (small q) and (2) estimation of an integral on small arcs (large q).

The methods used to solve the first problem, based on the theory of L-series, which are equally suitable both for mixed problems (the arguments of all terms are not necessarily primes) as well as for pure problems (arguments of all terms are primes), have been developed in full, just before the publication of my estimates for trigonometric sums with primes. These are the method based on Siegel's lemma and Esterman's method, which gives a much weaker order for the remainder term, not based on Siegel's lemma. Both the methods are based on certain results of Titchmarch and Page.

There is no need to use my estimates in mixed problems. My estimates are however essential in solving all pure problems.

Here it should be mentioned that to a considerable degree, my method can replace the theory of L-series in solving the first problem. For instance, using Theorem 1, we can easily derive my asymptotic expression for the number of representations of an integer N in the form

$$N = p_1^n + \dots + p_s^n; \qquad n > 1; \qquad s \geqq s_0$$

with the help of the usual formula for $\pi(x; q, l)$ proved only for the case $q < r^{\varepsilon}$ ($r = \ln N$), where ε is an arbitrarily small positive constant. If only Theorem 1 and Page's results are used, the remainder terms obtained in all the problems are almost as accurate as those derived with the help of the Esterman method. Thus, in my formula the remainder term is of the order of

$$N^{s\nu-1} r^{-s+4+\varepsilon}; \qquad \nu = \frac{1}{n},$$

i.e. it is of the same order as that which is obtained if the Esterman method is used correctly.

Lemma 5. Let N be an integer $> N_0$, where N_0 is a sufficiently large constant > 1, $r = \ln N$, D denote the product of all primes $\leqq \sqrt{N}$, h a positive number $\leqq 1/6$, $\lambda = 1 + h$. Then all the divisors of D which are less than or equal to N can be divided into

$$< r^{\ln r/\ln \lambda}$$

classes satisfying the condition so that for each of these classes there exists a positive integer τ and 3τ numbers $\gamma_1, \dots, \gamma_\tau$; $G_1, \dots, G_\tau$; $l_1, \dots, l_\tau$ having the following properties:

$$G_1 \geqq G_2 \geqq \dots \geqq G_\tau; \qquad \gamma_s^\lambda = G_s; \qquad l_s \geqq 0;$$

Every d belonging to a given class can be represented in the form of a product

$$d = g_1 \dots g_\tau,$$

where g_s is the product of l_s primes p satisfying the condition

$$\gamma_s \leqq p^{l_s} < G_s,$$

so that

$$\gamma_s \leqq g_s < G_s.$$

Proof. 1°. We shall divide all the prime divisors of D into groups as follows. Let τ denote the greatest integer satisfying the condition

$$2^{\lambda^{\tau-1}} \leqq \sqrt{N}.$$

Then we obtain τ groups of primes, if we include into the s-th group ($s = 1, 2, \dots, \tau$) the numbers p satisfying the condition:

$$2^{\lambda^{s-1}} \leqq p < 2^{\lambda^s}.$$

2°. By the definition of τ, it follows that

$$\tau < \frac{\ln r}{\ln \lambda}.$$

3°. For the number ν of prime factors of some $d \leqq N$, the upper bound is

$$\nu \leqq r-1.$$

4°. We shall divide the values of d under consideration into classes as follows: the values of d with different number of factors belonging to the same group shall be put into the same class. By virtue of 2° and 3°, the number of all such classes is

$$< r^{\ln r/\ln \lambda}.$$

5°. Let f_s denote the product of prime factors of the s-th group which occur in the number d of a chosen class, and let l_s denote the number of such factors ($f_s = 1, l_s = 0$, if there are no such factors). Thus we have

$$d = f_1 \ldots f_\tau.$$

Putting

$$\varphi_s = 2^{\lambda^{s-1} l_s}; \qquad F_s = 2^{\lambda^s l_s},$$

we obtain

$$\varphi_s \leqq f_s < F_s.$$

The following numbers

$$\varphi_1, \ldots, \varphi_\tau,$$
$$f_1, \ldots, f_\tau,$$
$$F_1, \ldots, F_\tau,$$

on being rearranged in such a way the F_s form a non-increasing sequence, shall be denoted by the symbols:

$$\gamma_1, \ldots, \gamma_\tau,$$
$$g_1, \ldots, g_\tau,$$
$$G_1, \ldots, G_\tau.$$

Accordingly, we shall change the numbers of the groups of p's as well. Thus, the proof of the lemma is complete.

Lemma 6. Let (u) and (v) be two sequences of positive integers coprime to q, $N \geqq N_0$, where N_0 is a sufficiently large constant >1, $r = \ln N$ and

$$1 < U_1 < U_2 \leqq N; \qquad 1 \leqq A \leqq N,$$
$$(a, q) = 1; \qquad 1 \leqq q \leqq N,$$

n is an integer $\geqq 1$,

$$S = \sum_u \varphi(u) \sum_v e^{2\pi i \frac{a}{q} u^n v^n},$$

where u runs through the numbers of the sequence (u) satisfying the condition

$$U_1 < u \leqq U_2$$

and v runs through, for a given u, the numbers of the sequence (v) satisfying the condition

$$\frac{N-A}{u} < v \leqq \frac{N}{u}.$$

Now let $\varphi(u)$ denote a function which for all u in the interval

$$U_1 < u \leqq U_2$$

take non-negative values, and which, for any u_1 and u_2 satisfying the condition

$$U_1 \leqq u_1 < u_2 \leqq U_2,$$

obeys the inequality

$$\sum_{u_1 < u \leqq u_2} (\varphi(u))^s \ll u_2 F_s.$$

Finally, let ϱ be an arbitrarily small positive constant. Then

$$S \ll Aq^{\varrho-\frac{1}{2}} \sqrt{\left(1+\frac{q}{U_1}\right)\left(1+\frac{U_2 q}{A}\right)\left(1+\frac{U_2}{A}\right)}\, rF_2^{1/2} + \left(\frac{Nq}{U_1}+U_2\right) rF_1.$$

Proof. 1°. Divide the interval

$$U_1 < u \leqq U_2$$

into $\ll r$ subintervals of the type:

$$U < u U_0; \qquad U_0 \leqq 2U.$$

2°. Consider that part S_0 of the sum S which corresponds to one such subinterval. After dividing this subinterval into $\ll (U/q+1)$ still smaller subintervals of the type

$$u_0 < u \leqq u'; \qquad u' - u_0 \leqq q,$$

we obtain

$$S_0 = \textstyle\sum_0 S_1; \qquad S_1 = \sum_{u_0 < u \leqq u'} \varphi(u) \sum_{\frac{N-A}{u} < v \leqq \frac{N}{u}} e^{2\pi i \frac{a}{q} u'' v''}.$$

The sum S_1 is approximately equal to

$$S_2 = \sum_{u_0 < u \leqq u'} \varphi(u) \sum_{\frac{N-A}{u_0} < v < \frac{N}{u_0}} e^{2\pi i \frac{a}{q} u'' v''}.$$

Since

$$\frac{N-A}{u} = \frac{N-A}{u_0} + O\left(\frac{Nq}{U^2}\right); \qquad \frac{N}{u} = \frac{N}{u_0} + O\left(\frac{Nq}{U^2}\right),$$

we obtain

$$S_1 - S_2 \ll \left(\frac{Nq^{3/2}}{U^2}+1\right) \sum_{u_0 < u \leqq u'} \varphi(u).$$

Therefore

$$\sum_0 S_1 - \sum_0 S_2 \ll \left(\frac{Nq}{U^2}+1\right) \sum_{U_1<u\leqq U_0} \varphi(u) \ll \left(\frac{Nq}{U}+U\right) F_1 .$$

Now we have

$$\left(\sum_0 S_2\right)^2 \ll \left(\frac{U}{q}+1\right) \sum_0 |S_2|^2 .$$

But each sum S_2 can be divided into

$$\ll \frac{A}{Uq}+1$$

subsums of the type

$$S_3 = \sum_{u_0<u\leqq u'} \varphi(u) \sum_{v_0<v\leqq v'} e^{2\pi i \frac{a}{q} u^n v^n}; \quad v'-v_0 \leqq q ,$$

where

$$S_2^2 \ll \left(\frac{A}{Uq}+1\right) \sum |S_3|^2 .$$

Summation is taken over all subsums S_3. Applying Lemma 3 to each subsum S_3, we obtain

$$S_3^2 \ll q^{1+\varrho} \sum_{u_0<u\leqq u'} (\varphi(u))^2 \sum_{v_0<v\leqq v} 1 ,$$

$$S_2^2 \ll q^{1+\varrho} \left(\frac{A}{Uq}+1\right) \sum_{u_0<u\leqq u'} (\varphi(u))^2 \left(\frac{A}{U}+1\right).$$

At the same time, we have

$$\left(\sum_0 S_2\right)^2 \ll \left(\frac{U}{q}+1\right) UF_2 \left(\frac{A}{Uq}+1\right)\left(\frac{A}{U}+1\right) q^{1+\varrho}$$

$$= A^2 q^{\varrho-1} \left(1+\frac{q}{U}\right)\left(1+\frac{Uq}{A}\right)\left(1+\frac{U}{A}\right) F_2 ,$$

$$S_0 \ll A q^{\varrho/2-1/2} \sqrt{\left(1+\frac{q}{U}\right)\left(1+\frac{Uq}{A}\right)\left(1+\frac{U}{A}\right)}\, F_2^{1/2} + \left(\frac{Nq}{U}+U\right) F_1 .$$

Hence, Lemma 6 follows immediately.

Theorem 2. Let ε, η and h be arbitrarily small positive constants <1, $N>2$, $r=\ln N$ $(a,q)=1$, $e^{r^\varepsilon}<q$, $N^{2/3+h} q^{3/2} \leqq A \leqq N$, and n a positive integer:

$$S = \sum_{N-A<p\leqq N} e^{2\pi i \frac{a}{q} p^n} .$$

Then

$$S \ll A q^{\eta-1/2} .$$

Proof.[1] 1°. Let $\eta = 2\varrho$. Then we have

$$S = \sum_d \sum_m \mu(d)\, e^{2\pi i \frac{a}{q} d^n m^n} + O(\sqrt{N}),$$

where d runs through the divisors of the product of all primes at most $\sqrt{N}$ that are not contained in q, while m runs through the natural numbers that are coprime to q, and the summation is taken over the range

$$N - A < dm \leqq N.$$

Then we find

$$\sum_d \sum_m \mu(d)\, e^{2\pi i d^n m^n} = T_0 - T_1,$$

where T_0 is the sum of terms which correspond to the values of d with even number of prime divisors, while T_1 is the sum of the terms which correspond to the values of d with an odd number of prime divisors. Now consider one of the sums T_0 or T_1, and denote it by T_2.

2°. Divide the numbers $d \leqq N$ into classes satisfying the conditions stated in Lemma 5 (the notations being as in this lemma).

3°. Divide the numbers m into $\ll r$ classes. Put all numbers m in a certain interval of the type

$$M < m \leqq M_0; \qquad M_0 \leqq 2M.$$

into one class.

4°. Denote by T_3 that part of the sum T_2, which corresponds to the chosen classes of d and m.

5°. Let

$$MG_1 \ldots G_\tau > N^{1/3},$$

where β is the first term in the sequence $1, 2, \ldots, \tau$ that satisfies the condition

$$MG_1 \ldots G_\beta > N^{1/3}.$$

6°. Let $\beta > 1$. Then, putting

$$u = mg_1 \ldots g_\beta; \qquad v = g_{\beta+1} \ldots g_\tau,$$

we obtain

$$MG_1 \ldots G_{\beta-1} \leqq N^{1/3}; \qquad G_\beta \leqq G_{\beta-1}; \qquad MG_1 \ldots G_\beta \leqq N^{2/3}.$$

Therefore for the case under consideration

$$T_3 = \sum_u \varphi(u) \sum_v e^{2\pi i \frac{a}{q} u^n v^n},$$

[1] An improved estimate for a trigonometrical sum containing primes; Izv. akad. nauk SSSR (ser. mat.) **2** (1938) 1–24.

where u runs through integers satisfying the condition

$$N^{1/3\lambda} < u \leqq 2N^{2/3},$$

while v, for a given u, runs through integers satisfying the condition

$$\frac{N-A}{u} < v \leqq \frac{N}{u}.$$

Here $\varphi(u)$ does not exceed the number $\tau(u)$ of divisors of u. Consequently, the numbers F_1 and F_2 stated in Lemma 6 will satisfy the conditions:

$$F_1 \ll r; \qquad F_2 \ll r^3.$$

Applying Lemma 6, we obtain

$$T_3 \ll Aq^{\varrho-1/2}\sqrt{\left(1+\frac{q}{N^{1/3\lambda}}\right)\left(1+\frac{N^{2/3}q}{A}\right)}\,r^{5/2} + (N^{1-1/3\lambda}q + N^{2/3})\,r^2$$
$$\ll Ar^{5/2}q^{\varrho-1/2}.$$

7°. Let $\beta = 1$. For the sake of brevity, assuming that $l_1 = \chi$ and denoting by χ_1 the least integer that satisfies the condition

$$MG^{\chi_1/\chi} > N^{1/3},$$

we shall represent each d in the form:

$$d = u_0 v,$$

where u_0 is the product of χ_1 prime factors of g_1, while v is the product of the remaining $(\chi-\chi_1)$ factors of g_1 and all $g_2, \ldots, g_\tau$. For $\chi_1 > 1$, we have

$$T_3 = \frac{1}{\binom{\chi}{\chi_1}} T_3'; \qquad T_3' = \sum_m \sum_{u_0} \sum_v e^{2\pi i \frac{a}{q} m^n u_0^n v^n},$$

where u_0 and v run through, independently of each other, the products of χ_1 and $(\chi-\chi_1)$ prime factors of g_1; here T_3' may contain only the terms with

$$(u_0, v) = 1.$$

Thus, we find

$$T_3' = \sum_\delta \mu(\delta) W_\delta; \qquad W_\delta = \sum_m \sum_{u_1} \sum_{v_1} e^{2\pi i \frac{a}{q} \delta^{2n} m^n u_1^n v_1^n},$$

where δ runs through all possible divisors of g_1, while u_1 and v_1, for a given δ, run through the quotients of division of u_0 and v by δ.
Putting $mu_1 = u$, we obtain

$$W_\delta = \sum_u \varphi(u) \sum_{v_1} e^{2\pi i \frac{a}{q} \delta^{2n} u^n v_1^n};$$

$$F_1 \ll r; \qquad F_2 \ll r^3.$$

The number u in the sum W_δ runs through the values satisfying the condition

$$\frac{N^{1/3\lambda}}{\delta} < u \leqq 2\frac{N^{2/3}}{\delta},$$

while v_1, for a given u, runs through the values satisfying the condition:

$$\frac{N-A}{\delta^2 u} < v_1 \leqq \frac{N}{\delta^2 u}.$$

8°. Evidently, $\delta \leqq \sqrt{N}$. In addition, let

$$\delta \leqq q.$$

Now we can apply Lemma 6. Thus, we have

$$W_\delta \ll \frac{A}{\delta^2} q^{\varrho-1/2} \sqrt{\left(1+\frac{q\delta}{N^{1/3\lambda}}\right)\left(1+\frac{N^{2/3}q\delta}{A}\right)\left(1+\frac{N^{2/3}\delta}{A}\right)}\, r^{5/2}$$
$$+\left(\frac{N^{1-1/3\lambda}q}{\delta}+\frac{N^{2/3}}{\delta}\right)r^2 \leqq \frac{A}{\delta} q^{\varrho-1/2} r^{5/2}.$$

9°. If, however, $\delta > q$, then W_δ does not exceed the number of systems of values of m, u_0/δ, v_1 satisfying the condition

$$\frac{N-A}{\delta^2} < m\frac{u_0}{\delta} v_1 \leqq \frac{N}{\delta^2}$$

and, consequently, by the well known property of $\tau(a)$, we have

$$W_\delta \ll \left(\frac{A}{\delta^2}+\frac{\sqrt{N}}{\delta}\right) r^2,$$

which is contained as a particular case in the inequality proved in 8°.

10°. Therefore, since $\delta \leqq \sqrt{N}$ and by virtue of 7°, we have

$$T_3 \ll T_3' \ll A r^{7/2} q^{\varrho-1/2}.$$

Similarly, we obtain the same estimate even when $\chi_1 = 1$.

11°. Now consider the case, where

$$M \geqq N^{1/3}.$$

Here $d < N^{2/3}$, $A > dq$. The sum

$$S_d = \sum_{\frac{N-A}{d} < m \leqq \frac{N}{a}}$$

can be divided into

$$\ll \frac{A}{dq}$$

subsums, each of which, according to Lemma 4, is

$$\ll q^{\varrho+1/2}.$$

Therefore

$$S_d \ll \frac{A}{d} q^{\varrho-1/2}; \qquad T_3 \ll Arq^{\varrho-1/2}.$$

12°. Finally, if

$$MG_1 \dots G_\tau < N^{1/3},$$

then evidently we have

$$T_3 \ll N^{1/3} r \ll Aq^{-1/2}.$$

13°. From the estimates derived for T_3 (in 6°, 10°, 11°, 12°), Lemma 5 and 1° and 4°, we find that our lemma is true.

More General Results For the Case of $n = 1$. In view of the special importance of the case $n = 1$, we shall demonstrate a more general theorem than Theorem 2.

Lemma 7.[1] Let $N \geqq N_0$, where N_0 is a sufficiently large constant >1, $r = \ln N$,

$$W \geqq 1; \qquad 1 \leqq V \leqq N; \qquad k - \text{integer} > 0,$$

$$\lambda = \frac{a}{q} + \frac{\theta}{q^2}; \qquad (a, q) = 1; \qquad 0 < q \leqq N,$$

t runs through (all or some) integers in a certain interval of length $\ll V$

$$S = \sum \min\left(W, \frac{1}{2(\lambda k t)}\right).$$

Then

$$S \ll \left(\frac{V}{q} + 1\right)(kW + qr).$$

Lemma 8. Let (u) and (v) be two sequences of positive integers, $N \geqq N_0$, where N_0 is a sufficiently large constant >1, $r = \ln N$;

$$1 \leqq U_1 < U_2 \leqq N, \qquad U_2 \leqq A \leqq N; \qquad k - \text{integer} > 0,$$

$$\lambda = \frac{a}{q} + \frac{\theta}{q^2}; \qquad (a, q) = 1; \qquad 0 < q \leqq N,$$

$$S = \sum_u \varphi(u) \sum e^{2\pi i \lambda k u v},$$

where u runs through the members of the sequence (u) satisfying the condition:

$$U_1 < u \leqq U_2,$$

while v runs, for a given u, through the members of the sequence (v) satisfying the condition:

$$\frac{N - A}{u} < v \leqq \frac{N}{u}.$$

---

[1] Estimates for trigonometrical sums Izv. akad. nauk SSSR (ser. mat.) **2** (1938) 505–524.

Here $\varphi(u)$ is a function which takes, for all u in the interval:

$$U_1 < u \leqq U_2$$

non-negative values; and for any u_1 and u_2 satisfying the condition

$$U_1 \leqq u_1 < u_2 \leqq U_2,$$

we always have

$$\sum_{u_1 < v \leqq u_2} (\varphi(u))^2 \ll u_2 F_2 .$$

Hence

$$S \ll A\sqrt{F_2}\sqrt{\frac{k}{q}+\frac{kU_2}{A}+\frac{N}{U_1 A}+\frac{Nq}{A^2}}\, r^{3/2}.$$

Proof. 1°. It suffices to consider only the case

$$A \geqq \max\left(U_2, \frac{N}{U_1}, \sqrt{Nq}\right).$$

Divide the interval

$$U_1 < u \leqq U_2$$

into $\ll r$ subintervals of the type

$$U < u \leqq U_0; \qquad U_0 \leqq 2U.$$

2°. Consider that part S_0 of the sum S which corresponds to one such subinterval. Now dividing this subinterval into

$$\ll \frac{N}{A}$$

new subintervals of the type

$$u_1 < u \leqq u_2; \qquad u_2 - u_1 \ll \frac{AU}{N},$$

we find that

$$S_0 = \sum\nolimits_1 S_1; \qquad S_1 = \sum_{u_1 < u \leqq u_2} \varphi(u) \sum_{\frac{N-A}{u} < v \leqq \frac{N}{u}} e^{2\pi i \lambda k u v},$$

$$S_0^2 \ll \frac{N}{A} \sum\nolimits_1 |S_1|^2; \qquad S_1^2 \ll \sum_{u_1 < u \leqq u_2} (\varphi(u))^2 R,$$

$$R = \sum_{u_1 < u \leqq u_2} \; \sum_{\frac{N-A}{u} < v \leqq \frac{N}{u}} \; \sum_{\frac{N-A}{u} < v_1 \leqq \frac{N}{u}} e^{2\pi i \lambda k u (v - v_1)},$$

where u runs through the integers in the above interval.

To estimate R, chenge the order of summation. Evidently, v and v_1 may take only those values which satisfy the condition

$$\frac{N-A}{u_2} < v \leqq \frac{N}{u_1}; \qquad \frac{N-A}{u_2} < v_1 \leqq \frac{N}{u_1},$$

and u, for the chosen v and v_1, runs through the values in the interval

$$u' < u \leqq u''; \qquad u' = \max\left(u_1, \frac{N-A}{v}, \frac{N-A}{v_1}\right),$$
$$u'' = \min\left(u_2, \frac{N}{v}, \frac{N}{v_1}\right).$$

Therefore, since $u'' - u' \leqq u_2 - u_1 \ll AU/N$, we obtain

$$R \ll \sum_v \sum_{v_1} \min\left(\frac{AU}{N}, \frac{1}{2(\lambda k(v-v_1))}\right).$$

Observing that v, for a given v_1, runs through the numbers in the interval of length

$$\frac{N}{u_1} - \frac{N-A}{u_2} = \frac{N(u_2-u_1)+Au_1}{u_1 u_2} \ll \frac{A}{U},$$

by Lemma 7, we find that

$$R \ll R_0; \qquad R_0 = \frac{A}{U}\left(\frac{A}{Uq}+1\right)\left(k\frac{AU}{N}+qr\right).$$

At the same time, we have

$$S_1^2 \ll \sum_{u_1 < u \leqq u_2} (\varphi(u))^2 R_0; \qquad S_0^2 \ll \frac{N}{A} R_0 \sum_{U < u \leqq U_0} (\varphi(u))^2$$
$$\ll \frac{NU}{A} F_2 R_0 \ll A^2 F_2 \left(\frac{1}{Uq}+\frac{1}{A}\right)\left(kU+\frac{Nqr}{A}\right),$$

$$S_1 \ll A\sqrt{F_2}\sqrt{\frac{k}{q}+\frac{kU}{A}+\frac{Nr}{UA}+\frac{Nqr}{A^2}}.$$

Hence Lemma 8 follows immediately.

Lemma 9[1]. Let k be an integer >0

$$N > 2; \qquad r = \ln N; \qquad 1 < d_0 < A \leqq N,$$
$$\lambda = \frac{a}{q} + \frac{\theta}{q^2}; \qquad (a,q) = 1; \qquad 0 < q \leqq N.$$

Then

$$\sum_{0 < d \leqq d_0} \min\left(\frac{A}{d}, \frac{1}{2(\lambda k d)}\right) \ll \left(d_0 + q + \frac{k^2 A}{q}\right) r.$$

Theorem 3. Let h be an arbitrarily small positive constant $\leqq 1/6$, $N > 2$, $r = \ln N$, k an integer >0,

$$\alpha = \frac{a}{q} + \frac{\theta}{q^2}; \qquad (a,q) = 1; \qquad 0 < q \leqq N, \qquad 1 \leqq A \leqq N,$$
$$S = \sum_{N-A < p \leqq N} e^{2\pi i \alpha k p}.$$

---

[1] An improved estimate for a trigonometrical sum containing primes; Izv. akad. nauk SSSR (ser. mat.) **2** (1938) 1–24.

Then

$(\lambda = 1 + h)$

$$S \ll A r^{\frac{\ln r}{\ln \lambda}+6} \sqrt{\frac{kN^{\frac{2+h}{3}}}{A} + \frac{Nq}{A^2} + \frac{k}{q} + \frac{k^4}{q^2}}.$$

Proof. 1°. We have

$$S = \sum_d \sum_m \mu(d)\, e^{2\pi i \alpha k d m} + O(\sqrt{N}),$$

where d runs through the divisors of the product of all primes at most $\sqrt{N}$ and m runs through natural numbers. Summation is taken over the range

$$N - A < dm \leqq N.$$

Now we find

$$\sum_d \sum_m \mu(d)\, e^{2\pi i \alpha k d m} = T_0 - T_1,$$

where T_0 is the sum of the terms corresponding to the values of d with even number of prime divisors, while T_1 is the sum of the terms corresponding to the values of d with odd number of prime divisors. Now consider one of the sums T_0 or T_1 and denote it by T_2.

2°. Divide the numbers $d \leqq N$ into classes satisfying the conditions stated in Lemma 5, all notations being as in Lemma 5.

3°. Also divide the numbers m into $\ll r$ classes. Put all the numbers contained in a certain interval of the type

$$M < m \leqq M_0; \qquad M_0 \leqq 2M$$

into the same class.

4°. Denote by T_3 that part of the sum T_2 which corresponds to the chosen classes of the values of d and m.

5°. First let

$$MG_1 \ldots G_\tau > N^{1/3},$$

here β is the first of the members in the sequence $1, 2, \ldots, \tau$ that satisfies the condition:

$$MG_1 \ldots G_\beta > N^{1/3}.$$

6°. Let $\beta > 1$. Then putting

$$u = mg_1 \ldots g_\beta; \qquad v = g_{\beta+1} \ldots g_\tau,$$

we obtain

$$MG_1 \ldots G_{\beta-1} \leqq N^{1/3}; \qquad G_\beta \leqq G_{\beta-1}; \qquad MG_1 \ldots G_\beta \leqq N^{2/3}.$$

Therefore for our case

$$T_3 = \sum_u \varphi(u) \sum_v e^{2\pi i \alpha k u v},$$

where u runs through the integers satisfying the condition

$$N^{1/3\lambda} < u \leqq 2N^{2/3},$$

while v, for a given u, runs through the integers satisfying the condition

$$\frac{N-A}{u} < v \leqq \frac{N}{u}.$$

Here $\varphi(u)$ does not exceed the number $\tau(u)$ of divisors of u. Consequently, the number F_2 stated in Lemma 8, will satisfy the condition

$$F_2 \ll r^3.$$

Applying Lemma 8, we obtain

$$T_3 \ll A\sqrt{r^3}\sqrt{\frac{k}{q} + \frac{kN^{2/3}}{A} + \frac{N^{1-1/3\lambda}}{A} + \frac{Nq}{A^2}}\, r^{3/2}.$$

7°. Let $\beta = 1$. For the sake of brevity, assuming that $l_1 = \chi$ and denoting by χ_1 the least integer satisfying the condition

$$MG_1^{\chi_1/\chi} > N^{1/3},$$

we shall represent each d in the form

$$d = u_0 v,$$

where u_0 is the product of χ_1 prime factors of g_1, while v is the product of the remaining $(\chi - \chi_1)$ factors of g_1 and all $g_2, \ldots, g_\tau$.

For $\chi_1 > 1$, we have

$$T_3 = \frac{1}{\binom{\chi}{\chi_1}} T_3'; \qquad T_3' = \sum_m \sum_{u_0} \sum_v e^{2\pi i \alpha k m u_0 v},$$

where u_0 and v run through, now independently of each other, the product of χ_1 and $(\chi - \chi_1)$ different prime divisors of g_1; only those terms for which $(u_0, v) = 1$ are contained in T_3'.

Just as before, we conclude that

$$N^{1/3\lambda} < u \leqq 2N^{2/3}.$$

Now we find

$$T_3' = \sum_\delta \mu(\delta) W_\delta; \qquad W_\delta = \sum_m \sum_{u_1} \sum_{v_1} e^{2\pi i \alpha k \delta^2 m u_1 v_1},$$

where δ runs through all possible divisors of g_1, whereas u_1 and v_1, for a given δ, run through the quotients of division of u_0 and v by δ.

Putting $mu_1 = u$, we obtain

$$W_\delta = \sum_u \varphi(u) e^{2\pi i \alpha \delta^2 u v_1}; \qquad F_2 \leqq r^3.$$

In the sum W_δ, u runs through the values satisfying the condition:

$$\frac{N^{1/3\lambda}}{\delta} < u \leqq \frac{2N^{2/3}}{\delta},$$

while v_1, for a given u, runs through the values satisfying the condition:

$$\frac{N-A}{\delta^2 u} < v_1 \leqq \frac{N}{\delta^2 u}.$$

8°. Evidently, we have $\delta \leqq \sqrt{N}$. Furthermore, let

$$\delta \leqq \delta_0; \qquad \delta_0 = \min(N^{1/3\lambda}, AN^{-2/3}).$$

Now we can apply Lemma 8. Thus, we have

$$W_\delta \ll \frac{A}{\delta^2}\sqrt{r^3}\sqrt{\frac{k}{q} + \frac{kN^{2/3}\delta}{A} + \frac{N^{1-1/3\lambda}\delta}{A} + \frac{Nq\delta^2}{A^2}} \cdot r^{3/2}.$$

9°. If, however, $\delta > \delta_0$, then W_δ does not exceed the number of systems of the values of m, u_0/δ, v_1 satisfying the condition

$$\frac{N-A}{\delta^2} < m\frac{u_0}{\delta}v_1 \leqq \frac{N}{\delta^2}.$$

Consequently, by the well known property of the function $\tau(a)$, we have

$$W_\delta \ll \left(\frac{A}{\delta^2} + \frac{\sqrt{N}}{\delta}\right) r^2,$$

which is a particular case of the inequality proved in 8°.

10°. By virtue of the statements proved above, on summing over all δ, we find

$$T_3 \ll A\sqrt{\frac{k}{q} + \frac{kN^{2/3}}{A} + \frac{N^{1-1/3\lambda}}{A} + \frac{Nq}{A^2}} \cdot r^5.$$

Similarly, we obtain the same estimate even when $\chi_1 = 1$.

11°. Now consider the case where

$$M > N^{1/3}.$$

Here we have $d < N^{2/3}$ and

$$T_3 = \sum S_d; \qquad S_d = \sum_m e^{2\pi i \alpha k d m},$$

where d and m run through the numbers in the chosen classes. Here m in the sum S_d runs through the values satisfying the condition

$$M < m \leqq M_0; \qquad \frac{N-A}{d} < m \leqq \frac{N}{d}.$$

For $A \geqq N^{2/3}$, we have

$$S_d \ll \min\left(\frac{A}{d}, \frac{1}{2(\alpha k d)}\right)$$

therefore, by Lemma 9,

$$T_3 \ll \left(N^{2/3} + q + \frac{k^2 A}{q}\right) r.$$

12°. Finally, for the case where

$$MG_1 \ldots G_\tau \leqq N^{1/3},$$

evidently we have

$$T_3 \ll N^{1/3} r.$$

13°. Collecting all the results proved in 1°, 2°, 4°, 6°, 11° and 12°, we obtain

$$S \ll A r^{\frac{\ln r}{\ln \lambda}+6} \sqrt{\frac{k}{q} + \frac{kN^{2/3}}{A} + \frac{N^{1-1/3\lambda}}{A} + \frac{Nq}{A^2} + \frac{k^4}{q^2}}.$$

Hence Theorem 3 follows directly.

Lemma 10[1]. Let ε be an arbitrarily small positive constant,

$$\lambda = \frac{a}{q} + \frac{\theta}{q^2}; \qquad (a, q) = 1; \qquad q > 0.$$

Let k and g run, independently of each other, through the values

$$1, 2, \ldots, K$$

$$1, 2, \ldots, G.$$

Then the number of values of the symbol (λkg) which do not exceed q^{-1} is

$$\ll (KGq^{-1} + 1)(KG)^\varepsilon.$$

Lemma 11. Let h be an arbitrarily small positive constant <1, $N > N_0$, where N_0 is a sufficiently large constant >2,

$$r = \ln N: \qquad \max(N^{2/3}, N^{1/2} q^{1/2}) \leqq A \leqq N,$$

$$\lambda = \frac{a}{q} + \frac{\theta}{q^2}; \qquad (a, q) = 1; \qquad 0 < q \leqq N.$$

Let k run through $1, \ldots, K$, where the number g satisfying the condition

$$\lambda k = \frac{c}{g} + \frac{\theta_1}{gq}; \qquad (c, g) = 1; \qquad 0 < g \leqq q; \qquad |\theta_1| < 1.$$

is put into correspondance with each k. Then, putting

$$D_k = \min\left(\frac{1}{k}\sqrt{\frac{N^{2+h/3}}{A} + \frac{Ng}{A^2} + \frac{1}{g}}, \ \frac{1}{k}\sqrt{\frac{kN^{2+h/3}}{A} + \frac{Nq}{A^2} + \frac{k^2}{q}}\right),$$

$$S = \sum_{0<k\leqq q} D_k,$$

[1] An analytic proof of the theorem on the distribution of the fractional parts of the values of an integral polynomial; Izv. akad. nauk SSSR (ser. mat.) (1927, no. 7–8) 567–578, Lemma 5.

we have

$$S \ll r^2 \sqrt{\frac{N^{2/3+h}}{A} + \frac{N^{1+h} q^{1-h}}{A^2} + \frac{1}{q^{1-h}}}.$$

Proof. 1°. First consider the case where

$$q > N^{5/6}, \qquad g > N^{1/6}.$$

Evidently, we have

$$D_k \ll \frac{1}{k} \sqrt{\frac{Nq}{A^2}}.$$

Consequently, that part of the sum S which corresponds to such a case is

$$\ll r \sqrt{\frac{Nq}{A^2}}.$$

2°. Now consider the cases where

$$q \leqq N^{5/6}; \qquad q > N^{5/6}, \qquad g \leqq N^{1/6}.$$

First we shall estimate that part of the sum S which includes the terms satisfying the condition

$$kg \geqq q.$$

For this purpose, we shall divide all values of k, and then in an exactly similar manner, all the values of g into groups as follows:

$$k = 1, \qquad g = 1,$$
$$k = 2, 3, \qquad g = 2, 3,$$
$$k = 4, 5, 6, 7, \qquad g = 4, 5, 6, 7,$$
$$\dots\dots\dots\dots$$
$$k = 2^s, 2^s + 1, \dots, q, \qquad g = 2^s, 2^s + 1, \dots, q.$$

Let $\chi, \chi + 1, \dots, \chi + \chi'$; $\gamma, \gamma + 1, \dots, \gamma + \gamma'$ be two such groups of k's and g's. The numbers k and g in these groups may satisfy the inequality $kg \geqq q$ if and only if

$$4\chi\gamma \geqq q.$$

Assume that this condition is fulfilled. Then, we have

$$\lambda k = \frac{c}{g} + \frac{\theta}{gq}; \qquad \lambda kg = c + \frac{\theta}{q}$$

and therefore, by Lemma 10, the sum of the terms of S satisfying the condition that k and g belong to the groups chosen, is equal to

$$\ll \chi^{\varepsilon} \gamma^{1+\varepsilon} q^{-1} \sqrt{\frac{N^{(2+h)/3}}{A} + \frac{N\gamma}{A^2} + \frac{1}{\gamma}}$$

$$\ll q^{2\varepsilon} \sqrt{\frac{N^{(2+h)/3}}{A} + \frac{N\gamma}{A^2} + \frac{1}{q}} \ll \sqrt{\frac{N^{2/3+h}}{A} + \frac{N^{1+h} q^{1-h}}{A^2} + \frac{1}{q^{1-h}}}.$$

Summing over all possible pairs of groups, we obtain a value

$$\ll r^2 \sqrt{\frac{N^{2/3+h}}{A} + \frac{N^{1+h} q^{1-h}}{A^2} + \frac{1}{q^{1-h}}}.$$

3°. Finally, we shall estimate that part of the sum S which corresponds to the case where

$$kg < q.$$

Comparing the two different expressions for λk, we obtain

$$\frac{c}{g} + \frac{\theta_1}{gq} = \frac{ak}{q} + \frac{\theta k}{q^2}; \qquad |akg - cq| < 2.$$

Consequently, either the difference $akg - cq$ is zero or $v = \pm 1$. The first case is impossible since $(a, q) = 1$, $kg < q$. Therefore

$$akg - cq = v.$$

Denoting by z_0 the least positive solution of the congruence

$$az \equiv v \pmod{q},$$

we obtain $kg = z_0$. Since the number of solutions in positive integral k and g of this equation is $\ll q^{\varepsilon}$, the number of terms in our sum satisfying the condition $kg < h$ will also be of the same order. Each term is

$$\ll \frac{1}{k} \cdot \sqrt{\frac{k N^{(2+h)/3}}{A} + \frac{Nq}{A^2} + \frac{k^2}{q}}.$$

Therefore, that part of the sum S which corresponds to our case is

$$\ll r \sqrt{\frac{N^{2/3+h}}{A} + \frac{N^{1+h} q^{1-h}}{A^2} + \frac{1}{q^{1-h}}}.$$

The lemma follows immediately from 1°, 2° and 3°.

Theorem 4. Let ε and h be arbitrarily small positive constants $\leqq 1/6$, $N > N_0$, where N_0 is a sufficiently large constant > 2, $r = \ln N$,

$$\lambda = \frac{a}{q} + \frac{\theta}{q^2}; \qquad (a, q) = 1; \qquad 1 \leqq q \leqq N, \qquad 1 \leqq A \leqq N; \qquad 0 < \delta \leqq 1.$$

Let T denote the number of all values of the fractional part

$$\{\lambda p\}; \qquad N - A < p \leqq N$$

and T_1 the number of values of the same fractional part satisfying the condition

$$0 \leqq \{\lambda p\} \leqq \delta.$$

Then

$$T_1 = \delta T + O(A\Delta),$$

$$\Delta = e^{r^{\varepsilon}} \sqrt{\frac{N^{2/3+h}}{A} + \frac{N^{1+h} q^{1-h}}{A^2} + \frac{1}{q^{1-h}}}.$$

Proof. 1°. According to Lemma 14 in Chap. I of my book "A New Method in Analytic Number Theory"[1] (assume $\Delta < 0.5$, otherwise Theorem 4 is trivial), there exists a periodic function $\psi(x)$ of period 1 with the following properties.
Let α and β be real numbers satisfying the condition

$$0 \leqq \beta - \alpha \leqq 1 - 2\Delta .$$

Then

1. $\psi(x) = 1$ in the interval $\alpha \leqq x \leqq \beta$.
2. $0 \leqq \psi(x) \leqq 1$ in the intervals $\alpha - \Delta \leqq x \leqq \alpha$, $\beta \leqq x \leqq \beta + \Delta$.
3. $\psi(x) = 0$ in the interval $\beta + \Delta \leqq x \leqq 1 + \alpha - \Delta$.
4. $\psi(x)$ can be expanded as a Fourier series of the type

$$\psi(x) = \beta - \alpha + \Delta + \sum_{k=1}^{\infty} (a_k \cos 2\pi kx + b_k \sin 2\pi kx),$$

where

$$a_k \ll \frac{1}{k}, \quad b_k \ll \frac{1}{k}, \quad \text{if} \quad k < \frac{1}{\Delta},$$

$$a_k \ll \frac{1}{\Delta k^2}, \; b_k \ll \frac{1}{\Delta k^2}, \quad \text{if} \quad k \geqq \frac{1}{\Delta}.$$

2°. Using 1° and Theorem 3, we find

$$\sum_{N-A<p\leqq N} \psi(\lambda p) = (\beta - \alpha + \Delta) T + AR,$$

$$\frac{1}{\Delta k^2}\sqrt{\frac{k^4}{q^2}} = \frac{1}{\Delta\sqrt{q}}\,\frac{1}{k}\sqrt{\frac{k^2}{q}} \ll \frac{1}{k}\sqrt{\frac{k^2}{q}},$$

$$R \ll r^{\frac{\ln r}{\ln \lambda}+6} \sum_{0<k\leqq q} D_k + \sum_{k>q} \frac{1}{\Delta k^2} \ll \Delta + \frac{1}{\Delta q} \ll \Delta .$$

Hence, arguing in the same way as in my book (Chap. VIII), we find that Theorem 4 is true.

Theorem 5. (Distribution of primes with respect to a given modulus). Let η be an arbitrarily small positive constant <1, $N > 2$,

$$1 \leqq A \leqq N; \qquad 1 \leqq Q \leqq N,$$
$$0 \leqq \delta \leqq 1 .$$

Replace all primes p in the interval

$$N - A < p \leqq N$$

by their remainders P on division by Q. Let T denote the number of all these remainders (i.e. the number of all primes in the interval $N - A < p \leqq N$) and T_1 the number of those primes which lie in the interval

$$0 \leqq P \leqq \delta Q .$$

---

[1] Trudy mat. inst. Steklov **10** (1937).

Then

$$T_1 = \delta T + O(A\Delta),$$

$$\Delta = N^\eta \sqrt{\frac{N^{2/3}}{A} + \frac{NQ}{A^2} + \frac{1}{Q}}.$$

Proof. In Theorem 4, let

$$\lambda = \frac{1}{Q}.$$

Then, expressing λ as

$$\lambda = \frac{1}{q} + \frac{\theta}{q^2}; \qquad q = [Q],$$

we arrive at Theorem 5 as a direct corollary of Theorem 4.

Particular case. The law of distribution of primes with respect to a given modulus Q which can be derived from Theorem 5 is of the simplest form when

$$A \geqq N^{2/3} Q.$$

Here we have

$$T_1 = \delta T + O\left(N^\eta \frac{A}{\sqrt{Q}}\right).$$

Example 1. Let $A = N^{5/6}$, $Q = N^{1/6}$, $\delta = 1/2$. On replacing all primes in the interval

$$N - N^{5/6} < p \leqq N$$

according to our theorem (particular case) by their remainders on division by $N^{1/6}$, and denoting by T the number of all remainders, by T_1 the number of all those remainders which are $\leqq \frac{1}{2} N^{1/6}$ we obtain

$$T_1 = \tfrac{1}{2} T + O(N^{5/6 + \varepsilon - 1/12}).$$

Example 2. Let $A = N$, $Q = N^{1/3}$, $\delta = 1/2$. On replacing all primes $p \leqq N$ by their remainders on division by $N^{1/3}$, and denoting by T the number of all remainders and by T_1 the number of remainders which are $\leqq N^{1/3}$, we obtain

$$T_1 = \tfrac{1}{2} T + O(N^{1+\eta-1/6}).$$

Note. Using my method, we can prove a theorem similar to Theorem 5 even in the case where, in place of all primes in the interval

$$N - A < p \leqq N$$

we take those primes which possess certain arithmetical properties, say, for instance, those which are the quadratic residues or non-residues to some modulus, or residues or non-residues of power $n > 2$, etc.

Certain Problems in Analytic Number Theory

In: Trudy tret'ego vsesoyuznogo matematicheskogo s'ezda (Proceedings of the Third All-Union Mathematical Congress), June-July 1956, vol. 3, Reviews, Moscow Izd. Akademii Nauk SSSR, (1958) 3–13.

Notations

n denotes a positive integer, $\nu = 1/n$;
ε is an arbitrarily small positive constant;
δ is a number such that $0 < \delta \leqq 1$;
θ is a number such that $|\theta| \leqq 1$;
$\{z\}$ denotes the fractional part of a real number z;
P is a sufficiently large positive integer;
p is an odd prime;
$C = C(a, \ldots, k)$ means that C depends only on $a, \ldots, k$;
$A \ll B$ for positive B means that $|A| \leqq cB$, where c is a constant.
A point is called integral if all its coordinates are integers.

1. In this report we shall endeavour to review the state-of-the-art of certain problems in analytic number theory, which may in short be called the problems of the distribution of the values of a function for integral values of the arguments. In each problem of this kind, some set Ω consisting of a large number of integral points $(x_1, \ldots, x_r)$ in an r-dimensional space is chosen, the values of a certain function $\Phi(x_1, \ldots, x_r)$ at these points are considered, and the laws governing the distribution of these values of this function are studied. Sometimes, the laws governing the joint distribution of several functions are investigated.

2. In solving this type of problems, a need arises for finding, as accurately as possible, the estimates for trigonometric sums of the type:

$$S = \sum_{\Omega} e^{2\pi i f(x_1, \ldots, x_r)} \tag{1}$$

under the assumption that the values of the function $f(x_1, \ldots, x_r)$ are real. Finding estimates for the sums (1) is an important problem by itself.

The absolute value of each term in the sum (1) is 1. Hence the "trivial" estimate of the sum is

$$|S| \leqq P,$$

where P is the number of points in the set Ω. But, by imposing certain restrictions on the set Ω and on the function $f(x_1, \ldots, x_r)$, we may try to derive, for the sum S, an estimate of the type:

$$|S| \leqq P^{\varkappa}$$

with γ as close to zero as possible. We shall show that such a "non-trivial" estimate is attainable for quite a large classes of sums S.

3. First we shall consider sums of the type:

$$\sum_{x=N+1}^{N+P} e^{2\pi i F(x)}, \tag{2}$$

which are particular cases of the sum (1) when $r = 1$ and under the condition that the set Ω consists of P points on the OX axis with the coordinates $N+1, \ldots, N+P$. We shall examine certain important classes of the sum (2); in addition to finding the estimates for each of these classes, we shall also point out the most important applications of these estimates.

4. The simplest non-trivial case of (2) is the sum of the type:

$$S = \sum_{x=1}^{P} e^{2\pi i \frac{a}{P} x^n}, \quad (a, P) = 1. \tag{3}$$

For $n = 2$, these sums were first studied by Gauss and are therefore called Gauss sums. If n is constant, the estimate of (3) is

$$S \ll P^{1-\nu},$$

which for a general case cannot be improved any further because there exist infinitely many values of P, for each of which the corresponding sum (3) is equal to $P^{1-\nu}$.

But for a prime $P = p$, this estimate can be replaced by a more accurate one:

$$|S| \leqq (n-1)\sqrt{p}. \tag{4}$$

From this estimate we can derive the basic law for the distribution of nth power residues and non-residues modulo $p > 2$ (reported by Schur and me in 1918).

Let n be a divisor of $(p-1)$, such that $1 < n < p-1$. We split the numbers x not divisible by p into n sets such the sth set ($s = 0, 1, \ldots, n-1$) consists of those x's which satisfy the condition $\operatorname{ind} x \equiv s \pmod{n}$ (consequently, the zeroth set consists of the nth power residues, while the remaining sets consist of non-residues). The total number T of numbers in the sth set that occur among p_1 ($p_1 < p$) consecutive integers can be represented as

$$T = \nu p_1 + \theta\sqrt{p}\ln p.$$

Under these conditions and for a constant n, the estimate (4) shows that the upper bound for the difference between a certain number and the nearest member of the sth set is of order $\sqrt{p}$. It is of great interest to examine whether this order can be reduced further, say, to p^{ε}. In particular, it is of considerable importance to continue the study of the conjecture that the least nonresidue is a number of order p^{ε}.

5. The sum (3) is a particular case of the sum of the type:

$$S = \sum_{x=1}^{P} e^{2\pi i f(x)}, \quad f(x) = \alpha_n x^n + \ldots + \alpha_1 x, \tag{5}$$

where $\alpha_n, \ldots, \alpha_1$ are rational irreducible fractions with positive denominators, for which the greatest common multiple is P. For the sake of simplicity, we shall take n constant.

The estimate obtained for this sum (by Mordell for a prime $P = p$ and then by Hua Loo-Keng for the general case) is

$$S \ll p^{1-\nu+\varepsilon},$$

which for the case of $P = p$ was subsequently improved (H. Weyl) as:

$$S \ll \sqrt{p}.$$

6. Now we shall consider still more general sums:

$$S_m = \sum_{x=1}^{P} e^{2\pi i m f(x)}, \quad f(x) = \alpha_n x^n + \ldots + \alpha_1 x, \tag{6}$$

where $\alpha_n, \ldots, \alpha_1$ are now real numbers and m is a positive integer. The first general method for the estimation of these sums was developed by Weyl in 1914; therefore these sums are called Weyl sums. The idea behind the Weyl method is quite simple. It is based on the fact that $|S_m|^2$ can be represented as the sum of $<2P$ sums of the same type as the sum S_m, but in each new sum x runs through $<P$ consecutive integers from the sequence $1, \ldots, P$ and $f(x)$ is replaced by a polynomial of degree $(n-1)$ or zero.

Van der Corput and others have also played a key role in the development of the Weyl method.

If

$$\alpha_n = \frac{a}{q} + \frac{\theta}{q^2}, \quad (a, q) = 1 \quad q > 0. \tag{7}$$

applying the Weyl method, we obtain

$$S_m \ll P^{1+\varepsilon}(P^{-1} + mq^{-1} + mp^{-n+1} + qP^{-n})^{\varrho}, \quad \varrho = \frac{1}{2^{n-1}}.$$

Using this estimate, we find that under condition (7), the number $T(\sigma)$ of fractions

$$\{f(x)\}, \quad x = 1, \ldots, P,$$

less than σ is given by the expression:

$$T(\sigma) = \sigma P + O(P^{1+\varepsilon}(P^{-1} + q^{-1} + qP^{-n})^{\varrho})$$

(the law of distribution of fractional parts of the values of a polynomial).

7. In 1934 I developed a new method for solving the problems in analytic number theory. It was applied in particular to deriving new estimates for the sums (6). Various versions of my method of deriving such estimates were also suggested by van der Corput, Linnik, Hua Loo-Keng and others.

Here we shall make an attempt to state the results obtained by the application of my method in evaluating the sums (6) in a form which, while not being the most accurate, is yet as clear as possible.

Let $n \geqq 12$. Divide the points of an n-dimensional space into two classes as follows. Class I contains a point of the type:

$$\left(\frac{a_n}{q_n} + z_n, \ldots, \frac{a_1}{q_1} + z_1\right),$$

where a_i/q_i are rational irreducible fractions with positive denominators whose greatest common multiple, Q is not greater than $P^{1/3-\nu/6}$, while z_i are such that

$$|z_s| \leqq P^{-s+\nu}, \qquad s = n, \ldots, 1.$$

Class II contains every point which is not a point of Class I. Now assuming

$$\varrho = \frac{1}{40n^2 \ln 12n^2}, \qquad 0 \leqq m \leqq P^{2\varrho},$$

we find that, if $(\alpha_n, \ldots, \alpha_1)$ is a point of Class II, the estimate is

$$|S_m| < n^{2n \ln 12n^2} P^{1-\varrho}.$$

If, however, $(\alpha_n, \ldots, \alpha_1)$ is a point of Class I, then

$$S_m = \frac{1}{Q} \sum_{x=1}^{Q} e^{2\pi i \left(\frac{ma_n}{q_n} x^n + \ldots + \frac{ma_1}{q_1} x\right)} \int_0^P e^{2\pi i (mz_n x^n + \ldots + mz_1 x)} dx + \theta n^{1.5n \ln 12n^2} P^{1-\varrho}.$$

Taking n to be constant, this formulation can be simplified as follows:
Let $n \geqq 12$. Divide the points of an r-dimensional space into two classes: Class I and Class II. Here Class I contains only those points

$$\left(\frac{a_n}{q_n} + z_n, \ldots, \frac{a_1}{q_1} + z_1\right),$$

which obey a more restrictive condition than in the previous case

$$Q \leqq P^\nu, \qquad |z_s| \leqq P^{-s+\nu} Q^{-1+n^\varepsilon},$$

Class II contains, as before, every point which is not a point of Class I. In addition, let

$$u_0 = \max(|z_n P^n|, \ldots, |z_1 P|),$$

retaining the previous value for ϱ and the previous inequality for m. Now, if $(\alpha_n, \ldots, \alpha_1)$ is a point of Class II, then

$$S_m \ll P^{1-\varrho}.$$

If, however, $(\alpha_n, \ldots, \alpha_1)$ is a point of Class I, then for $u_0 > 1$, we have

$$S_m \ll PQ^{-\nu+\varepsilon} u_0^{-\nu},$$

and for $u_0 \leqq 1$:

$$S_m \ll PQ^{-\nu+\varepsilon} (m, Q)^\nu.$$

With the help of these simplified results, we find that the number $T(\sigma)$ of fractions

$$\{f(x)\}, \qquad x = 1, \ldots, P,$$

less than σ is given by the expression:

$$T(\sigma) = \sigma P + O(\Delta),$$

where $\Delta = P^{1-\varrho} \ln P$ for the points of Class II and $\Delta = PQ^{-\nu+\varepsilon_1} u_0^{-\nu}$ for the points of Class I, if $u_0 > 1$, or $\Delta = PQ^{-\nu+\varepsilon_1}$ for the points of Class I, if $u_0 < 1$.

8. It should, however, be mentioned that the estimates derived for the sums (6) by my method are far from the best possible result. Indeed, denoting the sum (6) as a function of the coefficients of the polynomial $f(x)$ by the symbol $S(\alpha_n, \ldots, \alpha_1)$ (for a given m), we can find such a $C = C(n)$ that for all points in the n-dimensional cube:

$$0 \leqq \alpha_s \leqq 1, \qquad s = n, \ldots, 1,$$

except possibly certain points belonging to some domains of total volume $\leqq P^{-0.125n^2}$, the estimate is

$$S(\alpha_n, \ldots, \alpha_1) \leqq CP^{0.96}.$$

It is of great interest to further improve the estimate for the sum (6), at least, for the points of Class II obeying the condition:

$$Q > P^{1/3-\nu}, \qquad |z_s| \leqq P^{-s+1/3}.$$

9. Now we consider the sums

$$S = \sum_{x=N+1}^{N+P} e^{2\pi i f(x)}$$

under the condition that in the interval $N \leqq x \leqq N+P$ the function $f(x)$ has a continuous $(n+1)$-th derivative $f^{n+1}(x)$ satisfying the condition:

$$\frac{1}{A} \leqq \frac{f^{(n+1)}(x)}{n!} \leqq \frac{c}{A}, \tag{8}$$

where $A \geqq 2$ and c is a constant.

10. When $n = 1$, this condition takes the form

$$\frac{1}{A} \leqq \frac{f''(x)}{2} \leqq \frac{c}{A}. \tag{9}$$

In this case the estimate of S is

$$S \ll P\left(\frac{1}{\sqrt{A}} + \frac{\sqrt{A}}{P}\right). \tag{10}$$

If, besides the condition (9), certain additional restrictions are imposed on $f(x)$, and a function $\varphi(x)$ in the interval $N \leqq x \leqq N+P$ is considered under certain restrictions, then the sum

$$S_1 = \sum_{x=N+1}^{N+P} \varphi(x)\, e^{2\pi i f(x)}$$

can be approximated by the expression:

$$S_1 = \sum_{f'(N) \leqq s \leqq f'(N+P)} \frac{1+i}{\sqrt{2}} \frac{\varphi(x_s)}{\sqrt{f''(x_s)}} e^{2\pi i(-sx_s + f(x_s))} + \lambda, \tag{11}$$

where x_s can be found from the equation $f'(x_s) = s$ and λ has a good estimate (in 1918 I published the proof of this theorem for particular cases, and van der Corput in 1920 for the general case).

11. For $n > 1$, the estimates of the sum satisfying the condition (8) were derived by van der Corput with the help of the Weyl method.

12. These results for the sums satisfying the condition (8) have been applied in deriving asymptotic expressions for the number of integral points in plane domains. For instance, the number of integral points in

$$x^2 + y^2 \leqq r^2$$

can be expressed as

$$\pi r^2 + O(\Delta).$$

Moreover, using the estimate (10), one can show that the upper bound of Δ is $r^{2/3}$ (this was first demonstrated by Serpinsky in 1906, using the Voronoi method). Applying the formula (11) and the estimate obtained when $n > 1$ for the sums with the condition (8), van der Corput replaced 2/3 by a lesser exponent which was subsequently reduced still further to $13/20 + \varepsilon$ (Hua Loo-Keng). Later, it was shown (Hardy and Littlewood, 1914) that Δ cannot be replaced by a number of an order less than $r^{1/2}$.

It is of great interest to study the question: can this value of Δ not be replaced by a lesser magnitude, say, $\Delta = r^{1/2+\varepsilon}$?

The approximate expression (11) together with certain techniques of my method are used in deriving asymptotic formulas for the number of integral points in a three-dimensional domain. For example, (my result published in 1955) the number of integral points in the domain

$$x^2 + y^2 + z^2 \leqq r^2$$

is given by the expression:

$$\tfrac{4}{3}\pi r^3 + O(\Delta_1), \qquad \Delta_1 = r^{11/8+\varepsilon}.$$

It is known (Hardy and Littlewood's method) that Δ_1 cannot be replaced by a quantity of order less than r.

It is of great interest to investigate the question: can this value of Δ_1 not be replaced by a lesser quantity, say, $\Delta_1 = r^{1+\varepsilon}$?

13. My method can also be used in deriving the estimates of sums which satisfy the condition (8); but, for the sake of convenience, we shall formulate a result concerning a slightly more general sum: Let $n \geqq 11$; in the interval $N \leqq x \leqq N + P$, let a real-valued function $f(x)$ have a continuous derivative $f^{(n+1)}(x)$ satisfying the condition (8). Furthermore, let $P \leqq A \leqq CP^2, C > 1$. In the interval $N \leqq x \leqq N + P$, also let $\varphi(x)$ be a non-negative monotonic function such that $\varphi(x) \leqq \Phi$. Then, if P_1 denotes any integer, such that $0 < P_1 \leqq P$, we have

$$\left| \sum_{x=N+1}^{N+P_1} \varphi(x)\, e^{2\pi i f(x)} \right| < \Phi\, (8n)^{0.5n(\ln 120n - 1.6)}\, c\, C^{\varrho/3}\, P^{1-\varrho},$$

$$\varrho = \frac{1}{3n^2 \ln 120n}.$$

Estimates similar to the above one found with the help of my method are used in the theory of distribution of prime numbers (refinement for the remainder term in the asymptotic expression of the number $\pi(x)$ of primes not exceeding x, refinement for the upper bound of the difference between adjacent primes proved by N. G. Chudakov, Ingnam and others).

14. We shall not discuss in detail the estimates for other classes of sums (2) and their applications. We shall only consider the sums (2) with $F(x) = \alpha k^x$, where $\alpha > 0$ and $k > 1$.

For an integer k, considerable advances have been made regarding the estimates of such sums and the distribution of fractional parts of the values of a function $F(x)$ (N.M. Korobov and others); this case will be dealt with in special communications.

For the case non-integral k, almost no progress has been made in regard to these questions. For instance, when

$$F(x) = \left(\frac{a}{b}\right)^x, \qquad (a, b) = 1, \qquad a > b > 1,$$

so far neither a proof nor a disproof has been found for the hypothesis that the number T of fractions

$$\{F(x)\}, \qquad x = 1, \ldots, P,$$

less than σ can be represented in the form:

$$T = \sigma P + P\gamma, \qquad |\gamma| \leqq \Delta,$$

where Δ does not depend on σ and tends to zero in the limit as P increases unboundedly (a and b being constants). This question has also not been solved for the function $F(x) = e^x$.

15. Of particular interest are the distribution laws for the values of a function $\Phi(x_1, \ldots, x_r)$ which takes integral values at the points $(x_1, \ldots, x_r)$ of a set Ω. Here a question arises in respect of each integer N: for how many points in the set Ω will this N be the value of the function $\Phi(x_1, \ldots, x_r)$? In other words, what is the number $I(N)$ of representations of an integer N in the form:

$$N = \Phi(x_1, \ldots, x_r)$$

under the condition that the point $(x_1, \ldots, x_r)$ belongs to the set Ω. Here different questions may be formulated: sometimes the question of the existence of representations, i.e. the inequality $I(N) > 0$, and sometimes the question of the asymptotic expression for $I(N)$.

We shall only take up a particular case where $\Phi(x_1, \ldots, x_r) = x_1^n + \ldots + x_r^n$, and $x_1, \ldots, x_r$ run through non-negative integers. Here $I(N)$ stands for the number of representations of a number N in the form:

$$N = x_1^n + \ldots + x_r^n, \qquad x_1 \geqq 0, \ldots, x_r \geqq 0. \tag{12}$$

Evidently, one can restrict himself to the values of $x_1, \ldots, x_r$ which do not exceed N^ν. We can trivially prove that for $r \leqq n$, there exist an infinitely large number of positive integers N which cannot be represented in the form (12).

It is therefore natural to raise the question: What are the values of r for which Eq. (12) is solvable for any $N > 0$, or at least for $N > c$, where c is a sufficiently large constant?

Even Lagrange had shown that every non-negative N can be represented in the form:

$$N = x_1^2 + x_2^2 + x_3^2 + x_4^2$$

and in 1770 Waring conjectured that for any $n > 2$ there exists such an $r = r(n)$ for which Eq. (12) is solvable. This conjecture is known as the Waring problem. It was proved by Hilbert in 1909.

To simplify our discussion, we shall use the symbol $G(n)$ (used by Hardy and Littlewood) to denote the least r for which there exists a c satisfying the condition that for any $N > c$, Eq. (12) is solvable. From Hilbert's result, it follows that $G(n)$ exists. From what has been said before, we infer that $G(n) > n$.

In 1919 Hardy and Littlewood published a new method for solving the Waring problem. They found that the upper bound for $G(n)$ is $\sim n2^{n-2}$, and derived an asymptotic expression for the number $I(N)$ of representations:

$$I(N) = \frac{(\Gamma(1+v))^r}{\Gamma(r^v)} N^{rv-1}\sigma + O(N^{rv-1-c_0}),$$

when $r \geqq (n-2)2^{n-1} + 5$, where c_0 is a positive constant and σ is a special series whose sum exceeds a certain positive constant.

They used the Weyl method for estimating certain sums in their studies. Later, Hua Loo-Keng, after modifying the result of Hardy and Littlewood, demonstrated that the asymptotic expression derived by Hardy and Littlewood is also valid when

$$r \geqq {}^{2n} + 1 .$$

Application of my 1934 method yields the upper bound for $G(n)$:

$$G(n) < 3n(\ln n + 11) .$$

With the increasing n, this bound increases as a quantity of the order of $n \ln n$. In this respect this bound hardly differs from the unattainable bound of n.

The asymptotic expression of Hardy and Littlewood for the number of representations has been derived with the help of my method only for the case where

$$r \geqq [10n^2 \ln n]$$

(Hua Loo-Keng improved this bound to $r \geqq r_0$; $r_0 \sim 4n^2 \ln n$).

It is of great interest to study the question: can this bound not be replaced by a bound of lesser order, say, a bound of the order of $n^{1+\varepsilon}$?

16. Besides the Waring problem, of utmost importance is the question about the number $I(N_n, \ldots, N_1)$ of solutions to the system:

$$N_n = x_1^n + \ldots + x_r^n$$
$$\ldots\ldots\ldots\ldots$$
$$N_1 = x_1 + \ldots + x_r$$

under the condition that each $x_1, \ldots, x_r$ runs through the values $1, \ldots, P$. Here, for

$$n \geqq 12, \quad r \geqq r_0, \quad r_0 = [6.5n^2 \ln 12n^2]$$

we can establish the inequality:

$$\Gamma(N_n, \ldots, N_1) \leqq \int_0^1 \ldots \int_0^1 \left| \sum_{x=1}^{P} e^{2\pi i(\alpha_n x^n + \ldots + \alpha_1 x)} \right|^r d\alpha_n \ldots d\alpha_1 < CP^{r - \frac{n(n+1)}{2}},$$
$$C = (8n)^{4.3n^3 (\ln 12n^2)^2}.$$

The first assertion of this inequality is trivial, whereas the second is proved with the help of my method.

The question of the existence of solutions, just like the question of the asymptotic expression for $I(N_n, \ldots, N_1)$, can obviously be raised only when sufficiently stringent inequalities are stipulated to restrict the numbers $N_n, \ldots, N_1$.

With the aid of my method, assuming certain such inequalities, K.K. Mardzhanishvili and Hua Loo-Keng derived an asymptotic expression for the number of representations for an r of the order of $n^2 \ln n$ ($r \sim 4n^2 \ln n$ – Hua Loo-Keng). The existence of representations would follow as a corollary from this asymptotic expression, if we could show that the sum of some sequence, which is the factor of the principal term, is greater than a certain positive constant. The upper bound for the least r satisfying such a condition has been derived by Mardzhanishvili, but it is quite large. An important problem at present is to find the bounds of a still lower order, say, bounds $\sim 4n^2 \ln n$.

Now we shall take up the following problem. From the numbers $n, \ldots, 2, 1$, choose any m numbers:

$$n_m = n, \qquad m_{m-1}, \ldots, n_2, n_1.$$

Applying my method, we can show that the number $I(N_{n_m}, \ldots, N_{n_1})$ of solutions to the system:

$$N_{n_m} = x_1^{n_m} + \ldots + x_r^{n_m},$$
$$\ldots\ldots\ldots\ldots$$
$$N_{n_1} = x_1^{n_1} + \ldots + x_r^{n_1}$$

under the condition that each $x_1, \ldots, x_r$ runs through the values $1, \ldots, P$, for a constant n, is a quantity of the order of P^{n-k}; $k = n_m + \ldots + n_1$ for all $r \geqq r_0$, where r_0 is of the order of $n^2 \ln n$.

It would be great importance to establish some other order of growth for r_0, say, much closer to k (for example, $k^{1+\varepsilon}$).

17. Here we shall examine the questions relating to the laws of distribution of the values of a function $\Phi(x_1, \ldots, x_r)$ which does not necessarily take integral values at the points of Ω.

How specific are these questions is clear from the following fact. Let $x_1, \ldots, x_r$ run through positive integers. While the number of solutions of the equation

$$x_1 \ldots x_r = N$$

is a trivial question, the number of values of the product $(x_1 + \sigma) \ldots (x_r + \sigma)$, where $0 < \sigma < 1$, which lie in a given interval of length 1 is a much more complicated problem.

18. In 1937 I discovered that many of the sums over primes can be formed, only using addition and subtraction, from a comparatively small number of other sums, whose

estimates could be readily derived with the aid of techniques (described elsewhere in this report) which do not pertain to the theory of prime numbers. By such a technique, it is possible to estimate a large number of sums over primes: in particular, we can find an estimate for the sum:

$$\sum_{p \leqq N} e^{2\pi i F(p)} \tag{13}$$

under approximately the same conditions as those imposed earlier on the function $F(x)$. The rest of this communication is devoted to a consideration of two important classes of the sums (13) and the application of the estimates found for the sums in each class to questions relating to the distribution of primes.

19. First consider the simplest sum of the type (13) (when $F(x) = \alpha x$, which was previously regarded trivial)

$$S = \sum_{p \leqq N} e^{2\pi i \alpha p}.$$

Let

$$H = e^{0.5\sqrt{\ln N}}, \qquad \alpha = \frac{a}{q} + \frac{\theta}{q^2}, \qquad (a, q) = 1, \qquad 1 < q < N.$$

Then, for the sum S we obtain the bound

$$S \ll N(\ln N)^{4.5}\left(\sqrt{\frac{1}{q} + \frac{q}{N} + \frac{1}{H}}\right). \tag{14}$$

This bound together with certain theorems on the distribution of primes in arithmetic progressions (Page, Esterman, Siegel, Walfisz) have been used first in solving the Goldbach problem for sufficiently large odd N, i.e. in proving that any large number N can be represented as the sum of three primes:

$$N = p_1 + p_2 + p_3,$$

and in deriving an asymptotic expression for the number of such representations. A tentative proof based on certain hypotheses of the theory of L-series was published by Hardy and Littlewood in 1914. Later, in 1946 Linnik gave a rigorous proof based only on the established facts of the theory of L-series.

Furthermore, using this estimate, the questions relating to the representation of a sufficiently large integer N in the form:

$$N = K_1 p_1 + K_2 p_2 + K_3 p_3,$$

where K_1, K_2, K_3 are non-zero integers were studied in detail (van der Corput). It was shown (simultaneously by seven different authors) that "almost all" positive even numbers are the sums of two primes and that "almost all" such numbers are the differences of two primes (van der Corput).

What remains unsolved is the Goldbach problem for even numbers, i.e. the representation of all sufficiently large even numbers as the sum of two primes. Of great interest is also a more general question concerning the representation of numbers in the form $N = K_1 p_1 + K_2 p_2$, where K_1 is a positive integer, while K_2 is a positive or negative integer.

The bound (14) for the sum (13) can be improved by a far more accurate estimate with the help of which we can derive the following theorem.

Let q be an integer such that $1 < q \leqq N$ and l run through the least non-negative residues of primes $p \leqq N$ modulo q. Then, the number $T(\sigma)$ of the values of l less than σq can be represented in the form:

$$T(\sigma) = \sigma\pi(N) + O(\Delta), \qquad \Delta = N^{1+\varepsilon}\left(\sqrt{\frac{1}{q} + \frac{q}{N}} + N^{-0.2}\right).$$

20. Now we shall find an estimate for a sum of the type (13) for a more complicated case where the sum is similar to the Weyl sum. Let n be a constant $\geqq 12$,

$$f(x) = \alpha_n x^n + \ldots + \alpha_1 x,$$

and let each α_s be represented as

$$\alpha_s = \frac{a_s}{q_s} + \frac{\theta_s}{q_s \tau_s}, \qquad 0 < q_s \leqq \tau_s, \qquad \tau_s = P^{0.5s}.$$

Let Q denote the least common multiple of the numbers $q_n \ldots, q_1$, and χ_1 be given by the equation $Q = P^{\chi_1}$, and χ by the equation

$$\chi = \min\left(\tfrac{1}{9}, \chi_1\right).$$

Then, assuming

$$S_m = \sum_{p \leqq N} e^{2\pi i m f(p)}, \qquad \varrho = \frac{\chi v^2}{6.75 \ln 108 n^2},$$

for an integer m such that $0 < m \leqq P^{2\varrho}$, we obtain the bound

$$S_m \ll P^{1-\varrho+\varepsilon}. \tag{15}$$

Using this bound, for the number $T(\sigma)$ of fractions of the type:

$$\{f(p)\}, \qquad p \leqq N,$$

less than σ, we obtain the following expression:

$$T(\sigma) = \sigma\pi(N) + O(P^{1-\varrho_1+\varepsilon}), \qquad \varrho_1 = \frac{\chi v^2}{7 \ln 108 n^2}.$$

The estimate (15) can be supplemented with particular estimates for the cases where the bound is only a rough estimate. We can also supplement it with the estimates for the cases $n < 12$. All these estimates find application in the representation of an integer N in the form:

$$N = \Phi(p_1, \ldots, p_r),$$

where $p_1, \ldots, p_r$ run through primes and $\Phi(p_1, \ldots, p_r)$ takes only integral values. The simplest particular cases of this problem are the Goldbach problem and others which we have already mentioned.

Here we shall only dwell on the representation of N in the form:

$$N = p_1^n + \ldots + p_r^n \tag{16}$$

under the assumption that $n > 1$.

This question has been studied in detail for the case $n = 2$. It has been found that when

$$r > 5, \qquad r \equiv N \pmod{24}$$

any sufficiently large N can be represented in the form:

$$N = p_1^2 + \ldots + p_r^2,$$

and an asymptotic expression can be derived for the numbers of such representations. When $n > 2$, the part played by 24 will here be played by a certain K entirely determined by the number n; and if the condition

$$r > r_0, \qquad r \equiv N (\bmod K),$$

is satisfied, where r_0 is some number $\sim 4n \ln n$, then any sufficiently large N can be represented in the form (16) (Hua Loo-Keng).

For this case, an asymptotic expression can be derived only if r is a quantity of an order not less than $n^2 \ln n$.

As regards the question about the number of solutions of the system

$$\begin{aligned} N_n &= p_1^n + \ldots + p_r^n \\ &\ldots\ldots\ldots \\ N_1 &= p_1 + \ldots + p_r \end{aligned}$$

I should also like to say that as much progress has been made here as for a similar question discussed in the previous pages (Mardzhanishvili, Hua Loo-Keng).

Finally, we note that for some types of the functions $f(x)$, the asymptotic expressions for the number $T(\sigma)$ of fractions

$$\{f(p)\}, \qquad p \leqq N,$$

less than σ give rise to particular theorems, an example of which is the following.

The number T of primes $p \leqq N$ lying in intervals of the type

$$(2t-1)^2 < p \leqq (2t)^2, \qquad t = 1, 2, \ldots,$$

is given by the expression:

$$T = \tfrac{1}{2}\pi(N) + O(N^{0.9+\varepsilon}).$$

21. As an example of sums with primes, which are different from (13), and are estimated with the aid of my 1937 method, we can mention the sum

$$S = \sum_{p \leqq N} \chi(p+k),$$

where χ is the non-principal character with respect to an odd prime modulus q, and k is a constant not divisible by q. For this sum the following estimate

$$S \ll N^{1+\varepsilon}\Delta, \qquad \Delta = \sqrt{\frac{1}{q} + \frac{q}{N} + N^{-1/6}},$$

has been obtained, and if $q^{3/4} \ll N \ll q^{5/4}$, as Δ we can take

$$\Delta = (q^{3/4}/N)^{1/3}.$$

On the Distribution of the Fractional Parts of Values of a Polynomial

Izv. Akademii Nauk SSSR, Ser. Mat., **25** no. 6 (1961) 749–754.

Notations. Let n denote an integer $\geqq 4$; $\nu = 1/n$; m a positive integer; $\alpha_n, \ldots, \alpha_1$ real numbers; $f(x) = \alpha_n x^n + \ldots + \alpha_1 x$.

For the real numbers A and B satisfying the condition $0 \leqq B - A \leqq 1$ the notation $A \leqq z < B \pmod 1$ means that for some integer d, we have $A + d \leqq z < B + d$.

$p > 1$, x runs through an increasing sequence (x) of integers (for instance, the natural numbers, primes, etc.), p_1 is the number of all terms in the sequence (x) satisfying the condition $0 < x \leqq p$, while $p_1(A, B)$ is the number of those terms satisfying the same condition but for which $A \leqq f(x) < B \pmod 1$. Finally,

$$T_m = \sum_{0 < x \leqq p} e^{2\pi i m f(x)}.$$

The aim of this paper is to prove a certain general theorem which shows that for a sufficiently small positive δ less than 0.5, from an n-dimensional cube $0 < \alpha_n \leqq 1; \ldots;$ $0 < \alpha_1 \leqq 1$, it is possible to choose a domain which contains all the points $(\alpha_n, \ldots, \alpha_1)$, without any exception, of this n-dimensional cube with the condition that for any A and B, the inequality

$$|p_1(A, B) - (B - A)p_1| \geqq 3p^{1-\delta},$$

is satisfied and the volume of this domain rapidly tends to zero as p increases unboundedly.

Lemma 1. Let l be a positive integer,

$$D_l = 2^{2.9n^2 l + nl^2} n^{n^2 l}, \qquad b_l = \left[nl + \frac{n(n+1)}{4} + 1 \right].$$

Then

$$\int_0^1 \cdots \int_0^1 |T_1|^{2b_l} \, d\alpha_n \ldots d\alpha_1 < D_l p^{2b_l - \frac{n(n+1)}{2}(1 - (1-\nu)^l)}.$$

Proof. Let $P_l = (2n)^{2n(1-\nu)^{-l}}$. For $p \geqq P_l$, this lemma is a modification of the theorem established in [1]. It can be derived from the formulas stated in the last four lines on page 583 of [1]. Here we shall prove this lemma in full, using the method of induction.

Assume that the lemma is true for a certain l. Then for $p \geqq P_{l+1}$ we have

---

[1] On a multiple integral, Izv. akad. nauk SSSR (ser. mat.) **22** (1958) 577–584.

$$\int_0^1 \dots \int_0^1 |T_1|^{2b_{l+1}}\, d\alpha_n \dots d\alpha_1 \leqq p^{2n} \int_0^1 \dots \int_0^1 |T_1|^{2b_l}\, d\alpha_n \dots d\alpha_1$$
$$< D_l p^{2b_{l+1} - \frac{n(n+1)}{2}(1-(1-\nu)^l)},$$

where

$$\frac{D_l p^{2b_{l+1} - \frac{n(n+1)}{2}(1-(1-\nu)^l)}}{D_{l+1} p^{2b_{l+1} - \frac{n(n+1)}{2}(1-(1-\nu)^{l+1})}} < \frac{P_{l+1}^{((n+1)/2\,(1-\nu))\,(1-\nu)^{l+1}}}{2^{2.9n^2+n(2l+1)}\, n^{n^2}} < 1\,.$$

Therefore the lemma is true if $(l+1)$ is substituted for l. But the lemma is obviously true for $l=1$. Indeed, for $p \leqq P_1$, we find

$$\int_0^1 \dots \int_0^1 |T_1|^{2b_1}\, d\alpha_n \dots d\alpha_1 \leqq p^{2b_1},$$

and we have

$$\frac{p^{2b_1}}{D_1 p^{2b_1 - \frac{n(n+1)}{2}(1-(1-\nu))}} \leqq \frac{P_1^{(n+1)/2}}{2^{2.9n^2+n}\, n^{n^2}} < 1\,.$$

Consequently, the lemma is true for any l.

Lemma 2. Let $k \geqq 1$, $h = \ln k$, $\delta_0 = 1/4k$,

$$D' = (2^{2.9h+h^2} n^h)^{n^3}, \qquad \lambda' = 1 - \frac{h+1.3}{k}.$$

Then from an n-dimensional cube $0 < \alpha_n \leqq 1; \dots; 0 < \alpha_1 \leqq 1$, it is possible to choose a domain Ω' of volume not exceeding

$$D' p^{-\lambda' \frac{n(n+1)}{2}},$$

and containing all points $(\alpha_n, \dots, \alpha_1)$, without any exception, of this n-dimensional cube such that

$$|T_1| \geqq p^{1-\delta_0}. \tag{1}$$

Proof. For $k \leqq 1.96$ the lemma is trivial, because in this case we have $\lambda' < 0$. Therefore we shall assume that $k > 1.96$. Let $l = [nh]$, $\sigma = nh - l$. The n-dimensional cube stated in the lemma can be divided into identical n-dimensional parallelepipeds with such small length of sides that the modulus of the difference between the values $T_1^{2b_l}$ corresponding to two points in the same parallelepiped is always not greater than the difference between the right-hand and left-hand sides of the inequality of Lemma 1. We shall take Ω' to be a domain formed by all those small parallelepipeds which contain at least one point satisfying the condition (1). Representing the volume of the domain Ω' in the form of $D'R$, by Lemma 1, we find

$$D' R p^{2b_l - 2b_l \delta_0} \leqq D_l p^{2b_l - \frac{n(n+1)}{2}(1-(1-\nu)^l)},$$

$$R \leqq p^{-\frac{n(n+1)}{2}(1-G)}, \qquad G = (1-\nu)^l + \frac{4b_l \delta_0}{n(n+1)},$$

$$G < \frac{1}{k(1-\nu)^\sigma} + \left(\frac{4nh}{n+1} - \frac{4\sigma}{n+1} + 1 + \frac{1}{n+1}\right)\delta_0$$

$$< \frac{1}{k(1-\nu)} + \left(4h - \frac{3+4h}{n+1} + 1\right)\delta_0 < \frac{h+1.3}{k},$$

$$1 - G > \lambda', \qquad D'R < D'p^{-\lambda'\frac{n(n+1)}{2}}.$$

Example. Take $k = 2.5$. Then we find that the volume of the domain Ω' containing all points of the n-dimensional cube $0 < \alpha_n \leqq 1; \ldots; 0 < \alpha_1 \leqq 1$, satisfying the condition

$$|T_1| \geqq p^{0.9},$$

is equal to

$$< (12n)^{n^3} p^{-0.1\frac{n(n+1)}{2}}.$$

Lemma 3. Let α and β be real numbers,

$$0 < \Delta < 0.5, \qquad \Delta \leqq \beta - \alpha \leqq 1 - \Delta.$$

Then there exists a periodic function $\psi(z)$ of period 1 and satisfying the conditions:

1. $\psi(z) = 1$ in the interval $\alpha + 0.5\Delta \leqq z \leqq \beta - 0.5\Delta$;
2. $0 \leqq \psi(z) \leqq 1$ in the interval $\alpha - 0.5\Delta \leqq z \leqq \alpha + 0.5\Delta$; and in the interval $\beta - 0.5\Delta \leqq z \leqq \beta + 0.5\Delta$;
3. $\psi(z) = 0$ in the interval $\beta + 0.5\Delta \leqq z \leqq 1 + \alpha - 0.5\Delta$;
4. $\psi(z)$ can be expanded as a Fourier series of the type

$$\psi(z) = \beta - \alpha + \sum_{m=1}^{\infty} (a_m \cos 2\pi mz + b_m \sin 2\pi mz),$$

where

$$|a_m| < \frac{2}{\pi m}, \qquad |b_m| < \frac{2}{\pi m},$$

$$|a_m| < \frac{2}{\pi^2 m^2 \Delta}, \quad |b_m| < \frac{2}{\pi^2 m^2 \Delta}.$$

Proof. This lemma is a corollary of Lemma 12 proved in Chap. I of book[1] for the case of $r = 1$.

Theorem. Let $k \geqq 1$, $h = \ln k$, $\delta = 1/5k$

$$D = (2^{3h+h^2} n^h)^{n^3}, \qquad \lambda = 1 - \frac{h+1.35}{k}.$$

Then from an n-dimensional cube $0 < \alpha_n \leqq 1; \ldots; 0 < \alpha_1 \leqq 1$, it is possible to choose a domain Ω of volume not exceeding

$$Dp^{-\lambda\frac{n(n+1)}{2}},$$

[1] The method of trigonometrical sums in the theory of numbers, Trudy, mat. inst. Steklov **23** (1947).

and containing all points $(\alpha_n, \ldots, \alpha_1)$, without exception, of this n-dimensional cube with the following condition: for these points, for any A and B satisfying the condition $0 \leqq B - A \leqq 1$, the inequality

$$|p_1(A, B) - (B - A)p_1| \geqq 3p^{1-\delta}$$

is satisfied.

Proof. We shall use Lemmas 2 and 3 and the notations adopted therein. Now let

$$t = p^{\delta}, \qquad \Delta = \frac{1}{6t}, \qquad M = \frac{6t}{\pi}, \qquad M_1 = \frac{60t^2}{\pi}.$$

For $k \leqq 2$, the theorem is trivial, because in this case $\lambda < 0$. Therefore we shall take that $k > 2$. Construct the domain Ω as follows. For a given m, take a figure similar to an n-dimensional cube and divide it into n-dimensional elementary parallelepipeds as in the proof of Lemma 2, but here take the length of its sides to be $1/m$ of the previous length. The volumes of the new cube, of each new elementary parallelepiped and of the domain Ω_m into which the domain Ω' is converted will be $1/m^n$ of the previous volumes. The previous n-dimensional cube can be divided into m^n such new cubes. The set of the domains Ω_m of all these new cubes shall be called the domain Ω'_m. The volume of the domain Ω'_m is equal to the volume of the domain Ω'. The domain Ω shall be considered as the common part of the domains Ω'_m corresponding to the values of m which do not exceed M_1. The volume of the domain Ω is

$$\leqq D' p^{-\lambda' \frac{n(n+1)}{2}} M_1$$

and consequently, it will not exceed the bound stated in the lemma, because

$$\frac{60}{\pi} < 2^{4.3} < 2^{0.1hn^3}, \qquad t^2 \leqq p^{\frac{1}{20k}\frac{n(n+1)}{2}}.$$

Let $0 \leqq B - A \leqq 1 - 2\Delta$ and assume that the point $(\alpha_n, \ldots, \alpha_1)$ of the n-dimensional cube $0 < \alpha_n \leqq 1; \ldots; 0 < \alpha_1 \leqq 1$, does not belong to the domain Ω. Then, for $m \leqq M_1$, we have

$$|T_m| < p^{1-\delta_0}, \qquad \left|\sum_{0<x\leqq p} \cos 2\pi m f(x)\right| + \left|\sum_{0<x\leqq p} \sin 2\pi m f(x)\right| < \sqrt{2}p^{1-\delta_0},$$

and for $m > M_1$, the right-hand side of this inequality can be replaced by $\sqrt{2}p$. Taking $\alpha = A - 0.5\Delta$, $\beta = B + 0.5\Delta$, and since $\beta - \alpha = B - A + \Delta$, by Lemma 3, we obtain

$$\sum_{0<x\leqq p} \psi(f(x)) - (B-A)p_1$$

$$< p\Delta + \sqrt{2}p^{1-\delta_0}\left(\sum_{0<m\leqq M}\frac{2}{\pi m} + \sum_{M<m\leqq M_1}\frac{2}{\pi^2 m^2 \Delta} + \sum_{m>M_1}\frac{2t}{\pi^2 m^2 \Delta}\right)$$

$$< \frac{\sqrt{8}}{\pi}p^{1-\delta_0}\left(\frac{\pi}{6\sqrt{8}} + 1 + \int_1^M \frac{dm}{m} + \int_M^{M_1}\frac{dm}{\pi m^2 \Delta} + \frac{t}{\pi M_1^2 \Delta} + \int_{M_1}^{\infty}\frac{tdm}{\pi m^2 \Delta}\right)$$

$$< \frac{\sqrt{8}}{\pi}p^{1-\delta_0}\left(\frac{\pi}{6\sqrt{8}} 1 + \ln t + \ln\frac{6}{\pi} + 1 + \frac{1}{10}\right) < \frac{\sqrt{8}}{\pi}p^{1-\delta_0}(\ln t + 2.94).$$

But it is readily seen that

$$\frac{\sqrt{8}}{\pi}\,\frac{\ln t + 2.94}{t^{0.2}} \leqq \frac{\sqrt{8}}{\pi}\,\frac{6}{e^{0.412}} < 3\,, \qquad p^{-\delta_0}\,t^{0.2} = p^{-\delta}.$$

Hence

$$p_1(A, B) < (B - A)\,p_1 + 3p^{1-\delta}. \tag{2}$$

Similarly, for $2\Delta \leqq B - A \leqq 1$, assuming that $\alpha = A + 0.5\Delta$ and $\beta = B - 0.5\Delta$, we obtain

$$p_1(A, B) > (B - A)\,p_1 - 3p^{1-\delta}. \tag{3}$$

Therefore, for $2\Delta \leqq B - A \leqq 1 - 2\Delta$, we have

$$|p_1(A, B) - (B - A)\,p_1| < 3p^{1-\delta}. \tag{4}$$

For $0 \leqq B - A \leqq 2\Delta$, from the inequality $0 \leqq p_1(A, B)$ and (2), and for $1 - 2\Delta \leqq B - A \leqq 1$, from the inequality $p_1 \geqq p_1(A, B)$ and (3), we once again obtain the inequality (4).

Example. Taking $k = 3$, we find that the volume of the domain containing all the points $(\alpha_n, \ldots, \alpha_1)$ of the n-dimensional cube $0 < \alpha_n \leqq 1; \ldots; 0 < \alpha_1 \leqq 1$, for which, for any A and B satisfying the condition $0 \leqq B - A \leqq 1$, the inequality

$$p_1(A, B) - (B - A)\,p_1 \geqq 3p^{1-1/15},$$

is satisfied, does not exceed

$$(24n^{1.1})^{n^3}\,p^{-\frac{1}{6}\frac{n(n+1)}{2}}.$$

Note 1. Lemmas 1 and 2 as well as the theorem remain true even when the function $f(x)$ is replaced by the sum $f(x) + F(x)$, where $F(x)$ is any given function which takes real values for x's belonging to the sequence (x). This statement follows from the trivial fact that after the substitution used in Lemma 1, the multiple integral remains the same as before or even decreases.

Moreover, Lemmas 1 and 2 as well as the theorem, under a proper new interpretation of the symbols p_1 and $p_1(A, B)$, remain true even when the sum T_m is replaced by the sum:

$$\sum_{0<x\leqq p} \tau(x)\,e^{2\pi i m f(x)},$$

where $\tau(x)$ is any given function which takes values such that $|\tau(x)| \leqq 1$ for x's belonging to the sequence (x).

Note 2. A result similar to that obtained in the example for Lemma 2 was mentioned by me earlier in the Introduction to my book[1].

---

[1] The method of trigonometrical sums in the theory of numbers, Trudy mat. inst. Steklov **23** (1947).

Section II

The Method of Trigonometric Sums in Number Theory

Second edition: revised and enlarged. Moscow, Nauka (1980) 144p.

Contents

Notation

The reader is supposed to be acquainted with my course in the theory of numbers and the symbols used in it. In addition, we shall use the following symbols:
c – a positive constant,
θ – a number whose modulus is not greater than unity,
ε – an aribitrarily small positive constant less than unity,
P – an integer greater than unity.

For the constants $h, h_1, \ldots, h_k$, the notation $h = h(h_1, \ldots, h_k)$ means that the value of h is completely determined by the values of $h_1, \ldots, h_k$.

If A is dependent on other quantities we shall sometimes denote A in a more explicit form as $A(a_1, \ldots, a_n)$ showing the quantities in brackets which vary in the problem under consideration.

For positive B the expression $A \ll B$ means that $|A| \leqq cB$. The expression $A = O(B)$ has the same meaning.

The symbol (h) stands for the difference between a real number h and its nearest integer, i.e. the smaller of $\{h\}$, $1 - \{h\}$.

For real a and b the expression $a \equiv b$ denotes $a = b + l$, where l is an integer. Here, a is said to be congruent to b, or a and b are congruent.

n is a positive integer, and $\nu = 1/n$.

Two points of an n-dimensional space will be called congruent, if their coordinates are congruent.

A point is said to be integral, if all its coordinates are integers.

If $0 < B - A \leqq 1$, the notation $A \leqq g < B \pmod 1$ shows that g is congruent to a certain number g_1 satisfying the condition:

$$A \leqq g_1 < B.$$

The symbol $\sum_a$ shows that the sum is taken over the values of a shown.

The symbol $\overset{M}{\sum} H$ stands for the product of a positive number A and the sum of not more than B terms of H of the type shown, provided $AB = M$. This symbol, in particular, may be used to denote a sum of not more than M terms H of the type shown.

The punctuation mark ";" inserted between two formulas stands for the word "where".

Introduction

Gauss was the first to consider simple trigonometrical sums and to demonstrate the use of these sums as a tool in solving problems in the theory of numbers. In particular, he made a comprehensive study of the properties of the so-called Gauss sum:

$$S\left(\frac{ax^2}{P}\right) = \sum_{x=1}^{P} e^{2\pi i \frac{ax^2}{P}}; \qquad (a, P) = 1 .$$

Using these properties, he built up his proof of the reciprocity law for quadratic residues.

Subsequently, trigonometrical sums, naturally of a much more general type, became a powerful technique in solving several important problems in the theory of numbers. The central point in regard to such sums was then to find their estimates as accurately as possible (i.e. the most exact upper bound for their moduli).

For the Gauss sum, this problem was solved by Gauss himself; he gave the following accurate expressions for the modulus of this sum:

$$\left|S\left(\frac{ax^2}{P}\right)\right| = \sqrt{P}, \quad \text{if} \quad P \equiv 1 \pmod 2;$$

$$\left|S\left(\frac{ax^2}{P}\right)\right| = 0, \quad \text{if} \quad P \equiv 2 \pmod 4;$$

$$\left|S\left(\frac{ax^2}{P}\right)\right| = \sqrt{2P}, \quad \text{if} \quad P \equiv 0 \pmod 4 .$$

The idea underlying the proof of these expressions is very simple; to illustrate it let us consider the case $P \equiv 1 \pmod 2$. Here we find

$$\left|S\left(\frac{ax^2}{P}\right)\right|^2 = \sum_{y=1}^{P} \sum_{x=1}^{P} e^{2\pi i \frac{a(y^2-x^2)}{P}} = \sum_{h=0}^{P-1} S_h ,$$

where S_h denotes the sum of the terms of a double sum such that $y \equiv x + h \pmod P$ and therefore, can be expressed as

$$S_h = \sum_{x=1}^{P} e^{2\pi i \frac{a(2xh+h^2)}{P}} .$$

Obviously, $S_h = P$ if $h = 0$, and $S_h = 0$, if $h > 0$. Therefore

$$\left|S\left(\frac{ax^2}{P}\right)\right|^2 = P, \quad \left|S\left(\frac{ax^2}{P}\right)\right| = \sqrt{P} .$$

The Gauss sum is a particular of the more general "rational trigonometrical sum":

$$S\left(\frac{\varphi(x)}{P}\right) = \sum_{x=1}^{P} e^{2\pi i \frac{\varphi(x)}{P}}, \tag{1}$$

where $\varphi(x) = a_n x^n + \ldots + a_1 x$ is a polynomial of degree $n > 1$ under the condition $(a_n, \ldots, a_1, P) = 1$.

For a prime $P = p$, Mordell gave the following estimate for this sum

$$\left| S\left(\frac{\varphi(x)}{P}\right) \right| < np^{1-\underline{\nu}},$$

which, using one of Hasse's ideas, Weyl replaced by

$$\left| S\left(\frac{\varphi(x)}{P}\right) \right| < n\sqrt{p}.$$

Weyl's estimate, as regards the growth of the order (for a constant n) with increasing p, cannot be, in general, improved: we can show infinitely many cases where the modulus of the sum is not less than $\sqrt{p}$. Indeed, for $n = 2$, any Gauss sum under the condition $p \equiv 1 \pmod 2$ has a modulus equal to $\sqrt{p}$. And for any $n > 2$ and every p such that $n \backslash p - 1$, at least one of the sums of the type:

$$S\left(\frac{ax^n}{P}\right); \qquad 0 < a < p$$

has a modulus even greater than $\sqrt{p}$.

The proof of this assertion is quite simple: we can easily derive the identity:

$$\sum_{a=1}^{p-1} \left| S\left(\frac{ax^n}{p}\right) \right|^2 = \sum_{a=1}^{p} \sum_{y=1}^{p} \sum_{x=1}^{p} e^{2\pi i \frac{a(y^n - x^n)}{p}} - p^2,$$

hence, noting that the congruence $y^n \equiv x^n \pmod p$, for given x, has n solutions, if x is not divisible by p, and one solution, if x is divisible by p, we obtain:

$$\sum_{a=1}^{p-1} \left| S\left(\frac{ax^n}{p}\right) \right|^2 = p(p-1)n + p - p^2 = p(p-1)(n-1).$$

Consequently, for at least one value of a satisfying the condition $0 < a < p$, we have

$$\left| S\left(\frac{ax^n}{p}\right) \right|^2 \geqq p(n-1), \quad \left| S\left(\frac{ax^n}{p}\right) \right| \geqq \sqrt{(n-1)p} > \sqrt{p}.$$

The best estimate of the sum (1) for a composite number P has been found by Hua. He established that

$$\left| S\left(\frac{\varphi(x)}{P}\right) \right| \leqq c(n) P^{1-\underline{\nu}}.$$

This inequality is remarkable in that, for a constant n, as regards the order of the growth of the right-hand side with increasing P, it cannot be improved any further. This follows directly from the existence, for any prime p such that $(n, p) = 1$, of the sums

$$S\left(\frac{ax^n}{p^n}\right); \qquad (a, p) = 1,$$

each of which, as we shall show below, is equal to $P^{1-\nu}$; $P = p^n$. Indeed, assuming

$x = y + zp^{n-1}$, we obtain

$$S\left(\frac{ax^n}{p^n}\right) = \sum_{y=1}^{p^{n-1}} \sum_{z=0}^{p-1} e^{2\pi i \frac{a(y^n + ny^{n-1}zp^{n-1})}{p^n}} = \sum_{y=1}^{p^{n-1}} S_y;$$

$$S_y = \sum_{z=0}^{p-1} e^{2\pi i \left(\frac{ay^n}{p^n} + \frac{any^{n-1}z}{p}\right)}.$$

But, $S_y = p$ if y is divisible by p, and $S_y = 0$ in the contrary case. Therefore,

$$S\left(\frac{ax^n}{p^n}\right) = p^{n-2}p = p^{n-1} = P^{1-\nu}.$$

The rational trigonometrical sum is a particular case of a still more general class of sums of the type:

$$T = T(\alpha_n, \ldots, \alpha_1) = \sum_{0<x\leqq P} e^{2\pi i f(x)};$$
$$f(x) = \alpha_n x^n + \ldots + \alpha_1 x, \tag{2}$$

where $\alpha_n, \ldots, \alpha_1$ are arbitrary real numbers. The first general method for finding non-trivial estimates of sums (2) was given by H. Weyl (long before Mordell and Hua derived their results). These sums therefore are called Weyl sums.

The idea behind Weyl's method is quite simple. It is based on the following identity (under the assumption that $y = x + h$):

$$|T|^2 = \sum_{y=1}^{P} \sum_{x=1}^{P} e^{2\pi i (f(y) - f(x))} = \sum_{h=-P+1}^{P-1} S_h;$$

$$S_h = \sum_{\max(1, 1-h) \leqq x \leqq \min(P, P-h)} e^{2\pi i (f(x+h) - f(x))},$$

which expresses $|T|^2$ as the sum of less than $2P$ terms of S_h, where S_h is a sum similar to but simpler than T. Indeed, the number of terms in the sum S_h does not exceed P, and the exponent of each term contains, in place of $f(x)$, the difference

$$f(x+h) - f(x) = n\alpha_n h x^{n-1} + \ldots,$$

which is zero if $h = 0$, and a polynomial of degree $(n-1)$, if $h \neq 0$.

The estimate of the sum T obtained by the Weyl method can be formulated as the following theorem:

Let

$$\alpha_n = \frac{a}{q} + \frac{\theta t}{q^2}; \quad (a, q) = 1, \quad 0 < q \leqq P^n; \quad 1 \leqq t \leqq q.$$

Then, for a certain $c(n, \varepsilon)$ greater than unity, we have

$$|T| \leqq c(n, \varepsilon)\, P^{1+\varepsilon} (P^{-1} + tq^{-1} + tP^{-n+1} + qP^{-n})^{\varrho_0};$$
$$\varrho_0 = \frac{1}{2^{n-1}}. \tag{3}$$

The main demerit of this estimate is that its accuracy becomes poorer and poorer with increasing n. Indeed, the right-hand side of (3) is far greater than $P^{1-\varrho_0}$, whose exponent tends rapidly to unity with increasing n, since ϱ_0, being of the order of 2^{-n} in n, tends rapidly to zero.

Nonetheless, this estimate played a key role in the development of the theory of numbers: it was instrumental in obtaining the first (though far from perfect) solutions to several important problems in the number theory.

One such problem was the distribution of fractional parts of the values of a polynomial $f(x) = \alpha_n x^n + \ldots + \alpha_1 x$. Its solution was found in the form: Let

$$\alpha_n = \frac{a}{q} + \frac{\theta}{q^2}; \qquad (a, q) = 1, \quad 0 < q \leqq P^n,$$

and σ be an arbitrary number satisfying the condition $0 < \sigma \leqq 1$. On representing the number $D(\sigma)$ of the terms in the sequence $1, \ldots, P$, such that $0 \leqq \{f(x)\} < \sigma$, as

$$D(\sigma) = P\sigma + \lambda(\sigma), \tag{4}$$

we obtain an inequality of the type

$$|\lambda(\sigma)| \leqq c(n, \varepsilon_1)\, P^{1+\varepsilon_1} (P^{-1} + q^{-1} + qP^{-n})^{\varrho_0}; \qquad \varrho_0 = \frac{1}{2^{n-1}}.$$

for the number $\lambda(\sigma)$.

Another problem, which was solved with the help of estimate (3) is the Waring problem. In 1770 Waring asserted, that for every $n > 1$, there exists an $r = r(n)$ such that any positive integer N can be expressed in the form:

$$N = x_1^n + \ldots + x_r^n \tag{5}$$

with non-negative integers $x_1, \ldots, x_r$. This assertion was called Waring's problem and it was solved by D. Hilbert in 1909.

In 1919 Hardy and Littlewood discovered a new technique for solving the Waring problem incomparably more general and accurate than the Hilbert method. In this method the estimate of the sum T obtained by Weyl's method plays the key role. In the technique developed by Hardy and Littlewood Waring's problem is given a more complete and comprehensive formulation, than as a mere question of the existence of representations of a number N in the form (5). In particular, they studied the function $G(n)$, which is the least r for which all integers N beginning from a certain number can be represented in the form (5). For this function they derived the inequalities

$$n < G(n) \leqq n 2^{n-1} h; \qquad \lim_{n=\infty} h = 1. \tag{6}$$

The most important result is that, if

$$r > (n-2)\, 2^{n-1} + 5, \tag{7}$$

Hardy and Littlewood derived an asymptotic expression in the form:

$$I(N) = \frac{(\Gamma(1+\nu))^n}{\Gamma(r\nu)} N^{r\nu - 1}\, \mathfrak{S} + O(N^{r\nu - 1 - c(n,r)}), \tag{8}$$

for the number $I(N)$ of representations of N in the form (5), where $\mathfrak{S}$ is a certain "singular series" whose sum, as they showed, is greater than a certain number $c_1(n, r)$. Thereafter, they used a scheme quite similar in substance to my method but more simple in form. They investigated a number of new additive problems, which may be called "generalizations of Waring's problem", for example, the problem formulated by Hilbert in 1900 on the solvability of the system

$$\begin{aligned} x_1^n + \ldots + x_r^n &= N_n, \\ \ldots\ldots\ldots\ldots \\ x_1 + \ldots + x_r &= N_1 \end{aligned} \tag{9}$$

in non-negative integers $x_1, \ldots, x_r$ (for a sufficiently large $r = r(n)$ under more strigent natural inequalities bounding the numbers $N_n, \ldots, N_1$). In 1937 Mardzhanishvili was the first to derive an asymptotic expression for the number of solutions of this system for $r > 2^{2n} n!(n+1)^3$! The existence of solutions for a sufficiently large $r = r(n)$ was demonstrated by Kamke in 1921 by extending Hilbert's method for solving Waring's problem.

Finally, the estimate (3) for the sum T, and the estimates derived by J. van der Corput (by appropriately modifying the Weyl method) for the generalized sum T to a case where the function $f(x)$, though not a polynomial of degree n, is similar to such a polynomial in some respect, found application in other problems of the theory of numbers. For instance, these estimates were used in the theory of distribution of primes, viz., the existence of primes in a small interval (extension of the Bertrand postulate proved by Chebyshev that for $N > 1$, there always exists a prime between N and $2N$), and the problem of the number $\pi(N)$ of primes not exceeding N (improvement of the Poussin asymptotic expression for $\pi(N)$).

In 1934 I found a new method in the analytical theory of numbers. It not only introduced radical refinements in the solutions of problems already treated by other methods, but also paved the way to new problems. The method and its applications were worked out in cooperation with van der Corput, Chudakov, Hua and others. An elegant p-adic version of this method, which was later refined by Karatsuba, was developed by Linnik.

Two of the theorems obtained with the help of this method are of fundamental importance (Theorems 1 and 4, Chapter 4), since they give, for sufficiently large integer b, estimate for the integral:

$$I = \int_0^1 \ldots \int_0^1 |T|^{2b} \, d\alpha_n \ldots d\alpha_1 .$$

The first theorem gives the estimate for certain general values of b, the other for a particular value of b. Theorem 4 is quite simple: it is formulated as follows:
Let n be a constant $\geqq 3$, $b = 2r_1$, and

$$r_1 = [n^2(2\ln n + \ln\ln n + 4)],$$

then

$$I \ll P^{2b - \frac{n(n+1)}{2}} .$$

The first theorem gives new estimates for Weyl sums. In deriving these estimates there is no need to use my method at the points $(\alpha_n, \ldots, \alpha_1)$ belonging to certain small domains enveloping a point

$$\left(\frac{a_n}{q_n}, \ldots, \frac{a_1}{q_1}\right),$$

whose coordinates are irreducible fractions with a common denominator not exceeding P^ν, since at such points the sum $T = T(\alpha_n, \ldots, \alpha_1)$ can be estimated by simpler methods. For the points in a domain H that remains after these small domains are subtracted from an n-dimensional space, my method gives a general estimate of the type:

$$|T| \leqq c(n)\, P^{1-\gamma}; \quad \gamma = \frac{1}{8n^2 (\ln n + 1.5 \ln \ln n + 4.2)}.$$

Comparing this estimate with the estimate (3) obtained by Weyl's method, we find that with the increasing n the former estimate, though worsening (as regards the order of growth with increasing P), loses its accuracy much more slowly than the latter.

It should, however, be noted that my method begins to give better estimates for T than Weyl's method only from about $n = 12$. Naturally, the new estimates for Weyl sums have affected, to some extent, all previous solutions derived by Weyl's method.

In the first instance, this has touched upon the problem of the distribution of fractional parts of the values of a polynomial. For the remainder term $\lambda(\sigma)$ in the asymptotic expression (4) in the domain H, we now obtain an estimate of the type:

$$|\lambda(\sigma)| < c(n)\, P^{1-\varrho_1}; \quad \varrho_1 = \frac{1}{8n^2 (\ln n + 1.5 \ln \ln n + 4.2)}.$$

Using the simplest considerations of the new method, for the function $G(n)$ in Waring's problem we obtain an inequality of the type

$$G(n) < 2n \ln n (1 + \delta_n'); \quad \lim_{n=\infty} \delta_n' = 0.$$

This inequality, since $G(n) > n$, cannot be improved any further in the sense of the order of growth of the right-hand side with increasing n. On the other hand, using the second theorem formulated above for the estimate of the integral I and the new estimates for Weyl sums, we can show that the asymptotic expression derived by Hardy and Littlewood (8) holds, even if

$$r > 4n^2 \ln n (1 + \delta_n''); \quad \lim_{n=\infty} \delta_n'' = 0. \tag{10}$$

But even this condition, as regards the order of growth of the right-hand side with the increasing n, can hardly be believed to be the best possible result; it can probably be refined by a condition of the type $r > cn^{1+\varepsilon}$, or by a still more exact condition.

Under a condition of the type (10) my method also gives an asymptotic expression for the number of solutions of the system (9); however, this condition can no longer be improved in the sense of the order of growth of the right-hand side with increasing n (the right-hand side cannot be replaced by a quantity of an order less than n^2). It would be of interest to prove the existence of solutions under the same condition.

In the theory of distribution of primes, as regards the existence of primes in a small interval, with the aid of my method and the density theorems on the zeroes of the zeta-function it has been possible to establish the existence (with a weak asymptotic expression for the number) of primes p in the interval $N \leqq p \leqq N + N^\delta$ (N is sufficiently large) for $\delta = 3/5$, whereas in the Weyl method this was demonstrated only for $\delta = 32999/33000$. In the asymptotic expression

$$\pi(N) = \int_2^N \frac{dx}{\ln x} + R$$

my method gives $R = O(Ne^{-c_1 \lambda(N)(\ln N)^{0.6}})$, where $\lambda(N) = (\ln \ln N)^{-0.2}$, in place of $R = O(Ne^{-c\sqrt{\ln N}})$ found by Poussin (1895), while the Weyl method gives $R = O(Ne^{-c_2\sqrt{\ln N \ln \ln N}})$.

If, however, we assume the Riemann hypothesis, from which it follows that $R = O(\sqrt{N} \ln N)$, even this result is far from being the best and is in no way better than the first result obtained by Poussin. Besides, in Lemma 9, Chapter 4 of this book, if we had

$$|S| < 2a^{1-c/n} \quad \text{in place of} \quad |S| < 2a^{1-1/30000n^2},$$

we could only obtain (the rest of the proof being the same) the result

$$R = O(Ne^{-c_3 \lambda(N)(\ln N)^{2/3}}); \quad \lambda(N) = (\ln \ln N)^{-1/5},$$

which is only a slight improvement of the previous result. It is, most probably, rather difficult to make a significant step forward in solving this problem of the order of R (at least to find $R = O(N^{1-c})$ even for $c = 0.000001$) only by improving the estimates of Weyl sums without any advance in the theory of zeta functions.

In 1742, the correspondence between Goldbach and Euler gave rise to "Goldbach's problem", a hypothesis which asserts that any even integer not less than six can be represented as the sum of two odd primes (Goldbach's binary representation) and any odd integer not less than nine as the sum of three odd primes (Goldbach's ternary representation). Evidently, Goldbach's ternary representation theorem follows as a trivial corollary from the binary representation theorem.

In 1919, while attempting to solve the Goldbach binary representation problem, Brun developed a method which is a special modification of the Eratosthenes sieve. Using this method, he showed that any sufficiently large natural number can be represented as the sum of two numbers, each of which is the product of not more than nine primes. Later this result was refined, but Goldbach's binary representation problem could not be solved.

Nonetheless, Brun's method (and its various modifications) found wide application in the theory of distribution of primes and led to quite a few important theorems in this field.

Modifying the Brun method by introducing certain new ideas as to the density of a sequence of positive integers, in 1930 Schnirelman proved that any integer greater than unity can be represented in the form of a sum of a limited number of primes.

Hardy and Littlewood were the first to examine Goldbach's ternary representation in a broader formulation than as the mere existence of representation of an odd integer by the sum of three odd primes. They studied the asymptotic expressions for the number of representations. In 1923 they gave an intuitive version (based on certain unproved theorems on L-series) for deriving such formulas by a method similar to that used in the derivation of asymptotic expression in the Waring problem.

In 1937 I found that many sums of primes can be made up, only by addition and subtraction, of a comparatively small number of sums whose estimates could be obtained by my method and techniques not related to the theory of zeta functions (or L-series). In particular, one such sum is

$$T' = \sum_{p \leqq P} e^{2\pi i f(p)},$$

which is similar to the sum T, but summation is taken over only primes not greater than P. First an estimate was derived for the sum

$$\sum_{p \leqq P} e^{2\pi i \alpha p},$$

which is the simplest form (for $n=1$) of the sum T'. This estimate, together with the theorems on the distribution of primes in an arithmetic progression with a common difference not exceeding some function $\psi(P)$ slowly increasing with P (Page's theorem or Walfisz' theorem, which is more accurate but is based on Siegel's Lemma) made it possible to derive rigorously for the first time the asymptotic expression of Hardy and Littlewood for the number of representations of an odd number N in the form $N = p_1 + p_2 + p_3$. The existence of representations for all sufficiently large N has been demonstrated with the help of this formula, and has been derived as a corollary of this formula.

Then, in 1937, using certain considerations stated above and largely drawing upon Weyl's technique, I also derived an estimate of the sum T' (for $n > 1$) similar to the estimate of the sum T obtained by Weyl's method (which is somewhat less accurate). In 1948–1956, applying my technique instead of Weyl's method, I succeeded in deriving a general theorem on the estimate of the sum T' (for $n \geqq 12$) which is quite similar to my general theorem on the estimate of the sum T.

Applying the second theorem on the estimate of the integral I and the estimate of the sum T' obtained by the 1948 method (or by the 1937 method) under a condition of type (10), we can derive an asymptotic expression for the number of representations of a positive integer N in the form:

$$N = p_1^n + \ldots + p_r^n$$

by primes $p_1, \ldots, p_r$.

The aim of this monograph is to present my method in the analytical theory of numbers (found in 1934 and developed further in subsequent years). Without going into finer details and applications, the monograph explains the method in detail only as applied to a limited number of certain important problems in the analytical theory of numbers.

One such problem is the estimation of the sum T. Chapter 4 is devoted to this problem. It gives the proofs of two important theorems on the estimate of the integral I and the derivation of the general theorem on the estimate of the sum T. Chapter 5, closely linked with Chapter 4, outlines the derivation of an asymptotic expression for the distribution of the fractional parts of the values of a polynomial $f(x)$.

Chapter 6 is devoted to the application of the theorems proved in Chapter 4 to derive the asymptotic formula in Waring's problem on the representation of a positive integer N as the sum of numbers of the type x^n.

Chapter 7 deals with the derivation of a general theorem on the estimate of the sum T'. It is intimately connected with Chapter 8, which contains the derivation of an asymptotic expression for the distribution of fractional parts of the values of a polynomial $f(p)$, where p runs through primes.

Finally, Chapter 9 illustrates the application of the second theorem of Chapter 4 on an estimate for the integral I and of the Theorem 2 of Chapter 7, in the derivation of an asymptotic expression for the representation of a positive integer N as the sum of numbers of the type p^n, where p is a prime.

The lemmas and auxiliary theorems, which are freely used in this monograph, but whose proofs are not associated with my method, have been grouped together, wherever possible, in appropriate chapters. Assumptions relating to the function $\tau(a)$, which gives the number of divisors of a number a, are laid down in Chapter 1. Assumptions relating to trigonometrical series and integrals are stated in Chapter 2. Finally, Chapter 3 gives the proof of Hua's theorem on an estimate for the general rational trigonometrical sum and ceetain corollaries of this theorem.

The references include only those sources which deal with the questions discussed in this monograph. They also include some papers containing the proofs of the first, though rough, versions of important theorems, as well as papers containing the first formulations or applications of certain important considerations used in the text. A more detailed list of references on the problems (including those not given space in this monograph) which have been solved with the help of my method can be found in Hua's monograph [17].

In conclusion I express my indebtedness to A. Karatsuba, D.Sc. for his assistance in preparing the manuscript for publication.

Chapter 1. Number of Divisors

This chapter presents several lemmas relating to the function $\tau(a)$ which expresses the number of divisors of a natural number a. These lemmas are used in the subsequent chapters. Well known lemmas and their trivial corollaries are stated without proof.

Lemma 1. For $N \geqq 2$, we have

$$\sum_{0<a\leqq N} \tau(a) = N \ln N + (2E-1)N + O(\sqrt{N}),$$

where $E = 0.577\ldots$ is the Euler constant.

For $N \geqq 2$ and $N^{0.5} \leqq g \leqq N$, we have

$$\sum_{N-g<a\leqq N} \tau(a) \ll g \ln N.$$

The first assertion of the lemma is known, while the second is a trivial corollary of the first.

Lemma 2. For $N \geqq 2$ and a positive integer l,

$$\sum_{0<a\leqq N} (\tau(a))^l \ll N(\ln N)^{2^l-1}.$$

Proof. We shall first establish the inequality

$$\sum_{0<a\leqq N} \frac{(\tau(a))^l}{a} \ll (\ln N)^{2^l} \tag{1}$$

with the help of the method of mathematical induction.
Write down the sequence of fractions:

$$\begin{array}{llllll} 1, & \frac{1}{2}, & \frac{1}{3}, & \frac{1}{4}, & \frac{1}{5}, & \frac{1}{6}, \ldots \\ & \frac{1}{2}, & & \frac{1}{4}, & & \frac{1}{6}, \ldots \\ & & \frac{1}{3}, & & & \frac{1}{6}, \ldots . \end{array}$$

The s-th row from the top contains the fractions whose denominators are not greater than N and are multiples of s. The fraction $1/a$ $(0<a\leqq N)$ occurs in $\tau(a)$ sequences. Therefore the sum of all the terms contained in all the sequences is

$$\sum_{0<a\leqq N} \frac{\tau(a)}{a}.$$

On the other hand, the sum of the terms in the s-th sequence is

$$\ll \frac{1}{s} \ln N.$$

Therefore the sum of all the terms in all the sequences will be

$$\ll \ln N \sum_{0<s\leqq N} \frac{1}{s} \ll (\ln N)^2.$$

which completes the proof of the inequality (1) for $l = 1$.

Now assuming that the inequality (1) holds true for some l, we shall show that it is satisfied for $(l+1)$ as well. Write down new sequences of fractions as follows:

$$\begin{array}{cccccc} \frac{(\tau(1))^l}{1}, & \frac{(\tau(2))^l}{2}, & \frac{(\tau(3))^l}{3}, & \frac{(\tau(4))^l}{4}, & \frac{(\tau(5))^l}{5}, & \frac{(\tau(6))^l}{6}, \dots \\ & \frac{(\tau(2))^l}{2}, & & \frac{(\tau(4))^l}{4}, & & \frac{(\tau(6))^l}{6}, \dots \\ & & \frac{(\tau(3))^l}{3}, & & & \frac{(\tau(6))^l}{6}, \dots \\ \dots & \dots & \dots & \dots & \dots & \dots \end{array}$$

which contain only those fractions whose denominators are not greater than N. The term containing $\tau(a)$ occurs in $\tau(a)$ sequences and, consequently, the sum of all terms contained in all the sequences is

$$\sum_{0<a\leqq N} \frac{(\tau(a))^{l+1}}{a}.$$

On the other hand, by virtue of the familiar inequality $\tau(st) \leqq \tau(s)\,\tau(t)$, the sum of the terms in the s-th sequence is

$$\ll \frac{(\tau(s))^l}{s}\left(\frac{(\tau(1))^l}{1} + \frac{(\tau(2))^l}{2} + \dots + \frac{(\tau([Ns^{-1}]))^l}{[Ns^{-1}]}\right) \ll \frac{(\tau(s))^l}{s}(\ln N)^{2^l}.$$

Therefore the sum of all the terms in all the sequences is

$$\ll (\ln N)^{2^l} \sum_{0<s\leqq N} \frac{(\tau(s))^l}{l} \ll (\ln N)^{2^{l+1}}.$$

Thus, we find that the inequality holds true for $l+1$.

Now we shall prove the inequality of our lemma. For $l = 1$, this inequality is a trivial corollary of Lemma 1.

Now, assuming that the inequality is true for some l, we shall show that it holds good for $l+1$ as well. Write the sequences

$$\begin{array}{cccccc} (\tau(1))^l, & (\tau(2))^l, & (\tau(3))^l, & (\tau(4))^l, & (\tau(5))^l, & (\tau(6))^l, \dots \\ & (\tau(2))^l, & & (\tau(4))^l, & & (\tau(6))^l, \dots \\ & & (\tau(3))^l, & & & (\tau(6))^l, \dots \\ \dots & \dots & \dots & \dots & \dots & \dots \end{array}$$

which again include only those terms containing the values of $\tau(a)$ for $a \leqq N$. A term containing $\tau(a)$ occurs in $\tau(a)$ sequences. Therefore the sum of all terms contained in all the sequences is

$$\sum_{0<a\leqq N} (\tau(a))^{l+1}.$$

On the other hand, by virtue of the familiar inequality $\tau(st) \leqq \tau(s)\tau(t)$, the sum of the s-th sequence is

$$\ll (\tau(s))^l ((\tau(1))^l + (\tau(2))^l + \ldots + (\tau([Ns^{-1}]))^l) \ll (\tau(s))^l \frac{N}{s} (\ln N)^{2l-1}.$$

Therefore, by virtue of inequality (1), the sum of all the terms in all the sequences is

$$\ll N(\ln N)^{2^l-1} \sum_{0<s\leqq N} \frac{(\tau(s))^l}{s} \ll N(\ln N)^{2^{l+1}-1}.$$

Lemma 3. Let $n \geqq 2$, Q a natural number not exceeding $\sqrt{N}$, v satisfy the conditions $0 \leqq v < Q$, $(v, Q) = 1$, and g obey the inequality $Q\sqrt{N} < g \leqq N$. Then

$$\sum_{N-g<Qu+v\leqq N} \tau(Qu+v) \ll \frac{g}{Q} \ln N.$$

Proof. For every term in the sum on the left-hand side, we find

$$\tau(Qu+v) \leqq 2\tau_1(Qu+v),$$

where $\tau_1(Qu+v)$ is the number of divisors of $(Qu+v)$ not exceeding $\sqrt{N}$. Therefore

$$\sum_{N-g<Qu+v\leqq N} \tau(Qu+v) \ll \sum_{N-g<Qu+v\leqq N} \tau_1(Qu+v).$$

And, since each s not exceeding $\sqrt{N}$ is a divisor of not more than

$$\frac{g}{Qs} + 1$$

numbers of the type $Qu+v$ satisfying the condition $N-g < Qu+v \leqq N$, the right-hand side of the last inequality is

$$\ll \sum_{0<s\leqq\sqrt{N}} \left(\frac{g}{Qs} + 1\right) \ll \frac{g}{Q} \ln N,$$

which completes the proof of our inequality.

Lemma 4. Let σ be the number of different prime divisors of q, and let $\varepsilon < 1$. Then,

$$2^\sigma < 2^{2^{1/\varepsilon}} q^\varepsilon.$$

Proof. Let $q = p_1^{\alpha_1} \ldots p_\sigma^{\alpha_\sigma}$ be the canonical expansion of q. We find

$$\frac{2^\sigma}{q^\varepsilon} \leqq \frac{2}{p_1^\varepsilon} \cdots \frac{2}{p_\sigma^\varepsilon}.$$

Substitute 2 for each factor, if its denominator is <2, or 1, if its denominator is $\geqq 2$. Let p_r be the largest of $p_1, \ldots, p_\sigma$ for which the denominator is <2. Then

$$p_r < 2^{1/\varepsilon}, \quad 2^r < 2^{p_r} < 2^{2^{1/\varepsilon}}.$$

Thus the lemma has been proved.

Lemma 5. We have

$$\tau(q) < c(\varepsilon)\, q^{\varepsilon}.$$

Proof. Let $q = p_1^{\alpha_1} \dots p_\sigma^{\alpha_\sigma}$ be the canonical expansion of q. Then

$$\frac{\tau(q)}{q^{\varepsilon}} = \frac{\alpha_1 + 1}{p_1^{\varepsilon\alpha_1}} \cdots \frac{\alpha_\sigma + 1}{p_\sigma^{\varepsilon\alpha_\sigma}};$$

but the ratio

$$\frac{\alpha_\varepsilon + 1}{p_s^{\varepsilon\alpha_s}}$$

is

$$\leqq \frac{\alpha_s + 1}{2^{\alpha_s}} \leqq 1, \quad \text{if} \quad p_s \geqq 2^{1/\varepsilon}$$

and

$$\leqq \frac{\alpha_s + 1}{2^{\varepsilon\alpha_s}} \leqq \frac{\alpha_s + 1}{\varepsilon\alpha_s \ln 2} \leqq \frac{2}{\varepsilon \ln 2}, \quad \text{if} \quad p_s < 2^{1/\varepsilon}.$$

Therefore

$$\frac{\tau(q)}{q^{\varepsilon}} \leqq \left(\frac{2}{\varepsilon \ln 2}\right)^{2^{1/\varepsilon}}.$$

Thus, we have proved the lemma.

Chapter 2. Trigonometric Series and Integrals

This chapter contains those lemmas in the theory of trigonometrical series and integrals that are necessary for understanding the subsequent chapters.

Lemma 1 (Fourier Series). Let $F(x) = P(x) + iQ(x)$ be a periodic function of period 1 and let the interval $0 \leqq x \leqq 1$ be such that it can be divided into a finite number of subintervals in each of which $P(x)$ and $Q(x)$ are continuous and monotonic. Furthermore, at the points of discontinuity, let

$$F(x) = \frac{F(x-0)+F(x+0)}{2}.$$

Then

$$F(x) = \frac{a_0}{2} + \sum_{m=1}^{\infty} (a_m \cos 2\pi m x + b_m \sin 2\pi m x);$$

$$a_m = 2\int_0^1 F(\xi)\cos 2\pi m\xi\, d\xi, \quad b_m = 2\int_0^1 F(\xi)\sin 2\pi m\xi\, d\xi.$$

This lemma, as it can be found in university text-books, is given without proof.

Lemma 2. Let r be a positive integer, α and β be real numbers such that

$$0 < \Delta < 0.25, \quad \Delta \leqq \beta - \alpha \leqq 1 - \Delta.$$

Then there exists a periodic function $\psi(x)$ of period 1, such that
1. $\psi(x) = 1$ in the interval $\alpha + 0.5\Delta \leqq x \leqq \beta - 0.5\Delta$;
2. $0 < \psi(x) < 1$ in the intervals $\alpha - 0.5\Delta < x < \alpha + 0.5\Delta$ and $\beta - 0.5\Delta < x < \beta + 0.5\Delta$;
3. $\psi(x) = 0$ in the interval $\beta + 0.5\Delta \leqq x \leqq 1 + \alpha - 0.5\Delta$;
4. $\psi(x)$ can be expanded as a Fourier series of the type

$$\psi(x) = \beta - \alpha + \sum_{m=1}^{\infty} (g_m e^{2\pi i m x} + h_m e^{-2\pi i m x}),$$

where

$$|g_m| \leqq \frac{1}{\pi m}, \qquad |h_m| \leqq \frac{1}{\pi m},$$

$$|g_m| \leqq \beta - \alpha, \qquad |h_m| \leqq \beta - \alpha,$$

$$|g_m| < \frac{1}{\pi m}\left(\frac{r}{\pi m \Delta}\right)^r, \quad |h_m| < \frac{1}{\pi m}\left(\frac{r}{\pi m \Delta}\right)^r.$$

Proof. Consider a periodic function $\psi_0(x)$ of period 1 defined by the equations:

$$\psi_0(x) = 1 \quad \text{in} \quad \alpha < x < \beta,$$

$$\psi_0(x) = 0.5 \quad \text{at} \quad x = \alpha \quad \text{and} \quad x = \beta,$$

$$\psi_0(x) = 0 \quad \text{in} \quad \beta < x < 1 + \alpha.$$

From Lemma 1, we obtain

$$\psi_0(x) = \frac{a_{0,0}}{2} + \sum_{m=1}^{\infty} (a_{m,0}\cos 2\pi mx + b_{m,0}\sin 2\pi mx),$$

$$a_{0,0} = 2\int_{\alpha}^{\beta} d\xi = 2(\beta-\alpha),$$

$$a_{m,0} = 2\int_{\alpha}^{\beta} \cos 2\pi m\xi\, d\xi = \frac{\sin 2\pi m\beta - \sin 2\pi m\alpha}{\pi m},$$

$$b_{m,0} = 2\int_{\alpha}^{\beta} \sin 2\pi m\xi\, d\xi = \frac{\cos 2\pi m\alpha - \cos 2\pi m\beta}{\pi m}.$$

Now, taking $\Delta = 2r\delta$, consider the functions $\psi_1(x), \ldots, \psi_r(x)$ defined by the recurrence relationship:

$$\psi_\varrho(x) = \frac{1}{2\delta}\int_{-\delta}^{\delta} \psi_{\varrho-1}(x+z)\, dz.$$

Obviously, each of these functions has the following properties:

1. $\psi_\varrho(x) = 1$ in the interval $\alpha + \varrho\delta \leqq x \leqq \beta - \varrho\delta$;
2. $0 < \psi_\varrho(x) < 1$ in the intervals $\alpha - \varrho\delta < x < \alpha + \varrho\delta$ and $\beta - \varrho\delta < x < \beta + \varrho\delta$;
3. $\psi_\varrho(x) = 0$ in the interval $\beta + \varrho\delta \leqq x \leqq 1 + \alpha - \varrho\delta$;
4. $\psi_\varrho(x)$ can be expanded as a Fourier series of the type

$$\psi_\varrho(x) = \beta - \alpha + \sum_{m=1}^{\infty} (a_{m,\varrho}\cos 2\pi mx + b_{m,\varrho}\sin 2\pi mx),$$

where

$$a_{m,\varrho} = \frac{\sin 2\pi m\beta - \sin 2\pi m\alpha}{\pi m}\left(\frac{\sin 2\pi m\delta}{2\pi m\delta}\right)^\varrho,$$

$$b_{m,\varrho} = \frac{\cos 2\pi m\alpha - \cos 2\pi m\beta}{\pi m}\left(\frac{\sin 2\pi m\delta}{2\pi m\delta}\right)^\varrho.$$

Evidently, $\psi_0(x)$ has these four properties. For $\varrho > 0$, if the function $\psi_{\varrho-1}(x)$ has these four properties, the function $\psi_\varrho(x)$ has obviously the first three properties. And that it has the fourth property, too, follows from the relationships: from

$$\frac{a_{0,\varrho}}{2} = \int_0^1 \psi_\varrho(\xi)\, d\xi = \int_0^1 \left(\frac{1}{2\delta}\int_{-\delta}^{\delta} \psi_{\varrho-1}(\xi+z)\, dz\right) d\xi$$

$$= \frac{1}{2\delta}\int_{-\delta}^{\delta} dz \int_0^1 \psi_{\varrho-1}(\xi+z)\, d\xi$$

$$= \frac{1}{2\delta}\int_{-\delta}^{\delta} dz \int_0^1 \psi_{\varrho-1}(\xi)\, d\xi = \int_0^1 \psi_{\varrho-1}(\xi)\, d\xi = \beta - \alpha,$$

and if $m > 0$, from

$$a_{m,\varrho} = 2\int_0^1 \left(\frac{1}{2\delta}\int_{-\delta}^{\delta} \psi_{\varrho-1}(\xi+z)\,dz\right)\cos 2\pi m\xi\,d\xi$$

$$= \frac{1}{\delta}\int_{-\delta}^{\delta} dz \int_0^1 \psi_{\varrho-1}(\xi+z)\cos 2\pi m\xi\,d\xi$$

$$= \frac{1}{\delta}\int_{-\delta}^{\delta} dz \int_0^1 \psi_{\varrho-1}(\xi)\cos 2\pi m(\xi-z)\,d\xi$$

$$= \frac{1}{2\delta}\int_{-\delta}^{\delta} dz\,(a_{m,\varrho-1}\cos 2\pi mz + b_{m,\varrho-1}\sin 2\pi mz) = a_{m,\varrho-1}\frac{\sin 2\pi m\delta}{2\pi m\delta}$$

and from

$$b_{m,\varrho} = b_{m,\varrho-1}\frac{\sin 2\pi m\delta}{2\pi m\delta},$$

which can be derived in a similar manner for $m > 0$.

It is the last of these functions, $\psi_r(x)$, that can be used as the function $\psi(x)$ stated in the lemma. Indeed, since $r\delta = 0.5\Delta$, the function $\psi(x) = \psi_r(x)$ satisfies the first three conditions of the lemma. Transforming its expansion into a Fourier series of type (1) by the Euler formulae, we obtain

$$g_m = \frac{a_{m,\varrho}}{2} + \frac{b_{m,\varrho}}{2i} = \frac{1}{2\pi im}(e^{-2\pi im\alpha} - e^{-2\pi im\beta})\left(\frac{\sin 2\pi m\delta}{2\pi m\delta}\right)^r,$$

$$h_m = \frac{a_{m,\varrho}}{2} - \frac{b_{m,\varrho}}{2i} = \frac{1}{2\pi im}(e^{2\pi im\beta} - e^{2\pi im\alpha})\left(\frac{\sin 2\pi m\delta}{2\pi m\delta}\right)^r; \qquad \delta = \frac{\Delta}{2r},$$

which satisfy, as can be easily verified, all the inequalities of condition (4) of the lemma.

Lemma 3 (van der Corput's Lemma). Let $f(x)$ be a real differentiable function in the interval $M < x \leqq M'$; let its first derivative $f'(x)$ be monotonic and of constant sign in this interval, and for constant $\delta\,(0 < \delta < 1)$, let $|f'(x)| \leqq \delta$. Then

$$\sum_{M<x\leqq M'} e^{2\pi if(x)} = \int_M^{M'} e^{2\pi if(x)}\,dx + O\left(3 + \frac{2\delta}{1-\delta}\right).$$

Proof. Assume that the interval $M < x \leqq M'$ contains not less than two integers, otherwise the lemma is trivial. Let M_1 be the integer nearest to M exceeding M, and M_1' be the integer nearest to M not exceeding M'. Let x denote one of the numbers $M_1, \ldots, M_1' - 1$. Consider the periodic function $F(z)$ of period 1 and defined in the interval $0 < z < 1$ by the expression:

$$F(z) = e^{2\pi if(x+z)}.$$

Applying Lemma 1 to this function (here z plays the part of x in the lemma) and putting $z = 0$, we obtain

$$\frac{e^{2\pi if(x)} + e^{2\pi if(x+1)}}{2} = \frac{a_0}{2} + \sum_{m=1}^{\infty} a_m;$$

$$a_m = 2\int_0^1 e^{2\pi if(x+\xi)}\cos 2\pi m\xi\,d\xi.$$

Hence, we find

$$\frac{e^{2\pi i f(x)} + e^{2\pi i f(x+1)}}{2} - \int_0^1 e^{2\pi i f(x+\xi)}\, d\xi$$

$$= - \sum_{m=1}^{\infty} \frac{1}{m} \int_0^1 (e^{2\pi i m\xi} - e^{-2\pi i m\xi})\, e^{2\pi i f(x+\xi)} f'(x+\xi)\, d\xi .$$

Summing the last equality over all $x = M_1, \ldots, M_1' - 1$, we obtain

$$\sum_{M_1 \leqq x \leqq M_1'} e^{2\pi i f(x)} - \frac{1}{2} e^{2\pi i f(M_1)} - \frac{1}{2} e^{2\pi i f(M_1')} - \int_{M_1}^{M_1'} e^{2\pi i f(x)}\, dx = - \sum_{m=1}^{\infty} \frac{1}{m} U_m ;$$

$$U_m = \int_{M_1}^{M_1'} (e^{2\pi i m x} - e^{-2\pi i m x})\, e^{2\pi i f(x)} f'(x)\, dx$$

$$= \frac{1}{2\pi i} \int_{M_1}^{M_1'} \frac{f'(x)}{m + f'(x)}\, de^{2\pi i (mx + f(x))} + \frac{1}{2\pi i} \int_{M_1}^{M_1'} \frac{f'(x)}{m - f'(x)}\, de^{2\pi i (-mx + f(x))} .$$

Applying the second mean value theorem to the real and imaginary parts of each integral separately, we obtain

$$|U_m| < \frac{\sqrt{8}}{2\pi} \left(\frac{\delta}{m+\delta} + \frac{\delta}{m-\delta} \right) < \frac{\sqrt{8}}{\pi} \frac{\delta}{m-\delta} ,$$

$$\sum_{m=1}^{\infty} \frac{1}{m} |U_m| < \frac{\delta}{1-\delta} + \sum_{m=2}^{\infty} \frac{\delta}{m(m-1)} < \frac{2\delta}{1-\delta} ,$$

which completes the proof of the lemma.

Lemma 4. Let $\varphi(x) = u_n x^n + \ldots + u_1 x$, where $u_n, \ldots, u_1$ are all real numbers the greatest modulus of which is equal to u. Then for the integral

$$I = \int_0^1 e^{2\pi i \varphi(x)}\, dx$$

the estimate (which can obviously be replaced by the modulus of any other coefficient)

$$|I| \leqq \min(1{,}32 u^{-\nu}) .$$

holds true.

Proof. We shall assume that $n > 1$ and $u > (32)^n$, because otherwise the lemma is trivial. We have

$$I = U + iV , \quad U = \int_0^1 \cos 2\pi\varphi(x)\, dx , \quad V = \int_0^1 \sin 2\pi\varphi(x)\, dx .$$

Consider the integral U. Divide the range of integration $0 \leqq x \leqq 1$ into subintervals of two classes. Intervals of the first class are those intervals at every point of which the inequality

$$|\varphi'(x)| \leqq A ; \qquad A = \left(\frac{n-1}{4e} \right)^{1-\nu} u^{-\nu} .$$

is satisfied, while the remaining intervals are said to be intervals of the second class. Accordingly, we obtain

$$U = U_1 + U_2 .$$

Let μ be the sum of the lengths of the intervals belonging to the first class. Obviously, we have

$$|U_1| \leqq \mu .$$

To estimate μ, group together all the intervals of the first class into one interval (of length μ) and in this new interval take n points spaced at a distance of $\mu/(n-1)$. Now return the intervals to their original places: thus we obtain n points $x_1, \ldots, x_n$ belonging to the intervals of the first class, such that

$$|x_i - x_j| \geqq |i-j| \frac{\mu}{n-1} .$$

Now consider a linear system of equations in $u_1, 2u_2, \ldots, nu_n$:

$$\begin{aligned} u_1 + 2u_2 x_1 + \ldots + nu_n x_1^{n-1} &= \varphi'(x_1), \\ \ldots\ldots\ldots\ldots\ldots\ldots\ldots\ldots \\ u_1 + 2u_2 x_n + \ldots + nu_n x_n^{n-1} &= \varphi'(x_n) . \end{aligned} \tag{2}$$

Let $u = |u_{r+1}|$; then

$$(r+1)\,u = \left|\frac{\Delta'}{\Delta}\right| ,$$

where

$$\Delta = \begin{vmatrix} 1 & x_1 & \ldots & x_1^r & \ldots & x_1^{n-1} \\ \cdot & \cdot & \cdot & \cdot & \cdot & \cdot \\ 1 & x_k & \ldots & x_k^r & \ldots & x_k^{n-1} \\ \cdot & \cdot & \cdot & \cdot & \cdot & \cdot \\ 1 & x_n & \ldots & x_n^r & \ldots & x_n^{n-1} \end{vmatrix} ,$$

and Δ' is obtained from Δ on replacing the elements of the $(r+1)$-th column by the corresponding right-hand sides of the equations in the system (2).

On expanding Δ' with the elements of the $(r+1)$-th column, we obtain

$$\Delta' = \sum_{k=1}^{n} \varphi'(x_k)\, \Delta'_k ,$$

where Δ'_k is obtained from Δ by deleting the $(r+1)$-th column and the n-th row. The absolute value of the quotient obtained on dividing Δ'_k by the Vandermonde determinant formed from the numbers $x_1, \ldots, x_{k-1}, x_{k+1}, \ldots, x_n$ is the $(n-r-1)$-th symmetric function of these numbers and, consequently, does not exceed

$$\binom{n-1}{n-r-1} = \binom{n-1}{r} .$$

Therefore

$$(r+1)\,u \leqq \sum_{k=1}^{n} \frac{A\binom{n-1}{r}}{\sum\limits_{\substack{m=1\\ m\lessgtr k}} (x_k - x_m)} \leqq A\binom{n-1}{r}\left(\frac{n-1}{\mu}\right)^{n-1} \sum_{k=1}^{n} \frac{1}{(k-1)!\,(n-k)!}$$

$$= \frac{A}{(n-r-1)!\,r!}\left(\frac{2(n-1)}{\mu}\right)^{n-1},$$

$$\mu \leqq A^{\frac{1}{n-1}}\, 4eu^{-\frac{1}{n-1}} = 4e\left(\frac{n-1}{4e}\right)^{\nu} u^{-\nu}.$$

Now we shall estimate U_2. The intervals of the second class can be subdivided into $\leqq 2n-2$ subintervals, in each of which $\varphi'(x)$ is monotonic and of constant sign. Moreover, at each point of these subintervals we have $|\varphi'(x)| \geqq A$. Let $x_1 \leqq x \leqq x_2$ be one such subinterval and I' be that part of the integral U_2 which corresponds to this subinterval. Without loss of generality, we can restrict ourselves to a case where $\varphi'(x)$ is an increasing function in this interval. Assuming $\varphi(x) = v$, $\varphi(x_1) = v_1$, and $\varphi(x_2) = v_2$, we easily find

$$I' = \int_{v_1}^{v_2} \cos 2\pi v \, \frac{dv}{\varphi'(x)},$$

where $\varphi'(x)$ is regarded as a function of v. Using the numbers of the type $0.5l + 0.25$ (where l is an integer) lying in the interval, we divide the interval $v_1 \leqq v \leqq v_2$ into subintervals of length not exceeding 0.5. Then we can represent the integral I' by an alternating series. Hence it follows that for some v_0 and σ satisfying the condition $v_1 \leqq v_0 < v + \sigma \leqq v_2$, we have

$$|I'| \leqq \int_{v_0}^{v_2+\sigma} \frac{dv}{\varphi'(x)} = x'' - x';$$

$$v_0 = \varphi(x'), \quad v_0 + \sigma = \varphi(x'').$$

Now, applying Lagrange's formula, we obtain

$$\sigma = \varphi(x'') - \varphi(x') = (x'' - x')\,\varphi'(\xi); \quad x' \leqq \xi \leqq x'',$$

$$x'' - x' = \frac{\sigma}{\varphi'(x)} \leqq \frac{1}{2A}, \quad |U_2| \leqq \frac{n-1}{A} = 4e\left(\frac{n-1}{4e}\right)^{\nu} u^{-\nu},$$

$$|U| \leqq |U_1| + |U_2| \leqq 8e\left(\frac{n-1}{4e}\right)^{\nu} u^{-\nu}.$$

Arguing in a similar manner, we obtain the same upper bound for $|V|$ as well. Therefore

$$I < 8e\sqrt{2}\left(\frac{n-1}{4e}\right)^{\nu} u^{-\nu} < 32u^{-\nu}.$$

Lemma 5. For $P \geqq 2$, constant n greater than unity and real z, let

$$J(z) = \int_{2}^{P} \frac{e^{2\pi i z x^n}}{\ln x}\, dx.$$

Now, putting $z = \delta P^{-n}$, $u = \ln P$,

$$D_1 = \min(1, |\delta|^{-\nu}), \quad \text{if} \quad 0 \leqq |\delta| \leqq P^{n-1},$$
$$D_1 = u|\delta|^{-\nu}, \qquad \text{if} \quad P^{n-1} < |\delta|,$$

we obtain

$$J(z) \ll \frac{P}{u} D_1.$$

Proof. Without loss of generality, we shall consider only the case where $z \geqq 0$. For $\delta \leqq 1$ the lemma is trivial; therefore let $\delta > 1$. We find

$$J(z) = U(z) + iV(z);$$

$$U(z) = \int_{z2^n}^{zP^n} \frac{\cos 2\pi v}{\ln(vz^{-1})} z^{-\nu} v^{\nu-1}\, dv,$$

$$V(z) = \int_{z2^n}^{zP^n} \frac{\sin 2\pi v}{\ln(vz^{-1})} z^{-\nu} v^{\nu-1}\, dv.$$

Now consider the integral $U(z)$. The coefficient of $\cos 2\pi v$ in the integrand is a decreasing function of v. Divide up the range of integration, using the numbers of the type $0.5l + 0.25$ (where l is an integer) in it, into subintervals of length not exceeding 0.5. Thus the integral $U(z)$ can be expressed as an alternating series. Hence it follows that for some v_0 and σ satisfying the condition $z2^n \leqq v_0 \leqq z2^n + \sigma \leqq zP^n$; $\sigma \leqq 0.5$, we have

$$|U(z)| \leqq \int_{v_0}^{v_0+\sigma} \frac{z^{-\nu} v^{\nu-1}}{\ln(vz^{-1})}\, dv \leqq \int_{z2^n}^{z2^n+0.5} \frac{z^{-\nu} v^{\nu-1}}{\ln(vz^{-1})}\, dv = \int_{2}^{(2^n+0.5z^{-1})^\nu} \frac{dx}{\ln x}.$$

For $1 < \delta \leqq P^{n-1}$, the last integral is

$$\ll \frac{z^{-\nu}}{\ln P} = \frac{P}{u} D_1.$$

For $P^{n-1} < \delta \leqq P^n$, it is

$$\ll z^{-\nu} = \frac{P}{u} D_1.$$

Finally, for $P^n < \delta$, it is

$$\ll z^{-1} < z^{-\nu} = \frac{P}{u} D_1.$$

Consequently, for $\delta > 1$, we always have

$$U(z) \ll \frac{P}{u} D_1.$$

Arguing in a similar manner, we obtain the same upper bound for $V(z)$ as well. Therefore for $\delta > 1$, we have

$$J(z) \ll \frac{P}{u} D_1.$$

which completes the proof.

Lemma 6. Let N be an integer. Then

$$\int_0^1 e^{2\pi i N\alpha}\,d\alpha = \begin{cases} 1, & N=0, \\ 0, & N \neq 0. \end{cases}$$

Lemma 7. Let m be an integer greater than unity and a an integer. Then

$$\sum_{a=0}^{m-1} e^{2\pi i \frac{ax}{m}} = \begin{cases} m, & \text{if } a \text{ is divisible by } m, \\ 0, & \text{otherwise}. \end{cases}$$

Lemma 8. Let α be a real non-integer. Then for any N_1 and N_2 such that $N_1 < N_2$, we have

$$\left| \sum_{N_1 < x \leqq N_2} e^{2\pi i \alpha x} \right| < \frac{1}{2(\alpha)}.$$

Lemma 9. For an integer P greater than 1 and real numbers $\alpha_n, \ldots, \alpha_1$, let

$$T(\alpha_n, \ldots, \alpha_1) = \sum_{x=1}^{P} e^{2\pi i(\alpha_n x^n + \ldots + \alpha_1 x)}.$$

If, for given t, the sequence $\alpha'_n, \ldots, \alpha'_1$ can be derived from the sequence $\alpha_n, \ldots, \alpha_1$ on substituting some n_1 of its terms by new terms satisfying the condition $|\alpha'_s - \alpha_s| \leqq tP^{-s}$, then

$$|T(\alpha'_n, \ldots, \alpha'_1) - T(\alpha_n, \ldots, \alpha_1)| < 2\pi n_1 tP.$$

Proof. We have

$$|\alpha'_n x^n + \ldots + \alpha'_1 x - \alpha_n x^n - \ldots - \alpha_1 x| \leqq n_1 t.$$

Hence, using the familiar inequality $|e^{2\pi i l} - 1| \leqq 2\pi |l|$, which holds for any real l, we see the truth of the lemma.

Chapter 3. Rational Trigonometric Sums

This chapter deals with special "rational trigonometrical sums" of the type:

$$S\left(\frac{\varphi(x)}{q}\right)=\sum_{x=1}^{q} e^{2\pi i\frac{\varphi(x)}{q}},$$

where q is an integer greater than unity, and $\varphi(x)=a_n x^n+\ldots+a_1 x+a_0$ is polynomial of degree n with integral coeficients.

The central problem in this chapter is to estimate the upper bounds for the moduli of these sums with the accuracy necessary in subsequent chapters.

In this problem, certain restrictions can be imposed on the sum without any loss of generality. First, the coefficient $a_0=\varphi(0)$ in the polynomial $\varphi(x)$ can be taken to be zero, because the modulus of the sum does not depend on it. Second, we can assume that $(a_n,\ldots,a_1,q)=1$, as otherwise the fraction $\varphi(x)/q$ is reducible. Third, under these two restrictions, we can take $n>1$. Indeed, for $n=1$, $a_0=0$, $(a_1,q)=1$, we have

$$S\left(\frac{\varphi(x)}{q}\right)=S\left(\frac{a_1 x}{q}\right)=\sum_{x=1}^{q} e^{2\pi i\frac{a_1 x}{q}}=0.$$

The theorem which gives the solution to the upper bound problem is formulated as follows:

Let $(a_n,\ldots,a_1,q)=1$. Then

$$\left|S\left(\frac{\varphi(x)}{q}\right)\right|<c(n,\varepsilon_0)\,q^{1-\nu+\varepsilon_0}.$$

The complete proof of this theorem was given by Hua.

As already mentioned in the Introduction, Hua enunciated another version, more accurate, of this theorem which gives an estimate of the type:

$$\left|S\left(\frac{\varphi(x)}{q}\right)\right|<c(n)\,q^{1-\nu}.$$

But, as the second version does not introduce any improvements into the main contents of my book and Hua's proof is rather cumbersome, we shall not consider it.

Lemma 1. Let $q=q_1\ldots q_k$ be the product of positive, pairwise coprime integers and for each q_s, let a number Q_s be defined by the equality $q=q_sQ_s$. Then, for $a_0=0$,

$$S\left(\frac{\varphi(x)}{q}\right)=S\left(\frac{Q_1^{-1}\varphi(Q_1x)}{q_1}\right)\ldots S\left(\frac{Q_k^{-1}\varphi(Q_kx)}{q_k}\right).$$

Proof. If $x_1,\ldots,x_k$ run independently through a complete system of residues modulo $q_1,\ldots,q_k$, respectively, then $Q_1x_1+\ldots+Q_kx_k$ runs through a complete system of residues modulo q. Therefore

$$S\left(\frac{\varphi(x)}{q}\right)=\sum_{x_1}\dots\sum_{x_k} e^{2\pi i\frac{\varphi(Q_1x_1+\dots+Q_kx_k)}{q}}$$
$$=\sum_{x_1}\dots\sum_{x_k} e^{2\pi i\left(\frac{Q^{-1}\varphi(Q_1x_1)}{q_1}+\dots+\frac{Q_k^{-1}\varphi(Q_kx_k)}{q_k}\right)}.$$

Hence the lemma follows trivially.

Note. If the coefficients of the numerator and denominator in the sum on the left-hand side are coprime, so are the coefficients of the numerators and denominators in the sums on the right-hand side.

Lemma 2 (Mordell's Lemma). Let p be a prime and $a_0=0$ and $(a_n,\dots,a_1,p)=1$. Then

$$\left|S\left(\frac{\varphi(x)}{p}\right)\right|<np^{1-\nu}.$$

Proof. We shall assume that $p>n^n$, otherwise the lemma is trivial. Let each $a_n,\dots,a_1$, $x_1,\dots,x_n$, $y_1,\dots,y_n$ run through the values $1,\dots,p$. Taking

$$X_s=x_1^s+\dots+x_n^s,\quad Y_s=y_1^s+\dots+y_n^s,$$

we obtain

$$\sum_{a_n}\dots\sum_{a_1}\left|S\left(\frac{\varphi(x)}{p}\right)\right|^{2n}$$
$$=\sum_{a_n}\dots\sum_{a_1}\sum_{x_1}\dots\sum_{x_n}\sum_{y_1}\dots\sum_{y_n} e^{2\pi i\frac{a_n(X_n-Y_n)+\dots+a_1(X_1-Y_1)}{p}}=p^nN,$$

where N denotes the number of solutions $(x_1,\dots,x_n,y_1,\dots,y_n)$ of the system of congruences:

$$X_n\equiv Y_n\,(\mathrm{mod}\,p),\dots,X_1\equiv Y_1\,(\mathrm{mod}\,p). \tag{1}$$

Let ξ_s denote the s-th elementary symmetric function of $x_1,\dots,x_n$ and η_s, the s-th elementary symmetric function of $y_1,\dots,y_n$. According to Newton's formulae, the system (1) implies

$$\xi_n\equiv\eta_n\,(\mathrm{mod}\,p),\dots,\xi_1\equiv\eta_1\,(\mathrm{mod}\,p).$$

Therefore, the numbers $x_1,\dots,x_n$ form a permutation of $y_1,\dots,y_n$. Since each y_s runs, independently of the others, through p values, we have $N\leqq n!\,p^n$ and consequently,

$$\sum_{a_n}\dots\sum_{a_1}\left|S\left(\frac{\varphi(x)}{p}\right)\right|^{2n}\leqq n!\,p^{2n}. \tag{2}$$

We shall consider only those sums on the left-hand side in which the coefficient a_n of the highest term in the polynomial $\varphi(s)$ is not divisible by p. Two polynomials in which the corresponding coefficients are congruent modulo p are said to be identical to modulus p. The polynomial $\varphi(kx+l)-\varphi(l)$, where $k=1,2,\dots,p-1$ and $l=1,\dots,p$, is called a polynomial transform of $\varphi(x)$. The coefficient of the highest term in such a polynomial is a_nk^n and is therefore not divisible by p. The modulus of the sum $S(\varphi(x)/p)$ does not change if $\varphi(x)$ is replaced by a polynomial transform (if x

runs through a complete system of residues modulo p, then $kx + l$ also runs through a complete system of residues modulo p). It is easy to estimate the number h of polynomial transforms of $\varphi(x)$ identical to some one transform. Indeed, for the polynomial $\varphi(kx+l) - \varphi(l)$ to be identical with a given polynomial $\varphi(k_0x+l_0) - \varphi(l_0)$, the following two congruences should be satisfied:

$$a_n k^n \equiv a_n k_0^n \pmod p,$$

$$na_n k^{n-1} l + a_{n-1} k^{n-1} \equiv na_n k_0^{n-1} l_0 + a_{n-1} k_0^{n-1} \pmod p.$$

The first congruence is satisfied by not more than n values of k, while, for a given k, the second congruence is satisfied by only one value of l. Therefore, $h \leqq n$. Hence, we find that from among all the $p(p-1)$ polynomial transforms of $\varphi(x)$, the number of dissimilar polynomials, which as a consequence occur in different terms on the left-hand side of (2), is not less than

$$\frac{(p-1)p}{n}.$$

Therefore the equality (2) gives

$$\frac{(p-1)p}{n}\left|S\left(\frac{\varphi(x)}{p}\right)\right|^{2n} \leqq n!\,p^{2n}, \quad \left|S\left(\frac{\varphi(x)}{p}\right)\right| < c(n)\,p^{1-\nu},$$

$$c(n) = (2nn!)^{0.5\nu} < n.$$

Definition. Consider the coefficients

$$na_n, \ldots, sa_s, \ldots, a_1$$

of the function $\varphi'(x)$ and the exponents

$$t_n, \ldots, t_s, \ldots, t_1 \tag{3}$$

of the highest powers of a number p that divide these coefficients. Let t be the least of the numbers in the sequence (3) and let t_s be the first term from the left in the sequence (3) equal to t. The number s is called the *index* of the polynomial and is denoted by the symbol ind $\varphi(x)$.

Lemma 3. For an integer λ,

$$\operatorname{ind} \varphi(x+\lambda) = \operatorname{ind} \varphi(x).$$

Proof. Let

$$t'_n, \ldots, t'_s, \ldots, t'_1$$

be the exponents of the highest powers of p that divide the coefficients of the polynomial

$$\frac{d\varphi(x+\lambda)}{dx} = na_n(x+\lambda)^{n-1} + \ldots + sa_s(x+\lambda)^{s-1} + \ldots + a_1,$$

reduced to canonical form. The coefficient of $x^{\sigma-1}$ is obtained by adding σa_σ to the numbers each of which is necessarily divisible by one of the numbers $na_n, \ldots, (\sigma+1)\,a_{\sigma+1}$. Therefore, we have $t'_\sigma > t_s$, if $\sigma > s$; $t'_\sigma = t_s$ if $\sigma = s$, and $t'_\sigma \geqq t_s$ if $\sigma < s$. This proves the lemma.

Lemma 4. We have always

$$\operatorname{ind} \varphi(px) \leqq \operatorname{ind} \varphi(x).$$

If, however,

$$\operatorname{ind} \varphi(px) = \operatorname{ind} \varphi(x)$$

and some value of x satisfies the congruence:

$$\varphi'(x) \equiv 0 \pmod{p^{t+1}},$$

then

$$\operatorname{ind} \varphi(x) > 1$$

and this value of x is a multiple of p.

Proof. On substituting $\varphi(px)$ for $\varphi(x)$, we find that the sequence (3) is transformed as follows:

$$n + t_n, \ldots, s + t_s, \ldots, 1 + t_1, \tag{4}$$

and for $\sigma > s$, we have $\sigma + t_\sigma > s + t_s$. Therefore the first assertion of the lemma is true.

We shall now prove the second assertion. If $\sigma > s$, from the definition of index, it follows that $t_\sigma > t$. If $\sigma < s$, from the condition $\operatorname{ind} \varphi(px) = \operatorname{ind} \varphi(x)$ and from the sequence (4) it follows that

$$\sigma + t_\sigma \geqq s + t_s, \quad t \leqq t_\sigma + \sigma - s < t_\sigma.$$

Thus, all the coefficients of $\varphi'(x)$ different from sa_s are divisible by p^{t+1}. Therefore, divisibility of $\varphi'(x)$ by p^{t+1} implies the divisibility of $sa_s x^{s-1}$ by p^{t+1}, which is possible only when $s > 1$ and x is a multiple of p.

Lemma 5. Let $(a_n, \ldots, a_1, p) = 1$ and let p^μ be the highest power of p which divides all the coefficients of the polynomial $\varphi(\lambda + p^u x) - \varphi(\lambda)$. Then

$$u \leqq \mu \leqq nu.$$

Proof. We have

$$\varphi(\lambda + p^u x) - \varphi(\lambda) = a_n(\lambda + p^u x)^n$$
$$+ \ldots + a_\tau(\lambda + p^u x)^\tau + \ldots + a_1(\lambda + p^u x) - \varphi(\lambda).$$

The right-hand side is reducible to an n-th degree polynomial in x, all coefficients of which are divisible by p^u. Therefore, $u \leqq \mu$. Let a_τ be the first coefficient of $\varphi(x)$, which according to thecondition $(a_n, \ldots, a_1, p) = 1$, is not divisible by p. The coefficient of x^τ in $\varphi(\lambda + p^u x) - \varphi(\lambda)$ is obtained by adding $a_\tau p^{u\tau}$ to the numbers, each of which is necessarily divisible by one of the numbers $a_n p^{u\tau}, \ldots, a_{\tau+1} p^{u\tau}$, and, consequently, by $p^{u\tau+1}$. Therefore this coefficient is divisible by $p^{u\tau}$, but not by $p^{u\tau+1}$. Hence, $\mu \leqq \tau u \leqq nu$.

Lemma 6. Let p be a prime and l an integer greater than 1; also let $(a_n, \ldots, a_1, p) = 1$. Then

$$\left| S\left(\frac{\varphi(x)}{p^l}\right) \right| < n^{2n} p^{l(1-\nu)}.$$

Proof. Let t be the least of the numbers in (3). Assume that

$$x \equiv \lambda_1 \pmod{p^{t+1}}, \ldots, x \equiv \lambda_h \pmod{p^{t+1}}$$

are all different solutions of the congruence

$$\varphi'(x) \equiv 0 \pmod{p^{t+1}},$$

and that the representatives $\lambda_1, \ldots, \lambda_h$ of these solutions are chosen from the numbers $1, \ldots, p^{t+1}$.

Since the number of solutions of the congruence $p^{-t}\varphi'(x) \equiv 0 \pmod p$ does not exceed $(n-1)$, we have $h < p^t(n-1)$. By virtue of the condition $(a_n, \ldots, a_1, p) = 1$, among the coefficients $a_n, \ldots, a_1$, there exists some a_τ not divisible by p. Since $\tau\alpha_\tau$ is divisible by p^t, so τ is also divisible by p^t. Therefore $p^t \leqq n$ and, consequently, $h < n^2$. First consider the case where $l \leqq 2t+1$. We have

$$\left| S\left(\frac{\varphi(x)}{p^l}\right) \right| \leqq p^l \leqq p^{2t+1} \leqq n^2 p \leqq n^2 p^{l(1-\nu)}.$$

Then consider the case where $l > 2t+1$. We have

$$S\left(\frac{\varphi(x)}{p^l}\right) = \sum_{\xi=1}^{p^{t+1}} S_\xi; \qquad S_\xi = \sum_{\substack{1 \leqq x \leqq p^l \\ x \equiv \xi (mod\, p^{t+1})}} e^{2\pi i \frac{\varphi(x)}{p^l}}. \tag{5}$$

If ξ is not equal to any one of the numbers $\lambda_1, \ldots, \lambda_h$, we find

$$S_\xi = \sum_{\substack{1 \leqq y \leqq p^{l-t-1} \\ y \equiv \xi (\mathrm{mod}\, p^{t+1})}} \sum_{z=0}^{p^{t+1}-1} e^{2\pi i \frac{\varphi(y+p^{l-t-1}z)}{p^l}}$$

$$= \sum_{\substack{1 \leqq y \leqq p^{l-t-1} \\ y \equiv \xi (\mathrm{mod}\, p^{t+1})}} \sum_{z=0}^{p^{t+1}-1} e^{2\pi i \frac{\varphi(y)+\varphi'(y)p^{l-t-1}z}{p^l}}$$

$$= \sum_{\substack{1 \leqq y \leqq p^{l-t-1} \\ y \equiv \xi (\mathrm{mod}\, p^{t+1})}} e^{2\pi i \frac{\varphi(y)}{p^l}} \sum_{z=0}^{p^{t+1}-1} e^{2\pi i \frac{p^{-t}\varphi'(y)z}{p}} = 0.$$

Therefore, the equality (5) gives

$$S\left(\frac{\varphi(x)}{p^l}\right) = \sum_{r=1}^{h} S_{\lambda_r}. \tag{6}$$

First consider the case where $l \leqq n(t+1)$. Here we have

$$|S_{\lambda_r}| \leqq p^{l-t-1} \leqq p^{l(1-\nu)},$$
$$\left| S\left(\frac{\varphi(x)}{p^l}\right) \right| < n^2 p^{l(1-\nu)}. \tag{7}$$

Now consider the case where $l > n(t+1)$.

First assume that $\operatorname{ind} \varphi(px) = \operatorname{ind} \varphi(x)$. Let λ_r be any one of the numbers $\lambda_1, \ldots, \lambda_h$. Since in the sum S_{λ_r} summation is taken over all the values of x satisfying the congruence $\varphi'(x) \equiv 0 \pmod{p^{t+1}}$, by Lemma 4, all these values of x are multiples of p.

To the right-hand side of (6), on adding all those sums of S_ξ, where ξ is a multiple of p but is not a number from the sequence $\lambda_1, \ldots, \lambda_h$, we transform the right-hand side, without changing its value, into the sum of all terms in the sum $S(\varphi(x)/p^l)$ corresponding to the values of x which are multiples of p. Consequently,

$$S\left(\frac{\varphi(x)}{p^l}\right) = \sum_{x=1}^{p^{l-1}} e^{2\pi i \frac{\varphi(px)}{p^l}},$$

$$\left|S\left(\frac{\varphi(x)}{p^l}\right)\right| = \left|\sum_{x=1}^{p^{l-1}} e^{2\pi i \frac{\varphi(px)-\varphi(0)}{p^l}}\right|.$$

Let p^μ be the highest power of p dividing all the coefficients of the polynomial $\varphi(px) - \varphi(0)$. According to Lemma 5, we have $1 \leqq \mu \leqq n$. Putting $\varphi(px) - \varphi(0) = p^\mu g(x)$, we find that the greatest common divisor of the coefficients of $g(x)$ and p is unity and that

$$\left|S\left(\frac{\varphi(x)}{p^l}\right)\right| \leqq p^{\mu-1} \left|\sum_{x=1}^{p^{l-\mu}} e^{2\pi i \frac{g(x)}{p^{l-\mu}}}\right|,$$

or

$$\left|S\left(\frac{\varphi(x)}{p^l}\right)\right| \leqq p^{\mu(1-\nu)} \left|\sum_{x=1}^{p^{l-\mu}} e^{2\pi i \frac{g(x)}{p^{l-\mu}}}\right|. \tag{8}$$

Now assume that ind $\varphi(px) <$ ind $\varphi(x)$. Let S_{λ_r} be one of the sums $S_{\lambda_1}, \ldots, S_{\lambda_h}$ with the greatest value of the modulus. From the equality (6), it follows that

$$\left|S\left(\frac{\varphi(x)}{p^l}\right)\right| < n^2 \, |S_{\lambda_r}|.$$

Thus,

$$|S_{\lambda_r}| = \left|\sum_{x=1}^{p^{l-t-1}} e^{2\pi i \frac{\varphi(\lambda_r + p^{t+1}x) - \varphi(\lambda_r)}{p^l}}\right|.$$

Let p^{μ_r} be the highest power of p dividing all the coefficients of the polynomial $\varphi(\lambda_r + p^{t+1}x) - \varphi(\lambda_r)$. By Lemma 5, we have

$$t+1 \leqq \mu_r \leqq n(t+1).$$

Assuming that $\varphi(\lambda_r + p^{t+1}x) - \varphi(\lambda_r) = p^{\mu_r} g_r(x)$, we find that the greatest common divisor of the coefficients of $g_r(x)$ and p is 1 and that

$$|S_{\lambda_r}| = p^{\mu_r - t - 1} \left|\sum_{x=1}^{p^{l-\mu_r}} e^{2\pi i \frac{g_r(x)}{p^{l-\mu_r}}}\right| \leqq p^{\mu_r(1-\nu)} \left|S\left(\frac{g_r(x)}{p^{l-\mu_r}}\right)\right|.$$

Consequently,

$$\left|S\left(\frac{\varphi(x)}{p^l}\right)\right| < n^2 p^{\mu_r(1-\nu)} \left|S\left(\frac{g_r(x)}{p^{l-\mu_r}}\right)\right|. \tag{9}$$

Furthermore, applying Lemma 3, the first assertion of Lemma 4, and the inequality ind $\varphi(px) <$ ind $\varphi(x)$, we obtain the following relations:

$$\begin{aligned}\operatorname{ind} g_r(x) &= \operatorname{ind} p^{\mu_r} g_r(x) = \operatorname{ind}\,(\varphi(\lambda_r + p^{t+1}x) - \varphi(\lambda_r))\\ &= \operatorname{ind}\varphi(p^{t+1}x) \leqq \ldots \leqq \operatorname{ind}\varphi(p^2 x) \leqq \operatorname{ind}\varphi(px) < \operatorname{ind}\varphi(x),\end{aligned}$$

from which we find

$$\operatorname{ind} g_r(x) < \operatorname{ind}\varphi(x). \tag{10}$$

Applying either of the inequalities in (8) (if $\operatorname{ind}\varphi(px) = \operatorname{ind}\varphi(x)$) or the inequality (9) (if $\varphi(px) < \operatorname{ind}\varphi(x)$), we reduce the sum $S(\varphi(x)/p^l)$ to a similar sum but with smaller l until finally we arrive at a sum with an l not exceeding $n(t+1)$, to which we apply inequality (7).

Here, the coefficient n^2 will occur not more than $(n-1)$ times, because each application of the inequality (9) reduces the index by at least unity, and not more than once because of the application of the inequality (7).

Thus, the lemma follows trivially.

Theorem 1 (Hua). Let $(a_n, \ldots, a_1, p) = 1$. Then

$$\left|S\left(\frac{\varphi(x)}{q}\right)\right| < c(n, \varepsilon_0)\, q^{1-\nu+\varepsilon_0}.$$

Proof. Without loss of generality, we can assume that $\varphi(0) = 0$. Let

$$q = p_1^{\alpha_1} \ldots p_\sigma^{\alpha_\sigma}$$

be the canonical expansion of q. By Lemma 1, we have

$$S\left(\frac{\varphi(x)}{q}\right) = S\left(\frac{\varphi_1(x)}{p_1^{\alpha_1}}\right) \ldots S\left(\frac{\varphi_\sigma(x)}{p_\sigma^{\alpha_\sigma}}\right).$$

Here, according to the Note to Lemma 1, in the expression of the sum on the right-hand side the coefficients of the numerator and denominator are coprime. Applying Lemma 2 or Lemma 6 to each such sum, we obtain

$$\left|S\left(\frac{\varphi(x)}{q}\right)\right| < n^{2n\sigma} q^{1-\nu}.$$

But according to Lemma 4 of Chapter 1, we find

$$n^{2n\sigma} = 2^{\frac{2n\ln n}{\ln 2}\sigma} \leqq \left(2^{2^{1/\varepsilon}} q^{\varepsilon}\right)^{\frac{2n\ln n}{\ln 2}}.$$

Hence, putting

$$\varepsilon_0 = \varepsilon\,\frac{2n\ln n}{\ln 2},$$

we prove the theorem.

Theorem 2. Let m be an integer satisfying the condition $0 < m < q$. Let y_1 be coprime to q and let y run through a reduced system of residues modulo q. Then, for constant n and under the condition

$$(B_n, \ldots, B_1, q) = 1,$$

we have

$$\sum_y \left| \frac{1}{q} \sum_{x=1}^{q} e^{2\pi i \frac{mB_n(y^n - y_1^n)x^n + \ldots + mB_1(y-y_1)x}{q}} \right| \ll (m,q)^\nu q^{1-\nu+\varepsilon'} .$$

Proof. Putting

$$(m,q) = d,\ m = m_1 d,\ q = q_1 d,\ m_1 B_s = D_s ,$$

we find that the fraction in the exponent is given by the expression

$$\frac{D_n(y^n - y_1^n)x^n + \ldots + D_1(y-y_1)x}{q_1}; \quad (D_n, \ldots, D_1, q_1) = 1 .$$

Further, putting

$$(D_n(y^n - y_1^n), \ldots, D_1(y-y_1), q_1) = \delta ,$$
$$q_1 = q_2\delta,\ D_s(y^s - y_1^s) = E_s\delta_1$$

we find that the same fraction is also given by the expression

$$\frac{E_n x^n + \ldots + E_1 x}{q_2}; \quad (E_n, \ldots, E_1, q_2) = 1 .$$

Therefore, according to Theorem 1, the term corresponding to a given y will be

$$\ll q_2^{-\nu+\varepsilon_0} \ll d^\nu \delta^\nu q^{-\nu+\varepsilon_0} .$$

Since δ divides all numbers of the type $D_s(y^s - y_1^s)$; $s \leqq n$, it also divides the number $y^{s_0} - y_1^{s_0}$; $s_0 = n!$ And since the congruence $y^{s_0} - y_1^{s_0} \equiv 0 \pmod{\delta}$ has $\ll \delta^{\varepsilon''}$ solutions, it is satisfied by $\ll q\delta^{-1+\varepsilon''}$ values of y. Therefore the sum of all the terms corresponding to given δ is

$$\ll q\delta^{-1+\varepsilon''} d^\nu \delta^\nu q^{-\nu+\varepsilon_0} \ll d^\nu q^{1-\nu+\varepsilon_0} \delta^{\nu-1+\varepsilon''} .$$

And the sum of the terms corresponding to all δ dividing q is (by Lemma 5 of Chapter 1)

$$\ll d^\nu q^{1-\nu+\varepsilon_0} \tau(q) \ll d^\nu q^{1-\nu+\varepsilon'} .$$

Theorem 3. Let m be a natural number and

$$S' = \sum_x{}' e^{2\pi i \frac{m\varphi(x)}{q}} ,$$

where x runs through a reduced system of residues modulo q and let the condition $(a_n, \ldots, a_1, p) = 1$ be satisfied. Then, for constant n, we have

$$S' \ll (m,q)^\nu q^{1-\nu+\varepsilon'} .$$

Proof. We have

$$S' = \sum_{d \backslash q} \mu(d) S_d; \quad S_d = \sum_{0<dz\leqq q} e^{2\pi i \frac{m\varphi(dz)}{q}} .$$

Let δ be the greatest common divisor of all the coefficients of $m\,\varphi(dz)$ and q. Clearly, δ is divisible by d. We find that

$$S_d = \sum_{0<dz\leqq q} e^{2\pi i \frac{m\varphi(dz)\delta^{-1}}{q\delta^{-1}}}.$$

Hence applying Theorem 1, we get

$$S_d \ll \frac{\delta}{d}(q\delta^{-1})^{1-\nu+\varepsilon_0} \ll \frac{\delta^\nu}{d} q^{1-\nu+\varepsilon_0},$$

which, by virtue of $\delta\backslash(m, q)\, d^n$, is

$$\ll (m, q)^\nu q^{1-\nu+\varepsilon}.$$

Therefore, applying Lemma 5 of Chapter 1, we find that the theorem is proved.

Chapter 4. Weyl Sums

Notation. For $n \geqq 3$ and for real $\alpha_n, \ldots, \alpha_1$, we shall assume $f(x) = \alpha_n x^n + \ldots + \alpha_1 x$. Let P denote an integer greater than 1, and m, a positive integer. We shall suppose that

$$T = T_1 = T(\alpha_n, \ldots, \alpha_1) = \sum_{0<x\leqq P} e^{2\pi i f(x)},$$

$$T_m = T(m\alpha_n, \ldots, m\alpha_1).$$

For a positive integer b, integral β satisfying the condition $0 \leqq \beta \leqq 2b$, and arbitrary integers $A_n, \ldots, A_1$, the number of solutions of Diophantine equations

$$x_1^n + \ldots + x_\beta^n - x_{\beta+1}^n - \ldots - x_{2b}^n = A_n,$$

$$\cdots\cdots\cdots\cdots\cdots\cdots\cdots\cdots$$

$$x_1 + \ldots + x_\beta - x_{\beta+1} - \ldots - x_{2b} = A_1,$$

where each x_j runs independently of the others through the values $1, \ldots, P$ will be denoted by $I_{2B,\beta,A_n,\ldots,A_1}$. The value of this symbol for $\beta = b$, $A_n = \ldots = A_1 = 0$ will be denoted by I_{2b} or simply I.

An important problem in number theory lies in finding an upper bound for $I_{2b,\beta,A_n,\ldots,A_1}$. From the representations of $I_{2b,\beta,A_n,\ldots,A_1}$ and I in the form of integrals

$$I_{2b,\beta,A_n,\ldots,A_1} = \int_0^1 \ldots \int_0^1 T^\beta \, T^{2b-\beta} e^{-2\pi i(A_n\alpha_n + \ldots + A_1\alpha_1)} \, d\alpha_n \ldots d\alpha_1,$$

$$I = \int_0^1 \ldots \int_0^1 |T|^{2b} \, d\alpha_n \ldots d\alpha_1,$$

which can be easily derived, we trivially get

$$I_{2b,\beta,A_n,\ldots,A_1} \leqq I.$$

Therefore the problem of finding an upper bound for $I_{2b,\beta,A_n,\ldots,A_1}$ can be reduced to a simpler one, namely the problem of finding an upper bound for I. The solution of this upper bound problem and a closely interrelated problem, i.e. an upper bound for the modulus of the sum T, are dealt with in this chapter.

Obviously, without changing the value of the integral I, we can take any domain of the type

$$h_n \leqq \alpha_n < h_n + 1, \ \ldots, \ h_1 \leqq \alpha_1 < h_1 + 1,$$

as its range of integration. We shall denote this domain by the symbol Π_n, or in a more extended form by

$$\Pi_n(h_n, \ldots, h_1).$$

It should be noted that every point in an n-dimensional space is congruent to one and only one point of this domain.

We shall define two upper bounds for the integral I: "general upper bound" (Theorem 1) and "particular upper bound" (Theorem 4). We shall discuss the particular bound first as it is simpler.

If $n \geqq 3$ and

$$b \geqq b_0; \quad b_0 = [n^2(2\ln n + \ln\ln n + 4)],$$

then the particular upper bound is given by an inequality of the type

$$I \leqq c(n, b)\, P^{2b - \frac{n(n+1)}{2}}.$$

It is easy to see that the number b_0, in the sense of its order of growth with increasing n, cannot be improved any further. Since it is of order $n^2 \ln n$, it cannot be replaced even by $0.5n^2$, which is of order n^2.

Indeed, the integral I can be regarded as the number of solutions of the system of equations

$$\begin{gathered} x_1^n + \ldots + x_b^n = x_{b+1}^n + \ldots + x_{2b}^n, \\ \cdots\cdots\cdots\cdots\cdots\cdots \\ x_1 + \ldots + x_b = x_{b+1} + \ldots + x_{2b}, \end{gathered} \tag{1}$$

where each x_j, independently of the others, runs through $1, \ldots, P$.

To each set of $x_1, \ldots, x_b$, there corresponds at least one solution of this system. Therefore, $I \geqq P^b$, and if the bound of I is true for some $b < b_0$, then

$$P^b \leqq I \leqq c(n, b)\, P^{2b - \frac{n(n+1)}{2}}.$$

Selecting P so large that $c(n, b) < P^{0.5n}$, we get $b > 0.5n^2$.

It is easy to show that this bound is, in the sense of the order of growth with increasing P, exact.

Indeed, for the sake of clarity, selecting the domain $\Pi_n(-0.5, \ldots, -0.5)$ as the domain Π_n, consider the domain γ bounded by the inequalities:

$$-\frac{P^{-n}}{13n} \leqq \alpha_n < \frac{P^{-n}}{13n}, \ldots, -\frac{P^{-1}}{13n} \leqq \alpha_1 < \frac{P^{-1}}{13n}.$$

The volume of the domain γ is

$$c_1(n)\, P^{\frac{-n(n+1)}{2}}; \quad c_1(n) = (6.5n)^{-n}.$$

The domain γ contains the origin of coordinates with a value of T equal to P. Therefore (Lemma 9, Chapter 2) the value of T at any point of this domain differs from P by a number that does not exceed

$$2\pi n \frac{1}{13n} P < 0.5P,$$

in absolute value and, consequently, is numerically not greater than $0.5P$. Since the domain γ is a part of the domain Π_n, we readily get

$$I > c_2(n)\, P^{2b - \frac{n(n+1)}{2}}; \quad c_2(n) = c_1(n)\,(0.5)^{2b}.$$

While the particular upper bound is used in those cases where the number n remains unchanged as P increases to infinity, the general upper bound can be used in those cases where n may grow, though slowly, as P increases to infinity.

Let $n \geqq 3$ and, for a positive integer l, let b_l be defined by the equality

$$b_l = nl.$$

Then, for $b \geqq b_l$, the general upper bound is given by

$$I < (nl)^{3nl} (2n)^{\frac{3n(n+1)l}{2}} P^{2b - \frac{n(n+1)}{2} + \frac{n(n+1)}{2}(1-\nu)^l}.$$

A specific feature of this bound is that it contains a positive integer l, which can be chosen arbitrarily from the conditions of the problem, paying due allowance for the fact that an increase in l, while improving the factor dependent on P, simultaneously increases the lower bound b_l of the number b and also the factor dependent only on n.

This factor, dependent only on n, is a very rapidly increasing function of n. Its substitution by some other quantity, a much more slowly varying function of n, would be a great refinement of the general upper bound.

A direct corollary of Theorem 1 which gives the general upper bound of I is Theorem 2. In a simple form it is formulated as follows:

Let k be a constant $\geqq 8$. From an n-dimensional domain Π_n (of volume 1) we can choose such a small domain Ω of volume V, that

$$V < c(n,k)\, P^{-\frac{(n+1)n}{2} g_k}; \qquad g_k = 1 - \frac{4\ln k - 1}{k},$$

where $c(n,k)$ is a certain constant dependent only on n and k, and that for the complementary region of the domain Π_n we have

$$|T| \leqq P^{1-1/k}.$$

We can easily show that it is impossible to replace the inequality for V by some other inequality $V < c'(n,k)\, P^{-\frac{n(n+1)}{2} g'}$ of the same type satisfying the condition $g' = g(n,k) > 1$. Indeed, let us consider the domain γ once again. We have shown that for any point in the domain γ, the inequality $|T| > 0.5P$ is satisfied: this inequality, when $P > 2^k$, gives the inequality $|T| > P^{1-1/k}$. Therefore, for $P > 2^k$, the domain γ will be a part of the domain Ω and, consequently,

$$V \geqq c_2(n)\, P^{-\frac{n(n+1)}{2}},$$

from which our assertion follows immediately.

It is useful to compare the following elementary lemma with Theorem 2:

Lemma. Let λ_1 and λ be any numbers satisfying the condition:

$$0 < \lambda_1 < \lambda < 0.5.$$

Then, from the domain Π_n it is possible to choose a domain Ω' of volume V' not exceeding

$$V_0' = P^{2\lambda - 1},$$

such that, for the points in the remaining domain, we have

$$|T| \leqq P^{1-\lambda_1}.$$

Proof. We find, trivially

$$\int_0^1 \cdots \int_0^1 |T|^2 \, d\alpha_n \ldots d\alpha_1 = P.$$

The domain Π_n can be divided into identical n-dimensional cubes with such small sides that the modulus of the difference between the values of T corresponding to two points belonging to the same cube is less than $P^{1-\lambda_1} - P^{1-\lambda}$. By the domain Ω' we mean a domain covered with cubes, each containing at least one point at which the estimate stated in the lemma is not true. Then for any point in the domain Ω', the inequality $|T| > P^{1-\lambda}$ is satisfied. For the volume V' of this domain, we have $V' P^{2-2\lambda} \leqq P$; hence we find $V' \leqq V_0'$.

Theorem 2 differs radically from the lemma in that with unbounded increase in P the number V_0 in Theorem 2 tends to zero much more rapidly than V_0' in the lemma (V_0 tends to zero as a quantity of order $P^{-\frac{n(n+1)}{2}k}$, whereas V_0' approaches zero only as a quantity of order as small as $P^{2\lambda-1} > P^{-1}$). Therefore, though the upper bound of $|T|$ in Theorem 2 is much worse than in the lemma, Theorem 2 is a powerful tool in solving many important problems in number theory; unlike the lemma, which is of no use in this respect.

The upper bound for the modulus of the sum T will be considered, with due regard for the requirements of Chapter 5, as a particular case of an upper bound for the modulus of the sum T_m. Here we shall only discuss in detail the general solution of this problem contained in Theorem 3 and meant to be applied to cases where the number n remains constant as P increases to infinity. This general theorem is quite sufficient for the purposes of subsequent chapters. It is stated as follows:

Let $n \geqq 3$. Divide the points of an n-dimensional space into two classes as follows. Class I contains all points of the type

$$\left(\frac{a_n}{q_n} + z_n, \ldots, \frac{a_1}{q_1} + z_1\right),$$

where the first factors in the coordinates are rational irreducible fractions with positive denominators whose least common multiple is a number Q not exceeding P^ν, while the second factors satisfy the conditions:

$$|z_s| \leqq P^{-s+\nu}.$$

Class II contains all points which are not the points of Class I.
Now assuming

$$\varrho = \frac{1}{8n^2(\ln n + 1.5 \ln \ln n + 4.2)},$$

if $m \leqq P^{2\varrho}$, for the points of Class II we have

$$T_m \ll P^{1-\varrho}.$$

And assuming

$$\delta_0 = \max(|z_n| P^n, \ldots, |z_1| P),$$

if $m \leqq P^{4\nu^2}$ for the points of Class I we have

$$T_m \ll P(m, Q)^\nu Q^{-\nu+\varepsilon}$$

or

$$T_m \ll PQ^{-\nu+\varepsilon_0}\delta_0^{-\nu}, \quad \text{if } \delta_0 \geqq 1 .$$

The main advantage of this general solution is that the domain of the points belonging to Class I is almost a negligible part of the domain Π_n. Simple calculation shows that the volume of this part is $\ll P^{1+2\nu-n(n+1)/2}$. For the remaining part of the domain Π_n (the domain of the points belonging to Class II), the upper bound of the modulus of the sum T is uniformly bounded by the inequality $T \ll P^{1-\varrho}$.

My method gives non-trivial estimates for sums which are extensions of the sum T to a case where the function $f(x)$, while not being a polynomial of degree n, is close, in some respects, to such a polynomial. In view of the needs of the modern theory of distribution of primes, I have included Lemma 9, which gives the estimate for one such sum, at the end of this chapter. This lemma is stated as follows:

Let $n \geqq 20$, a and a_1 be integers $(a < a_1 \leqq 2a)$ and let $t = a^{n-\theta}$ $(0 \leqq \theta < 1)$. Then for the sum

$$S = \sum_{a < u \leqq a_1} e^{2\pi i F(u)}; \qquad F(u) = -\frac{t \ln u}{2\pi},$$

we have

$$|S| < 2a^{1 - \frac{1}{30{,}000 n^2}}.$$

Of the two proofs of this lemma which I published in 1957 and 1965, I give here the second, as it is simpler.

Lemma 9 has given a new estimate for the function $\zeta(1+it)$ and, on its basis, a new estimate for the remainder term in the asymptotic expression for $\pi(N)$ (see Introduction).

Now we shall proceed to present the main matter of the chapter.

Lemma 1. Let r and m be positive integers, and $x_1, \ldots, x_r$ be non-negative integers. Then

a) $(x_1 + \ldots + x_r)^m \leqq r^{m-1}(x_1^m + \ldots + x_r^m)$,

b) $x_1 \ldots x_r \leqq \dfrac{x_1^r + \ldots + x_r^r}{r}$.

Lemma 2. Let n and b be positive integers and let each $x_1, \ldots, x_b, x_{b+1}, \ldots, x_{2b}$ run through its own sequence of integers such that x_{b+j} runs through the same sequence of values as x_j. Then the numerical value of the integral

$$J_{2b} = \int_0^1 \ldots \int_0^1 \sum_{x_1} e^{2\pi i f(x_1)} \ldots \sum_{x_b} e^{2\pi i f(x_b)}$$
$$\times \sum_{x_{b+1}} e^{-2\pi i f(x_{b+1})} \ldots \sum_{x_{2b}} e^{-2\pi i f(x_{2b})}\, d\alpha_n \ldots d\alpha_1$$

does not change, if, for an integer a, $f(a+x_j)$ is substituted for each $f(x_j)$.

Proof. Indeed, the integral J_{2b} is equal to the number of solutions of the system

$$X_n = \ldots = X_1 = 0;$$
$$X_s = x_1^s + \ldots + x_b^s - x_{b+1}^s - \ldots - x_{2b}^s.$$

The transformed integral is equal to the number of solutions of the system

$$X_n + \binom{n}{1} X_{n-1} a + \ldots + \binom{n}{1} X_1 a^{n-1} = 0,$$
$$\ldots\ldots\ldots\ldots\ldots\ldots\ldots\ldots$$
$$X_2 + \binom{2}{1} X_1 a = 0,$$
$$X_1 = 0,$$

This number (as can be readily verified by considering a new sequence taken in reverse order) is equal to the number of solutions of the first system.

Lemma 3. Let $n > 2$, $P > (2n)^{4n}$, $H = (2n)^3$, R be the least integer satisfying the condition $HR \geqq P$, and finally, let $v_1, \ldots, v_n$ run through the integers in the intervals:

$$X_1 < v_1 \leqq Y_1, \ \ldots, \ X_n < v_n \leqq Y_n, \tag{1}$$

where for some ω satisfying the condition $0 \leqq \omega < P$, we have

$$-\omega < X_1, \quad X_1 + R = Y_1,$$
$$Y_1 + R \leqq X_2, \ \ldots, \ X_n + R = Y_n, \quad Y_n \leqq -\omega + P.$$

Then the number E_1 of system of values of $v_1, \ldots, v_n$ such that the sums

$$V_1 = v_1 + \ldots + v_n, \ \ldots, \ V_n = v_1^n + v_2^n + \ldots + v_n^n$$

lie in some intervals of lengths, respectively,

$$1, \ \ldots, \ P^{n-1}, \tag{2}$$

satisfies the inequality:

$$E_1 < H^{\frac{n(n-1)}{2}} e^{1.15n + 0.95}.$$

If $v_1', \ldots, v_n'$ run through the same values as $v_1, \ldots, v_n$ (independently of the latter), then the number E of cases where the differences $V_1 - V_1', \ldots, V_n - V_n'$ lie in given intervals of lengths

$$P^{1-\nu}, \ \ldots, \ P^{n(1-\nu)}, \tag{3}$$

satisfies the inequality

$$E < H^{\frac{n(n-1)}{2}} H^{-n} e^{1.72n} P^{\frac{3n-1}{2}}.$$

Proof. First we shall evaluate E_1. Let s be an integer such that $1 < s \leqq n$. If, for given $v_{s+1}, \ldots, v_n$, the sums $V_1, \ldots, V_n$ lie, respectively, within certain intervals of lengths (2), then the sums

$$v_1 + \ldots + v_s, \ \ldots, \ v_1^s + \ldots + v_s^s$$

lie, respectively, in certain intervals of lengths

$$1, \ldots, P^{s-1}.$$

Let $\eta_1, \ldots, \eta_s$ and $\eta_1 + \xi_1, \ldots, \eta_s + \xi_s$ be two systems of values of $v_1, \ldots, v_s$ having such a property and least η_s (consequently, $\xi_s > 0$). Then

$$\frac{(\eta_1+\xi_1)-\eta_1}{\xi_1}\xi_1 + \ldots + \frac{(\eta_s+\xi_s)-\eta_s}{\xi_s}\xi_s = \theta_0,$$

$$\ldots\ldots\ldots\ldots\ldots\ldots\ldots\ldots\ldots\ldots\ldots$$

$$\frac{(\eta_1+\xi_1)^s-\eta_1^s}{s\xi_1}\xi_1 + \ldots + \frac{(\eta_s+\xi_s)^s-\eta_s^s}{s\xi_s}\xi_s = \frac{\theta_{s-1}}{s}P^{s-1}.$$

Hence we find

$$\Delta\xi_s - \Delta' = 0;$$

$$\Delta = \begin{vmatrix} \frac{(\eta_1+\xi_1)-\eta_1}{\xi_1} & \ldots & \frac{(\eta_s+\xi_s)-\eta_s}{\xi_s} \\ \ldots & \ldots & \ldots \\ (\eta_1+\xi_1)^s-\eta_1^s & \ldots & \frac{(\eta_s+\xi_s)^s-\eta_s^s}{s\xi_s} \end{vmatrix},$$

$$\Delta' = \begin{vmatrix} \frac{(\eta_1+\xi_1)-\eta_1}{\xi_1} & \ldots & \frac{(\eta_{s-1}+\xi_{s-1})-\eta_{s-1}}{\xi_s} & \theta_0 \\ \ldots & \ldots & \ldots & \ldots \\ \frac{(\eta_1+\xi_1)^s-\eta_1^s}{s\xi_1} & \ldots & \frac{(\eta_{s-1}+\xi_{s-1})^s-\eta_{s-1}^s}{s\xi_{s-1}} & \frac{\theta_{s-1}}{s}P^{s-1} \end{vmatrix}. \tag{4}$$

Now apply the following transformation to equality (4). Expand both its determinants in terms of the elements of the first column and then apply the Lagrange formula to the result, regarding it as the difference of the values of a certain function of v_1 for $v_1 = \eta_1 + \xi_1$ and $v_1 = \eta_1$. Thus we obtain a new equality in which the elements of the first column of each determinant are replaced by $1, \ldots, x_1^{s-1}$, respectively, where x_1 is a number satisfying the condition $X_1 < x_1 < Y_1$. Carrying out similar transformations with respect to subsequent columns up to the penultimate and finally the last column for which only the first determinant is subjected to transformation, we obtain

$$\Delta_s\xi_s - \Delta_s' = 0, \quad \Delta_s = \begin{vmatrix} 1 & \ldots & 1 \\ \ldots & \ldots & \ldots \\ x_1^{s-1} & \ldots & x_s^{s-1} \end{vmatrix},$$

$$\Delta_s' = \begin{vmatrix} 1 & \ldots & 1 & \theta_0 \\ \ldots & \ldots & \ldots & \ldots \\ x_1^{s-1} & \ldots & x_{s-1}^{s-1} & \frac{\theta_{s-1}}{s}P^{s-1} \end{vmatrix},$$

$$X_1 < x_1 < Y_1, \ \ldots, \ X_s < x_s < Y_s.$$

Hence we find

$$\Delta_s' = \sum_{r=0}^{s-1} \frac{\theta_r}{r+1} P^r U_r,$$

where U_r is the coefficient of x_s^r in the expansion of

$$\Delta_s = (x_s - x_1) \dots (x_s - x_{s-1}) \Delta_{s-1}$$

in powers of x_s and, consequently, is equal to the product of Δ_{s-1} and the sum of the products of the numbers $-x_1, \dots, -x_{s-1}$ taken $(s-1-r)$ factors at a time. Therefore, we have

$$U_r \leqq \Delta_{s-1} \binom{s-1}{r} P^{s-1-r},$$

$$\xi_s < \sum_{r=1}^{s-1} \frac{\binom{s-1}{r} P^{s-1}}{(r+1)(x_s - x_1) \dots (x_s - x_{s-1})},$$

Hence, by virtue of the inequality $x_{j+t} - x_j \geqq (2t-1) R$, which is true for $t \geqq 1$, we obtain

$$\xi_s < \sum_{r=1}^{s} \frac{\binom{s}{r} H^{s-1}}{1 \cdot 3 \cdot \dots \cdot (2s-3) s} < \frac{(2^{s+1}-2) H^{s-1}}{3 \dots (2s-1)} < L_s H^{s-1} - 1;$$

$$L_s = \frac{4}{(2-0.5) \dots (s-0.5)},$$

where, since

$$\ln(2-0.5) + \dots + \ln(s-0.5) \geqq \int_1^s \ln x \, dx = s \ln s - s + 1 .$$

we have

$$L_s < 4e^{s-1} s^{-s}.$$

Thus, from the foregoing, it follows that v_s, for $s > 1$ and given $v_{s+1}, \dots, v_n$, can have only less than $4e^{s-1} s^{-s} R^{s-1}$ different values. And since v_1, for given $v_2, \dots, v_n$, lies in an interval of length 1 and, consequently, cannot have more than two different values, we have

$$E_1 < 2 \prod_{s=2}^{n} (4e^{s-1} s^{-s} H^{s-1}) = 2 \cdot 4^{n-1} (eH)^{\frac{n(n-1)}{2}} \prod_{s=2}^{n} s^{-s},$$

from which, since

$$\sum_{s=2}^{n} s \ln s > \int_1^n s \ln s \, ds + \frac{1}{2}(n-1) = \frac{n^2}{2} \ln n - \frac{n^2}{4} + \frac{n}{2} - \frac{1}{4},$$

we find

$$E_1 < (e^{1.5} H n^{-1})^{\frac{n(n-1)}{2}} n^{-\frac{n}{2}} e^{1.15n+0.95} < H^{\frac{n(n-1)}{2}} e^{1.15n+0.95} .$$

From this inequality, since

$$\left(\frac{P^{1-\nu}}{1}+1\right)\dots\left(\frac{P^{(n-1)(1-\nu)}}{P^{n-1}}+1\right)<e^{0.05}P^{\frac{n-1}{2}},$$

we obtain for the number E' of the systems of values of $v_1, \dots, v_n$ such that the sums $V_1, \dots, V_n$ lie, respectively, in given intervals of lengths (3), the following inequality

$$E'<H^{\frac{n(n-1)}{2}}e^{1.15n+1}P^{\frac{n-1}{2}}.$$

Finally, noting that the number of all systems $v_1, \dots, v_n$ is less than $2P^n H^{-n}$, we find

$$E<H^{\frac{n(n-1)}{2}}H^{-n}e^{1.72n}P^{\frac{3n-1}{2}}.$$

Theorem 1. Let n be an integer >2, $\nu=n^{-1}$, $f(x)=\alpha_n x^n+\dots+\alpha_1 x$, $P\geqq 1$, l a positive integer, $b\geqq nl$,

$$J_b(P)=\int_0^1\dots\int_0^1\left|\sum_{0<x\leqq P}e^{2\pi i f(x)}\right|^{2b}d\alpha_n\dots d\alpha_1.$$

Then

$$J_b(P)<D_l P^{2b-\Delta(l)};$$

$$D_l=(nl)^{3nl}(2n)^{\frac{3n(n+1)}{2}},\qquad \Delta(l)=\frac{n(n+1)}{2}(1-(1-\nu)^l).$$

Proof. Obviously, it suffices to demonstrate the theorem only for the case $b=nl$. For $l=1$ and any P, the theorem is true. Indeed, the integral $J_n(P)$ is equal to the number of solutions of the Diophantine system

$$x_1^n+\dots+x_n^n-x_{n+1}^n-\dots-x_{2n}^n=0,$$
$$\dots\dots\dots\dots\dots\dots\dots\dots$$
$$x_1+\dots+x_n-x_{n+1}-\dots-x_{2n}=0,$$

which is not greater than $n!\,P^n<D_1P^{2n-2}$. For $l>1$ and $P<D_l^{1/\Delta(l)}$ the theorem is obvious. Therefore, we shall only consider the case where $l>1$ and $P\geqq D_l^{1/\Delta(l)}$.

Let m and P_0 be natural numbers and assume the theorem true for $l\leqq m$ and $P\leqq P_0$ and also when $l\leqq m+1$ and $P<P_0$. We shall show that the theorem is also true when $l\leqq m+1$ and $P=P_0$; accordingly, by the mathematical induction principle, the theorem is always true.

In the proof, we shall use the following notation:

$$b=n(m+1),\qquad H=(2n)^3,\qquad R=[PH^{-1}+1].$$

Thus, we have $P\leqq RH$, and consequently, $J_b(P)\leqq J_b(RH)$.

Let y be an integer in the interval $0\leqq y<H$. Divide the interval $0<x\leqq RH$ into H small subintervals, such that for given y's, all values of x can be expressed in the form $x=Ry+z$, $(0<z\leqq R)$. Denote the corresponding part of the sum $\sum_{0<x\leqq RH}e^{2\pi i f(x)}$ by the symbol Z_y. Thus we obtain

$$J_b(P)\leqq\int_0^1\dots\int_0^{1}\left(\left|\sum_{y_1}\dots\sum_{y_b}Z_{y_1}\dots Z_{y_b}\right|\right)^2d\alpha_n\dots d\alpha_1. \tag{5}$$

A sequence of the type $\eta_1, \ldots, \eta_r$ and a product $Z_{\eta_1} \ldots Z_{\eta_r}$ associated with it are said to be regular, if from the numbers $\eta_1, \ldots, \eta_r$, an increasing sequence of n numbers can be so chosen that the difference between any largest and the least neighbouring terms in them is greater than unity. A sequence and its associated product are called irregular, if no such sequence can be chosen.

Let J_1 and J_2 be integrals similar to the integral on the right-hand side of the inequality (5); the first integral only contains regular products $Z_{y_1} \ldots Z_{y_b}$, while the second only irregular products $Z_{y_1} \ldots Z_{y_b}$. We find (Lemma 1, a) that

$$J_b(P) < 2J_1 + 2J_2. \tag{6}$$

We shall first evaluate $2J_1$; it does not, by Lemma 1, a, exceed $2H^{2b}$ multiplied by the maximum of an integral of the type

$$J' = \int_0^1 \ldots \int_0^1 |Z_1 \ldots Z_b|^2 \, d\alpha_n \ldots d\alpha_1,$$

where $Z_1 \ldots Z_b$ is one of the regular products $Z_{y_1} \ldots Z_{y_b}$ but with factors so rearranged that $Z_1 \ldots Z_n$ is a product of those factors which characterize $Z_{y_1} \ldots Z_{y_b}$, as a regular product and $Z_{n+1} \ldots Z_b$ is the product of the remaining factors. Divide each term in the latter into $\leqq k$ small sums; $k = [RP^{-1+\nu} + 1]$, each with a length of summation $P^{1-\nu}$, or possibly less for the last sum. Then $Z_{n+1} \ldots Z_b$ can be expressed as the sum of not more than k^{b-n} terms of the type $U_{n+1} \ldots U_b$, where U_s is one of the sums obtained in the subdivision of Z_s. The integral J' (Lemma 1, b) is not greater than $k^{2(b-n)}$ multiplied by the maximum of the integral

$$J'' = \int_0^1 \ldots \int_0^1 |Z_1 \ldots Z_n|^2 \, |U|^{2(b-n)} \, d\alpha_n \ldots d\alpha_1,$$

where U is one of the sums $U_{n+1}, \ldots, U_b$. After replacing the summation variables (let these be $z_1, \ldots z_n, u$) of the sums $Z_1, \ldots, Z_n, U$ by new variables $\zeta_1, \ldots, \zeta_n, w$ defined by the equalities $z_1 = a + \zeta_1, \ldots, z_n = a + \zeta_n$, $u = a + w$ (where a is the least value of u), we obtain (Lemma 2)

$$J'' = \int_0^1 \ldots \int_0^1 \sum_{S_1, \ldots, S_n} \; \sum_{W_1, \ldots, W_n} \exp \{2\pi i (\alpha_n (S_n + W_n) + \ldots + \alpha_1 (S_1 + W_1))\} \, d\alpha_n \ldots d\alpha_1,$$

$$S_j = \zeta_1^j + \ldots + \zeta_n^j - \zeta_{n+1}^j - \ldots - \zeta_{2n}^j,$$

$$W_j = w_1^j + \ldots + w_{b-n}^j - w_{b-n+1}^j - \ldots - w_{2b-n}^j.$$

Obviously, the numerical value of the integral J'' can only be altered by the values of $S_1, \ldots, S_n$ which do not go ourside certain intervals of lengths

$$2bP^{1-\nu} - 1, \ \ldots, \ 2bP^{n(1-\nu)} - 1$$

(because $W_1, \ldots, W_n$ do not lie outside such intervals of such lengths, respectively). Therefore (Lemma 3)

$$J'' < (2b)^n H^{\frac{n(n-1)}{2}} H^{-n} e^{1.72n} P^{\frac{3n-1}{2}} J_{b-n}(P^{1-\nu}).$$

Furthermore, we find ($l = m+1$)

$$l > \Delta(l) > \frac{l(n+1)}{4} \quad \text{and} \quad P > (2n)^{6n}, \quad \text{if} \quad l \leqq n;$$

$$l > \Delta(l) > \frac{n(n+1)}{4} \quad \text{and} \quad P > (2n)^{6l}, \quad \text{if} \quad l > n;$$

$$R < PH^{-1}\left(1 + \frac{H}{P}\right), \qquad R^{2b-\Delta(l)} < 2P^{2b-\Delta(l)}\,H^{-2b+\Delta(l)},$$

$$k < P^{\nu} H^{-1}\left(1 + \frac{2H}{P^{\nu}}\right), \qquad k^{2(b-n)} < 2P^{2\frac{b-n}{n}}\,H^{-2(b-n)},$$

$$2J_1 < 2H^{2b}\,2H^{-2b+2n}\,P^{2\frac{b-n}{n}}\,(2b)^n \times H^{\frac{n(n-1)}{2}}\,H^{-n}\,e^{1.72n}\,P^{\frac{3n-1}{2}}\,J_{b-n}(P^{1-\nu})$$

$$< b^{3n}(2n)^{\frac{3n(n+1)}{2}}\,P^{2\frac{b-n}{n}+\frac{3n-1}{2}}\,J_{b-n}(P^{1-\nu}), \tag{7}$$

$$J_{b-n}(P^{1-\nu}) < (b-n)^{3(b-n)}\,(2n)^{\frac{3n(n+1)n}{2}}\,P^{(1-\nu)\left(2(b-n)-\frac{n(n+1)}{2}(1-(1-\nu)^m)\right)},$$

$$2J_1 < \tfrac{1}{2}D_{m+1}\,P^{2b-\Delta(m+1)}.$$

Now we shall estimate $2J_2$. From the numbers $0, \ldots, H-1$, we can choose an increasing sequence of $n-1$ numbers in not more than $H^{n-1}/n!$ ways. A total number of $(2n-2)^b$ sequences of $y_1, \ldots, y_b$ are associated with each such a sequence. Therefore the total number of irregular sequences $y_1, \ldots, y_b$ is not greater than

$$\frac{H^{n-1}}{(n-1)!}(2n-2)^b < \frac{H^{n-1}}{2}(2n)^b.$$

Consequently, the integral $2J_2$ does not exceed $H^{n-1}(2n)^b$ multiplied by the maximum of the integral of the type

$$\int_0^1 \ldots \int_0^1 |Z_1 \ldots Z_b|^2\, d\alpha_n \ldots d\alpha_1 .$$

The last integral does not exceed the integral $J_b(R)$ (Lemmas 1b and 2), for which, by induction, we obtain

$$J_b(R) < D_{m+1}\,R^{2b-\Delta(m+1)} < D_{m+1}\,P^{2b-\Delta(m+1)}\,2H^{-2b+\Delta(m+1)},$$

from which, since $H^{-1}(2n)^b\,2H^{-2b+\Delta(m+1)} < \frac{1}{2}$, we find

$$2J_2 < \tfrac{1}{2}D_{m+1}\,P^{2b-\Delta(m+1)}. \tag{8}$$

The proof of the theorem follows from (6)–(8).

Theorem 1 proved above also holds for certain more general sums T_0. This follows from the lemma given below.

Lemma 4. Let

$$I' = \int_0^1 \ldots \int_0^1 |T_0|^{2b}\, d\alpha_n \ldots d\alpha_1; \quad T_0 = \sum_{0<x\leqq P} \psi(x)\, e^{2\pi i f(x)},$$

where $\psi(x)$, for the values of x under consideration, satisfies the condition $|\psi(x)| \leqq 1$. Then

$$I' \leqq I.$$

Proof. The integral I' is equal to the sum

$$\sum_{x_1} \dots \sum_{x_{2b}} \psi(x_1) \dots \psi(x_b) \overline{\psi(x_{b+1})} \dots \overline{\psi(x_{2b})},$$

taken over all the solutions of the system (1), the number of solutions of which is the integral I.

Theorem 2. Let $C \geqq 1$, $k \geqq 10$, $n \geqq 8$. From an n-dimensional domain Π_n bounded by the inequalities $0 \leqq \alpha_n < 1, \dots, 0 \leqq \alpha_1 < 1$, such a domain Ω of volume R can be chosen, where

$$R < (nl)^{3nl} (2n)^{\frac{3n(n+1)l}{2}} C^{-2nl} P^{-\frac{n(n+1)}{2} + \frac{n(n+1)}{2} \frac{4\ln k - 1}{k}};$$

$$l = \left[(n+1) \ln \frac{k}{4} + 1 \right],$$

that for the points $(\alpha_n, \dots, \alpha_1)$ of the remaining domain,

$$|T| < CP^{1-1/k}. \tag{9}$$

Proof. The domain Π_n can be divided into identical n-dimensional cubes of such small sides that the modulus of the difference of the values of $|T|^{2nl}$ corresponding to two points in the same cube is less than the difference between the right-hand and left-hand sides of the inequality of Theorem 1. By Ω we mean the domain formed by those cubes which contain at least one point that does not satisfy the inequality (9). By Theorem 1, we have

$$RC^{2nl} P^{(1-1/k)2nl} < D_l P^{2nl - \Delta(l)}.$$

Hence, we obtain

$$R < C^{-2nl} D_l P^{-\frac{n(n+1)}{2} + \frac{n(n+1)}{2}\lambda}; \qquad \lambda = \frac{2nl}{k} + \frac{n(n+l)}{2}(1-\nu)^l,$$

which, as $\lambda < (4\ln k - 1)/k$, proves the theorem.

Note. By virtue of Lemma 4, Theorem 2 and its proof remain true, even if the sum T is replaced by the sum T_0 in Lemma 4.

Lemma 5. To each integer y there corresponds a point $(Y_{n-1}, \dots, Y_1)$ of an $(n-1)$-dimensional space defined by the expansion

$$f(x+y) - f(y) = \alpha_n x^n + Y_{n-1} x^{n-1} + \dots + Y_1 x$$

of the difference $f(x+y) - f(x)$ in powers of x.

Let Y be a positive integer less than P. A necessary condition that a point corresponding to a certain number y in the sequence $0, \dots, Y$, may be made, by adding the terms which do not numerically exceed

$$L_{n-1} = P^{-n+1}, \ \dots, \ L_1 = P^{-1},$$

to its coordinates, congruent to a point corresponding to some definite number y_0 of the same sequence, is that the inequality

$$(n \dots s\alpha_s(y-y_0)) \leqq n \dots (s+1)\,(\tfrac{3}{2}n)^{n-s} L_{s-1}$$

should be satisfied for $s = n, \dots, 2$.

Proof. We find that

$$Y_{n-1} = \binom{n}{1}\alpha_n y + \alpha_{n-1},$$

$$Y_{n-2} = \binom{n}{2}\alpha_n y^2 + \binom{n-1}{1}\alpha_{n-1} y + \alpha_{n-2},$$

$$Y_{n-3} = \binom{n}{3}\alpha_n y^3 + \binom{n-1}{2}\alpha_{n-1} y^2 + \binom{n-2}{1}\alpha_{n-2} y + \alpha_{n-3},$$

$$\dots\dots\dots\dots\dots\dots\dots\dots\dots\dots\dots\dots\dots$$

$$Y_{s-1} = \binom{n}{n-s+1}\alpha_n y^{n-s+1} + \binom{n-1}{n-s}\alpha_{n-1} y^{n-s}$$
$$+ \binom{n-2}{n-s-1}\alpha_{n-2} y^{n-s-1} + \dots + \binom{s+1}{2}\alpha_{s+1} y^2 + \binom{s}{1}\alpha_s y + \alpha_{s-1}.$$

Assuming that

$$j\alpha_j(y-y_0) = A_j, \quad y^{j-1} + y^{j-2} y_0 + \dots + y_0^{j-1} = h_{j-1},$$

we find that, by the condition stated in the lemma, it follows

$$\theta_{n-1} L_{n-1} \equiv A_n,$$

$$\theta_{n-2} L_{n-2} \equiv \binom{n}{2}\frac{h_1}{n} A_n + A_{n-1},$$

$$\theta_{n-3} L_{n-3} \equiv \binom{n}{3}\frac{h_2}{n} A_n + \binom{n-1}{2}\frac{h_1}{n-1} A_{n-1} + A_{n-2},$$

$$\dots\dots\dots\dots\dots\dots\dots\dots\dots\dots\dots\dots\dots$$

$$\theta_{s-1} L_{s-1} \equiv \binom{n}{n-s+1}\frac{h_{n-s}}{n} A_n + \binom{n-1}{n-s}\frac{h_{n-s-1}}{n-1} A_{n-1}$$
$$+ \binom{n-2}{n-s-1}\frac{h_{n-s-2}}{n-2} A_{n-2} + \dots + \binom{s+1}{2}\frac{h_1}{s+1} A_{s+1} + A_s.$$

On multiplying these congruences by the numbers

$$D_n = 1, \quad D_{n-1} = n, \quad \dots, \quad D_s = n \dots (s+1),$$

respectively, we get

$$\theta_{n-1} D_n L_{n-1} \equiv D_n A_n,$$

$$\theta_{n-2} D_{n-1} L_{n-2} \equiv \binom{n-1}{1} \frac{h_1}{2} \frac{D_{n-1}}{D_n} D_n A_n + D_{n-1} A_{n-1},$$

$$\theta_{n-3} D_{n-2} L_{n-3} \equiv \binom{n-1}{2} \frac{h_2}{3} \frac{D_{n-2}}{D_n} D_n A_n + \binom{n-2}{1} \frac{h_1}{2} \frac{D_{n-2}}{D_{n-1}} D_{n-1} A_{n-1} + D_{n-2} A_{n-2},$$

. . . .

$$\theta_{s-1} D_s L_{s-1} \equiv \binom{n-1}{n-s} \frac{h_{n-s}}{n-s+1} \frac{D_s}{D_n} D_n A_n + \binom{n-2}{n-s-1} \frac{h_{n-s-1}}{n-s} \frac{D_s}{D_{n-1}} D_{n-1} A_{n-1} + \binom{n-3}{n-s-2} \frac{h_{n-s-2}}{n-s-1} \frac{D_s}{D_{n-2}} D_{n-2} A_{n-2} + \ldots + \binom{s}{1} \frac{h_1}{2} \frac{D_s}{D_{s+1}} D_{s+1} A_{s+1} + D_s A_s.$$

Here the coefficients of the products $D_j A_j$ are integers ($D_s D_t^{-1}$ is divisible by $t-s+1$, if $s<t$)). Therefore, our congruences hold true when these products are replaced by numbers congruent to them. This fact along with the relationships:

$$\frac{h_j}{j+1} \leqq P^j \ (j \geqq 0), \qquad P^i L_s = L_{s-i} \ (s>i)$$

shows a way to prove the inequality of the lemma. It is not difficult to verify that this inequality can be written as

$$(D_s A_s) \leqq (\tfrac{3}{2} n)^{n-s} D_s L_{s-1}.$$

Indeed, the first congruence immediately gives

$$(D_n A_n) \leqq D_n L_{n-1}.$$

Applying this inequality, from the second congruence we obtain

$$(D_{n-1} A_{n-1}) \leqq D_{n-1} L_{n-2} + \binom{n-1}{1} P D_{n-1} L_{n-1}$$

$$= D_{n-1} L_{n-2} + \binom{n-1}{1} D_{n-1} L_{n-2} = n D_{n-1} L_{n-2} < \tfrac{3}{2} n D_{n-1} L_{n-2}.$$

Consequently, assuming that the lemma is true for all $(D_{s'} A_{s'})$, with $s'>s$, when $s<n-1$, we can, by induction, show that the lemma is also true for all $(D_s A_s)$. But, under this assumption, the last congruence gives

$$\begin{aligned}(D_s A_s) &\leqq D_s L_{s-1} + \binom{n-1}{n-s} P^{n-s} D_s L_{n-1} \\ &+ \binom{n-2}{n-s-1} P^{n-s-1}\, \tfrac{3}{2} n D_s L_{n-2} \\ &+ \binom{n-3}{n-s-2} P^{n-s-2} \left(\tfrac{3}{2} n\right)^2 D_s L_{n-3} \\ &+ \ldots + \binom{s}{1} P \left(\tfrac{3}{2} n\right)^{n-s-1} D_s L_s < D_s L_{s-1} \left(\tfrac{3}{2} n\right)^{n-s} \\ &\times \left(\frac{\left(\frac{2}{3}\right)^{n-s}}{(n-s)!} + \frac{\left(\frac{2}{3}\right)^{n-s-1}}{(n-s-1)!} + \frac{\left(\frac{2}{3}\right)^{n-s-2}}{(n-s-2)!} + \ldots + \frac{\frac{2}{3}}{1!}\right) \\ &< D_s L_{s-1} \left(\tfrac{3}{2} n\right)^{n-s} (e^{2/3} - 1) < \left(\tfrac{3}{2} n\right)^{n-s} D_s L_{s-1}.\end{aligned}$$

Lemma 6. Let m be a positive integer, λ a real,

$$\Phi(y) = m \frac{ay + \lambda y}{q}; \quad (a, q) = 1,\ q > 0,$$

where y runs through $\leqq Y$ consecutive integers. Then, for $V \geqq 0$, the number of values of y, which satisfy the condition

$$(\Phi(y)) \leqq V q^{-1} \tag{10}$$

does not exceed

$$\lambda Y m + m + 2V, \qquad \text{if} \quad Y \leqq q,$$

$$(\lambda Y m + m + 2V) \frac{2Y}{q}, \qquad \text{if} \quad Y > q.$$

Proof. We shall assume that $\lambda \geqq 0$, because if $\lambda < 0$ we can take a function $-\Phi(y)$, possessing the property $(-\Phi(y)) = (\Phi(y))$.

First consider the first case only for $m = 1$. Let y_0 be the least value of y, B be the integral part and β the fractional part of the number λy_0. Denoting by z the least non-negative residue of the number $ay + B$ modulo q, and assuming that $\beta + \lambda(y - y_0) = \varphi(z)$, we obtain

$$(\Phi(y)) = \left(\frac{z + \varphi(z)}{q}\right); \quad \beta \leqq \varphi(z) \leqq \beta + \lambda Y.$$

We can easily demonstrate that the number of values of y satisfying the condition (10) is

$$\leqq \lambda Y + 1 + 2V.$$

Indeed, this assertion is obvious for $q \leqq \lambda Y + 1 + 2V$; for the opposite inequality, this assertion follows from the fact that the cases where the inequality (10) is possible are only those for which

$$z = q - [\beta + \lambda Y + V],\ \ldots,\ q - 1,\ 0,\ 1,\ \ldots,\ [V - \beta].$$

Now, consider the first case for any m. Let $(m, q) = d$; $m = m_1 d$, $q = q_1 d$. We find

$$m\frac{ay+\lambda y}{q} = \frac{m_1 ay + m_1 \lambda y}{q_1}.$$

Among $Y_1 \leqq q_1$ consecutive values of y, the number of values which satisfy the condition

$$\left(\frac{m_1 aay + m_1 \lambda y}{q_1}\right) \leqq Vd^{-1} q_1^{-1},$$

according to the lemma being proved, is

$$\leqq m_1 \lambda Y_1 + 1 + 2Vd^{-1}.$$

Consequently, among all the values of y, the number of values satisfying this condition is

$$\leqq m_1 \lambda Y + d + 2V \leqq \lambda Ym + m + 2V.$$

The second statement of the lemma follows trivially from the first.

Lemma 7. Let $P \geqq n^n$, m a positive integer not exceeding P^{ν^2}, and to each integer y let there correspond a point $(mY_{n-1}, \dots, mY_1)$ defined by the expansion

$$mf(x+y) - mf(y) = m\alpha_n x^n + mY_{n-1} x^{n-1} + \dots + mY_1 x$$

of the polynomial $mf(x+y) - mf(y)$ in powers of x. Put a number $\tau_s = P^{s-0.5}$ into correspondence with each $s = n, \dots, 2$, and let α_s be represented (which is always possible) in the form:

$$\alpha_s = \frac{a_s}{q_s} + \frac{\theta_s}{q_s \tau_s}; \quad (a_s, q_s) = 1, \quad 0 < q_s \leqq \tau_s.$$

Let Q_0 denote the least common multiple of the numbers $q_n, \dots, q_2$.

Let G be the number of those points corresponding to the number y in the sequence $0, \dots, P-1$, which can be made, by adding numbers numerically not exceeding

$$L_{n-1} = P^{-n+1}, \dots, L_1 = P^{-1},$$

respectively, to their coordinates, congruent to a point corresponding to some definite number y_0 from the same sequence.

Then, for $Q \geqq P^{0.5-0.4\nu}$, we have

$$G < mn^{2n-2} P^{0.5+0.4\nu}.$$

Proof. First assume that there exists at least one q_s that satisfies the condition $q_s \geqq P^{0.5-0.4\nu}$ (then we have $Q_0 \geqq P^{0.5-0.4\nu}$). By Lemma 5, the number G does not exceed the number of values of $y' = y - y_0$ satisfying the condition:

$$\left(\frac{n \dots sma_s y' + n \dots sm\theta_s \tau_s^{-1} y'}{q_s}\right) \leqq n \dots (s+1) \left(\tfrac{3}{2}n\right)^{n-s} L_{s-1}.$$

Therefore we can use Lemma 8 to determine an upper bound for G. Thus, we obtain

$$G < (n \dots sm(P^{1-s+0.5}+1) + 2n \dots (s+1)\,(\tfrac{3}{2}n)^{n-s} q_s L_{s-1})\,H;$$
$$H = 1, \quad \text{if} \quad P \leqq q_s,$$
$$H = \frac{2P}{q_s}, \quad \text{if} \quad P > q_s.$$

Since in the first case, we have

$$q_s L_{s-1} \leqq P^{0.5},$$

and in the second case

$$q_s L_{s-1} < 1, \quad H \leqq 2P^{0.5+0.4\nu},$$

for both cases, we find

$$G < mn!\,(2 + (\tfrac{3}{2}n)^{n-2})\,P^{0.5+0.4\nu} < mn^{2n-2}\,P^{0.5+0.4\nu}.$$

Now assume that every q_s does not exceed $P^{0.5-0.4\nu}$. Then for $s = n, \dots, 2$, we find

$$|n \dots sm\theta_s \tau_s^{-1} y'| < 0,\,1,\, n \dots (s+1)\,(\tfrac{3}{2}n)^{n-2} q_s P^{-s+1} < 0.1\,.$$

Therefore, $n \dots smy'$ is divisible by each q_s under consideration and, consequently, by Q_0. Let

$$(n \dots sm, Q_0) = d, \quad Q_0 = Q_1 d.$$

Since y' is necessarily divisible by Q_1, we have

$$G < \frac{P}{Q_1} + 1 = \frac{Pd}{Q_0} + 1 < 2dP^{0.5+0.4\nu} < mn^{2n-2}\,P^{0.5+0.4\nu}.$$

Lemma 8. Let $n \geqq 3$, and to each $s = n, \dots, 1$, let there correspond its number $\tau_s = P^{s-0.5}$, and α_s is represented in the form:

$$\alpha_s = \frac{a_s}{q_s} + \frac{\theta_s}{q_s \tau_s}; \quad (a_s, q_s) = 1,\; 0 < q_s \leqq \tau_s.$$

Let the least common multiple Q_0 of the numbers $q_n, \dots, q_2$ obey the condition $Q_0 > P^{0.5-0.4\nu}$, and let

$$\varrho = \frac{1}{k}; \quad k = 8n^2\,(\ln n + 1.5 \ln\ln n + 4.2)\,.$$

Then, for a positive integer m not exceeding $P^{2\varrho}$, we have

$$|T_m| < CP^{1-\varrho}; \quad C = (n(n+1)\ln k)^{3/2}\,(2n)^{2(n+1)}.$$

Proof. We shall restrict ourselves to the case $P \geqq C^k$, because in the contrary case the lemma is trivial. Assuming $Y = [P^{1-\varrho}]$, we shall put a point $(m\alpha_n, mY_{n-1}, \dots, mY_1)$ of an n-dimensional space into correspondence with each y taken from the sequence

0, 1, ..., Y (notation as in Lemma 7), and, accordingly, also a domain (y) of points $(\eta_n, \dots, \eta_1)$ of the n-dimensional space bounded by the inequalities:

$$m\alpha_n - 0.5L_n P^{-\varrho} \leqq \eta_n < m\alpha_n + 0.5L_n P^{-\varrho},$$
$$mY_{n-1} - 0.5L_{n-1} P^{-\varrho} \leqq \eta_{n-1} < mY_{n-1} + 0.5L_{n-1} P^{-\varrho},$$
$$\dots\dots\dots\dots\dots\dots\dots\dots$$
$$mY_1 - 0.5L_1 P^{-\varrho} \leqq \eta_1 < mY_1 + 0.5L_1 P^{-\varrho}$$

into correspondence with such a point. The domain of points congruent to the points of this domain that lie in the domain Π_n (see the Introduction of this chapter) is called the domain (y). By Lemma 7, the number of domains (y) intersecting a given domain (y_0) is equal to

$$< mn^{2n-2} P^{0.5+0.4\nu}.$$

Therefore, among all $Y+1$ domains, the number of domains which do not intersect one another is equal to

$$> \frac{Y+1}{mn^{2n-2} P^{0.5+0.4\nu}} \geqq \frac{P^{0.5-3\varrho-0.4\nu}}{n^{2n-2}}.$$

But, for the total volume V of a domain W covered by such domains, we have

$$V > \frac{P^{0.5-3\varrho-0.4\nu}}{n^{2n-2}} \frac{P^{-n\varrho}}{P^{\frac{n(n+1)}{2}}} > P^{-\frac{n(n+1)}{2}+0.5-0.4\nu}.$$

Assuming that

$$T_m^{(y)} = \sum_{x=1}^{P} e^{2\pi i (mf(x+y) - mf(y))} \quad \text{(consequently, } T_m^{(0)} = T_m\text{)},$$

for any y taken from the sequence 0, ..., Y, we have

$$|T_m - T_m^{(y)}| < 2Y.$$

Thus, since for the sum $T(m\eta'_n, \dots, m\eta'_1)$ corresponding to any point $(\eta'_n, \dots, \eta'_1)$ in the domain $(y)'$, by Lemma 9 of Chapter 2, we have

$$|T_m^{(y)} - T(m\eta'_n, \dots, m\eta'_1)| < \pi n P^{1-\varrho},$$

whereas for any point $(\eta_n, \dots, \eta_1)$ of the domain W, the following inequality

$$|T_m - T(m\eta_n, \dots, m\eta_1)| < (\pi n + 2) P^{1-\varrho}.$$

holds true. Now, putting

$$C_0 = (nl)^{3/2} (2n)^{3(n+1)/4}; \quad l = \left[(n+1) \ln \frac{k}{4} + 1 \right],$$

we find that $C - C_0 > \pi n + 2$. Now, assume that

$$|T_m| > CP^{1-\varrho}, \tag{11}$$

Then, for any point of the domain W, we have

$$|T_m| > C_0 P^{1-\varrho}.$$

But, from Theorem 1, it follows that

$$P^{-\frac{n(n+1)}{2}+0.5-0.4\nu} C_0^{2nl} P^{(1-\varrho)2nl} < (nl)^{3nl} (2n)^{\frac{3n(n+1)l}{2}} \times P^{2nl-\frac{n(n+1)}{2}(1-(1-\nu)^l)},$$

which implies

$$0 < 2nl\varrho + \frac{n(n+1)}{2}(1-\nu)^l + 0.5\nu - 0.5,$$

which is obtained from a comparison of the powers of P on the left-hand and right-hand sides.

From the last inequality, with the help of simple calculations, we obtain:

$$0 < \frac{4n(n+1)(\ln k - 1.55)}{k} + \nu - 1,$$

which (as it is not difficult to verify) is not true. Therefore, the assumption (11) is also not true, that means the contrary is true. Thus, this completes the proof of our lemma.

Theorem 3. Let n be a constant greater than or equal to 3. Divide the points of an n-dimensional space into two classes: Class I and Class II. A point

$$\left(\frac{a_n}{q_n} + z_n, \ldots, \frac{a_1}{q_1} + z_1\right),$$

where a_i/q_i $(i = 1, \ldots, n)$ are rational irreducible fractions with positive denominators having a least common multiple, Q, not exceeding P^ν, while z_i $(i = 1, \ldots, n)$ are numbers satisfying the condition:

$$|Z_s| \leqq P^{-s+\nu}$$

is said to be a point of Class I. A point which is not a point of Class I is said to be a point of Class II.

Then, assuming

$$\varrho = \frac{1}{8n^2(\ln n + 1.5 \ln\ln n + 4.2)},$$

for $m \leqq P^{2\varrho}$, for the points of Class II, we find that

$$|T_m| \ll P^{1-\varrho}.$$

And assuming

$$\delta_s = Z_s P^s, \quad \delta_0 = \max(|\delta_n|, \ldots, |\delta_1|),$$

for $m \leqq P^{4\nu^2}$, for the points of Class I, we find that

$$|T_m| \ll P(m, Q)^\nu Q^{-\nu+\varepsilon_0}$$

or

$$|T_m| \ll PQ^{-\nu+\varepsilon_0} \delta_0^{-\nu}, \quad \text{if} \quad \delta_0 \geqq 1.$$

Proof. Without any loss of generality, we shall assume that $P \geqq n^{2n^2}$.

First, we shall prove the theorem for the points of Class I. With the help of the substitution

$$x = Q\xi + \eta,$$

where η runs through the values 1, ..., Q, while ξ, for a given η, runs through the integers in the interval

$$-\frac{\eta}{Q} < \xi \leqq \frac{P-\eta}{Q},$$

we obtain

$$T_m = \sum_{\eta} e^{2\pi i m\left(\frac{a_n}{q_n}\eta^n + \ldots + \frac{a_1}{q_1}\eta\right)} W_\eta;$$

$$W_\eta = \sum_{\xi} e^{2\pi i(mz_n(Q\xi+\eta)^n + \ldots + mz_1(Q\xi+\eta))}.$$

The ξ-derivative of the polynomial within the brackets in the exponent in the expression for W_η does not numerically exceed

$$\frac{n(n+1)}{2} mQP^{-1+\nu} < 0.01.$$

Therefore the sum W_η can be divided into not more than $2n$ subsums, to each of which Lemma 3 of Chapter 2 can be applied. Thus we obtain

$$W_\eta = \int_{-\eta/Q}^{P-\eta/Q} e^{2\pi i(mz_n(Q\xi+\eta)^n + \ldots + mz_1(Q\xi+\eta))}\, d\xi + O(1)$$

$$= \frac{P}{Q}\int_0^1 e^{2\pi i(m\delta_n t^n + \ldots + m\delta_1 t)}\, dt + O(1),$$

from which we readily derive

$$T_m = UV + O(Q); \quad U = \frac{1}{Q}\sum_{\eta=1}^{Q} e^{2\pi i m\left(\frac{a_n}{q_n}\eta^n + \ldots + \frac{a_1}{q_1}\eta\right)},$$

$$V = P\int_0^1 e^{2\pi i(m\delta_n t^n + \ldots + m\delta_1 t)}\, dt.$$

When $\delta_0 \geqq 1$, applying Theorem 1 of Chapter 3 and Lemma 4 of Chapter 2, we find that the theorem is proved for the points of Class I.

Now we shall prove the theorem for the points of Class II. To each $s = n, \ldots, 1$, put into correspondence an appropriate number $\tau_s = P^{s-0.5}$, and represent the number α_s as

$$\alpha_s = \frac{a_s}{q_s} + z_s; \quad (a_s, q_s) = 1, \quad 0 < q_s \leqq \tau_s, \quad |z_s| \leqq \frac{1}{q_s \tau_s}.$$

Denote the least common multiple of $q_n, \ldots, q_1$ by Q, and the least common multiple of $q_n, \ldots, q_2$ by Q_0.

For $Q_0 > P^{0.5-0.4\nu}$, the theorem follows as a corollary from Lemma 8.

If $Q_0 \leqq P^{0.5-0.4\nu}$, transform the sum T_m by the substitution $x = Q_0\xi + \eta$, where η runs through the values $1, \ldots, Q_0$, and ξ runs, for given η, through the integers in the interval:

$$-\frac{\eta}{Q_0} < \xi \leqq \frac{P-\eta}{Q_0}.$$

Thus, we obtain

$$T_m = \sum_{\eta} e^{2\pi i m\left(\frac{a_n}{q_n}\eta^n + \ldots + \frac{a_1}{q_1}\eta\right)} W_\eta;$$

$$W_\eta = \sum_{\xi} e^{2\pi i\left(mz_n(Q_0\xi+\eta)^n + \ldots + mz_1(Q_0\xi+\eta)^2 + \left(\frac{a_1'}{q_1} + mz_1 Q_0\right)\xi + mz_1\eta\right)},$$

where a_1' is the number nearest to zero congruent to $ma_1 Q_0 \pmod{q_1}$. The ξ-derivative of the polynomial within the brackets in the exponent in the expression for W_η does not exceed

$$0.5 + \frac{n(n+1)}{2} mQ_0 P^{-1+\nu} < 0.51.$$

Therefore the sum W_η can be divided into not more than $2n$ subsums, to each of which Lemma 3 of Chapter 2 is applicable. Thus, we obtain

$$W_\eta = \int_{-\eta/Q_0}^{P-\eta/Q_0} e^{2\pi i\left(mz_n(Q_0\xi+\eta)^n + \ldots + mz_1(Q_0\xi+\eta)^2 + \left(\frac{a_1'}{q_1} + mz_1 Q_0\right)\xi + mz_1\eta\right)} d\xi + O(1)$$

$$= \frac{P}{Q_0} H + O(1);$$

$$H = \int_0^1 e^{2\pi i\left(mz_n P^n t^n + \ldots + mz_2 P^2 t^2 + \left(\frac{a_1'}{q_1} + mz_1 Q_0\right)\frac{Pt-\eta}{Q_0} + mz_1\eta\right)} dt.$$

Now assume that mQ_0 is not divisible by q_1. Then the coefficient of t in the polynomial within the brackets in the exponent is numerically greater than

$$\left(\frac{1}{q_1} - \frac{mQ_0}{q_1 P^{0.5}}\right)\frac{P}{Q_0} > \frac{P}{2q_1 Q_0} \geqq \frac{P^{0.4\nu}}{2}.$$

Therefore, applying Lemma 4 of Chapter 2, we obtain:

$$|H| < 32\left(\frac{P^{0.4\nu}}{2}\right)^{-\nu} \ll P^{-0.4\nu^2} \ll P^{-\varrho}.$$

Hence, it follows that

$$W_\eta \ll \frac{P^{1-\varrho}}{Q_0} + O(1) \ll \frac{P^{1-\varrho}}{Q_0}, \qquad |T_m| \ll P_1^{1-\varrho}.$$

Finally, assume that mQ_0 is divisible by q_1. Then we easily find

$$T_m = UV + O(Q_0);$$

$$U = \frac{1}{Q_0}\sum_{\eta=1}^{Q_0} e^{2\pi i m\left(\frac{a_n}{q_n}\eta^n + \ldots + \frac{a_1}{q_1}\eta\right)}, \qquad V = P\int_0^1 e^{2\pi i(m\delta_n t^n + \ldots + m\delta_1 t)} dt.$$

When $m \leqq P^{2\varrho}$ and $Q \geqq P^{\nu}$, applying Theorem 1 of Chapter 3 to U, we find that $|T| \ll P(m, Q_0)^{\nu} Q_0^{-\nu+\xi_0}$, from which, since mQ_0 is divisible by q_1 and, consequently, by Q, we find $mQ_0 \geqq Q$, $Q_0 > Qm^{-1}$. Hence we readily obtain

$$|T_m| \ll Pm^{\nu}(Qm^{-1})^{-\nu+\varepsilon_0} \ll P^{1-\varrho}.$$

When $m \leqq P^{2\varrho}$, $\delta_0 \geqq P$, applying Lemma 4 of Chapter 2 to V, we obtain

$$|T_m| \ll P^{1-\varrho}.$$

Thus, the proof of the theorem is complete.

Theorem 4. Let n be a constant $\geqq 3$, r an integer $\geqq 2r_1$, where

$$r_1 = [n^2(2\ln n + \ln\ln n + 4)].$$

Then, assuming

$$I_r = \int_0^1 |T|^r \, d\alpha_n \ldots d\alpha_1,$$

we have

$$I_r \ll P^{r - \frac{n(n+1)}{2}}.$$

Proof. Without loss of generality, we shall assume that $P \geqq C^k$ (notation as in Lemma 8). Consider the interval

$$I_{2r_0}; \quad r_0 = \frac{n(n+1)}{2} + b_l; \quad b_l = nl,$$

where l obeys the condition:

$$\frac{n(n+1)}{2}(1-\nu)^l \leqq n(n+1)\varrho; \quad \varrho = \frac{1}{8n^2(\ln n + 1.5\ln n + 4.2)}.$$

But this condition is a corollary of the condition:

$$e^{-l/n} \leqq 2\varrho, \quad \text{or} \quad l \geqq n \ln\frac{1}{\varrho}.$$

In particular, we can take

$$l > [n(2\ln n + \ln\ln n + 3) + 1].$$

Hence we have

$$r_0 = \frac{n(n+1)}{2} + nl < n^2(2\ln n + \ln\ln n + 4).$$

Changing the range of integration as follows:

$$-P^{-n+\nu} \leqq \alpha_n < -P^{n+\nu} + 1, \ldots,$$
$$-P^{-1+\nu} \leqq \alpha_1 < -P^{-1+\nu} + 1,$$

divide the new range into two subranges as follows. The range of Class I is an interval bounded by the inequalities:

$$\frac{a_n}{q_n} - P^{-n+\nu} \leqq \alpha_n < \frac{a_n}{q_n} + P^{-n+\nu}, \ldots, \frac{a_1}{q_1} - P^{-1+\nu} \leqq \alpha_1 < \frac{a_1}{q_1} + P^{-1+\nu},$$

where

$$(a_n, q_n) = 1, \quad q_n > 0, \ldots, \quad (a_1, q_1) = 1, \quad q_1 > 0,$$

and the least common multiple, Q, of the numbers $q_n, \ldots, q_1$ does not exceed P^ν. It can be easily verified that the intervals of Class I do not intersect each other. The range that remains after the intervals of Class I have been chosen is called the range of Class II.

In accordance with this division of the integration range, the integral I_{2r_0} is divided into the sum of two subintegrals:

$$I_{2r_0} = I'_{2r_0} \pm I''_{2r_0}.$$

First we shall evaluate I''_{2r_0}. According to Theorem 3 (when $m=1$), in the range of Class II, we have

$$|T|^{n(n+1)} \ll P^{n(n+1)(1-\varrho)}.$$

And according to Theorem 1, we have

$$\int_0^1 \ldots \int_0^1 |T|^{2bl} \ll P^{2bl - \frac{n(n+1)}{2} + \frac{n(n+1)}{2}(1-\nu)^l}.$$

Therefore

$$I''_{2r_0} \ll P^{2r_0 - \frac{n(n+1)}{2}}.$$

Now we shall estimate the integral I'_{2r_0}. First we shall evaluate that part H which corresponds to some definite interval of Class I with a given value of Q. According to Theorem 3, in this interval

$$|T| \ll PQ^{-\nu+\varepsilon_0} \min(1, \delta_0^{-\nu}).$$

Replace this inequality by a more rough but convenient inequality:

$$|T| \ll PQ^{-\nu+\varepsilon_0} \min(1, |\delta_n|^{-\nu^1}) \ldots \min(1, |\delta_1|^{-\nu^2}); \quad \delta_s = z_s P^s,$$

from which we easily find that

$$|H| \ll P^{2r_0} Q^{-2\nu r_0 + 2\varepsilon_0 r_0} \int_0^{P^{-n+\nu}} \min(1, \delta_n^{-2\nu^2 r_0})\, dz_n \ldots \int_0^{P^{-1+\nu}} \min(1, \delta_1^{-2\nu^2 r_0})\, dz_1$$

$$= P^{2r_0} Q^{-2\nu r_0 + 2\varepsilon_0 r_0} P^{-\frac{n(n+1)}{2}} \int_0^{P^\nu} \min(1, \delta_n^{-2\nu^2 r_0})\, d\delta_n \ldots \int_0^{P^\nu} \min(1, \delta_1^{-2\nu^2 r_0})\, d\delta_1$$

$$\ll P^{2r_0 - \frac{n(n+1)}{2}} Q^{-2\nu r_0 + 2\varepsilon_0 r_0}.$$

because

$$\int_0^{P^\nu} \min(1, \delta_s^{-2\nu^2 r_0})\, d\delta_s = \int_0^1 d\delta_s + \int_1^{P^\nu} \delta_s^{-2\nu^2 r_0}\, d\delta_s \ll 1.$$

On summing this bound for $|H|$ over all intervals of Class I with a given Q, we obtain

$$\ll P^{2r_0-\frac{n(n+1)}{2}}Q^{-2\nu r_0+2\varepsilon_0 r_0+1}.$$

Therefore

$$I'_{2r_0} \ll \sum_{0<Q\leqq P^\nu} P^{2r_0-\frac{n(n+1)}{2}}Q^{-2\nu r_0+2\varepsilon_0 r_0+1} \ll P^{2r_0-\frac{n(n+1)}{2}}.$$

The proof of the theorem is complete.

Lemma 9. Let $n \geqq 20$, a and a_1 be integers ($a<a_1\leqq 2a$), $t=a^{n-\theta}$, $0\leqq\theta<1$. Then, for the sum

$$S=\sum_{a<u\leqq a_1} e^{2\pi i F(u)}; \quad F(u)=-\frac{t\ln u}{2\pi},$$

we have

$$|S|<2a^{1-\frac{1}{30{,}000n^2}}.$$

Proof. We shall restrict ourselves to the case $a>e^{20{,}000n^2}$, because in the contrary case the lemma is trivial. Let

$$P=[a^{1/3}], \quad n_0=3n, \quad \nu_0=\frac{1}{n_0},$$

$$k=100, \quad \chi=\frac{1}{k}, \quad l_0=\left[(n+1)\ln\frac{k}{4}+1\right].$$

Put the numbers

$$d_s=\frac{1}{C_s n_0 P^{s+\chi}},$$

into correspondence with each $s=n_0,\ldots,1$, where C_s, satisfying the condition $12.6<C_s\leqq 13$, is so chosen that the denominator of d_s is an integer. Divide the domain Π_{n_0} into elementary parallelepipeds of sides $d_{n_0},\ldots,d_1$. Let there be a point $(\alpha_{n_0},\ldots,\alpha_1)$ in one of these parallelepipeds, such that

$$|T(\alpha_{n_0},\ldots,\alpha_1)|>P^{1-\chi}. \qquad (12)$$

Then for any point $(\alpha'_{n_0},\ldots,\alpha'_1)$ in this parallelepiped, by Lemma 9 of Chapter 2, we obtain

$$|T(\alpha'_{n_0},\ldots,\alpha'_1)-T(\alpha_{n_0},\ldots,\alpha_1)|<2\pi n_0\frac{P}{12.6n_0P^\chi}<0.5P^{1-\chi},$$

$$|T(\alpha'_n,\ldots,\alpha'_1)|>0.5P^{1-\chi}.$$

Therefore, if M denotes the number of different elementary parallelepipeds, each having at least one point which satisfies the condition (12), by means of Theorem 1, we obtain:

$$Md_{n_0}\ldots d_1(0.5P^{1-\chi})^{2n_0l_0}<(n_0l_0)^{3n_0l_0}(2n_0)^{\frac{3n_0(n_0+1)l_0}{2}}P^{2n_0l_0-\frac{n_0(n_0+1)}{2}+\frac{n_0(n_0+1)}{2}(1-\nu_0)^{l_0}},$$

from which, after a simple calculation, we find

$$M < (2n_0)^{6.44n_0(n_0+1)^2} P^{0.185\frac{n_0(n_0+1)}{2}}.$$

To evaluate S, substitute $u + xy$ for u, where x and y run, independently of each other, through the values $1, \ldots, P$. Thus, we get

$$|S| < \frac{1}{P^2} \sum_{\sigma < u \leqq a_1} |S_u| + 2a^{2/3},$$

$$S_u = \sum_x \sum_y e^{2\pi i(A_1 xy + \ldots + A_{n_0} x^{n_0} y^{n_0})}, \quad A_s = \frac{(-1)^s t}{2\pi s u^s}.$$

Now, putting $b = n_0 l$; $l = 5n_0$, and using the notation

$$Y_s = y_1^s + \ldots + y_b^s - y_{b+1}^s - \ldots - y_{2b}^s,$$

we obtain

$$|S_u|^{2b} \leqq P^{2b-1} \sum_{y_1, \ldots, y_{2b}} \left| \sum_{x=1}^{P} e^{2\pi i(A_1 Y_1 x + \ldots + A_{n_0} Y_{n_0} x^{n_0})} \right|. \tag{13}$$

Here Y_s can take only integer values z in the interval $-bP^s < z < bP^s$, and, by Theorem 1 (see also the introduction to this chapter), the number of solutions of the system $Y_{n_0} = z_{n_0}, \ldots, Y_1 = z_1$ does not exceed

$$U = (5n_0^2)^{15n_0^2} (2n_0)^{\frac{15n_0(n_0+1)_2}{2}} P^{10n_0^2 - \frac{n_0(n_0+1)}{2} + \frac{n_0(n_0+1)}{2}(1-v_0)^{5n_0}}$$
$$< (2n_0)^{sn_0(n_0+1)^2} P^{10n_0^2 - \frac{n_0(n_0+1)}{2} + \frac{n_0(n_0+1)}{2} 0.003}.$$

When $1.5n < s \leqq 3n - 4$, all the values of $A_s z_s$ lie within an interval of length $2b$. Therefore the number of values of $\{A_s z_s\}$, which lie in an interval of length d_s, does not exceed

$$2b(d_s |A_s|^{-1} + 1) < 2bP^s P^{s-3n+3}.$$

When $n < s \leqq 1.5n - 2$, all the values of $A_s z_s$ lie in an interval of length $2bP^s |A_s|$, where $|A_s| > d_s$. Therefore, the number of values of $\{A_s z_s\}$ which lie in an interval of length d_s does not exceed

$$2bP^s |A_s| + 1 < 2bP^s P^{3n-3\varepsilon}.$$

Finally, for any one of the remaining values of s, we shall trivially estimate the number of values of $\{A_s z_s\}$ which lie in an interval of length d_s as a number that does not exceed $2bP^s$.

Consequently, the total number of points $\{A_{n_0} z_{n_0}, \ldots, A_1 z_1\}$ which lie in a given elementary parallelepiped does not exceed

$$(2b)^{n_0} P^{\frac{n_0(n_0+1)}{2}} \prod_{1.5n < s \leqq 3n-1} P^{s-3n+3} \prod_{n < s \leqq 1.5n-2} P^{3n-2s}$$
$$< (2b)^{n_0} P^{\frac{n_0(n_0+1)}{2} - \frac{3}{2}n^2 + \frac{15}{2}n - 3} < (2b)^{n_0} P^{\frac{n_0(n_0+1)}{2}(1-0.2b)}.$$

Multiplying this expression by the upper bound of M and by U, we find that the number W of systems $(Y_{n_0}, \ldots, Y_1)$ satisfying the condition

$$|T(A_{n_0} Y_{n_0}, \ldots, A_1 Y_1)| > P^{1-\chi}, \tag{14}$$

is bounded by the inequality

$$W < (2n_0)^{14.5} P^{10n_0^2 - 0.06 \frac{n_0(n_0+1)}{2}} < P^{10n_0^2 - 100}.$$

On substituting P for the modulus of the sum $T(A_n Y_{n_0}, \ldots, A_1 Y_1)$ in each of the W cases of (14) and $P^{1-\chi}$ in each of $< P^{10n_0^2}$ remaining cases, from the inequality (13) we obtain

$$|S_u|^{10n_0^2} < P^{10n_0^2 - \chi}, \quad |S_u| < P^{2 - \frac{k}{10n_0^2}},$$

$$|S| < aP^{-\frac{\chi}{10n_0^2}} + 2a^{2/3} < 2a^{1 - \frac{1}{30{,}000n^2}}.$$

Chapter 5. Distribution of Fractional Parts of the Values of a Polynomial

Notation. $n\,(n \geqq 3)$ is a constant;
$\alpha_n, \ldots, \alpha_1$ are real numbers;
$f(x) = \alpha_n x^n + \ldots + \alpha_1 x$;
P is a positive integer;
$\Pi = \Pi_n$ has the same meaning as in Chapter 4.

The problem of an upper bound for the modulus of the sum T_m considered in Chapter 4 is closely related to the problem of the distribution of fractional parts $\{f(x)\}$ of the values of the polynomial $f(x)$ for $x = 1, \ldots, P$. The general solution of this problem is given in Theorem 1.

If we retain the same division of points $(\alpha_n, \ldots, \alpha_1)$ of an n-dimensional space into classes as was adopted in Theorem 3 of Chapter 4, this solution can be formulated as follows.

Let σ be any number such that $0 < \sigma \leqq 1$, and let $D(\sigma)$, the number of terms in the sequence $x = 1, \ldots, P$, which satisfy the condition $\{f(x)\} < \sigma$, be represented in the form:

$$D(\sigma) = P\sigma + \lambda(\sigma).$$

Then, $\lambda(\sigma) \leqq P\Delta_0$, where putting

$$\varrho_0 = \frac{1}{8n^2(\ln n + 1.5 \ln\ln n + 4.4)},$$

we find that for the points of Class II:

$$\Delta_0 = P^{-\varrho_0}.$$

And putting $\delta_0 = \min(|z_n| P^n, \ldots, |z_1| P)$, for the points of Class I, we find

$$\Delta_0 = Q^{-\nu+\varepsilon''}, \quad \text{if} \quad \delta_0 < 1,$$
$$\Delta_0 = Q^{-\nu+\varepsilon''}\delta_0^{-\nu+\varepsilon''}, \quad \text{if} \quad \delta_0 \geqq 1.$$

The main advantage of this general solution is that the domain of the points of Class I forms an extremely negligible part of the domain Π_n. For the remaining part of the domain Π_n, i.e. the domain of the points of Class II, the upper bound of the modulus of the number $\lambda(\sigma)$ is given by the uniform inequality $|\lambda(\sigma)| \ll P^{1-\varrho_0}$.

An important supplement to Theorem 1 is Theorem 2. In particular from the latter it follows that for $k_0 \geqq 12$ and for some $c(n, k_0)$ and a certain $j(k_0)$ satisfying the condition $0.02 < j(k_0) < 1$, which tends to unity as k_0 increases to infinity, we can eliminate from the domain Π_n a region Ω' of volume not exceeding

$$c(n, k_0)\, P^{-\frac{n(n+1)}{2} j(k_0)},$$

such that for any point of the remaining region, we have

$$|\lambda(\sigma)| < 3.4 P^{1-\frac{1}{k_0}} \ln P$$

regardless of the choice of σ.

Theorem 1. Let $n \geqq 3$. Divide the points of an n-dimensional space into two classes: Class I and Class II. A point of Class I is such a point

$$\left(\frac{a_n}{q_n} + z_n, \ldots, \frac{a_1}{q_1} + z_1\right),$$

where a_i/q_i $(i=1, \ldots, n)$ are rational irreducible fractions with positive denominators having a least common multiple Q not greater than P^ν, and z_i $(i=1, \ldots, n)$ satisfy the condition:

$$|z_s| \leqq P^{-\varepsilon+\nu}.$$

A point of Class II is a point, which is not a point of Class I.

Let σ be any number such that $0 < \sigma \leqq 1$, and let $D(\sigma)$, the number of terms in the sequence $x = 1, \ldots, P$ satisfying the condition $\{f(x)\} < \sigma$, be represented in the form:

$$D(\sigma) = P\sigma + \lambda(\sigma).$$

Then we have

$$\lambda(\sigma) \ll P\Delta_0,$$

where, putting

$$\varrho_0 = \frac{1}{8n^2(\ln n + 1.5 \ln \ln n + 4.4)},$$

we find that for the points of Class II:

$$\Delta_0 = P^{-\varrho_0}$$

and putting

$$\delta_0 = \max(|z_n| P^n, \ldots, |z_1| P),$$

we find that for the points of Class I:

$$\Delta_0 = Q^{-\nu+\varepsilon''}$$

or

$$\Delta_0 = Q^{-\nu+\varepsilon''} \delta_0^{-\nu+\varepsilon''}, \quad \text{if} \quad \delta_0 \geqq 1.$$

Proof. For the points of Class II, let

$$\Delta = P^{-\varrho}; \quad \varrho = \frac{1}{8n^2(\ln n + 1.5 \ln \ln n + 4.2)}$$

and for the points of Class I, let

$$\Delta = Q^{-\nu+\varepsilon_0}, \qquad \text{if} \quad \delta_0 < 1,$$

$$\Delta = Q^{-\nu+\varepsilon_0} \delta_0^{-\nu}, \quad \text{if} \quad \delta_0 \geqq 1.$$

Assuming

$$T_m = \sum_{x=1}^{P} e^{2\pi i m f(x)},$$

we have, for $m \leqq \Delta^{-2}$, according to Theorem 3 of Chapter 4,

$$|T_m| \ll P\Delta \quad \text{for the points of Class II,}$$
$$|T_m| \leqq (mQ)^{\nu} P\Delta \quad \text{for the points of Class I, or also}$$
$$|T_m| \ll P\Delta \quad \text{for the points of Class I, if } \delta_0 \geqq 1.$$

Without loss of generality, we shall assume that $\Delta \leqq 0.1$, otherwise the theorem is trivial.

Taking arbitrary α and β such that $\Delta \leqq \beta - \alpha \leqq 1 - \Delta$, and putting $r = 1$, consider the periodic function $\psi(x)$ of period 1 defined in Lemma 2 of Chapter 2. Then we have

$$\psi(f(x)) = 1, \quad \text{if} \quad \alpha + 0.5\Delta \leqq f(x) \leqq \beta - 0.5\Delta \pmod 1,$$
$$0 < \psi(f(x)) < 1, \quad \text{if} \quad \alpha - 0.5\Delta < f(x) < \alpha + 0.5\Delta \pmod 1,$$
$$\text{or} \quad \beta - 0.5\Delta < f(x) < \beta + 0.5\Delta \pmod 1,$$
$$\psi(f(x)) = 0, \quad \text{if} \quad \beta + 0.5\Delta \leqq f(x) \leqq 1 + \alpha - 0.5\Delta \pmod 1,$$

$$\psi(f(x)) = \beta - \alpha + \sum_{m=1}^{\infty} (g_m e^{2\pi i m f(x)} + h_m e^{-2\pi i m f(x)}), \tag{1}$$

where

$$|g_m| \leqq \frac{1}{\pi m}, \quad |h_m| \leqq \frac{1}{\pi m}, \qquad \text{if} \quad m \leqq \Delta^{-1},$$

$$|g_m| \leqq \frac{1}{\pi^2 \Delta m^2}, \quad |h_m| \leqq \frac{1}{\pi^2 \Delta m^2}, \quad \text{if} \quad m > \Delta^{-1}.$$

For the sake of brevity, assuming

$$U(\alpha, \beta) = \sum_{0 < x \leqq P} \psi(f(x)),$$

from (1) we obtain

$$U(\alpha, \beta) = P(\beta - \alpha) + H;$$

$$H \ll \sum_{0 < m \leqq \Delta^{-1}} \frac{|T_m|}{m_1} + \sum_{\Delta^{-1} < m \leqq \Delta^{-2}} \frac{|T_m|}{\Delta m^2} + \sum_{\Delta^{-1} < m} \frac{P}{\Delta m^2}.$$

If $(\alpha_n, \ldots, \alpha_1)$ is a point of Class II, then we readily find

$$H \ll \sum_{0 < m \leqq \Delta^{-1}} \frac{P\Delta}{m} + \sum_{\Delta^{-1} < m \leqq \Delta^{-1}} \frac{P\Delta}{\Delta m^2} + P\Delta \ll P\Delta \ln(\Delta^{-1}) \ll P\Delta_0.$$

If $(\alpha_n, \ldots, \alpha_1)$ is a point of Class I, when $\delta_0 \geqq 1$, obviously we obtain the same result.

Finally, if $(\alpha_n, \ldots, \alpha_1)$ is a point of Class I such that $\delta_0 < 1$, we get

$$H \ll \sum_{0 < m \leqq \Delta^{-1}} \frac{P\Delta (m, Q)^{\nu}}{m} + \sum_{\Delta^{-1} < m \leqq \Delta^{-2}} \frac{P\Delta (m, Q)^{\nu}}{\Delta m^2} + P\Delta.$$

The part of the sum of the first two terms on the right which corresponds to the case $(m, Q) = d$, is

$$\ll P\Delta\left(d^{\nu-1} \sum_{0<m_1\leqq \Delta^{-1}d^{-1}} \frac{1}{m_1} + d^{\nu-2} \sum_{\Delta^{-1}d^{-1}<m_1\leqq \Delta^{-2}d^{-2}} \frac{1}{\Delta m_1^2}\right) \ll P\Delta d^{\nu-1}\ln(\Delta^{-1}).$$

And since d can only be a divisor of Q, then, applying Lemma 5 of Chapter 1, we obtain

$$H \ll P\Delta_0.$$

Hence the following formula

$$U(\alpha, \beta) = P(\beta - \alpha) + O(P\Delta_0) \tag{2}$$

holds for all the cases.

For $0 \leqq \beta - \alpha \leqq 1$, let $D(\alpha, \beta)$ denote the number of terms in the sequence $x = 1, \ldots, P$ such that $\alpha \leqq f(x) < \beta \pmod 1$. If $2\Delta \leqq \beta - \alpha \leqq 1 - 2\Delta$, from the obvious inequalities

$$U(\alpha + 0.5\Delta, \beta - 0.5\Delta) \leqq D(\alpha, \beta) \leqq U(\alpha - 0.5\Delta, \beta + 0.5\Delta)$$

and from (2), it follows that

$$D(\alpha, \beta) = P(\beta - \alpha) + O(P\Delta_0), \tag{3}$$

which is readily extended to the remaining cases, too: to the case $0 \leqq \beta - \alpha \leqq 2\Delta$ with the aid of the identity

$$D(\alpha, \beta) = D(\alpha, \alpha + 1 - 2\Delta) - D(\beta, \alpha + 1 - 2\Delta),$$

and to the case $1 - 2\Delta \leqq \beta - \alpha \leqq 1$ with the aid of the identity:

$$D(\alpha, \beta) = D(\alpha, \alpha + 2\Delta) + D(\alpha + 2\Delta, \beta).$$

Putting $\alpha = 0$, $\beta = \sigma$ in (3), we complete the proof of the theorem.

Theorem 2. Let an increasing sequence (x) consist of P_1 terms from the sequence $1, \ldots, P$. Let σ be any number such that $0 < \sigma \leqq 1$ and let $D(\sigma)$, the number of terms in the sequence (x) satisfying the condition $\{f(x)\} < \sigma$, be represented in the form:

$$D(\sigma) = P_1\sigma + \lambda(\sigma).$$

Then, for $k \geqq 12$, we can eliminate from the domain Π_n a region Ω of volume not exceeding

$$(nl)^{3nl}(2n)^{\frac{3n(n+1)l}{2}} P^{-\frac{n(n+1)}{2}\left(1-\frac{4\ln k - 0.4}{k}\right)};$$

$$l = \left[(n+1)\ln\frac{k}{4} + 1\right].$$

such that for any point $(\alpha_n, \ldots, \alpha_1)$ in the remaining region, we have, for $n \geqq 8$,

$$|\lambda(\sigma)| < 3.4P^{1-1/k}\ln P$$

regardless of the choice of σ.

Proof. Let $P > (2n)^n$, otherwise the theorem is trivial. Consider the sum:

$$T' = \sum_{(x)} e^{2\pi i f(x)},$$

where x runs through the numbers in the sequence (x). Since this sum is a particular case of Lemma 6 of Chapter 4, we can apply Theorem 2 of Chapter 4 (according to the note given at the end of that theorem) to our sum. From the domain Π_n of an n-dimensional cube $0 \leqq \alpha_n < 1, \ldots, 0 \leqq \alpha_1 < 1$, we can eliminate a region Ω of volume not exceeding

$$(nl)^{3nl}(2n)^{\frac{3n(n+1)l}{2}} P^{-\frac{n(n+1)}{2}\left(1-\frac{4\ln k-1}{k}\right)},$$

such that for any point in the remaining region, we have

$$|T'| \leqq P^{1-1/k}.$$

Taking a positive integer m, for integral $g_n, \ldots, g_1$, consider the regions $\Pi_{g_n,\ldots,g_1}$ and $\Omega_{g_n,\ldots,g_1}$ obtained from the regions Π_n and Ω by a transformation, which maps the point $(\alpha_n, \ldots, \alpha_1)$ into the point $((\alpha_n + g_n)/m, \ldots, (\alpha_1 + g_1)/m)$. Making each g_s run, independently of the others, through the values $0, \ldots, m-1$, we obtain m^n regions $\Pi^{(m)}_{g_n,\ldots,g_1}$, which will completely cover (without overlapping) the region Π_n. The region covered by the corresponding $\Omega^{(m)}_{g_n,\ldots,g_1}$ regions will be called the region $\Omega(m)$. The volume of the region $\Omega(m)$ is obviously equal to the volume of the region Ω.

Put $M = P^{1/k}$. We shall call the region which contains all the subregions of $\Omega(m)$ satisfying the condition $m \leqq M^2$ the region Ω'. The volume of the region Ω' does not exceed the product of M^2 and the upper bound stated above for the volume of the region Ω. As is not difficult to verify, this volume does not exceed the bounds stated in the theorem.

From the method by which the region Ω' was constructed it is clear that, if the point $(\alpha_n, \ldots, \alpha_1)$ does not belong to Ω', any sum of the type

$$T'_m = \sum_{(x)} e^{2\pi i m f(x)},$$

where m is an integer numerically not greater than M^2, satisfies the inequality

$$|T'_m| < P^{1-1/k}.$$

Put $\Delta = M^{-1} = P^{-1/k}$. Then we have $\Delta < 1/12$.

Taking any arbitrary α and β such that $\Delta \leqq \beta - \alpha \leqq 1 - \Delta$ and assuming $r = 1$, consider the periodic function $\psi(x)$ of period 1 defined in Lemma 2 of Chapter 2 (see the proof of Theorem 1). Then we have

$$\psi(f(x)) = \beta - \alpha + \sum_{m=1}^{\infty} (g_m e^{2\pi i m f(x)} + h_m e^{-2\pi i m f(x)}).$$

Hence, assuming, for the sake of brevity, that

$$U(\alpha, \beta) = \sum_{\alpha \leqq f(x) < \beta \pmod 1} \psi(f(x)),$$

where x runs through the numbers in the sequence (x), we find, if the point $(\alpha_n, \ldots, \alpha_1)$ does not belong to the region Ω', that

$$|U(\alpha,\beta)-(\beta-\alpha)P_1|$$

$$\leqq \sum_{0<m\leqq M}\frac{2}{\pi m}P\Delta+\sum_{M<m\leqq M^2}\frac{2}{\pi^2 m^2\Delta}P\Delta+\sum_{M^2<m}\frac{2}{\pi^2 m^2\Delta}P$$

$$< P\Delta\left(\frac{2}{\pi}+\frac{2}{\pi}\int_1^M\frac{dm}{m}+\frac{2}{\pi^2\Delta}\int_M^{M^2}\frac{dm}{m^2}+\frac{2}{\pi^2\Delta^2}\int_{M^2}^{\infty}\frac{dm}{m^2}\right) \tag{4}$$

$$< P\Delta\left(\frac{2}{\pi}+\frac{2}{\pi}\ln M+\frac{4}{\pi^2}\right)$$

$$< P\Delta\frac{2}{\pi}(\ln M+1.64) < 0.25P^{1-1/k}\ln P.$$

For $0<\beta-\alpha\leqq 1$, let $D(\alpha,\beta)$ denote the number of terms in the sequence (x) satisfying the condition $\alpha\leqq\{f(x)\}<\beta\,(\mathrm{mod}\,1)$.

If $2\Delta\leqq\beta-\alpha\leqq 1-2\Delta$, from the obvious inequalities

$$U(\alpha+0.5\Delta,\beta-0.5\Delta)\leqq D(\alpha,\beta)\leqq U(\alpha-0.5\Delta,\beta+0.5\Delta)$$

and from (4), it follows that

$$|D(\alpha,\beta)-(\beta-\alpha)P_1|<1.7P^{1-1/k}\ln P.$$

If $0\leqq\beta-\alpha\leqq 2\Delta$, using the identity

$$D(\alpha,\beta)=D(\alpha,\alpha+1-2\Delta)-D(\beta,\alpha+1-2\Delta)$$

and, if $1-2\Delta\leqq\beta-\alpha\leqq 1$, using the identity

$$D(\alpha,\beta)=D(\alpha,\alpha+2\Delta)+D(\alpha+2\Delta),\beta)$$

we obtain:

$$|D(\alpha,\beta)-(\beta-\alpha)P_1|<3.4P^{1-1/k}\ln P.$$

Putting $\alpha=0$, $\beta=\sigma$, we complete the proof of the theorem.

Chapter 6. The Asymptotic Formula in Waring's Problem

Notation. $n\,(n \geqq 4)$ is a constant; r is a fixed integer, N is an integer.

In this chapter we shall illustrate the application of the results obtained in Chapter 4 to Waring's problem. We shall prove (Theorem 1) that the asymptotic formula:

$$I_N = \frac{(\Gamma(1+\nu))^r}{\Gamma(r\nu)} N^{r\nu-1}\,\mathfrak{S} + O\left(N^{r\nu-1-\frac{1}{20n\ln n}}\right)$$

established by Hardy and Littlewood for the number I_N of representations of a positive integer N in the form

$$N = x_1^n + \ldots + x_r^n$$

with positive integers $x_1, \ldots, x_r$ is valid for

$$r \geqq r_2; \quad r_2 = 2[n^2(2\ln n + \ln\ln n + 5)].$$

Further, we shall apply the Hardy and Littlewood method to demonstrate that $\mathfrak{S}$ exceeds a certain positive constant. Then we shall derive a corollary to show the existence of a number $N_0 = N_0(n, r)$ such that, for $N \geqq N_0(n, r)$, there are $\gg N^{r\nu-1}$ representations of N in the form stated above.

The results of Chapter 4 can also be successfully applied to solve other additive problems which may be called "generalizations of Waring's problem". Among these, for instance, are the representation of N as the sum of the values of a polynomial of degree $n \geqq 4$ which takes integral values for integral values of its argument, simulataneous representation of $N_n, \ldots, N_1$ in the form:

$$N_n = x_1^n + \ldots + x_r^n,$$

$$\ldots\ldots\ldots\ldots$$

$$N_1 = x_1 + \ldots + x_r$$

with positive integers $x_1, \ldots, x_r$, and others.

Lemma 1. Let

$$r \geqq r_0; \quad r_0 = 2r_1; \quad r_1 = [n^2(2\ln n + \ln\ln n + 4)].$$

Then

$$\int_0^1 |S_\alpha|^r\, d\alpha \leqq P^{r_0-n}; \quad S_\alpha = \sum_{x=1}^{P} e^{2\pi i\alpha x^n}.$$

Proof. Let each $x_1, \ldots, x_{r_0}$ run, independently of the others, through the values $1, \ldots, P$. The integral in the lemma expresses the number of solutions of the equation:

$$x_1^n + \ldots + x_{r_1}^n - x_{r_1+1}^n - \ldots - x_{r_0}^n = 0. \tag{1}$$

In addition to this integral, we shall also consider, for given integers $N_{n-1}, \ldots, N_1$, the integral

$$\int_0^1 \ldots \int_0^1 |S|^{r_0} e^{-2\pi i\alpha_{n-1}N_{n-1} - \ldots - 2\pi i\alpha_1 N_1} d\alpha_n \ldots d\alpha_1;$$

$$S = \sum_{x=1}^{P} e^{2\pi i(\alpha_n x^n + \ldots + \alpha_1 x)},$$

which expresses the number of solutions of the system

$$\left.\begin{aligned} x_1^n + \ldots + x_{r_1}^n - x_{r_1+1}^n - \ldots - x_{r_0}^n &= 0, \\ x_1^{n-1} + \ldots + x_{r_1}^{n-1} - x_{r_1+1}^{n-1} - \ldots - x_{r_0}^{n-1} &= N_{n-1}, \\ \ldots\ldots\ldots\ldots\ldots\ldots\ldots\ldots\ldots \\ x_1 + \ldots + x_{r_1} - x_{r_1+1} - \ldots - x_{r_0} &= N_1. \end{aligned}\right\} \quad (2)$$

This integral does not exceed the integral

$$\int_0^1 \ldots \int_0^1 |S|^{r_0} d\alpha_n \ldots d\alpha_1$$

in absolute value and, consequently, according to Theorem 4 of Chapter 4, it is

$$\ll P^{r_0 - \frac{n(n+1)}{2}}.$$

But the system (2) is solvable only when each N_s satisfies the condition:

$$-r_0 P^s < N_0 < r_0 P^s.$$

Therefore the number of different systems of values of $N_{n-1}, \ldots, N_1$ for which the system is solvable is $\ll P^{n(n-1)/2}$. And since the number of solutions of Eq. (1) is the sum of the numbers of solutions of all different systems (2), it is

$$\ll P^{r_0 - \frac{n(n+1)}{2}} P^{\frac{n(n-1)}{2}} = P^{r_0 - n}.$$

This completes the proof of our lemma.

Note. Lemma 1 obviously remains true if the sum S_α is replaced by a partial sum taken over only a certain part of the numbers of the sequence $1, \ldots, P$: say, over the primes in this sequence.

Lemma 2. Let r be a positive integer, $N > 0$, and $K_r(N)$ be the number of solutions, in positive integers $x_1, \ldots, x_r$, of the inequality:

$$x_1^n + \ldots + x_r^n \leqq N.$$

Then

$$K_r(N) = T_r N^{rv} - \theta r N^{rv-v}; \quad T_r = \frac{(\Gamma(1+v))^r}{\Gamma(rv+1)}, \quad \theta \geqq 0.$$

Proof. Obviously, we have

$$K_1(N) = N^v - \theta'; \quad \theta' \geqq 0,$$

and, thus, the lemma is true for $K_1(N)$.

We shall use induction to prove the lemma. Assume that, for some $r \geqq 1$, the lemma holds true for $K_r(N)$. We find $(\theta'' > 0,\ \theta''' > 0)$ that

$$
\begin{aligned}
K_{r+1}(N) &= \sum_{0<x\leqq N^{\nu}} K_r(N-x^n) \\
&= T_r \sum_{0<x\leqq N^{\nu}} (N-x^n)^{r\nu} - \theta'' r N^{r\nu} \\
&= T_r \int_0^{N^{\nu}} (N-x^n)^{r\nu}\, dx - \theta'''(r+1) N^{r\nu} \\
&= T' N^{(r+1)\nu} - \theta'''(r+1) N^{r\nu};
\end{aligned}
$$

$$
T' = T_1 \nu \int_0^1 (1-z)^{r\nu} z^{\nu-1}\, dz = T_1 \nu \frac{\Gamma(r\nu+1)\,\Gamma(\nu)}{\Gamma(r\nu+1+\nu)} = T_{r+1}.
$$

Hence, the lemma holds true also for $K_{r+1}(N)$.

Lemma 3. Let

$$
r \geqq r_0; \quad r_0 = 2\,[n^2(2\ln n + \ln\ln n + 4)], \quad P_0 = N^{\nu},
$$

$$
I_\alpha = \int_0^{P_0} e^{2\pi i \alpha x^n}\, dx, \quad J = \int_{-\infty}^{\infty} I_\alpha^r e^{-2\pi i \alpha N}\, d\alpha .
$$

Then

$$
J = \frac{(\Gamma(1+\nu))^r}{\Gamma(r\nu)} P_0^{r-n} + O(P_0^{r-n-0.25}).
$$

Proof. Let $P = [P_0]$, $N_1 = [NP^{-0.25}]$. The number G of solutions, in positive integers $x_1, \ldots, x_r$, of the inequality:

$$
N - N_1 < x_1^n + \ldots + x_r^n \leqq N,
$$

can be represented, by virtue of Lemma 2, in the form:

$$
\begin{aligned}
K_r(N) - K_r(N-N_1) &= T_r r\nu N^{r\nu-1} N_1 + O(N^{r\nu-1} N_1 P_0^{-0.25}) \\
&= \frac{(\Gamma(1+\nu))^r}{\Gamma(r\nu)} P_0^{r-n} N_1 + O(P_0^{r-n} N_1 P_0^{-0.25}).
\end{aligned}
$$

But, G can also be represented (Lemma 6, Chapter 2) in the following form:

$$
\int_{-P_0^{0.5-n}}^{1-P_0^{0.5-n}} \sum_{m=0}^{N_1-1} S_\alpha^r e^{-2\pi i \alpha(N-m)}\, d\alpha; \quad S_\alpha = \sum_{x=1}^{P} e^{2\pi i \alpha x^n}.
$$

First applying Lemma 8 of Chapter 2 and then Lemma 1, we find

$$
\begin{aligned}
\int_{P^{0.5-n}}^{1-P^{0.5-n}} \sum_{m=0}^{N_1-1} S_\alpha^r e^{-2\pi i \alpha(N-m)}\, d\alpha &\ll \int_{P^{0.5-n}}^{1-P^{0.5-n}} |S_\alpha|^r \frac{d\alpha}{2\,(\alpha)} \\
&\ll P^{n-0.5} \int_{P^{0.5-n}}^{1-P^{0.5-n}} |S_\alpha|^r\, d\alpha \ll P_0^{r-n} N_1 P_0^{-0.25}.
\end{aligned}
$$

Now comparing the two expressions for G, we obtain

$$
\frac{(\Gamma(1+\nu))^r}{\Gamma(r\nu)} P_0^{r-n} N_1 = \int_{-P_0^{0.5-n}}^{P_0^{0.5-n}} \sum_{m=0}^{N_1-1} S_\alpha^r e^{-2\pi i \alpha(N-m)}\, d\alpha + O(P_0^{r-n} N_1 P_0^{-0.25}). \tag{3}
$$

Apply Lemma 3 of Chapter 2 to the sum S_α. Since, for $|\alpha| \leqq P_0^{0.5-n}$, in the interval $0 \leqq x \leqq P_0$ we have $|(\alpha x^n)'| \leqq nP_0^{-0.5}$ (which does not exceed 0.5, if $P_0 \geqq 4n^2$), we find

$$S_\alpha = I_\alpha + O(1); \quad I_\alpha = \int_0^{P_0} e^{2\pi i \alpha x^n}\, dx. \tag{4}$$

Putting $\alpha = \delta P_0^{-n}$ and then transforming the integral I_α by the substitution $x = P_0 \xi$, apply Lemma 4 of Chapter 2. Thus, we obtain

$$I_\alpha \ll P_0 D; \quad D = \min(1, |\delta|^{-\nu}).$$

Now, from (4) it follows that

$$S_\alpha^r e^{-2\pi i \alpha (N-m)} = I_\alpha^r e^{-2\pi i \alpha (N-m)} + O(P_0^{r-1} D^{r-1}),$$

hence

$$\int_{-P_0^{0.5-n}}^{P_0^{0.5-n}} P_0^{r-1} D^{r-1}\, d\alpha \ll P_0^{r-n-1} \int_0^{P_0^{0.5}} \min(1, \delta^{-\nu(r-1)})\, d\delta \ll P_0^{r-n-1},$$

because $\alpha = \delta P_0^{-n}$, and so we find

$$\int_{-P_0^{0.5-n}}^{P_0^{0.5-n}} S_\alpha^r e^{-2\pi i \alpha (N-m)}\, d\alpha = \int_{-P_0^{0.5-n}}^{P_0^{0.5-n}} I_\alpha^r e^{-2\pi i \alpha (N-m)}\, d\alpha + O(P_0^{r-n-1}).$$

But the difference between the integral on the right-hand side and the integral

$$\int_{-P_0^{0.5-n}}^{P_0^{0.5-n}} I_\alpha^r e^{-2\pi i \alpha N}\, d\alpha,$$

is obviously

$$\ll \int_0^{P_0^{0.5-n}} P_0^r D^r |e^{2\pi i \alpha m} - 1|\, d\alpha \ll P_0^{r-n} \int_0^{P_0^{0.5}} \min(1, \delta^{-r\nu}) \min(1, \delta P_0^{-0.25})\, d\delta,$$

because $\alpha = \delta P_0^{-n}$, where the integrand

$$\begin{aligned} &\ll P_0^{-0.25}, && \text{if } 0 \leqq \delta \leqq 1, \\ &\ll \delta^{-r\nu+1} P_0^{-0.25}, && \text{if } 1 < \delta \leqq P_0^{0.25}, \\ &\ll \delta^{-r\nu}, && \text{if } P_0^{0.25} < \delta \leqq P_0^{0.5}. \end{aligned}$$

Therefore, this difference $\ll P_0^{r-n-0.25}$. Thus, the right-hand side of (3) takes the form:

$$N_1 \int_{-P_0^{0.5-n}}^{P_0^{0.5-n}} I_\alpha^r e^{-2\pi i \alpha N}\, d\alpha + O(N_1 P_0^{r-n-0.25}).$$

Since each of the integrals

$$\int_{-\infty}^{-P_0^{0.5-n}} I_\alpha^r e^{-2\pi i \alpha N}\, d\alpha, \quad \int_{P_0^{0.5-n}}^{\infty} I_\alpha^r e^{-2\pi i \alpha N}\, d\alpha,$$

is obviously

$$\ll \int_{P_0^{0.5-n}}^{\infty} P_0^r D^r\, d\alpha \ll P_0^{r-n} \int_{P_0^{0.5}}^{\infty} \delta^{-r\nu}\, d\delta \ll P_0^{r-n-1},$$

the right-hand side is transformed into

$$N_1 \int_{-\infty}^{\infty} I_\alpha^r e^{-2\pi i \alpha N} d\alpha + O(N_1 P_0^{r-n-0.25}),$$

which completes the proof of our lemma.

Theorem 1. Let

$$n \geqq 4, \quad r \geqq r_2; \quad r_2 = 2[n^2(2\ln n + \ln\ln n + 5)].$$

Then the number I_N of representations of a positive integer N in the form

$$N = x_1^n + \ldots + x_r^n$$

with positive integers $x_1, \ldots, x_r$, is given by the expression:

$$I_N = \frac{(\Gamma(1+\nu))^r}{\Gamma(r\nu)} N^{r\nu-1} \mathfrak{S} + O(N^{r\nu-1-\frac{1}{20n\ln n}}); \qquad \mathfrak{S} = \sum_{Q=1}^{\infty} A(Q),$$

$$A(Q) = \sum_{\substack{0 \leqq a < Q \\ (a,Q)=1}} \left(\frac{S_{a,Q}}{Q}\right)^r e^{-2\pi i \frac{a}{Q} N}, \quad S_{a,Q} = \sum_{1 \leqq l < Q} e^{2\pi i \frac{a}{Q} l^n}.$$

Proof. Taking $P = [N^\nu]$, we find

$$I_N = \int_{-P^{-n+\nu}}^{1-P^{-n+\nu}} S_\alpha^r e^{-2\pi i \alpha N} d\alpha; \quad S_\alpha = \sum_{x=1}^{P} e^{2\pi i \alpha x^n}.$$

Divide the range of integration into intervals of two classes as follows. Representing α as

$$\alpha = a/Q + z; \quad z = \delta P^{-n}; \quad (a, Q) = 1, \ 0 \leqq a < Q,$$

into Class I put those intervals containing values of α such that

$$Q \leqq P^\nu; \qquad |\delta| \leqq P^\nu.$$

Evidently, the intervals of Class I do not intersect.

Put those intervals that remain after the intervals of Class I have been selected, into Class II. In compliance with our division of the range of integration, the integral I_N can be expressed as the the sum of two subintegrals:

$$I_N = I_{N,1} + I_{N,2}.$$

If α belongs to an interval of Class II, then, according to Theorem 2 of Chapter 4, we find

$$S_\alpha \ll P^{1-\varrho}; \quad \varrho = \frac{1}{8n^2(\ln n + 1.5\ln\ln n + 4.2)}.$$

Putting $r_0 = 2[n^2(2\ln n + \ln\ln n + 4)]$, and using the trivial inequality $r - r_0 > 2n^2$, we obtain

$$S_\alpha^{r-r_0} \ll P^{r-r_0-2n^2\varrho}.$$

Therefore, applying Lemma 1, we find

$$I_{N,2} \ll P^{r-r_0-2n^2\varrho} \int_0^1 |S_\alpha|^{r_0} d\alpha \ll P^{r-n-2n^2\varrho} \ll P^{r-n-\frac{1}{20\ln n}}.$$

If α belongs to an interval of Class I that contains the fraction a/Q, then we find

$$S_\alpha = \sum_{l=1}^{Q} S^{(l)}; \quad S^{(l)} = e^{2\pi i \frac{a}{Q} l^n} \sum_{0<Qu+l\leqq P} e^{2\pi i z (Qu+l)^n}.$$

But, we have

$$\left| \frac{d}{du} z(Qu+l)^n \right| \ll \frac{\delta Q}{P} \ll P^{2\nu-1} < P^{-1/3}.$$

Therefore, according to Lemma 3 of Chapter 2, we find

$$S^{(l)} = e^{2\pi i \frac{a}{Q} l^n} \int_{-lQ-1}^{(P-l)Q-1} e^{2\pi i z (Qu+l)^n} du + O(1)$$

$$= \frac{e^{2\pi i \frac{a}{Q} l^n}}{Q} I(z) + O(1);$$

$$I(z) = \int_0^{P_0} e^{2\pi i z x^n} dx, \quad P_0 = N^\nu.$$

Consequently,

$$S_\alpha = \frac{S_{a,Q}}{Q} I(z) + O(P^\nu).$$

But, according to Lemma 4 of Chapter 2, we have

$$I(z) \ll Z; \quad Z = P_0 \min(1, |\delta|^{-\nu}).$$

Furthermore, we find

$$S_\alpha^r e^{-2\pi i \alpha N} = \left(\frac{S_{a,Q}}{Q}\right)^r e^{-2\pi i \frac{a}{Q} N} (I(z))^r e^{-2\pi i z N} + O(Z^{r-1} P^\nu),$$

hence, noting that

$$\int_{-P^{\nu-n}}^{P^{\nu-n}} Z^{r-1} P^\nu dz \ll P^{r-1-n+\nu} \int_0^{P^\nu} \min(1, \delta^{-\nu(r-1)}) d\delta \ll P^{r-n-1+\nu},$$

we find that the part of the integral $I_{N,1}$ over the interval chosen is given by the expression:

$$\left(\frac{S_{a,Q}}{Q}\right)^r e^{-2\pi i \frac{a}{Q} N} \int_{-P^{\nu-n}}^{P^{\nu-n}} (I(z))^r e^{-2\pi i z N} dz + O(P^{r-n-1+\nu}).$$

Summing the last expression over all intervals for a given Q, and then over all Q's not exceeding P^ν, we obtain

$$I_{N,1} = \sum_{Q\leqq P^\nu} A(Q) \int_{-P^{\nu-n}}^{P^{\nu-n}} (I(z))^r e^{-2\pi i z N} dz + O(P^{r-n-0.25}).$$

According to Theorem 1 of Chapter 3, we find

$$A(Q) \ll Q^{1-\nu r+\varepsilon_0 r}, \quad \mathfrak{S} \ll 1, \quad \sum_{Q\geqq P^\nu} A(Q) \ll P^{-4\ln n}.$$

Besides, we find

$$\int_0^{P^{\nu-n}} Z^r\,dz \ll P^{r-n}, \qquad \int_{P^{\nu-n}}^{\infty} Z^r\,dz \ll P^{r-n-4\ln n}.$$

Therefore, applying Lemma 3, we obtain

$$I_{N,1} = \frac{(\Gamma(1+\nu))^r}{\Gamma(r\nu)}\,\mathfrak{S}N^{r\nu-1} + O(P^{r-n-0.25}),$$

which, together with what has been established for $I_{N,2}$, completes the proof of our theorem.

Notation. Henceforth we shall retain the notation of Theorem 1, but use them under much weaker restrictions. We shall assume that $N > 0$, $n \geqq 4$, $r > 0$ and consider the number $\mathfrak{S}$ when the series defining it converges. Further, we shall use the symbol $M(q)$ to denote the number of solutions of the congruence:

$$x_1^n + \ldots + x_r^n \equiv N \pmod q,$$

where each x runs through the values $0, \ldots, q-1$.
We shall use the letter p to denote a prime, and the symbol $\psi(p)$ to denote the number (when the series defining it converges) given by the equality:

$$\psi(p) = \sum_{s=0}^{\infty} A(p^s).$$

The letter τ stands for the exponent of p in the canonical expansion of n; γ denotes the number defined by the equalities:

$$\gamma = \begin{cases} \tau+1, & \text{if} \quad p > 2, \\ \tau+2, & \text{if} \quad p = 2. \end{cases}$$

For $s \geqq \gamma$, the symbol h_s is used to denote the greatest exponent to which numbers not divisible by p belong modulo p^s and is defined by the following relationship:

$$h_s = \begin{cases} \varphi(p^s) & \text{if} \quad p > 2 \quad (p^\tau(p-1) \quad \text{for} \quad s = \gamma), \\ \frac{1}{2}\varphi(2^s), & \text{if} \quad p = 2 \quad (2^\tau(2-1) \quad \text{for} \quad s = \gamma). \end{cases}$$

The letter g is used to denote a number belonging to h_s modulo p^s.

We can easily show that g belongs to the exponent h_s modulo p^{s_1} for any arbitrary s_1 taken from the sequence $s-1, \ldots, \gamma$. Indeed, assume that g belongs to the exponent σ modulo p^{s_1}. We find ($t, t_1, \ldots$ being integers)

$$\begin{aligned} &g^\sigma \equiv 1 \pmod{p^{s_1}}, \quad g^\sigma \equiv 1 + p^{s_1} t, \\ &g^{\sigma p} = (1 + p^{s_1} t)^p = 1 + p^{s_1+1} t_1, \\ &g^{\sigma p^2} = (1 + p^{s_1+1} t_1)^p = 1 + p^{s_1+2} t_2, \\ &\ldots\ldots\ldots\ldots\ldots\ldots\ldots\ldots \\ &g^{\sigma p^{s-s_1}} = 1 + p^s t_{s-s_1} \equiv 1 \pmod{p^s}. \end{aligned}$$

Consequently, σp^{s-s_1} is divisible by h_s. Since $h_s = h_{s_1} p^{s-s_1}$, σ is divisible by h_{s_1}, and since σ is a divisor of h_{s_1}, we have $\sigma = h_{s_1}$, which proves our assertion.

For $p > 2$, to each N not divisible by p there corresponds a unique b from the sequence $0, \ldots, h_s - 1$ such that

$$N \equiv g^b \pmod{p^s}.$$

And for $p = 2$, to each N of the type $4m + 1$ there corresponds a unique b from the sequence $0, \ldots, h_s - 1$ such that

$$N \equiv g^b \pmod{p^s}.$$

Our next problem is to prove that, for $r \geqq 4n$, the number $\mathfrak{S}$ exceeds a certain positive constant.

Lemma 4. For $r \geqq 4n$, the congruence

$$x_1^n + \ldots + x_r^n \equiv N \pmod{p^\gamma} \tag{5}$$

is solvable, provided not all $x_1, \ldots, x_r$ are divisible by p.

Proof. Without loss of generality, we shall consider only the case $0 < N \leqq p^\gamma$.

If $p = 2$, then $2^\gamma = 2^{\tau+2} \leqq 4n$, and the congruence can be satisfied by taking the first N values of x_j equal to 1 and the remaining $(r - N)$ values equal to 0.

If $p > 2$, first we shall show that the congruence (5) is solvable, assuming that N is not divisible by p and that r is a certain number from the sequence $1, \ldots, 2n - 1$. For this purpose, consider the least $r = r(N)$ for which the congruence (5) is solvable. Determining b from the conditions (see notation) $0 \leqq b < h_\gamma$, $N \equiv g^b \pmod{p^\gamma}$, we shall relate N to the least non-negative residue v of the number b modulo n. If N' is another number related to v, then determining b_1 from the conditions $0 \leqq b_1 < h_\gamma$, $N' \equiv g^{b_1} \pmod{p^\gamma}$, we find that for a certain integer k, we have $b_1 = b + kn$. Hence

$$N' \equiv N z^n \pmod{p^\gamma}; \quad z = \begin{cases} g^k, & \text{if } k \geqq 0, \\ g^{k+h_\gamma}, & \text{if } k < 0. \end{cases}$$

Now, multiplying the congruence (5), term by term, by z^n, we obtain

$$(x_1 z)^n + \ldots + (x_r z)^n \equiv N' \pmod{p^\gamma},$$

hence we find that $r(N') \leqq r(N)$. Similarly, we find that $r(N) \leqq r(N')$. Hence we conclude that $r(N) = r(N')$. Consequently, all the values of N under consideration can be distributed over $m \leqq n$ sets, the values corresponding to the same $r(N)$ in the same set. Let the least representative of these sets, arranged in increasing order, be

$$N_1, \ldots, N_m.$$

The least value of N under consideration is $N = 1 \equiv 1^n \pmod{p^\gamma}$. Therefore, obviously we have

$$N_1 = 1, \quad r(N_1) = 1 = 2 - 1.$$

The equality $r(N_1) = 2 \cdot 1 - 1$ is a particular case of the relation:

$$r(N_j) \leqq 2j - 1,$$

which holds true also for all N_j under consideration (this can be proved by induction).

Indeed, assume that the above inequality has been demonstrated for $N_1, \ldots, N_j$ less than N_m. The number N_{j+1} is the least representative of its set, and one of the numbers

$N_{j+1}-1, N_{j+1}-2$ is definitely not divisible by p and, consequently, our inequality has been proved to hold true for it. Therefore

$$r(N_{j+1}) \leqq 2j-1+2 = 2(j+1)-1,$$

and this inequality is valid for N_{j+1} as well. In particular, we have $r(N_m) \leqq 2m-1 \leqq 2n-1$, which proves our assertion.

We can easily show that this assertion will still hold, if the condition $r \leqq 2n-1$ is replaced by the weaker condition $r \leqq 2n$. Indeed, if N is divisible by p, then $N-1$ is not, and from

$$x_1^n + \ldots + x_{r_1}^n \equiv N-1 \pmod{p^\gamma}; \quad r_1 \leqq 2n-1,$$

it follows that

$$x_1^n + \ldots + x_{r_1}^n + 1^n \equiv N \pmod{p^\gamma}.$$

This proves our lemma, since any number of terms, each equal to $0 = 0^n$, can be added to the left-hand side of (5).

Lemma 5. If

$$y^n \equiv a \pmod{p^\gamma}$$

is solvable when y is not divisible by p, then the congruence

$$x^n \equiv a \pmod{p^s}; \quad s > \gamma,$$

is also solvable.

Proof. For $p=2$, and y not divisible by 2, the congruence $y^n \equiv a \pmod{2^\gamma}$ is solvable only when a is of the form $4m+1$. Determining b from the congruence (see the notation)

$$a \equiv y^n g^b \pmod{p^s} \tag{6}$$

and using the first congruence of the lemma, we obtain

$$g^b \equiv 1 \pmod{p^\gamma}.$$

Therefore, for a certain non-negative integer b', we have (see the notation)

$$b = p^\tau (p-1) b'.$$

But, since for any positive integer k

$$g^{p^{s-1}(p-1)k} \equiv 1 \pmod{p^s},$$

from (6) we get

$$a \equiv y^n g^{p^\tau (p-1)(b' + kp^{s-1-\tau})} \pmod{p^s}.$$

Assuming that $n = p^\tau n_1$, and, since $(n_1, p) = 1$, the number k can be chosen from the condition that the second expression in brackets in the exponent is divisible by n_1 and the exponent itself is divisible by n, we find that the last congruence takes the form:

$$a \equiv (yg^f)^n \pmod{p^s},$$

where f is an integer. This completes the proof of our lemma.

Lemma 6. Let $s > \gamma$ and $r \geqq 4n$. Then

$$M(p^s) \geqq p^{(s-\gamma)(r-1)}.$$

Proof. According to Lemma 4, there exist an integer y not divisible by p and integers $y_2, \ldots, y_r$ such that

$$y^n \equiv N - y_2^n - \ldots - y_r^n \pmod{p^\gamma}.$$

Making each x_j of the sequence $x_2, \ldots, x_r$ run through those numbers of the system of least non-negative residues modulo p^s which are congruent to y_j modulo p^γ, we obtain $p^{(s-\gamma)(r-1)}$ systems of values of $x_2, \ldots, x_r$. For each of these systems the following congruence holds:

$$y^n \equiv N - x_2^n - \ldots - x_r^n \pmod{p^\gamma}$$

and, consequently (Lemma 5), the congruence

$$x^n \equiv N - x_2^n - \ldots - x_r^n \pmod{p^s}$$

is solvable. This proves the lemma.

Lemma 7. Let n be a positive integer. Then

$$\sum_{q \setminus m} A(q) = m^{-(r-1)} M(m).$$

Proof. Applying Lemma 7 of Chapter 2, we find

$$mM(m) = \sum_{a=0}^{m-1} \sum_{x_1=0}^{m-1} \ldots \sum_{x_r=0}^{m-1} e^{2\pi i \frac{a(x_1^n + \ldots + x^n - N)}{m}},$$

where the sum of the terms on the right-hand side corresponding to the same value of $m(a, m)^{-1} = q$ is equal to $m^r A(q)$. This trivially proves the lemma.

Lemma 8. We have

$$A(q) \ll q^{1 - \nu r + \varepsilon_0 r}.$$

Proof. This lemma follows as a corollary from Theorem 1 of Chapter 3.

Lemma 9. For $q_1, \ldots, q_k$, which are pairwise coprimes, we have

$$A(q_1 \ldots q_k) = A(q_1) \ldots A(q_k).$$

Proof. For each q_s, let Q_s be defined by the equality $q_1 \ldots q_k = q_s Q_s$. Let $a_1, \ldots, a_k$ run through reduced systems of residues modulo $q_1, \ldots, q_k$, respectively. Then, $Q_1 a_1 + \ldots + Q_k a_k$ will run through the reduced system of residues modulo $q_1 \ldots q_k$, and we get

$$A(q_1 \ldots q_k) = \sum_{a_1} \ldots \sum_{a_k} \left(S\left(\frac{(Q_1 a_1 + \ldots + Q_k a_k) x^n}{q_1 \ldots q_k} \right) \right)^r e^{-2\pi i \frac{Q_1 a_1 + \ldots + Q_k a_k}{q_1 \ldots q_k} N}.$$

But, according to Lemma 1 of Chapter 3, we find

$$S\left(\frac{(Q_1 a_1 + \ldots + Q_k a_k) x^n}{q_1 \ldots q_k} \right) = S\left(\frac{a_1 Q_1^n x^n}{q_1} \right) \ldots S\left(\frac{a_k Q_k^n x^n}{q_k} \right)$$
$$= S\left(\frac{a_1 x^n}{q_1} \right) \ldots S\left(\frac{a_k x^n}{q_k} \right).$$

Moreover, we easily obtain

$$e^{-2\pi i \frac{Q_1 a_1 + \ldots + Q_k a_k}{q_1 \ldots q^k} N} = e^{-2\pi i \left(\frac{a_1}{q_1} N + \ldots + \frac{a_k}{q_k} N\right)}.$$

Therefore the expression found for $A(q_1, \ldots, q_k)$ trivially reduces to $A(q_1) \ldots A(q_k)$.

Lemma 10. For $r \geqq 2n + 1$, the series $\mathfrak{S}$ and $\psi(p)$ converge absolutely, and

$$\mathfrak{S} = \prod_p \psi(p).$$

Proof. The absolute convergence of $\mathfrak{S}$ and $\psi(p)$ follows from Lemma 8. Now, let $\xi > 2$. According to Lemma 9, we find

$$\prod_{p \leqq \xi} \psi(p) = \prod_{p \leqq \xi} \sum_{s=0}^{\infty} A(p^s) = \sum_{q \leqq \xi} A(q) + \sum_{q > \xi}' A(q),$$

where the second sum only contains the terms corresponding to values of q not divisible by any prime exceeding ξ. As ξ increases to infinity, the second sum tends to zero, while the first sum tends to $\mathfrak{S}$.

Lemma 11. For $r \geqq 4n$, we have

$$\mathfrak{S} \gg 1.$$

Proof. By virtue of Lemmas 6 and 7, we have

$$\sum_{q \backslash p^s} A(q) \geqq p^{-s(r-1)+(s-\gamma)(r-1)} = p^{-\gamma(r-1)}.$$

Taking the limit as $s \to \infty$, we obtain

$$\psi(p) \geqq p^{-\gamma(r-1)}.$$

Moreover, according to Lemma 8, we have

$$\psi(p) - 1 \ll \sum_{s=1}^{\infty} p^{s(1-\nu r+\varepsilon_0 r)} \ll p^{-3+\varepsilon_0 r}.$$

Consequently, there exists a large constant p_0 greater than 2, such that for $p > p_0$, we always have

$$|\psi(p) - 1| \leqq p^{-2}, \ \psi(p) \geqq 1 - p^{-2}.$$

Therefore, applying Lemma 10, we obtain

$$\mathfrak{S} = \prod_{p \leqq p_0} \psi(p) \prod_{p > p_0} \psi(p) \geqq \prod_{p \leqq p_0} p^{-\gamma(r-1)} \prod_{p > p_0} (1 - p^{-2}) \gg 1.$$

Corollary (the Statement or Problem of Waring). For any integer $r \geqq r_0$; $r_0 = [2n^2(2\ln n + \ln\ln n + 5)]$, there exists a large positive constant $N_0 = N_0(n, r)$, such that any integer $N \geqq N_0$ can be represented in the form:

$$N = x_1^n + \ldots + x_r^n$$

with positive integers $x_1, \ldots, x_r$.

Proof. In the asymptotic expression of our theorem, according to Lemma 11, for some $c = c(n, r)$, we have

$$\mathfrak{S} \geqq c(n, r),$$

and thus the main term in this expression is

$$\geqq \frac{(\Gamma(1+v))^r}{\Gamma(rv)} N^{rv-1} c(n, r) \gg N^{rv-1},$$

Since the remainder term approaches zero, as it is a quantity of order less than N^{rv-1}, there does exist an $N_0 = N_0(n, r)$, such that for $N \geqq N_0$ the expression for I_N given by our formula is positive. This completes the proof of the corollary.

Chapter 7. Sums over Prime Numbers

The method applied in Chapter 4 to find an upper bound for the modulus of the trigonometric sum T_m over all the integers in the interval $0 < x \leqq P$ is also suitable for a wider class of sums taken over a certain subset of these integers. An excellent example is the sum:

$$T'_m = \sum_{p \leqq P} e^{2\pi i m f(p)},$$

which is similar to the sum T_m, but is taken only over prime numbers.

In this chapter we shall derive an upper bound for the modulus of the sum T'_m. As regards accuracy, this bound hardly differs from the bound found in Chapter 4 for the modulus of the sum T_m.

Notation. In deriving the following lemmas, we shall use the notation given below. Assume that n is a constant equal to or greater than 5 and that

$$r = 2\,[n^2(2\ln n + \ln\ln n + 4)], \quad \varrho = \frac{1}{20n^2(2\ln n + \ln\ln n + 4)}.$$

The coordinates of a point $(\alpha_n, \ldots, \alpha_1)$ of an n-dimensional space shall be represented in the form:

$$\alpha_s = \frac{a_s}{q_s} + z_s; \quad (a_s, q_s) = 1, \quad 0 < q_s,$$

where some upper bound will be specified for q_s and for $|z_s|$. For instance, it is known that for $\tau_s \geqq 1$, the number α_s can be represented in the above form, if

$$q_s \leqq \tau_s, \quad |z_s| \leqq \frac{1}{q_s \tau_s}.$$

The symbol Q will be used to denote the least common multiple of the numbers $q_n, \ldots, q_1$; the symbol Q_0 to denote the least common multiple of the numbers $q_n, \ldots, q_2$.

The number z_s will be represented as $z_s = \delta_s P^{-s}$. Moreover, we shall put

$$\delta_0 = \max(|\delta_n|, \ldots, |\delta_1|).$$

We shall assume that P_0 is greater than 1 and is chosen so large that for $P \geqq P_0$ all inequalities containing P are satisfied.

Let m denote a positive integer for which some upper bound shall be specified. Consider a sum of the type:

$$S = \sum_{y_1} \sum_{u} \sum_{v} \sum_{x_1} e^{2\pi i m f(y_1 u v x_1)},$$

where each variable y_1, u, v, x_1 runs through its own increasing sequence of positive numbers coprime to Q, while v runs through the numbers which are coprime to each value of x_1. Those numbers for which $(u, v) > 1$ shall be omitted, and summation will be taken only over the range:

$$C' P^{0.245} < y_1 u \leqq C'' P^{0.5}; \quad y_1 uvx_1 \leqq P.$$

The sum S can obviously be represented as

$$S = \sum_t \mu(t) S^{(t)}; \quad S^{(t)} = \sum_{y_1} \sum_{u_1} \sum_{v_1} \sum_{x_1} e^{2\pi imf(t^2 y_1 u_1 v_1 x_1)}, \tag{1}$$

where t runs through those numbers which divide simultaneously at least one value of u and at least one value of v. In the sum $S^{(t)}$ the variables u_1 and v_1 run through the quotients obtained in the division of those values of u and v that are multiples of t, by t, and summation is taken over the range:

$$C' P^{0.245} < t y_1 u_1 \leqq C'' P^{0.5}, \quad t^2 y_1 u_1 v_1 x_1 \leqq P.$$

Further, putting $y_1 u_1 = y$, $v_1 x_1 = x$, and denoting the number of solutions of the equation $y_1 u_1 = y$ by the symbol $\psi(y)$, we obtain

$$S^{(t)} = \sum_y \sum_x \psi(y)\, e^{2\pi imf(t^2 yx)},$$

where y and x run, independently of each other, through certain increasing sequences of natural number which are coprime to Q, summation being taken over the range:

$$C' P^{0.245} t^{-1} < y \leqq C'' P^{0.5} t^{-1}, \quad t^2 yx \leqq P. \tag{2}$$

In particular, we will also encounter degenerate sums S, where u takes a unique value, $u = 1$. Then t will take a unique value, $t = 1$, and equality (1) reduces to

$$S = S^{(1)}.$$

Lemma 1. For each $s = n, \ldots, 1$, let

$$q_s \leqq P^{0.5s}, \quad |z_s| < \frac{1}{q_s P^{0.5s}}.$$

Furthermore, let

$$m \leqq P^{2\varrho}, \quad \varrho = \frac{1}{20n^2 (2 \ln n + \ln \ln n + 4)}$$

and assume that one of the conditions:

$$1°.\ Q > P^{0.125},$$

$$2°.\ Q \leqq P^{0.125}$$

is satisfied. But for at least one s, we have

$$|z_s| > \frac{P^{0.5}}{q_s P^s}.$$

Then the inequality

$$S \ll P^{1-\varrho}$$

holds valid.

Proof. The interval (2) can be divided (for sufficiently large P_0) into $\ll \ln P$ intervals of the type:

$$cY < y \leqq Y; \quad 0.5 < c \leqq 0.75. \tag{3}$$

Let $S(Y)$ denote that part of the sum $S^{(t)}$ which corresponds to the interval (3).

When $t > t_0$; $t_0 = P^{\varrho}(\ln P)^2$, we find (Lemma 1, Chapter 1):

$$S(Y) \ll \sum_{cY<y\leqq Y} \psi(y)\frac{P}{t^2 Y} \ll \frac{P\ln P}{t^2}, \quad S^{(t)} \ll \frac{P(\ln P)^2}{t^2},$$

$$\sum_{t>t_0} |S^{(t)}| \ll P^{1-\varrho}.$$

Now consider the case of $t \leqq t_0$. Putting

$$H = [P^{(2n+3)\varrho_1}], \quad Y_0 = (1-c)YH^{-1}; \quad \varrho_1 = 1.2\varrho,$$

divide the interval (3) into $\ll H$ small intervals of the type:

$$Y_1 - Y_0 < y \leqq Y_1. \tag{4}$$

Let S' denote that part of the sum $S(Y)$ which corresponds to the interval (4), and let S'_y denote that part of the sum S' which corresponds to a given y. Putting

$$X = \frac{P}{t^2 Y_1},$$

we find that for each y in the interval (4):

$$\frac{P}{t^2 y} - X \ll \frac{PY_0}{t^2 Y_1^2} \ll XH^{-1}, \quad S_y = \psi(y)(S'_y + O(XH^{-1}));$$

$$S'_y = \sum_{0<x\leqq X} e^{2\pi i f(t^2 yx)}.$$

At the same time, we obtain

$$S' \ll P^{\varepsilon}\left(\sum_{Y_1 - Y_0 < y \leqq Y_1} |S'_y| + Y_0 XH^{-1}\right).$$

Now, putting $\beta_j = \beta_j(y) = m\alpha_j t^{2j} y^j$, we obtain

$$S'_y = S'(\beta_n, \ldots, \beta_1) = \sum_{0<x\leqq X} e^{2\pi i(\beta_n x^n + \ldots + \beta_1 x)}.$$

For the sum $S'' = S''(\eta_n, \ldots, \eta_1)$ corresponding to a certain point $(\eta_n, \ldots, \eta_1)$ in the domain (y) of an n-dimensional space bounded by the inequalities:

$$\beta_n - 0.5X^{-n}P^{-\varrho_1} < \eta_n \leqq \beta_n + 0.5X^{-n}P^{-\varrho_1},$$

$$\cdots\cdots\cdots\cdots\cdots\cdots\cdots\cdots$$

$$\beta_1 - 0.5X^{-1}P^{-\varrho_1} < \eta_1 \leqq \beta_1 + 0.5X^{-1}P^{-\varrho_1},$$

according to Lemma 9 of Chapter 2, we find

$$S'_y - S'' \ll XP^{-\varrho_1},$$

hence we obtain

$$|S'_y|^r \ll |S''|^r + X^r P^{-r\varrho_1}. \tag{5}$$

Furthermore, we have

$$|S'|^r \leqq P^{\varepsilon r}\left(Y_0^{r-1} \sum_{Y_1 - Y_0 < y \leqq Y_1} |S'_y|^r + Y_0^r X^r H^{-r}\right). \tag{6}$$

Multiplying (5) term by term by $X^{n(n+1)/2} P^{n\varrho_1}$, and then integrating over the domain (y), we obtain

$$|S'_y|^r \ll X^{\frac{n(n+1)}{2}} P^{n\varrho_1} \int \underset{(y)}{\ldots} \int |S''|^r d\eta_n \ldots d\eta_1 + X^r P^{-n\varrho_1},$$

hence we find

$$\sum_{Y_1 - Y_0 < y \leqq Y_1} |S'_y|^r \ll X^{\frac{n(n+1)}{2}} P^{n\varrho} G \int_0^1 \ldots \int_0^1 |S''|^r d\eta_n \ldots d\eta_1 + Y_0 X^r P^{-r\varrho_1},$$

where G denotes the maximum number of domains (y) corresponding to the values of y from the interval, which intersect with one (y_0) of the domains. Hence, applying Theorem 4 of Chapter 4, we obtain

$$\sum_{Y_1 - Y_0 < y \leqq Y_1} |S'_y|^r \ll X^r P^{n\varrho_1} G + Y_0 X^r P^{-r\varrho_1}$$

and, finally, applying inequality (6), we find

$$|S'|^r \ll P^{\varepsilon r} Y_0^r X^r \left(P^{n\varrho_1} \frac{G}{Y_0} + P^{-r\varrho_1}\right). \tag{7}$$

1°. We shall estimate G under the assumption that there exists at least one q_s greater than $P^{0.125}$. Assuming that (y) has a common point with (y_0), we find

$$\frac{ma_s t^{2s}(y^s - y_0^s)}{q_s} \equiv -z_s m t^{2s}(y^2 - y_0^s) + \theta' X^{-s} P^{-\varrho_1} \equiv \frac{h_s}{q_s}, \tag{8}$$

where h_s is an integer such that $h_s \ll P^{(n-1)\varrho_1}$. But for each h_s, the congruence

$$m(y^s - y_0^s) \equiv h_s \pmod{q_s}$$

in the reduced system of residues modulo q_s is satisfied by $\ll mP^{\varepsilon'}$ values of y. Therefore

$$G \ll mP^{\varepsilon''}\left(\frac{Y_0}{q_s} + 1\right) P^{(n-1)\varrho_1} \ll Y_0 P^{-0.125} P^{(n+1)\varrho_1} \ll Y_0 P^{-0.125}.$$

Now we shall consider the case, where

$$q_s \leqq P^{0.125},$$

for each s, but $Q > P^{0.125}$. Assuming that (y) has a common point with (y_0), for each s, we obtain the congruence (8), where h_s is less than 1 and, therefore, is zero. Therefore, the number $m(y^s - y_0^s)$ is divisible by q_s, while the number $m(y^{s_0} - y_0^{s_0})$, where s_0 is the least common multiple of the numbers s, is divisible by Q. In other words, the congruence

$$m(y^{s_0} - y_0^{s_0}) \equiv 0 \pmod{Q},$$

holds true, which is satisfied by $\ll mP^{\varepsilon''}$ values of y in the reduced system of residues modulo Q. Therefore we have

$$G \ll mP^{\varepsilon}\left(\frac{Y_0}{Q}+1\right) \ll Y_0 P^{-0.12}.$$

2°. Finally, consider the case where $Q \leqq P^{0.125}$, but

$$|z_s| > \frac{P^{0.5}}{q_s P^s}.$$

for at least one s. Assuming that for y and y_0 belonging to the same class with respect to modulus q_s, the domains (y) and (y_0) have a common point, we find

$$0 \equiv -z_s m t^{2s}(y^s - y_0^s) - \theta' X^{-s} P^{-\varrho_1},$$

where the right-hand side is less than 1 and, therefore, is zero. Hence we readily find

$$\frac{P^{0.5} m t^{2s} Y^{s-1}(y-y_0)}{q_s P^s} \ll X^{-s} P^{-\varrho_1}, \qquad \frac{y-y_0}{q_s} \ll P^{-\varrho_1},$$

which is possible only if $y = y_0$, because $(y-y_0)$ is divisible by q_s. Consequently, not more q_s domains (y) taken not more than one at a time from q_s different sets can intersect with (y_0), and thus we obtain

$$G \leqq q_s < Y_0 P^{-0.115}.$$

From (7), for both the cases 1° and 2°, we have

$$|S'|^r \ll P^{\varepsilon r} Y_0^r X^r (P^{n\varrho_1 - 0.115} + P^{-r\varrho_1}) \ll Y_0^r X^r (P^{-0.113} + P^{-1.2r\varrho}),$$

hence it follows that

$$|S'| \ll Y_0 X (P^{-\frac{0.113}{r}} + P^{-1.2\varrho}),$$

$$S(Y) \ll \frac{P}{t^2}(P^{-\frac{0.113}{r}} + P^{-1.2\varrho}),$$

$$S^{(t)} \ll \frac{P^{1-\varrho}}{t^2}, \quad S \ll P^{1-\varrho}.$$

Lemma 2. For each $s = n, \ldots, 1$, let

$$|z_s| \leqq \frac{P^{0.5}}{q_s P^s}$$

and also let $Q \leqq P^{0.125}$, $m \leqq P^{2\varrho}$.
If one of the conditions:

1°. $Q > P^{0.1\nu}$,

2°. $Q \leqq P^{0.1\nu}$, but $\delta_0 > P^{0.1\nu}$

is satisfied, then

$$S \ll P^{1-\varrho}.$$

Proof. The interval (2) can be divided into $\ll \ln P$ intervals of the type:

$$Y' < y \leqq Y; \quad 0.5\,Y < Y' \leqq 0.75\,Y. \tag{9}$$

Let $S(Y)$ denote that part of the sum $S^{(t)}$ which corresponds to the interval (9). If $t > t_0$; $t_0 = P^{\varrho}(\ln P)^2$, we find (Lemma 1, Chapter 1)

$$S(Y) \ll \sum_{cY<y\leqq Y} \psi(y)\,\frac{P}{t^2 Y} \ll \frac{P\ln P}{t^2},$$

$$S^{(t)} \ll \frac{P(\ln P)^2}{t^2}, \qquad \sum_{t>t_0} S^{(t)} \ll P^{1-\varrho}.$$

Now consider the case where $t \leqq t_0$. Divide the interval (9) into small intervals of the type:

$$Y_1 - Y_0 < y \leqq Y_1, \tag{10}$$

where, assuming

$$H = \left[\frac{Y-Y'}{Q} + 1\right], \qquad \text{if} \quad Q > P^{0.1\nu},$$

$$H = \left[\left(1 - \frac{Y'}{Y}\right)P^{0.4\nu} + 1\right], \quad \text{if} \quad Q \leqq P^{0.1\nu},$$

take

$$Y_0 = (Y - Y')\,H^{-1}.$$

Let S' denote that part of the sum $S(Y)$ which corresponds to the interval (10). Putting

$$X = \frac{P}{t^2 Y_1},$$

we find that for each y in the interval (10):

$$\frac{P}{t^2 y} - X \ll \frac{PY_0}{t^2 Y_1^2} \ll XH^{-1}.$$

At the same time, we obtain

$$S' = \sum_{Y-Y_0<y\leqq Y_1}\ \sum_{0<x\leqq X} \psi(y)\, e^{2\pi i m f(t^2 yx)} + O(P^{\varepsilon} Y_0 X H^{-1}).$$

Hence, after changing the order of summation, we find

$$|S'|^2 \ll X \sum_{0<x\leqq X}\ \sum_{y_1}\sum_{y} \psi(y_1)\,\psi(y)\, e^{2\pi i \Phi(y_1,y,x)} + P^{2\varepsilon} Y_0^2 X^2 H^{-2},$$

where $\Phi(y_1, y, x) = mf(t^2 y_1 x) - mf(t^2 yx)$, x runs through the integers in the interval (10), while y_1 and y run independently through values from the interval (10). Rewriting the last inequality as

$$|S'|^2 \ll X \sum_{y_1}\sum_{y} \psi(y_1)\,\psi(y)\, W_{y_1,y} + P^{2\varepsilon} Y_0^2 X^2 H^{-2};$$
$$W_{y_1,y} = \sum_{0<x\leqq X} e^{2\pi i \Phi(y_1,y,x)}, \tag{11}$$

we shall proceed to estimate the sum $W_{y_1,y}$. We have

$$W_{y_1,y} = \sum_{v=1}^{Q} S_v; \quad S_v = \sum_{0<Qu+v\leqq X} e^{2\pi i \Phi(y_1, y, Qu+v)};$$

$$\Phi(y_1, y, Qu+v) = F(y_1, y, v) + \lambda(y_1, y, Qu+v),$$

$$F(y_1, y, v) = \frac{ma_n t^{2n}(y_1^n - y^n)}{q_n} v^v + \ldots + \frac{ma_1 t^2(y_1 - y)}{q_1} v,$$

$$\lambda(y_1, y, Qu+v) = mz_n t^{2n}(y_1^n - y^n)(Qu+v)^n + \ldots + mz_1 t^2(y_1 - y)(Qu+v).$$

Now, from

$$\frac{d}{du} mz_s t^{2s}(y_1^s - y^s)(Qu+v)^s \ll \frac{mP^{0.5} t^{2s} Y^{s-1} Y_0 X^{s-1} Q}{q_s P^s} \ll \frac{mt^2 Y_0 Q}{q_s P^{0.5}} < P^{-0.2v}$$

it follows that

$$\frac{d}{du} \lambda(y_1, y, Qu+v) < 0.5.$$

Therefore, by Lemma 3 of Chapter 9, we find

$$S_v = e^{2\pi i F(y_1, y, v)} \int_{-v/Q}^{X-v/Q} e^{2\pi i \lambda(y_1, y, Qu+v)} du + O(1)$$
$$= e^{2\pi i F(y_1, y, v)} \frac{X}{Q} \int_0^1 e^{2\pi i \lambda(y_1, y, X\xi)} d\xi + O(1).$$

Now we have

$$W_{y_1,y} = UV + O(Q);$$

$$U = \frac{1}{Q} \sum_{v=1}^{Q} e^{2\pi i F(y_1, y, v)}, \quad V = X \int_0^1 e^{2\pi i \lambda(y_1, y, X\xi)} d\xi,$$

and from (11) we obtain

$$|S'|^2 \ll P^{2\varepsilon} X \sum_{y_1} \sum_{y} |U|\,|V| + P^{2\varepsilon} Y_0^2 (XQ + X^2 H^{-2}).$$

First, let $Q > P^{0.1v}$. Thus, we easily find that $0.5Q < Y_0 < Q$. For a given y_1, by Theorem 2 of Chapter 3, we have

$$\sum_{y} |U| \ll (m, Q)\, Q^{1-v+\varepsilon'}.$$

Consequently,

$$|S'|^2 \ll P^{2\varepsilon} X^2 Y_0^2 P^{2\varrho v - 0.1v^2 + \varepsilon'} \ll X^2 Y_0^2 P^{-0.097v^2},$$
$$|S'| \ll XY_0 P^{-0.048v^2}, \quad S(Y) \ll XYP^{-0.048v^2},$$
$$S^{(t)} \ll \frac{P^{1-0.047v^2}}{t^2} \ll \frac{P^{1-\varrho}}{t^2}, \quad S \ll P^{1-\varrho}.$$

Now, let $Q < P^{0.1\nu}$, but $\delta_0 > P^{0.1\nu}$. Hence, for some s, we have $\delta_0 = |\delta_s|$. For the modulus of the coefficient of ξ^s in the polynomial $\lambda(y_1, y, X\xi)$, we have the following inequality

$$\frac{m\delta t^{2s}(y_1^s - y^s)}{P^s} \gg \frac{P^\nu |y_1 - y|}{Y}.$$

Therefore, by Lemma 4 of Chapter 2, we have

$$V \ll X \min\left(1, \left(\frac{P^\nu |y_1 - y|}{Y}\right)^{-\nu}\right).$$

But $|y_1 - y|$, for a given y_1, runs through the integers η in the interval $0 \leqq \eta < Y_0$, and takes each value not more than twice.
Therefore

$$\sum_y V \ll X + X \sum_{0 < \eta < Y_0} \left(\frac{Y}{P^\nu \eta}\right)^\nu$$

$$\ll X + XP^{-\nu^2} Y^\nu Y_0^{1-\nu} \ll Y_0 XP^{-0.6\nu^2},$$

$$|S'|^2 \ll Y_0^2 X^2 P^{-0.58\nu^2}, \quad S' \ll Y_0 XP^{-0.29\nu^2},$$

$$S(Y) \ll YXP^{-0.29\nu^2}, \quad S^{(t)} \ll \frac{P^{1-\varrho}}{t^2}, \quad S \ll P^{1-\varrho}.$$

Lemma 3. Let $Q \leqq P^{0.1\nu}$ and $\delta_0 \leqq P^{0.1\nu}$. Then, assuming that

$$\Delta = (\ln P)^2 Q^{-0.5\nu+\varepsilon'}, \quad \mu = (m, Q)^{0.5\nu},$$

or, if $\delta_0 > 1$, that

$$\Delta = (\ln P)^2 Q^{-0.5\nu+\varepsilon'} \delta_0^{-0.5\nu}, \quad \mu = 1,$$

when $m \leqq \Delta^{-2}$, we have

$$S \ll P\Delta\mu.$$

Proof. The interval (2) can be divided into $\ll \ln P$ intervals of the type

$$Y' < y \leqq Y; \quad 0.5Y < Y' \leqq 0.75Y. \tag{12}$$

Let $S(Y)$ denote that part of the sum $S^{(t)}$ which corresponds to the interval (12). If $t \geqq \Delta^{-1}(\ln P)^2$, we find (Lemma 1, Chapter)

$$S(Y) \ll \sum_{Y' < y \leqq Y} \psi(y) \frac{P}{t^2 Y} \ll \frac{P \ln P}{t^2};$$

$$S^{(t)} \ll \frac{P(\ln P)^2}{t^2}, \quad \sum_{t \geqq \Delta^{-1}(\ln P)^2} S^{(t)} \ll P\Delta.$$

Now consider the case where $t < \Delta^{-1}(\ln P)^2$. Denoting by S_x that part of the sum $S(Y)$ which corresponds to a given x, we obtain

$$S(Y) = {\sum_{0 < x \leqq P/t^2 Y}}' S_x; \quad S_x = \sum_y{}' \psi(y) e^{2\pi i m f(t^2 yx)},$$

where the summation denoted by the symbol $\sum_y'$ is taken over those values of y, which lie in the interval

$$Y' < y \leqq \min\left(\frac{P}{t^2 x}, Y\right). \tag{13}$$

Hence we find that

$$|S(Y)|^2 \ll \frac{P}{t^2 Y} \sum_{0 < x \leqq P/t^2 Y'} |S_x|^2$$

$$\ll \frac{P}{t^2 Y} \sum_{0 < x \leqq P/t^2 Y'} \sum_{y_1}' \sum_{y}' \psi(y_1)\, \psi(y)\, e^{2\pi i \Phi(y_1, y, x)},$$

where x runs through the integers in the range stated above, $\Phi(y_1, y, x) = mf(t^2 y_1 x) - mf(t^2 yx)$, where y_1 runs, independently of y, through the same values as y.

Now change the order of summation. Each pair of the values of y_1 and y taken from the interval (13) may be contained in some $|S_x|^2$, if and only if

$$x \leqq X_{y_1, y} = \min\left(\frac{P}{t^2 y_1}, \frac{P}{t^2 y}\right).$$

Thus, as a result of the change of order of summation, we obtain

$$|S(Y)|^2 \ll \frac{P}{t^2 Y} \sum_{Y' < y \leqq Y} \sum_{Y' < y_1 \leqq Y} \psi(y_1)\, \psi(y) \sum_{0 < x \leqq X_{y_1, y}} e^{2\pi i \Phi(y_1, y, x)}$$

$$= S_1 + S_2 + S_3,$$

where S_1, S_2 and S_3 contain the terms satisfying the conditions $y_1 = y$, $y_1 > y$ and $y_1 < y$, respectively.

According to Lemma 2 of Chapter 1, we have

$$S_1 \ll \frac{P}{t^2 Y} \sum_y (\psi(y))^2 \frac{P^2}{t^2 Y} \ll \frac{P^2}{t^2 Y} (\ln P)^3.$$

Therefore, since S_2 and S_3 are complex conjugate, we obtain

$$\begin{aligned} &|S(Y)|^2 \ll \frac{P}{t^2 Y} \sum_{Y' < y_1 \leqq Y} \psi(y_1) W_{y_1} + \frac{P^2}{t^4 Y} (\ln P)^3; \\ &W_{y_1} = \sum_{Y' < y \leqq y_1} \psi(y)\, K; \quad K = \sum_{0 < x \leqq X} e^{2\pi i \Phi(y_1, y, x)}; \quad X = \frac{P}{t^2 y_1}. \end{aligned} \tag{14}$$

The sum K can be represented in the form:

$$K = \sum_{0 < v \leqq Q} K_v; \quad K_v = \sum_{0 < Qu + v \leqq X} e^{2\pi i \Phi(y_1, y, Qu + v)},$$

where

$$\Phi(y_1, y, Qu + v) = F(y_1, y, v) + \lambda(y_1, y, Qu + v);$$

$$F(y_1, y, v) = \frac{m a_n t^{2n} (y_1^n - y^n)}{q_n} v^n + \ldots + \frac{m a_1 t^2 (y_1 - y)}{q_1} v,$$

$$\lambda(y_1, y, Qu + v) = \frac{m \delta_n t^{2n} (y_1^n - y^n)}{P^n} (Qu + v)^n + \ldots + \frac{m \delta_1 t^2 (y_1 - y)}{P} (Qu + v).$$

For each s, we find that

$$\frac{d}{du}\frac{m\delta_s t^{2s}(y_1^s - y^s)}{P^s}(Qu+v)^s \ll \frac{m|\delta_s|t^{2s}Y^sX^{s-1}}{P^s}Q$$
$$\ll m|\delta_s|tP^{-0.5}Q \ll P^{-0.4}.$$

Therefore

$$\frac{d}{du}\lambda(y_1, y, Qu+v) < 0.5$$

and by Lemma 3 of Chapter 2, we have

$$K_v = e^{2\pi i F(y_1, y, v)} \int_{-v/Q}^{(X-v)/Q} e^{2\pi i\lambda(y_1, y, Qu+v)}\,du + O(1)$$
$$= e^{2\pi i F(y_1, y, v)}\frac{X}{Q}\int_0^1 e^{2\pi i\lambda(y_1, y, Xz)}\,dz + O(1).$$

Furthermore, we have

$$K = UV + O(Q);$$
$$U = U_{y_1, y} = \frac{1}{Q}\sum_{0<v\leqq Q} e^{2\pi i F(y_1, y, v)} \qquad V = X\int_0^1 e^{2\pi i\lambda(y_1, y, Xz)}\,dz.$$

Now, by Lemma 1 of Chapter 1, we have

$$W_{y_1} = \sum_{Y'<y<y_1}\psi(y)\,UV + O(QY\ln P),$$

hence, we find

$$W_{y_1} = \sum_{\substack{0<v\leqq Q\\(v,Q)=1}} L_v + O(QY\ln P); \qquad L_v = U_{y_1, v}\sum_{\substack{Y'<y<y_1\\y\equiv v\,(\mathrm{mod}\,Q)}}\psi(y)\,V.$$

First we shall prove the first assertion of Lemma 3. By Lemma 3 of Chapter 1, we have

$$L_v \ll |U_{y_1, v}|\sum_{\substack{Y'<y<y_1\\y\equiv v\,(\mathrm{mod}\,Q)}}{}' \psi(y)\,X \ll |U_{y_1, v}|\frac{XY\ln P}{Q},$$

$$W_{y_1} \ll \sum_{\substack{0\leqq v<Q\\(v,Q)=1}}{}' |U_{y_1, v}|\frac{XY\ln P}{Q} + QY\ln P,$$

hence, by Theorem 2 of Chapter 3, we obtain

$$W_{y_1} \ll XY\ln P\,(m, Q)^{\nu} Q^{-\nu+\varepsilon'}.$$

Now, from (14) we obtain

$$|S(Y)|^2 \ll \frac{P}{t^2 Y} \sum_{y_1} \psi(y_1)\, XY \ln P\, (m,Q)^\nu Q^{-\nu+\varepsilon'} + \frac{P^2}{t^4 Y} (\ln P)^3$$

$$\ll \frac{P^2}{t^4} (\ln P)^2 (m,Q)^\nu Q^{-\nu+\varepsilon'},$$

$$S(Y) \ll \frac{P}{t^2} \ln P\, (m,Q)^{0.5\nu} Q^{-0.5\nu+\varepsilon'},$$

$$S^{(t)} \ll \frac{P}{t^2} (\ln P)^2 (m,Q)^{0.5\nu} Q^{-0.5\nu+\varepsilon'},$$

$$S \ll P (\ln P)^2 (m,Q)^{0.5\nu} Q^{-0.5\nu+\varepsilon'}.$$

Now we shall prove the second assertion of Lemma 3, using the equality established for L_0. Let $\delta_0 = |\delta_s|$, where $\delta_0 > 1$. The modulus of the coefficient of z^s in the polynomial $\lambda(y, y_1, Xz)$ is

$$\gg \frac{m\delta_0 t^{2s} (y_1^s - y^s) X^s}{P^s} \gg m\delta_0 \frac{y_1 - y}{Y}.$$

Therefore (Lemma 4 of Chapter 2) we have

$$L_v \ll |U_{y_1,v}|\, XR;$$

$$R = \sum_{\substack{Y' < y < y_1 \\ y \equiv v \,(\mathrm{mod}\, Q)}} \psi(y) \min\left(1, \left(\frac{Y}{m\delta_0 (y_1 - y)}\right)^\nu\right).$$

Putting $T = Q\sqrt{y_1}$, we find that when $y_1 - Y' \leqq 2T$, we have $R \ll Q\sqrt{y_1}\, P^\varepsilon$. If $y_1 - Y' > 2T$, to estimate R, divide the interval $Y' < y < y_1$ into smaller intervals:

$$Y' < y \leqq y_1 - T2^b, \ \ldots, \ y_1 - T2^2 < y \leqq y_1 - T2,$$

$$y_1 - T2 < y \leqq y_1 - T, \quad y_1 - T < y < y_1,$$

where b is the largest integer such that $T2^{b+1}$ does not exceed the length of the above interval. That part of R which correponds to the interval with an upper bound y_1 is $\ll Q\sqrt{y_1}\, P^\varepsilon$, while that part which corresponds to the interval with upper bound $y_1 - T2^\gamma$, (by Lemma 1 of Chapter 2) is

$$\ll \frac{T2^\gamma}{Q} \ln P \min\left(1, \left(\frac{Y}{m\delta T2^\gamma}\right)^\nu\right).$$

Hence, we easily find that

$$R \ll \frac{y_1 \ln P}{Q (m\delta_0)^\nu}, \qquad L_v \ll |U_{y_1,v}| \frac{XY \ln P}{Q} (m\delta_0)^{-\nu},$$

$$W_{y_1} \ll \sum_{\substack{0 \leqq v < Q \\ (v,Q)=1}} |U_{y_1,v}| \frac{XY \ln P}{Q} (m\delta_0)^{-\nu} + QY \ln P,$$

and, consequently (Theorem 3 of Chapter 3), we have

$$W_{y_1} \ll XY \ln P Q^{-\nu+\varepsilon} m\delta_0^{-\nu}.$$

Now, from (14) it follows that (Lemma 1 of Chapter 1)

$$|S(Y)|^2 \ll \frac{P}{t^2}(\ln P)^2 XYQ^{-\nu+\varepsilon'''}\delta_0^{-\nu} + \frac{P^2(\ln P)^3}{t^4 Y}$$

$$\ll \frac{P^2(\ln P)^2}{t^4} Q^{-\nu+\varepsilon'''}\delta_0^{-\nu},$$

$$S(Y) \ll \frac{P\ln P}{t^2} Q^{-0.5\nu+\varepsilon'}\delta_0^{-0.5\nu},$$

$$S^{(t)} \ll \frac{P}{t^2}(\ln P)^2 Q^{-0.5\nu+\varepsilon'}\delta_0^{-0.5\nu},$$

$$S \ll P(\ln P)^2 Q^{-0.5\nu+\varepsilon'}\delta_0^{-0.5\nu}.$$

Lemma 4. Divide the points of an n-dimensional space into two classes: Class I and Class II. If $Q \leqq P^{0.1\nu}$ and $\delta_0 \leqq P^{0.1\nu}$, then the point is said to be a point of Class I. A point which is not a point of Class I is said to be a point of Class II. Then assuming $\Delta = P^{1-\nu}$, $\mu = 1$ for the points of Class II, and

$$\Delta = (\ln P)^2 Q^{-0.1\nu+\varepsilon'}, \quad \mu = (m, Q)^\nu$$

for the points of Class I, and if $\delta_0 > 1$, assuming that

$$\Delta = (\ln P)^2 Q^{-0.1\nu+\varepsilon'}\delta_0^{-0.1\nu}, \quad \mu = 1$$

for the points of Class I, if $m \leqq \Delta^2$, we always have

$$S \ll P\Delta\mu.$$

Proof. The validity of the theorem for the points of Class II follows from Lemmas 1 and 2, whereas for the points of Class I from Lemma 3.

Notation. In deriving the lemmas on the following pages we shall use the same notation as before, but with slight modifications. We shall use the symbol S to denote the sum

$$S = \sum_k \sum_d e^{2\pi i m f(dk)},$$

where d runs through a certain increasing sequence of numbers coprime to Q, while k runs, for a given d, through successive numbers coprime to Q, and summation is taken over the range:

$$K < k \leqq K_0, \quad 0 < dk \leqq P,$$

where K and K_0 satisfy the conditions

$$P^{0.75} < K < P, \quad 2K \leqq K_0 \leqq 4K.$$

We have

$$S = \sum_{0<d\leqq P^{0.25}} S(d); \quad S(d) = \sum_{K<k\leqq K'} e^{2\pi i m f(dk)};$$

$$K' = \min\left(K_0, \frac{P}{d}\right),$$

and $S(d)$ can be represented as

$$S(d) = \sum_{t \setminus Q}' k(t)\, S^{(t)};$$

$$S^{(t)} = S^{(t)}(b_n, \ldots, b_1) = \sum_{W_0 < w \leqq W} e^{2\pi i \Phi(w)};$$

$$\Phi(w) = b_n w^n + \ldots + b_1 w; \quad b_s = m\alpha_s d^s t^s;$$

$$W_0 = \frac{K}{t}; \quad W = \frac{K'}{t},$$

where w runs through all integers in the range indicated above.

Lemma 5. For each $s = n, \ldots, 1$, let

$$q_s \leqq P^{0.5s}, \quad |z_s| < \frac{1}{q_s P^{0.5s}}.$$

Furthermore, let $m \leqq P^{2\varrho}$, and let one of the conditions

$$1°.\ Q > P^{0.125},$$

$$2°.\ Q \leqq P^{0.125}$$

be satisfied, and for at least one $s > 1$, suppose

$$|z_s| > \frac{P^{0.5}}{q_s P^s}.$$

Then

$$S \ll P^{1-\varrho}.$$

Proof. Putting $t_0 = P^{\varrho + \varepsilon'}$, we obtain

$$\sum_{t > t_0} |S^{(t)}| \ll \sum_{\substack{t > t_0 \\ t \setminus Q}} \frac{P}{dt} \ll \frac{P^{1-\varrho}}{dP^{0.5\varepsilon'}}, \qquad \sum_{0 < d < P^{0.125}} \frac{P^{1-\varrho}}{dP^{0.5\varepsilon'}} \ll P^{1-\varrho}.$$

Therefore, we shall henceforth confine curselves to the case where $t \leqq t_0$.

Taking $Y = [P^{0.495}]$ for the case 1°, and $Y = [P^{0.745} d^{-1}]$ for the case 2° (in both cases we have $Y \ll WP^{-\varrho_1}$; $\varrho_1 = 1.2\varrho$) and then finding $Y_n, Y_{n-1}, \ldots, Y_1$ from the expansion

$$\Phi(y + w) - \Phi(y) = Y_n w^n + Y_{n-1} w^{n-1} + \ldots + Y_1 w,$$

we shall consider the sum

$$S_y = S(Y_n, Y_{n-1}, \ldots, Y_1) = \sum_{W_0 < w \leqq W} e^{2\pi i (Y_n w^n + \ldots + Y_1 w)}$$

for $y = 0, \ldots, Y$, and the domain $(y)'$ of the points $(\gamma_n, \ldots, \gamma_1)$, of an n-dimensional space bounded by the inequalities:

$$Y_n - 0.5 W^{-n} P^{-\varrho_1} \leqq \gamma_n \leqq Y_n + 0.5 W^{-n} P^{-\varrho_1},$$

$$\cdots\cdots\cdots\cdots\cdots\cdots\cdots\cdots\cdots\cdots$$

$$Y_1 - 0{,}5 W^{-1} P^{-\varrho_1} \leqq \gamma_1 \leqq Y_1 + 0.5 W^{-1} P^{-\varrho_1},$$

as well as a part (y) of the domain Π_n (see the proof of Lemma 8, Chapter 4) which corresponds to the above domain.

It is seen that for any of the values of y specified above, we have

$$|S^{(t)}| - |S_y| \ll Y.$$

And for a given y, for any point $(\gamma_n, \ldots, \gamma_1)$ of the domain (y) assuming that $S' = S(\gamma_n, \ldots, \gamma_1)$, we obtain (Lemma 9, Chapter 2):

$$|S_y| - |S'| \ll WP^{-\varrho_1}$$

and, consequently, we have

$$|S^{(t)}| \ll |S'| + WP^{-\varrho_1}; \quad |S^{(t)}|^r \ll |S'|^r + W^r P^{-r\varrho_1}.$$

Multiplying the last inequality by

$$W^{\frac{n(n+1)}{2}} P^{n\varrho_1} d\gamma_n \ldots d\gamma_1$$

and then integrating over the domain $(y)'$, we obtain

$$|S^{(t)}|^r \ll W^{\frac{n(n+1)}{2}} P^{n\varrho_1} \int \underset{(y)'}{\ldots} \int |S'|^r d\gamma_n \ldots d\gamma_1 + W^r P^{-r\varrho_1}.$$

Writing down similar inequalities for all $y = 0, \ldots, Y$, and then adding them, we obtain:

$$Y|S^{(t)}|^r \ll W^{\frac{n(n+1)}{2}} P^{n\varrho_1} G \int_0^1 \ldots \int_0^1 |S'|^r d\gamma_n \ldots d\gamma_1 + YW^r P^{-r\varrho_1},$$

where G is the least number of domains (y) which intersect a given domain (y_0). Hence, applying Theorem 4 of Chapter 4, we find

$$|S^{(t)}|^r \ll P^{n\varrho_1} \frac{G}{Y} W^r + W^r P^{-r\varrho_1} \ll W^r \left(P^{n\varrho_1} \frac{G}{Y} + P^{-r\varrho_1} \right). \tag{15}$$

To estimate G, we shall apply Lemma 5 of Chapter 4, according to which G does not exceed the number of solutions of the system of inequalities

$$(C_1 b_s(y - y_0)) \leqq C_1 C_2 W^{-\varepsilon+1}; \quad C_1 = n \ldots s,$$
$$C_2 = s^{-1} (\tfrac{3}{2} n)^{n-s}$$
$$(s = n, \ldots, 2),$$

where, since

$$W \gg \frac{K}{t} \gg \frac{P^{1-\varrho}}{dt},$$

the inequality corresponding to a given s implies

$$\left(\frac{C_1 a_s m d^s t^s (y - y_0)}{q_s} + \frac{C_1 \theta_s P^{-0.5s} m d^s t^s (y - y_0)}{q_s} \right) < \frac{C_1 C_2 q_s (P^{1-\varrho} d^{-1} t^{-1})^{-s+1}}{q_s}. \tag{16}$$

It is not difficult to verify that the numerator of the second term within brackets is numerically less than 0.1 in all cases.

Let us consider case 1°. Assume that from among the numbers $q_n, \ldots, q_2$ there exists at least one q_s exceeding $P^{0.125}$. Now, assuming that $q_s \leqq P^{0.495}$, we find the numerator of the right-hand side of (16) is less than 0.1 and, consequently, the numerator of the first term in brackets on the left-hand side should be divisible by q_s. Therefore the product $C_1 mt^s(y-y_0)$ should also be divisible by q_s, and thus we have

$$G \ll \frac{mt^s Y}{q_s} \ll \frac{Y}{P^{0.123}}, \quad \frac{G}{Y} \ll P^{-0.123}.$$

Now assume that $q_s > P^{0.495}$. For the numerator of the right-hand side of (16) (which is greatest when $s=2$), we obtain

$$C_1 C_2 q_s (P^{1-\varrho} d^{-1} t^{-1})^{-s+1} \ll P^{0.251}.$$

Consequently, $C_1 mt^s(y-y_0)$ is congruent modulo q_s to not more than $\ll P^{0.261}$ numbers, and thus we obtain

$$G \ll P^{0.261}, \quad \frac{G}{Y} \ll P^{-0.234}.$$

Now suppose that none of the numbers $q_n, \ldots, q_2$ exceeds $P^{0.125}$. Then we find that the numerator of the right-hand side of (16) is less than 0.1 and, consequently, the numerator of the first term within brackets on the left-hand side is divisible by q_s. Therefore the product $C_1 mt^s(y-y_0)$ should also be divisible by q_s. But the product $C_1 mt^n(y-y_0)$, being divisible by every $q_n, \ldots, q_2$, should be divisible by Q_0, and thus we obtain

$$G \ll \frac{mt^n Y}{Q_0} \ll \frac{Y}{P^{0.123}}, \quad \frac{G}{Y} \ll P^{-0.123}.$$

Now consider case 2°. Here the numerator of the right-hand side of (16) is less than 0.1 and, consequently, the numerator of the first term within brackets can be replaced by zero. Thus we obtain

$$P^{-0.5s} md^s t^s |y-y_0| < C_q q_s (P^{1-\varrho} d^{-1} t^{-1})^{-s+1},$$

$$|y-y_0| \ll q_s P^{\varrho} d^{-1} \ll \frac{P^{0.125}}{d}, \quad \frac{G}{Y} \ll P^{-0.369}.$$

In both cases 1° and 2°, inequality (15) gives

$$|S^{(t)}|^r \ll W^r (P^{n\varrho_1} P^{-0.123} + P^{r\varrho_1}), \quad |S^{(t)}| \ll WP^{-\varrho-c_1} \ll \frac{P^{1-\varrho-c_1}}{d},$$

where c_1 is a positive constant. Hence we easily find that

$$S(d) \ll \frac{P^{1-\varrho-0.5c_1}}{d}, \quad S \ll P^{1-\varrho}.$$

Lemma 6. For each $s = n, \ldots, 1$, let

$$|z_s| \leqq \frac{P^{0.5}}{q_s P^s}.$$

And also let

$$Q_0 \leqq P^{0.125}, \quad m \leqq P^{2\varrho}.$$

Then, if one of the conditions

$$Q > P^{0.1\nu}, \quad \delta_0 > P^{0.1\nu}$$

is satisfied, we have

$$S \ll P^{1-\varrho}.$$

Proof. Just as in the proof of Lemma 5, we can restrict ourselves to the case $t \leqq t_0$; $t_0 = P^{\varrho+\varepsilon'}$.

We find

$$S^{(t)} = \sum_{0 \leqq v < Q_0} \sum_{W_0 < Q_0 u + v \leqq W} e^{2\pi i \left(\Phi(v) + \lambda(Q_0 u + v) + \frac{a_1' u}{q_1}\right)};$$

$$\Phi(v) = \frac{ma_n}{q_n} d^n t^n v^n + \ldots + \frac{ma_1}{q_1} dw,$$

$$\lambda(Q_0 u + v) = mz_n d^n t^n (Q_0 u + v)^n + \ldots + mz_1 dt (Q_0 u + v),$$

where a_1' stands for the integer nearest to zero that is congruent to $ma_1 dtQ_0$ modulo q_1. For each $s = n, \ldots, 1$, we find

$$\frac{d}{du} mz_s d^s t^s (Q_0 u + v)^s \ll \frac{mQ_0 t}{P^{0.25}} \ll P^{-0.124}.$$

Therefore

$$\left| \frac{d}{du} \left(\lambda(Q_0 u + v) + \frac{a_1'}{q_1} u \right) \right| < 0.6.$$

Consequently, we can apply Lemma 3 of Chapter 2 to that part of the sum $S^{(t)}$ which corresponds to any value of v. We obtain

$$S^{(t)} = \sum_{0 \leqq v < Q_0} e^{2\pi i \Phi(v)} \int_{(W_0 - v)/Q_0}^{(W - v)/Q_0} e^{2\pi i \left(\lambda(Q_0 u + v) + \frac{a_1'}{q_1} u\right)} du + O(Q_0)$$

$$= \frac{1}{Q_0} \sum_{0 \leqq v < Q_0} e^{2\pi i \Phi(u)} \int_{W_0}^{W} e^{2\pi i \left(\lambda(\xi) + \frac{a_1'}{q_1} \frac{\xi - v}{Q_0}\right)} d\xi + O(Q_0).$$

Thus, we find

$$\int_{W_0}^{W} e^{2\pi i \left(\lambda(\xi) + \frac{a_1}{q_1} \frac{\xi - v}{Q_0}\right)} d\xi = V - V_0;$$

$$V = W \int_0^1 e^{2\pi i \left(\lambda(W\eta) + \frac{a_1}{q_1} \frac{W\eta - v}{Q_0}\right)} d\eta;$$

$$V_0 = W_0 \int_0^1 e^{2\pi i \left(\lambda(W_0\eta) + \frac{a_1}{q_1} \frac{W_0\eta - v}{Q_0}\right)} d\eta.$$

If mtQ_0 is not divisible by q_1, then a'_1 is not zero and each of the coefficients of η in brackets of the exponents of the expressions for V and V_0 is

$$\gg \frac{W}{q_1 Q_0}.$$

Therefore, by Lemma 4 of Chap. 2, each of the products V and V_0 is

$$\ll W\left(\frac{W}{q_1 Q_0}\right)^{-\nu} \ll \frac{P}{dt}\left(\frac{P}{dtq_1 Q_0}\right)^{-\nu} \ll \frac{P^{1-0.124\nu}}{d},$$

Hence it follows that

$$S^{(t)} \ll \frac{P^{1-0.124\nu}}{d} \ll \frac{P^{1-\varrho-c_1}}{d}, \quad S \ll P^{1-\varrho}.$$

If $mQ_0 t$ is divisible by q_1, then a'_1 is zero, and we obtain

$$V = W\int_0^1 e^{2\pi i\lambda(W\eta)}\,d\eta, \quad V_0 = W_0\int_0^1 e^{2\pi i\lambda(W_0\eta)}\,d\eta,$$

$$S^{(t)} = U(V - V_0) + O(Q_0); \quad U = \frac{1}{Q_0}\sum_{0 \leqq v < Q_0}' e^{2\pi i\Phi(v)}.$$

The coefficient of η^s in the polynomial $\lambda(W\eta)$ is equal to $mz_s d^s t^s W^s$. Therefore, by Lemma 4 of Chapter 2, we get (if $|\delta_s| = \delta_0$)

$$V \ll W\min(1, m|z_s|\,d^s t^s W^s)^{-\nu} \ll \frac{P}{dt}\min(1, m^{-\nu}\delta_0^{-\nu}).$$

We obtain the same inequality for V_0. Besides, evaluating U with the help of Theorem 2 of Chapter 3 we find

$$S^{(t)} \ll \frac{P}{dt}(m, Q_0)^\nu Q_0^{-\nu+\varepsilon_0}\min(1, m^{-\nu}\delta_0^{-\nu}). \tag{17}$$

If $Q > P^{0.1\nu}$, from the divisibility of $mQ_0 t$ by q_1, it follows that $Q_0 \geqq Q/mt > P^{0.095\nu}$ Therefore, from (17), we obtain

$$S^{(t)} \ll \frac{P^{1-0.092\nu^2}}{dt} \ll \frac{P^{1-\varrho-c_1}}{d}, \quad S \ll P^{1-\varrho}.$$

If $\delta_0 > P^{0.1\nu}$, from (17) we obtain

$$S^{(t)} \ll \frac{P^{1-0.1\nu^2}}{dt} \ll \frac{P^{1-\varrho-c_1}}{d}, \quad S \ll P^{1-\varrho}.$$

This completes the proof of the lemma.

Lemma 7. Let

$$Q \leqq P^{0.1\nu}, \quad \delta_0 \leqq P^{0.1\nu}.$$

Then, assuming that

$$\Delta = \ln PQ^{-\nu+\varepsilon''}, \quad \mu = (m, Q),$$

or, if $\delta_0 > 1$, that

$$\Delta = \ln PQ^{-\nu+\varepsilon''}\delta_0^{-\nu}, \quad \mu = 1,$$

we have, for $m \leqq \Delta^{-2}$,

$$S \ll P\Delta\mu.$$

Proof. Just as in the previous lemmas, it suffices to consider only the case $t \leqq \Delta^{-1} Q^{\varepsilon}$. We find

$$S^{(t)} = \sum_{0 < v \leqq Q} \sum_{W_0 \leqq Qu+v \leqq W} e^{2\pi i(\Phi(v) + \lambda(Qu+v))};$$

$$\Phi(v) = \frac{ma_n}{q_n} d^n t^n v^n + \ldots + \frac{ma_1}{q_1} dtv;$$

$$\lambda(Qu+v) = mz_n d^n t^n (Qu+v)^n + \ldots + mz_1 dt(Qu+v).$$

For each s, we find

$$\frac{d}{du} mz_s d^s t^s (Qu+v)^s \ll \frac{mP^{0.1\nu} Qt}{P^{0.75}} \ll P^{-0.6}.$$

Therefore

$$\frac{d}{du} \lambda(Qu+v) \ll P^{-0.6}.$$

Consequently, we can apply Lemma 3 of Chapter 2 to that part of the sum $S^{(t)}$ which corresponds to a given v. Thus, we obtain

$$S^{(t)} = \sum_{0<v\leqq Q} e^{2\pi i\Phi(v)} \int_{(W_0-v)/Q}^{(W-v)/Q} e^{2\pi i\lambda(Qu+v)} du + O(Q) = U(V-V_0) + O(Q);$$

$$U = \frac{1}{Q} \sum_{0<v\leqq Q} e^{2\pi i\Phi(v)},$$

$$V = W \int_0^1 e^{2\pi i\lambda(W\eta)} d\eta, \quad V_0 = W_0 \int_0^1 e^{2\pi i\lambda(W_0\eta)} d\eta.$$

The coefficient of η^s in the polynomial $\lambda(W\eta)$ is equal to $mz_s d^s t^s W^s$. Therefore, by Lemma 4 of Chapter 2, we have

$$V \ll W \min(1, (m|z_s| d^s t^s W^s)^{-\nu}) \ll \frac{P}{dt} \min(1, m^{-\nu}\delta_0^{-\nu}).$$

The same inequality is obtained for V_0. Besides, estimating U with the help of Theorem 1 of Chapter 3, we find

$$S^{(t)} \ll \frac{P}{dt}(m, Q)\, Q^{-\nu+\varepsilon'} \min(1, m^{-\nu}\delta_0^{-\nu}).$$

Hence, both the assertions of our lemma follow without any difficulty.

Lemma 8. Divide the points of an n-dimensional space into two classes Class I and Class II. A point is said to be a point of Class I, if

$$Q \leqq P^{0.1\nu}, \quad \delta_0 \leqq P^{0.1\nu}.$$

A point which is not a point of Class I is said to be a point of Class II.

If $m \leqq P^{2\varrho}$, for the points of Class II, we have

$$S \ll \frac{P^{1-\varrho}}{d}.$$

Assuming that

$$\Delta = \ln P Q^{-\nu+\varepsilon'}, \quad \mu = (m, Q)^{-\nu},$$

or, if $\delta_0 > 1$, that

$$\Delta = \ln P Q^{-\nu+\varepsilon'} \delta_0^{-\nu}, \quad \mu = 1,$$

we find that, if $m \leqq \Delta^{-2}$, for the points of Class I

$$S \ll P \Delta \mu.$$

Proof. The assertion concerning the points of Class II follows from Lemmas 5 and 6 the assertion concerning the points of Class I from Lemma 7.

Lemma 9. Let $\varepsilon_2 \leqq 0.001$ and F be the product of primes $p \leqq P^{0.25}$. Then, assuming

$$D = (\ln P)^{\frac{\ln \ln P - 1}{\ln(1+\varepsilon^2)}},$$

the divisors d of F which are not greater than P can be distributed among $< D$ sets with the following properties:

(a) Numbers d, belonging to the same set, have the same number β of prime factors and, hence, the same value of $\mu(d) = (-1)^{\beta}$.

(b) One of the sets, called the "simplest" set, consists of only one number $d = 1$. For this set, let $\varphi = 1$, and thus, we have $d = \varphi = 1$. To each of the remaining sets, there corresponds a definite φ such that all numbers of this set satisfy the condition:

$$\varphi < d \leqq \varphi^{1+\varepsilon_1}.$$

(c) For every set other than the simplest set, and for any U $(0 \leqq U < \varphi)$, there exist two such sets of numbers d, viz. the set of d' and the set of d'' with respective φ' and φ'' that satisfy the conditions

$$U < \varphi' \leqq U P^{0.25}, \quad \varphi' \varphi'' = \varphi$$

(the second set may prove to be the simplest set) such that, for a certain natural number B, each number d of a chosen set is repeated B times, if, from among all the products $d'd''$, only those satisfying the condition $(d', d'') = 1$ are chosen.

Proof. Distribute all odd prime divisors of F over the $\tau + 1$ intervals of the type

$$e^{(1+\varepsilon_2)^t} < p \leqq e^{(1+\varepsilon_2)^{t+1}},$$

which are obtained by making t (the ordinal number of an interval) run through the values $t = 0, 1, \ldots, \tau$, where τ is the largest integer satisfying the condition:

$$e^{(1+\varepsilon_2)^\tau} < P^{0.25}.$$

From this condition, we easily find that $\tau+1$, the number of intervals, satisfies the inequality:

$$\tau+1<\frac{\ln\ln P-\ln 4}{\ln(1+\varepsilon_2)}+1<\frac{\ln\ln P-1}{\ln(1+\varepsilon_2)}.$$

We shall relate each odd divisor $d(1<d\leqq P)$ of F to a non-decreasing sequence $l_0, l_1, \ldots, l_\tau$, where l_t denotes the number of prime factors of d in the interval with ordinal number t. The set of values of d related to the same sequence is precisely the set of numbers d stated in the lemma.

Since each d is the product of less than $\ln P-1$ prime factors, we have $l_t\leqq \ln P-1$ for each $t=0, \ldots, \tau$ and, consequently, the total number of different sets does not exceed

$$(\ln P)^{\tau+1}<D.$$

From the definition of the set of numbers d, it follows that the numbers d belonging to the same set indeed have the same number β of prime factors and, consequently, the same $\mu(d)=(-1)^\beta$.

Now consider some set of the numbers d other than the simplest set. Let $d=p_1\ldots p_\beta$ be the canonical expansion of some of its term written in an increasing order of factors. If t_s denotes the ordinal number of that set which contains p_s, and the number

$$e^{(1+\varepsilon_2)^{t_s}}$$

is denoted by φ_s, we find

$$\varphi_s<p_s\leqq\varphi_s^{1+\varepsilon_2}.$$

Putting the number $\varphi=\varphi_1\ldots\varphi_\beta$ into correspondence with the chosen set, we obtain

$$\varphi<d\leqq\varphi^{1+\varepsilon_2}.$$

Let U be a number such that $0\leqq U<\varphi$. Denoting by λ the least number satisfying the inequality $U<\varphi_1\ldots\varphi_\lambda$, consider two auxiliary sets: the set of numbers d', which run through product of the type $p_1\ldots p_\lambda$ and the set of numbers d'', which run through products of the type $p_{\lambda+1}\ldots p_\beta$ (the set of numbers d'' may prove to be the simplest set). Then the number $\varphi'=\varphi_1\ldots\varphi_\lambda$ will correspond to the set of numbers d', while the number $\varphi''=\varphi_{\lambda+1}\ldots\varphi_\beta$ to the set of numbers d''. Thus, we will have

$$U<\varphi'\leqq UP^{0.25};\quad \varphi'\varphi''=\varphi.$$

If d is any number from the chosen set, then the equality $d=d'd''$ is possible if and only if $(d', d'')=1$, and it then has

$$B=\binom{k_1+k_2}{k_1}$$

solutions, where k_1 is the number of factors φ_s in the product φ', which are equal to φ_λ, while k_2 is the number of factors φ_s in the product φ'' which are equal to φ_λ.

Theorem 1. Divide the points $(\alpha_n, \ldots, \alpha_1)$ of an n-dimensional space into two classes: Class I and Class II. A point, for which

$$Q\leqq P^{0.1\nu},\quad \delta_0\leqq P^{0.1\nu}$$

is said to belong to Class I. A point which is not a point of Class I is said to belong to Class II.

For the points of Class II assume that

$$\Delta_1 = P^{-\varrho}, \quad \mu = 1,$$

while for the points of Class I put

$$\Delta_1 = Q^{-0.5\nu+\varepsilon}, \quad \mu = (m, Q)^{0.5\nu}$$

or $\quad \Delta_1 = Q^{-0.5\nu+\varepsilon}\delta_0^{-0.5\nu}, \quad \mu = 1, \text{ if } \delta_0 > 1.$

Then, assuming

$$H = (\ln P)^{\frac{\ln\ln P}{\ln(1+\varepsilon_0)}},$$

when $m \leqq \Delta_1^{-2}$, we have

$$T'_m \ll HP\Delta_1\mu.$$

Proof. We shall use Lemmas 4 and 8. And, for the sake of uniformity we shall replace the estimate given in Lemma 8 by an approximate one to match that given in Lemma 4. For a better understanding of the proof, we may note that always

$$P^{0.996} \ll P\Delta_1.$$

Let F be the product of all primes not exceeding $P^{0.25}$ which do not divide Q. Let g run through all positive integers coprime to FQ, and d through the divisors of F (consequently through numbers coprime to Q). Finally, let k run through all positive numbers coprime to Q in the interval specified above. We find

$$\sum_{g \leqq P} e^{2\pi i m f(g)} = \sum_{dk \leqq P}\sum \mu(d)\, e^{2\pi i m f(dk)}. \tag{18}$$

Let g_j run through the values of g having exactly j prime factors. Then the left-hand side of (18) can be represented as

$$T'_m + T'_{m,2} + T'_{m,3} + O(P^{0.25});$$

$$T'_{m,2} = \sum_{g_2 \leqq P} e^{2\pi i m f(g_2)}, \quad T'_{m,3} = \sum_{g_3 \leqq P} e^{2\pi i m f(g_3)}.$$

We shall first estimate $T'_{m,2}$. Obviously, we have

$$T'_{m,2} = \tfrac{1}{2} T''_{m,2} + O(P^{3/4}); \quad T''_{m,2} = \sum_{g_1 g'_1 \leqq P}\sum e^{2\pi i m f(g_1 g'_1)},$$

where g'_1 runs through the same values as g_1. We find

$$T''_{m,2} = \sum_{\substack{g_1 \leqq \sqrt{P} \\ g_1 g'_1 \leqq P}}\sum e^{2\pi i m f(g_1 g'_1)} + \sum_{\substack{g_1 > \sqrt{P} \\ g_1 g'_1 \leqq P}}\sum e^{2\pi i m f(g_1 g'_1)}.$$

Now we can apply Lemma 4 to the sums on the right-hand side (even in degenerate cases with $y = g_1$, $x = g'_1$ for the first sum, and $x = g_1$, $y = g'_1$ for the second sum). Thus, we obtain the estimates stated in the lemma (in accordance with the class of the points).

Now, we shall estimate $T'_{m,3}$. Obviously, we have

$$T'_{m,3} = \tfrac{1}{3} T''_{m,3} + O(P^{3/4}); \quad T''_{m,3} = \sum_{g_1 g_2 \leqq P}\sum e^{2\pi i m f(g_1 g_2)}.$$

Here also we can apply Lemma 4 to the sum $T''_{m,3}$ (as in the degenerate case with $y=g_1$, $x=g_2$). Once again we obtain the estimates stated in the lemma.

Now, using Lemma 9, we shall estimate the right-hand side of (18). Without any loss of generality, we can restrict ourselves only to a part E of this sum which contains the set of values of d satisfying the condition $\mu(d)=1$.

Divide the interval $0<k\leqq P$ into $\ll \ln P$ intervals of the type

$$K<k\leqq K'; \quad 2K\leqq K'<4K, \tag{19}$$

Let U_K denote that part of the sum E which corresponds to the interval (19). And let $U_{K,\varphi}$ stand for that part of the sum U_K which contains those d that belong to the set with the corresponding number φ.

First consider the case $K<P^{0.245}$ and assume that $K_\varphi \geqq P^{0.995}$. Then, we have $\varphi>P^{0.75}$. Assuming $U=P^{0.245}K^{-1}$, construct the two sets of numbers d' and of d'' having the properties stated in Lemma 9. In particular, the number φ' satisfying the condition

$$P^{0.245}K^{-1}\leqq \varphi'<P^{0.495}K^{-1},$$

will correspond to d' and, consequently, all the values of d will lie in the interval:

$$P^{0.245}K^{-1}<d<(P^{0.495}K^{-1})^{1+\varepsilon_0}.$$

Therefore we can write

$$U_{K,\varphi}=\frac{1}{B}\sum_k\sum_{d'}\sum_{d''}e^{2\pi i m f(kd'd'')},$$

where k runs through values lying in the interval (19), except that values satisfying the condition $(d',d'')>1$ are omitted. And summation is taken over the interval

$$P^{0.245}<kd'<4P^{0.496}, \quad kd'd''\leqq P.$$

The sum on the right-hand side is a particular case of the sum S stated in Lemma 4. Therefore the estimates given in that lemma hold valid for this sum.

Now consider the case

$$P^{0.245}<K\leqq P^{0.5}.$$

Here all values of k lie in the interval

$$P^{0.245}<k<4P^{0.75}$$

and, consequently, the sum $U_{K,\varphi}$ is a degenerate case of the sum S of Lemma 4 $(y=k, x=d)$: the estimates stated there hold for this sum.

Further, consider the case

$$P^{0.5}<K\leqq P^{0.75}.$$

Here all values of k lie in the interval

$$P^{0.5}<k\leqq 4P^{0.75}.$$

Consequently, it is sufficient to consider the case where φ satisfies the condition

$$P^{0.5}>\varphi\geqq P^{0.25}.$$

Thus, we can once again apply Lemma 4 ($y=d$, $x=k$) and the estimates given the lemma are found to be true for this case, too.

Finally, consider the case

$$P^{0.75} < K < P.$$

Here the approximate estimates given in Lemma 8 hold for $U_{K,\varphi}$. By what has been proved above, and since there are only

$$\ll D \ln P = H(\ln P)^{1-\frac{1}{\ln(1+\varepsilon^0)}}$$

sums $U_{K,\varphi}$, we find that our theorem is true.

Lemma 10. Let $\varepsilon \leqq 0.01$, $u = \ln P$, $b_0 = e^{u^{1-\varepsilon}}$, $b = e^{u^{1-2\varepsilon}}$, $0 < q \leqq e^{\sqrt{u}}$, $0 \leqq l < q$, $(l, q) = 1$, $U \geqq 0$, $W \geqq b_0$, let $p_1, \ldots, p_\sigma$ be primes not greater than b that do not divide q, and let F be their product.

Then the number T of numbers of the type $qx + l$ which are coprime to F lying in the interval

$$U < qx + l \leqq U + W,$$

satisfies the condition:

$$T \ll \frac{Wu^{2\varepsilon}}{u\varphi(q)}.$$

Proof. We shall apply a simplied version of the Brun method (see the Introduction). Let $\Omega(s)$ denote the number of different prime divisors of the number s. Assuming that

$$m = 2[2\ln u + 1],$$

by a well-known theorem of number theory we find that

$$T \ll \sum_{\substack{d \backslash F \\ \Omega(d) \leqq m}} \mu(d) S_d,$$

where S_d is the number of numbers of the type $qx + l$ which are multiples of d and lie in the interval under consideration. But we easily find

$$S_d = \frac{W}{qd} + \theta_d.$$

Therefore

$$T \ll \left| \sum_{\substack{d \backslash F \\ \Omega(d) \leqq m}} \mu(d) \frac{W}{qd} \right| + \sum_{s=0}^{m} \binom{\sigma}{s}.$$

Furthermore, we have

$$\sum_{s=0}^{m} \binom{\sigma}{s} \leqq \sum_{s=0}^{m} \sigma^s \ll b^m \ll b^{5\ln u} \ll \frac{W}{uq} \frac{uq}{b_0} b^{5\ln u} \ll \frac{W}{uq},$$

$$\sum_{d \backslash F} \mu(d) \frac{W}{qd} = \frac{W}{q} \frac{\prod_{p \leqq b} \left(1 - \frac{1}{P}\right)}{\prod_{p \backslash q} \left(1 - \frac{1}{P}\right)} \ll \frac{W}{\ln b\varphi(q)} \ll \frac{Wu^{2\varepsilon}}{u\varphi(q)},$$

$$\sum_{\substack{d \backslash F \\ \Omega(d) > m}} \mu(d) \frac{W}{qd} \ll \frac{W}{q} \sum_{s=m+1}^{\sigma} \sum_{\substack{d \backslash F \\ \Omega(d)=s}} \frac{1}{d} \ll \frac{W}{q} \sum_{s=m+1}^{\sigma} \frac{\left(\frac{1}{P_1} + \ldots + \frac{1}{P_\sigma}\right)^s}{s!}$$

$$\ll \frac{W}{q} \sum_{s=m+1}^{\sigma} \left(\frac{e(c+\ln u)}{s}\right)^s \ll \frac{W}{q} \sum_{s=m+1}^{\sigma} \left(\frac{3}{4}\right)^s \ll \frac{W}{uq},$$

which completes the proof of our lemma.

Theorem 1a. Let $\varepsilon < 0.01$,

$$u = \ln P, \quad \delta_0 \leqq e^{u^\varepsilon}, \quad Q \leqq e^{u^\varepsilon}, \quad m \leqq e^{2\nu u^\varepsilon}.$$

Then, for the sum

$$S' = \sum_{p \leqq P} e^{2\pi i m f(p)}$$

we have

$$S' \ll P u^{-1+9\varepsilon} (m, Q)^{0.5\nu} Q^{-0.5\nu+\varepsilon'}$$

or

$$S' \ll P u^{-1+9\varepsilon} (m\delta_0)^{-0.5\nu}, \quad \text{if} \quad \delta_0 \geqq 1.$$

Proof. Let $b_0 = e^{u^{1-\varepsilon}}, p_1, \ldots, p_{\sigma_0}$ be primes not greater than b_0 which do not divide Q, and F_0 be their product. Let r run through positive numbers coprime to $F_0 Q$ and k through positive numbers coprime to Q. We find

$$\sum_{r \leqq P} e^{2\pi i m f(r)} = \sum_{d \backslash F} k(d) S_d; \quad S_d = \sum_{k \leqq P d^{-1}} e^{2\pi i m f(dk)}. \tag{20}$$

First, we shall estimate the sum D_1 of the terms on the right-hand side satisfying the condition $d > P^{0.8}$. Thus, we easily find

$$\Omega(d) > 0.8 u^\varepsilon.$$

Therefore (see the proof of Lemma 10), we have

$$D_1 \ll \sum_{\Omega(d) > 0.8 u^\varepsilon} \frac{P}{d} \ll P \sum_{s > 0.8 u^\varepsilon} \sum_{\Omega(d) = s} \frac{1}{d}$$

$$\ll P \sum_{s > 0.8 u^\varepsilon} \frac{\left(\frac{1}{P_1} + \ldots + \frac{1}{P_{\sigma_0}}\right)^\varepsilon}{s!}$$

$$\ll P \sum_{s > 0.8 u^\varepsilon} \left(\frac{e(c+\ln u)}{s}\right)^s \ll P \left(\frac{3 \ln u}{0.8 u^\varepsilon}\right)^{0.8 u^\varepsilon} \ll P e^{-0.5 u^\varepsilon}.$$

Now we shall estimate the sum D_2 of the terms on the right-hand side satisfying the condition $d \leqq P^{0.8}$. We find

$$S_d = \sum_{\substack{0 \leqq l < Q \\ (l,Q)=1}} e^{2\pi i m \Phi(dl)} S^{(l)}; \quad S^{(l)} = \sum_{-\frac{dl}{dQ} < v \leqq \frac{P-dl}{dQ}} e^{2\pi i m \lambda(d(Qv+l))}$$

$$\Phi(dl) = \frac{ma_n}{q_n} d^n l^n + \ldots + \frac{ma_1}{q_1} dl,$$

$$\lambda(d(Qv+l)) = mz_n d^n (Qv+l)^n + \ldots + mz_1 d(Qv+l).$$

Furthermore, we find

$$\frac{d}{dv} mz_s d^s (Qv+l)^s \ll P^{-0.1}, \quad \frac{d}{dv} \lambda(d(Qv+l)) \ll P^{-0.1}.$$

Therefore, applying Lemma 3 of Chapter 2, we obtain

$$S^{(l)} = \int_{\frac{-dl}{dQ}}^{\frac{P-dl}{dQ}} e^{2\pi i \lambda(d(Qv+l))} dv + O(1) = \frac{P}{dQ} \int_0^1 e^{2\pi i \lambda(P\xi)} d\xi + O(1).$$

Now, assuming that

$$P \int_0^1 e^{2\pi i \lambda(P\xi)} d\xi = V, \quad \frac{1}{Q} \sum_{\substack{0 \leqq l < Q \\ (l,Q)=1}} e^{2\pi i \Phi(dl)} = U,$$

we find $S_d = UV/d + O(Q)$. Therefore

$$D_2 = UV \sum_{d \leqq P^{0.8}} \frac{\mu(d)}{d} + O(P^{0.8} Q).$$

But, from what was proved at the beginning, it follows that

$$\sum_{d > P^{0.8}} \frac{\mu(d)}{d} \ll \sum_{\Omega(d) > 0.8u^\varepsilon} \frac{1}{d} \ll e^{-0.5u^\varepsilon}.$$

Moreover, (see the proof to Lemma 10), we have

$$\sum_{d \setminus F_0} \frac{\mu(d)}{d} = \frac{\prod_{p \leqq b_0} \left(1 - \frac{1}{p}\right)}{\prod_{p \setminus Q} \left(1 - \frac{1}{p}\right)} \ll \frac{Q}{\varphi(Q) \ln b_0} \ll \frac{u^{2\varepsilon}}{u}.$$

Therefore

$$\sum_{d \leqq P^{0.8}} \frac{\mu(d)}{d} \ll \frac{u^{2\varepsilon}}{u}, \quad D_2 \ll |UV| \frac{u^{2\varepsilon}}{u} + Pe^{-0.5u^\varepsilon},$$

$$D_1 + D_2 \ll |UV| \frac{u^{2\varepsilon}}{u} + Pe^{-0.5u^\varepsilon}.$$

Hence, applying Theorem 3 of Chapter 3 to U, we obtain

$$D_1 + D_2 \ll Pu^{-1} u^{2\varepsilon} (m, Q)^\nu Q^{-\nu+\varepsilon'}.$$

Applying Lemma 4 of Chapter 2 to V, we obtain

$$D_1 + D_2 \ll Pu^{-1} u^{2\varepsilon} (m\delta_0)^{-\nu}, \quad \text{if} \quad \delta_0 \geqq 1.$$

Now we shall consider the left-hand side of (20). Let r_κ run through the values of r having exactly κ different prime factors, and let $S_\kappa = \sum_{r_\kappa \leqq P} e^{2\pi i m f(r_\kappa)}$. For $\kappa > 1$, we easily find

$$S_\kappa = \frac{1}{\kappa} \sum_{r_1 r_{\kappa-1} \leqq P}\sum e^{2\pi i m f(r_1 r_{\kappa-1})} + O(Pb_0^{-1})$$

$$= \frac{1}{\kappa} \sum_{\substack{r_1 \leqq \sqrt{P} \\ r_1 r_{\kappa-1} \leqq P}}\sum e^{2\pi i m f(r_1 r_{\kappa-1})} + \frac{1}{\kappa} \sum_{\substack{r_1 > \sqrt{P} \\ r_1 r_{\kappa-1} \leqq P}}\sum e^{2\pi i m f(r_1 r_{\kappa-1})} + O(Pb_0^{-1}).$$

Now we shall only estimate the first sum on the right-hand side. For the sake of simplicity, putting $r_1 = y$, $r_{\kappa-1} = x$, write this sum as

$$S_0 = \sum_{\substack{y \leqq \sqrt{P} \\ yx \leqq P}}\sum e^{2\pi i m f(yx)}.$$

The second sum on the right-hand side is evaluated in a similar way and the same result is obtained in this case, too.

Divide the interval $b_0 < y \leqq \sqrt{P}$ into $\ll u$ intervals of the type:

$$R < y \leqq R_1; \quad 2R \leqq R_1 \leqq 4R. \tag{21}$$

Let $S(R)$ denote that part of the sum S_0 which corresponds to the interval (21). Now putting $b_1 = e^{u^{1-2\varepsilon}}$, divide the first quadrant into the following squares:

$$b_1 \xi < x \leqq b_1(\xi + 1), \quad b_1 \eta < y \leqq b_1(\eta + 1) \tag{22}$$

with integral ξ and η. The number of integral points in each square is $\ll b_1^2$. The number of integral points in all those squares which intersect the contour of the domain (G) of the range of summation of $S(R)$ is

$$\ll Pb_1 R^{-1} \ll Pe^{-0.5u^{1-\varepsilon}}.$$

Let the square (22) be completely contained in the domain (G) and $S_{\xi,\eta}$ be that part of the sum $S(R)$ which corresponds to this square. Finally, for given l and h coprime to Q, taken from the sequence $0, \ldots, Q-1$, let $S(l, h)$ denote the sum of those terms in the sum $S_{\xi,\eta}$ which satisfy the condition:

$$x \equiv l \pmod{Q}, \quad y \equiv h \pmod{Q}, \tag{23}$$

For a term of the sum $S(l, h)$, we have

$$mf(xy) = F(lh) + \lambda(xy);$$

$$F(lh) = \frac{ma_n}{q_n} l^n h^n + \ldots + \frac{ma_1}{q_1} lh,$$

$$\lambda(xy) = mz_n x^n y^n + \ldots + mz_1 xy.$$

In the expression for $\lambda(xy)$, if $b_1\xi$ and $b_1\eta$ are substituted for x and y, respectively, then for $mz_s x^s y^s$ we get a number

$$\ll m|\delta_s| \frac{b_1}{R} \ll e^{2\nu u^\varepsilon + u^\varepsilon + u^{1-2\varepsilon}} R^{-1} \ll e^{2u^{1-2\varepsilon}} R^{-1},$$

and for $mf(xy)$ we get a number $\ll e^{2u^{1-2\varepsilon}} R^{-1}$. Accordingly, for $S(l, h)$ we get a number

$$\ll \frac{b_1^2}{Q^2} e^{2u^{1-2\varepsilon}} R^{-1} \ll \frac{b_1^2}{Q^2} e^{-0.5u^{1-\varepsilon}}$$

and it is equal to

$$S(l, h) = e^{2\pi i\lambda(b_1^2\xi\eta)} S_1(l, h) + O\left(\frac{b_1^2}{Q^2} e^{-0.5u^{1-\varepsilon}}\right);$$

$$S_1(l, h) = L_{\xi,l} H_{\eta,h} e^{2\pi i F(lh)},$$

where $L_{\xi,l}$ and $H_{\eta,h}$ are the numbers of values of x and y which satisfy the conditions (22) and (23), respectively. Thus, by Lemma 9 (2ε in place of ε), we obtain $L_{\xi,l} \ll K$ and $H_{\eta,h} \ll K$; $K = b_1 u^{4\varepsilon}/u\varphi(Q)$. Now we find

$$S_{\xi,\eta} = S'_{\xi,\eta} + O(b_1^2 e^{-0.5u^{1-\varepsilon}});$$

$$S'_{\xi,\eta} = e^{2\pi i\lambda(b_1^2\xi\eta)} \sum_l \sum_h L_{\xi,l} H_{\eta,h} e^{2\pi i F(lh)},$$

$$S(R) = S_1(R) + O(Pe^{-0.5u^{1-\varepsilon}}); \quad S_1(R) = \sum_{\substack{R < b_1\eta \leqq R_1 \\ 0 < b_1^2\xi\eta \leqq P}}\sum S'_{\xi,\eta}.$$

First consider the general case. Here we find

$$|S'_{\xi,\eta}|^2 \ll K^2 Q \sum_h \sum_{h_1} H_{\eta,h} H_{\eta,h_1} \left|\sum_{l=0}^{Q-1} e^{2\pi i(F(lh) - F(lh_1))}\right|,$$

Hence, by Theorem 2 of Chapter 3, it follows

$$|S'_{\xi,\eta}|^2 \ll K^4 Q^4 (m, Q)^\nu Q^{-\nu+\varepsilon'},$$

$$|S'_{\xi,\eta}| \ll K^2 Q^2 (m, Q)^{0.5\nu} Q^{-0.5\nu+0.5\varepsilon'} \ll \frac{b_1^2 u^{8\varepsilon}}{u^2} (m, Q)^{0.5\nu} Q^{-0.5\nu+\varepsilon'}.$$

Therefore

$$S(R) \ll P \frac{u^{8\varepsilon}}{u^2} (m, Q)^{0.5\nu} Q^{-0.5\nu+\varepsilon'}, \quad S_0 \ll P \frac{u^{8\varepsilon}}{u} (m, Q)^{0.5\nu} Q^{-0.5\nu+\varepsilon'}.$$

Multiplying the right-hand side of this inequality by κ^{-1} and then summing over all $\ll u^\varepsilon$ values of κ exceeding 1, we obtain a number

$$\ll P \frac{u^{9\varepsilon}}{u} (m, Q)^{0.5\nu} Q^{-0.5\nu+\varepsilon'}.$$

Now consider the case $\delta_0 \geqq 1$. Here we find

$$S_1(R) \leqq \sum_l \sum_h |B_{l,h}|; \quad B_{l,h} = \sum_{\substack{R_2 < \eta \leqq R_3 \\ 0 < \xi\eta \leqq P_1}}\sum L_\xi H_\eta e^{2\pi i\lambda_1(\xi\eta)},$$

and for the sake of brevity, put

$$L_{\xi,l} = L_\xi, \quad H_{\eta,h} = H_\eta, \quad \frac{R}{b_1} = R_2, \quad \frac{R_1}{b_1} = R_3,$$

$$\frac{P}{b_1^2} = P_1, \quad \lambda_1(\xi\eta) = m\frac{\delta_n}{P_1^n}\xi^n\eta^n + \ldots + m\frac{\delta_1}{P_1}\xi\eta.$$

Let B_ξ be that part of the sum $B_{l,h}$ which corresponds to a given ξ. We find $B_\xi \ll K\left|\sum' H_\eta e^{2\pi i\lambda_1(\xi,\eta)}\right|$, where summation is taken over those values of η that lie in the interval $R_2 < \eta \leqq \min\left(\frac{P_1}{\xi}, R_3\right)$. Hence we obtain

$$|B_\xi|^2 \ll K^2 \sum_\eta{}' \sum_{\eta_1}{}' H_\eta H_{\eta_1} e^{2\pi i\Phi(\eta,\eta_1,\xi)};$$
$$\Phi(\eta,\eta_1,\xi) = \lambda_1(\xi\eta) - \lambda_1(\xi\eta_1).$$

We further find that

$$|B_{l,h}|^2 \ll \frac{P_1}{R_2} \sum_{0<\xi\leqq P_1/R_2} |B_\xi|^2,$$

where ξ runs through all integers between the limits shown. But each pair of η' and η'' may serve as the values of η and η_1 contained in some B_ξ, if and only if ξ does not exceed the least of the numbers P_1/η' and P_1/η''. Therefore, changing the order of summation, we obtain

$$|B_{l,h}|^2 \ll \frac{P_1 K^2}{R_2} \sum_\eta \sum_{\eta_1} H_\eta H_{\eta_1} \sum_{0<\xi\leqq\xi_{\eta,\eta_1}} e^{2\pi i\Phi(\eta,\eta_1,\xi)};$$
$$\xi_{\eta,\eta_1} = \min\left(\frac{P_1}{\eta}, \frac{P_1}{\eta_1}\right).$$

The right-hand side of this inequality can be represented in the form $S_1 + S_2 + S_3$, where S_1 corresponds to the case $\eta = \eta_1$, S_2 to the case $\eta < \eta_1$, and finally S_3 to the case $\eta > \eta_1$. But, we have

$$S_1 < \frac{P_1 K^2}{R_2} \sum_\eta K^2 \frac{P_1}{R_2} \ll \frac{P_1^2 K^4}{R_2}.$$

Noting that S_2 and S_3 are complex conjugate, we obtain

$$|B_{l,h}|^2 \ll \frac{P_1 K^4}{R_2} \sum_{\eta_1} \sum_{R_2<\eta<\eta_1} |W_{\eta,\eta_1}| + \frac{P_1^2 K^4}{R_2};$$
$$W_{\eta,\eta_1} = \sum_{0<\xi\leqq\xi_0} e^{2\pi i\Phi(\eta,\eta_1,\xi)}, \quad \xi_0 = \frac{P_1}{\eta_1}.$$

Furthermore, we find

$$\frac{d}{d\xi} m \frac{\delta_s}{P_1^s}(\eta^s - \eta_1^s)\xi^s \ll m|\delta_s| P_1^{-1} R_2 \ll P^{-0.4},$$
$$\frac{d}{d\xi} \Phi(\eta,\eta_1,\xi) \ll P^{-0.4}.$$

Therefore, by Lemma 3 of Chapter 2, we have

$$W_{\eta,\eta_1} = \int_0^{\xi_0} e^{2\pi i\Phi(\eta,\eta_1,\xi)}\, d\xi + O(1) = \xi_0 \int_0^1 e^{2\pi i\Phi(\eta,\eta_1,\xi_0 t)}\, dt + O(1).$$

Now let $\delta_0 = |\delta_s|$. The modulus of the coefficient of t^s in the polynomial $\Phi(\eta, \eta_1, \xi_0 t)$ is

$$\frac{m\delta_0(\eta_1^s - \eta^s)\xi_0^s}{P_1^s} \gg m\delta_0 \frac{\eta_1 - \eta}{R_2}.$$

Consequently, by Lemma 4 of Chapter 2, we have

$$W_{\eta,\eta_1} \ll \xi_0 \left(m\delta_0 \frac{\eta_1 - \eta}{R_2}\right)^{-\nu}.$$

For a given η_1, summing the right-hand side over all values of η in the interval $R_2 < \eta < \eta_1$, we obtain a number

$$\ll \xi_0 \left(\frac{m\delta_0}{R_2}\right)^{-\nu} R_2^{1-\nu} \ll P_1 (m\delta_0)^{-\nu}.$$

Therefore

$$|B_{l,h}|^2 \ll P_1 K^4 R_2^{-1} R_2 P_1 (m\delta_0)^{-\nu} + P_1^2 K^4 R_2^{-1} \ll P_1^2 K^4 (m\delta_0)^{-\nu},$$
$$|B_{l,h}| \ll P_1 K^2 (m\delta_0)^{-0.5\nu},$$
$$S_1(R) \ll (\varphi(Q))^2 P_1 K^2 (m\delta_0)^{-0.5\nu} \ll Pu^{-2} u^{8\varepsilon} (m\delta_0)^{-0.5\nu},$$
$$S(R) \ll Pu^{-2} u^{8\varepsilon} (m\delta)^{-0.5\nu},$$
$$S_0 \ll Pu^{-1} u^{8\varepsilon} (m\delta_0)^{-0.5\nu}.$$

Multiplying the right-hand side of the last inequality by κ^{-1}, and then summing over all $\ll u^{\varepsilon}$ values of κ exceeding 1, we obtain a number $\ll Pu^{-1} u^{9\varepsilon} (m\delta_0)^{-0.5\nu}$.

From what has been proved above for the right-hand and left-hand sides of (20), we find that our theorem is true.

Theorem 2. Let $\varepsilon \leqq 0.01$. Divide the points $(\alpha_n, \ldots, \alpha_1)$ of an n-dimensional space into classes as follows.

In Class 1a put those points which satisfy the condition

$$Q \leqq e^{u^{\varepsilon}}, \quad \delta_0 \leqq e^{u^{\varepsilon}}.$$

In Class 1b put those points which are not points of Class 1a but satisfy the condition

$$Q \leqq P^{0.2\nu}, \quad \delta_0 \leqq P^{\nu}.$$

Finally, put all remaining points in Class 2.

For the points of Class 1a, assume that

$$\Delta = u^{9\varepsilon} Q^{-0.5\nu+\varepsilon'}, \quad \mu = (m, Q)^{0.5\nu}$$

or

$$\Delta = u^{9\varepsilon} \delta_0^{-0.5\nu}, \quad \mu = m^{-0.5\nu}, \quad \text{if} \quad \delta_0 \geqq 1.$$

For the points of Class 1b, putting $\varepsilon_3 = 2\varepsilon'$, assume that

$$\Delta = Q^{-0.5\nu+\varepsilon_3}, \ \mu = (m, Q)^{0.5\nu}, \quad \text{if} \quad Q > e^{u^{\varepsilon}},$$
$$\Delta = Q^{-0.5\nu+\varepsilon_3} \delta_0^{-0.5\nu+\varepsilon_3}, \ \mu = 1, \quad \text{if} \quad \delta_0 > e^{u^{\varepsilon}}$$

(if $Q > e^{u^{\varepsilon}}$ and $\delta_0 > e^{u^{\varepsilon}}$, we can take either pair for the values of Δ and μ).

Finally, for the points of Class 2, assume that

$$\Delta = P^{-\varrho_1}; \quad \varrho_1 = \frac{1}{20n^2(2\ln n + \ln\ln n + 5)}, \quad \mu = 1.$$

Then, if $m \leqq \Delta^{-2}$, we always have

$$T'_m \ll \frac{P}{u}\Delta\mu.$$

Proof. We shall assume that $P \geqq P_0$, where P_0 is a sufficiently large positive constant.

The assertion stated in the theorem relating to the points of Class 1 a follows directly from Theorem 1 a.

For the points of Class 1 b, if $Q > e^{u^\varepsilon}$, Theorem 1 (Class I) gives $T'_m \ll \frac{P}{u} uHQ^{-0.5\nu+\varepsilon'}$ $(m, Q)^{0.5\nu}$. Hence, as $uH \ll Q^{\varepsilon'}$, we find that the theorem true for this case.

If $\delta_0 > e^{u^\varepsilon}$, for the points of Class 1 b, Theorem 1 (Class I) gives

$$T'_m \ll \frac{P}{u} uHQ^{-0.5\nu+\varepsilon'} \delta_0^{-0.5\nu},$$

hence, as $uH \ll \delta_0^{\varepsilon_3}$, we find that the theorem holds for this case as well.

Finally, for the points of Class 2, Theorem 1 (Class II) gives

$$T'_m \ll \frac{P}{u} uHP^{-\varrho},$$

hence, as $uH \ll P^{\varrho - \varrho_1}$, we find that the theorem holds also for the points of Class 2.

Chapter 8. Distribution of Fractional Parts of the Values of a Polynomial when the Argument Runs Through Prime Numbers

Notation. We shall retain the notation of Theorem 2 of Chapter 7 and, in addition, shall take

$$\varrho_2 = \frac{1}{20n^2(2\ln n + \ln\ln n + 5)}.$$

The problem on the upper bound of the modulus of the sum T'_m considered in Chapter 7 is closely related to the problem of the distribution of fractional parts $\{f(p)\}$ of the values of the polynomial $f(p)$ for prime numbers p not exceeding P. The general solution of this problem is given in the following theorem.

Theorem. Divide the points $(\alpha_n, \ldots, \alpha_1)$ of an n-dimensional space into three classes as follows. In Class 1a put those point which satisfy the condition:

$$Q \leqq e^{u^\varepsilon}, \quad \delta_0 \leqq e^{u^\varepsilon}.$$

In Class 1b, put those points which are not points of Class 1a but satisfy the condition:

$$Q \leqq P^{0.2\nu}, \quad \delta_0 \leqq P^\nu.$$

Finally, put all the remaining points in Class 2.

For the points of Class 1a, assume ($\varepsilon_3 = 2\varepsilon'$) that

$$\Delta_1 = u^{10\varepsilon} Q^{-0.5\nu+\varepsilon_3}$$

or

$$\Delta_1 = u^{10\varepsilon} \delta_0^{-0.5\nu}, \quad \text{if} \quad \delta_0 \geqq 1.$$

If $\varepsilon_4 = 3\varepsilon'$, for the points of Class 1b, assume that

$$\Delta_1 = Q^{-0.5\nu+\varepsilon_4}, \qquad \text{if} \quad Q > e^{u^\varepsilon},$$

$$\Delta_1 = Q^{-0.5\nu+\varepsilon_4}\delta^{-0.5\nu+\varepsilon_4}, \quad \text{if} \quad \delta_0 > e^{u^\varepsilon}.$$

Finally, for the points of Class 2, assume that

$$\Delta_1 = P^{-\varrho_2}.$$

Let σ be some number such that $0 < \sigma < 1$, and that the number D of primes not exceeding P, which satisfy the condition $\{f(p)\} < \sigma$, be represented in the form:

$$D = \sigma\pi(P) + \lambda.$$

Then

$$\lambda \ll Pu^{-1}\Delta_1.$$

Proof. Put

$$T'_m = \sum_{p \leqq P} e^{2\pi i m f(p)}.$$

Without loss of generality, we can take $\Delta_1 \leqq 0.1$, otherwise the theorem is trivial.

Taking any arbitrary α and β such that $\Delta \leqq \beta - \alpha \leqq 1 - \Delta$ and putting $r = 1$, consider the periodic function $\psi(x)$ of period 1 defined in Lemma 2 of Chapter 2. Then we have

$$\psi(f(x)) = 1, \quad \text{if} \quad \alpha + 0.5\Delta \leqq f(x) \leqq \beta - 0.5\Delta \pmod 1,$$

$$0 < \psi(f(x)) < 1, \quad \text{if} \quad \alpha - 0.5\Delta < f(x) < \alpha + 0.5\Delta \pmod 1$$

$$\text{or} \quad \beta - 0.5\Delta < f(x) < \beta + 0.5\Delta \pmod 1,$$

$$\psi(f(x)) = 0, \quad \text{if} \quad \beta + 0.5\Delta \leqq f(x) \leqq 1 + \alpha - 0.5\Delta \pmod 1,$$

$$\psi(f(x)) = \beta - \alpha + \sum_{m=1}^{\infty} (g_m e^{2\pi i m f(x)} + h_m e^{-2\pi i m f(x)}), \tag{1}$$

$$g_m \ll \frac{1}{m}, \quad h_m \ll \frac{1}{m}, \qquad \text{if} \quad m \leqq \Delta^{-1},$$

$$g_m \ll \frac{1}{\Delta m^2}, \quad h_m \ll \frac{1}{\Delta m^2}, \quad \text{if} \quad m > \Delta^{-1}.$$

Assuming $U(\alpha, \beta) = \sum_{p \leqq P} \psi(f(p))$, from (1) we obtain

$$U(\alpha, \beta) = (\beta - \alpha)\pi(P) + H;$$

$$H \ll \sum_{0 < m \leqq \Delta^{-1}} \frac{|T_m|}{m} + \sum_{\Delta^{-1} < m \leqq \Delta^{-2}} \frac{|T_m|}{\Delta m^2} + \sum_{\Delta^{-2} < m} \frac{Pu^{-1}}{\Delta m^2}. \tag{2}$$

Hence, applying Theorem 2 of Chapter 7, if $\mu = 1$, we obtain

$$H \ll \sum_{0 < m \leqq \Delta^{-1}} \frac{Pu^{-1}\Delta}{m} + \sum_{\Delta^{-1} < m \leqq \Delta^{-2}} \frac{Pu^{-1}\Delta}{\Delta m^2} + \sum_{\Delta^{-2} < m} \frac{Pu^{-1}}{\Delta m^2}$$

$$\ll Pu^{-1}\Delta \ln \frac{1}{\Delta} \ll \frac{P}{u}\Delta_1.$$

If, however, $\mu = (m, Q)^{0.5\nu}$, the sum of the terms in the first two sums on the right-hand side of (2) which satisfy the condition $(m, Q) = d$, is equal to

$$Pu^{-1}\Delta \left(d^{0.5\nu - 1} \sum_{0 < m_1 \leqq \Delta^{-1} d^{-1}} \frac{1}{m_1} + d^{0.5\nu - 2} \sum_{\Delta^{-1} d^{-1} < m_1 \leqq \Delta^{-2} d^{-1}} \frac{1}{\Delta m_1^2} \right) \ll \frac{P}{u} \Delta d^{0.5\nu - 1} \ln \frac{1}{\Delta}.$$

Since the total number of values of d is not greater than $\tau(Q) \ll Q^{\varepsilon''}$ (Lemma 5, Chapter 1), we find

$$H \ll Pu^{-1}\Delta_1.$$

Finally, when $\mu = m^{-0.5\nu}$, we have

$$H \ll \sum_{0 < m \leqq \Delta^{-1}} \frac{Pu^{-1}\Delta}{m^{1+0.5\nu}} + \sum_{\Delta^{-1} < m \leqq \Delta^{-2}} \frac{Pu^{-1}\Delta}{\Delta m^{2+0.5\nu}} + \sum_{\Delta^{-2} < m} \frac{Pu^{-1}}{\Delta m^2}$$

$$\ll Pu^{-1}\Delta \ll Pu^{-1}\Delta_1.$$

Consequently, in all the cases, we have

$$U(\alpha, \beta) = (\beta - \alpha)\pi(P) + O(Pu^{-1}\Delta_1). \tag{3}$$

In the interval $0 \leqq \beta - \alpha \leqq 1$, we shall use the symbol $D(\alpha, \beta)$ to denote the number of values of p which satisfy the condition:

$$\alpha \leqq f(p) < \beta \,(\mathrm{mod}\, 1).$$

If $2\Delta \leqq \beta - \alpha \leqq 1 - 2\Delta$, from the obvious inequalities

$$U(\alpha + 0.5\Delta, \beta - 0.5\Delta) \leqq D(\alpha, \beta) \leqq U(\alpha - 0.5\Delta, \beta + 0.5\Delta)$$

and from (3), it follows that

$$D(\alpha, \beta) = (\beta - \alpha)\,\pi(P) + O(Pu^{-1}\Delta_1), \tag{4}$$

which we shall extend to the remaining cases as well: to the case where $0 < \beta - \alpha \leqq 2\Delta$ by the identity:

$$D(\alpha, \beta) = D(\alpha, \alpha + 1 - 2\Delta) - D(\beta, \alpha + 1 - 2\Delta),$$

and to the case where $1 - 2\Delta \leqq \beta - \alpha \leqq 1$ by the identity:

$$D(\alpha, \beta) = D(\alpha, \alpha + 2\Delta) + D(\alpha + 2\Delta, \beta).$$

In (4), putting $\alpha = 0$ and $\beta = \sigma$, we find that our theorem is true.

Chapter 9. The Asymptotic Formula in Waring's Problem with Prime Numbers

In this chapter we shall derive an asymptotic expression for the number I_N of representations of a positive integer N in the form:

$$N = p_1^n + \ldots + p_r^n,$$

where $p_1, \ldots, p_r$ are prime numbers.

In structure the proof is similar to the procedure used in Chapter 6 to derive the asymptotic formula in the Waring problem. Theorem 2 of Chapter 7 plays the key role in our proof. To obtain the principal term, we shall use the following simple version of Page's theorem which I state here without proof.

Lemma 1 (Page's Lemma). Let c be an arbitrarily large constant greater than unity. The total number $\pi(N, Q, l)$ of primes not exceeding N contained in the arithmetic progression:

$$Qx + l; \quad 0 < Q \leqq (\ln N)^{2-\varepsilon_0};$$

$$(l, Q) = 1, \quad 0 \leqq l < Q,$$

is given by the expression:

$$\pi(N, Q, l) = \frac{1}{Q_1} \int_2^N \frac{dx}{\ln x} + O\left(\frac{N}{Q_1 \ln N} (\ln N)^{-c}\right);$$

$$Q_1 = \varphi(Q).$$

Theorem. Let $n \geqq 5$, r be a constant satisfying the condition $r \geqq r_2$;

$$r_2 = [2n^2(2\ln n + \ln\ln n + 5)].$$

Then, the number I_N' of representations of a positive integer N in the form

$$N = p_1^n + \ldots + p_r^n$$

is given by the expression:

$$I_N' = H(N)\, \mathfrak{S}' + O\left(\frac{N^{r\nu-1}}{u_0^r} u_0^{-10n}\right);$$

$$H(N) = \int_{-\infty}^{\infty} (J(z))^r e^{-2\pi i z N}\, dz,$$

$$J(z) = \int_2^{P_0} \frac{e^{2\pi i z x^n}}{\ln x}\, dx, \quad P_0 = N^{\nu}, \quad u_0 = \ln P_0,$$

$$\mathfrak{S}' = \sum_{Q=1}^{\infty} A(Q); \quad A(Q) = \sum_{\substack{(a, Q) = 1 \\ 0 \leqq a < Q}} \left(\frac{S_{a,Q}'}{Q_1}\right)^r e^{-2\pi i \frac{a}{Q} N},$$

$$S_{a,Q}' = \sum_{\substack{(l, Q) = 1 \\ 0 \leqq l < Q}} e^{2\pi i \frac{a}{Q} l^n}, \quad Q_1 = \varphi(Q),$$

where

$$H(N) = K(N)\frac{N^{rv-1}}{u_0^r}; \quad K(N) = \frac{(\Gamma(1+v))^r}{\Gamma(rv)} + O\left(\frac{\ln u_0}{u_0}\right).$$

Proof. Assuming $P = [N^v]$, $u = \ln P$, we find that

$$I_N' = \int_{-P^{-n+v}}^{1-P^{-n+v}} S_\alpha^r e^{-2\pi i \alpha N}\,d\alpha; \quad S_\alpha = \sum_{p \leqq P} e^{2\pi i \alpha p^n}.$$

Divide the range of integration into intervals of three classes as follows. Express each α as

$$\alpha = (a/Q) + z; \quad z = \delta P^{-n}, \quad (a, Q) = 1, \quad 0 \leqq a < Q.$$

In Class I put those intervals which contain the values of α satisfying the condition $Q \leqq u^{2-\varepsilon_0}$, $|\delta| \leqq u^2$.

Those intervals containing the values of α not contained in the intervals of Class I but satisfying the condition $Q \leqq e^{u^\varepsilon}$, $|\delta| \leqq e^{u^\varepsilon}$ are put into Class II.

Finally, the intervals which contain all the remaining values of α are put into Class III. In accordance with this division of the integration range, the integral I_N' can be expressed as the following sum of three subintegrals:

$$I_N' = I_{N,1}' + I_{N,2}' + I_{N,3}'. \tag{1}$$

We shall estimate $|S_\alpha|$, using Theorem 2 of Chapter 7 (for $m = 1$), where we shall now take $\alpha_n = \alpha$ and $\alpha_s = 0$: if $s < n\delta$ we shall in addition put that $\delta_0 = |\delta|$.

Therefore for the intervals of Class I and Class II, we have

$$S_\alpha \ll P u^{-1} u^{9\varepsilon} Q^{-0.5v+\varepsilon'}$$

or $$S_\alpha \ll P u^{-1} u^{9\varepsilon} \delta_0^{-0.5v}, \quad \text{if } \delta_0 \geqq 1.$$

But, for the intervals of Class III, we have

$$S_\alpha \ll P u^{-1} e^{-0.5 v u^\varepsilon + \varepsilon_3 u^\varepsilon}.$$

Putting $r_0 = 2[n^2(2\ln n + \ln\ln n + 5)]$, in the intervals of Class III we obtain

$$S_\alpha^r \ll \left(\frac{P}{u}\right)^{r-r_0} e^{(r-r_0)(-0.5vu^\varepsilon + \varepsilon_3 u^\varepsilon)} |S_\alpha|^{r_0}.$$

Therefore, since $r_2 - r_0 > 2n^2$, applying Lemma 1 of Chapter 6 (see the note to that lemma), we obtain

$$I_{N,3} \ll \left(\frac{P}{u}\right)^{r-r_0} e^{2n^2(-0.5vu^\varepsilon + \varepsilon_3 u^\varepsilon)} P^{r_0 - n}$$

$$\ll \frac{P^{r-n}}{u^r} u^{r_0} e^{-4u^\varepsilon} \ll \frac{P^{r-n}}{u^r} u^{-10n}.$$

Now let α belong to an interval of Class I which contains the fraction a/Q. Then, taking ε', ε_0 (Lemma 1) and ε sufficiently small ($\varepsilon' \leqq 0.05v$, $\varepsilon_0 \leqq 0.1$, $9\varepsilon \leqq (2-\varepsilon_0) \times (0.1v - \varepsilon')$), we have

$$S_\alpha \ll \frac{P}{u} Q^{-0.4v'}, \text{ if } \delta_0 \leqq Q, \quad S_\alpha \ll \frac{P}{u} \delta_0^{-0.4v}, \text{ if } \delta_0 \geqq Q.$$

Consequently, that part of the integral $I_{N,2}$ which corresponds to the chosen interval is

$$\ll \int_0^{QP^{-n}} |S_\alpha|^r\, dz + \int_{QP^{-n}}^{e^{u^\varepsilon} P^{-n}} |S_\alpha|^r\, dz \ll P^{-n} \int_0^{Q} |S_\alpha|^r\, d\delta_0 + P^{-n} \int_Q^{e^{u^\varepsilon}} |S_\alpha|^r\, d\delta_0$$

$$\ll \frac{P^{r-n}}{u^r} Q^{-0.4\nu r+1}.$$

Hence, first summing over all a for given Q, and then over all Q satisfying the condition $u^{2-\varepsilon_0} \leqq Q \leqq e^{u^\varepsilon}$, we obtain:

$$I_{N,2} \ll \frac{P^{r-n}}{u^r} (u^{2-\varepsilon_0})^{-0.4\nu r+3} \ll \frac{P^{r-n}}{ur} u^{-10n}.$$

Now let α belong to an interval of Class I which contains the fraction a/Q. Taking an arbitrarily large constant h greater than 1, divide the sum S_α into $\ll u^h$ subsums of the type

$$S_{\alpha, M} = \sum_{M < p \leqq M_1} e^{2\pi i \left(\frac{a}{Q} + z\right) p^n}; \quad 0 < M_1 - M \leqq Pu^{-h},\ M \geqq 2.$$

Furthermore, we find

$$S_{\alpha, M} = \sum_{\substack{0 \leqq l < Q \\ (l, Q) = 1}} S^{(l)}; \quad S^{(l)} = \sum_{\substack{M < p \leqq M_1 \\ p \equiv l \pmod{Q}}} e^{2\pi i \left(\frac{a}{Q} l^n + zp^n\right)}.$$

But, as $0 < x \leqq Pu^{-h}$, we have

$$z((M+x)^n - M^n) \ll u^{2-h}$$

and, besides, by Lemma 1, the number of terms in the sum $S^{(l)}$ is equal to

$$\frac{1}{Q_1} J_M + O\left(\frac{Pu^{-c}}{Q_1 u}\right); \quad J_M = \int_M^{M_1} \frac{dx}{\ln x}.$$

Therefore, we obtain

$$S^{(l)} = \frac{e^{2\pi i \frac{a}{Q} l^n}}{Q_1} \int_M^{M_1} \frac{e^{2\pi i z x^n}}{\ln x}\, dx + O\left(\frac{Pu^{-c-1}}{Q_1} + \frac{J_M}{Q_1} u^{2-h}\right),$$

$$S_{\alpha, M} = \frac{S'_{a,Q}}{Q_1} \int_M^{M_1} \frac{e^{2\pi i z x^n}}{\ln x}\, dx + O(Pu^{-c-1} + J_M u^{2-h});$$

$$S'_{a,Q} = \sum_{\substack{0 \leqq l < Q \\ (l, Q) = 1}} e^{2\pi i \frac{a}{Q} l^n}.$$

Hence, summing over all $\ll u^h$ values of M, and putting $c = 2h - 2$, we obtain (because $P - P_0 \ll 1$):

$$S_\alpha = \frac{S'_{a,Q}}{Q_1} J(z) + O(P_0 u^{1-h}).$$

But, according to Lemma 5 of Chapter 2, we have

$$J(z) \ll Z; \quad Z = \frac{P_0}{u_0} D_1;$$
$$D_1 = \min(1, \delta_0^{-\nu}), \quad \text{if} \quad \delta_0 \leqq P_0^{n-1},$$
$$D_1 = u_0 \delta_0^{-\nu}, \qquad \text{if} \quad \delta_0 > P_0^{n-1}.$$

Consequently,

$$S_\alpha^r e^{-2\pi i \alpha N} = \left(\frac{S'_{a,Q}}{Q_1}\right)^r e^{-2\pi i \frac{a}{Q} N} (J(z))^r e^{-2\pi i z N} + O(Z^{r-1} P_0 u^{1-h}).$$

Therefore, that part of the integral $I_{N,1}$ which corresponds to the chosen interval is

$$\left(\frac{S'_{a,Q}}{Q_1}\right)^r e^{-2\pi i \frac{a}{Q} N} \int_{-u^2 P^{-n}}^{u^2 P^{-n}} (J(z))^r e^{-2\pi i z N} dz + O\left(\int_0^{u^2 P^{-n}} Z^{r-1} P_0 u^{1-h} dz\right).$$

Summing this expression over all values of a satisfying the condition

$$0 \leqq a < Q; \quad (a, Q) = 1,$$

and then over all values of Q satisfying the condition $Q \leqq u^{2-\varepsilon_0}$, we obtain

$$I'_{N,1} = \sum_{Q \leqq u^{2-\varepsilon_0}} A(Q) \int_{-u^2 P^{-n}}^{u^2 P^{-n}} (J(z))^r e^{-2\pi i z N} dz + R;$$

$$R \ll u^4 \int_0^{n^2 P - u} Z^{r-1} P_0 u^{1-h} dz \ll u^4 P^{-n} \int_0^{u^2} \frac{P_0^r}{u_0^r} D_1^{r-1} u_0^{2-h} d\delta_0$$

$$\ll \frac{P_0^{r-n}}{u_0^r} u^{6-h} \ll \frac{P_0^{r-n}}{u_0^r} u_0^{-10n}, \tag{2}$$

if we take $h = 10n + 6$.

According to Theorem 3 of Chapter 3, we find

$$A(Q) \ll Q^{1-\nu r+\varepsilon' r} \ll Q^{-17n};$$

$$\sum_{Q=1}^{\infty} Q^{-17n} \ll 1, \qquad \sum_{Q \geqq u^{2-\varepsilon_0}} Q^{-17n} \ll u^{-10n}.$$

Therefore the first factor in the main term of (2) is equal to

$$\mathfrak{S}' + O(u^{-10n}); \quad \mathfrak{S}' \ll 1.$$

Furthermore, we find

$$\int_2^{\infty} \left(\frac{P}{u}\right)^r D_1^r\, dz \ll \frac{P^{r-n}}{u^r},$$

$$\int_{u^2 P^{-n}}^{\infty} \left(\frac{P}{u}\right)^r D_1^r\, dz \ll \frac{P^{r-n}}{u^r} u^{-10n}.$$

Therefore the second factor in the main term of (2) is equal to

$$H(N) + O\left(\frac{P^{r-n}}{u^r} u^{-10n}\right); \qquad H(N) \ll \frac{P^{r-n}}{u^r}.$$

On the basis of these facts, and from (2) we obtain

$$I'_{N,1} = H(N)\ \mathfrak{S}' + O\left(\frac{P_0^{r-n}}{u_0^r} u_0^{-10n}\right).$$

Now, formula (1) gives

$$I'_N = H(N)\ \mathfrak{S}' + O\left(\frac{P^{r-n}}{u_0^r} u_0^{-10n}\right).$$

Now we prove the equality asserted for $H(N)$ in our theorem. Assuming

$$I_1(z) = \frac{1}{u}\int_0^{P_0} e^{2\pi i z x^n}\,dx,$$

for $0 \leqq z \leqq u^2 P^{-n}$, we find

$$J(z) - I_1(z) = \int_2^{Pu^{-2}} e^{2\pi i z x^n}\,\frac{u-\ln x}{u\ln x}\,dx + \int_{Pu^{-2}}^{P_0} e^{2\pi i z x^n}\,\frac{u-\ln x}{u\ln x}\,dx - \frac{1}{u}\int_0^2 e^{2\pi i z x^n}\,dx$$

$$\ll Pu^{-2} + P\frac{\ln u}{u^2} + \frac{1}{u} \ll P\frac{\ln u}{u^2} \ll \frac{P}{u}\frac{\ln u}{u}.$$

Furthermore, we find

$$\int_{-u^2P^{-n}}^{u^2P^{-n}} (J(z))^r e^{-2\pi i z N}\,dz - \int_{-u^2P^{-n}}^{u^2P^{-n}} (I_1(z))^r e^{-2\pi i z N}\,dz$$

$$\ll \int_0^{u^2P^{-n}} \frac{P^r}{u^r} D_1^{r-1}\frac{\ln u}{u}\,dz \ll \frac{P^{r-n}}{u^r}\frac{\ln u}{u}.$$

But, by Lemma 3 of Chapter 6, we have

$$\int_{-\infty}^{\infty} (I_1(z))^r e^{-2\pi i z N}\,dz = \frac{(\Gamma(1+\nu))^r}{\Gamma(r\nu)}\frac{P_0^{r-n}}{u_0^r} + O\left(\frac{P_0^{r-n-0.25}}{u_0^r}\right).$$

Moreover, applying the inequalities

$$J(z) \ll \frac{P_0}{u} D_1,\quad I_1(z) \ll \frac{P_0}{u} D;\quad D = \min(1, |\delta|^{-\nu})$$

(the first of which was already used in this proof, while the second follows from the inequality $I_\alpha \ll P_0 D$ used in the proof of Lemma 3 of Chapter 6), we find that

$$\int_{u^2P^{-n}}^{\infty} |J(z)|^r\,dz \ll \frac{P_0^{r-n}}{u_0^r} u_0^{-10n},$$

$$\int_{u^2P^{-n}}^{\infty} |I_1(z)|^r\,dz \ll \frac{P_0^{r-n}}{u_0^r} u_0^{-10n}.$$

From these facts, we easily derive the equality for $H(N)$ asserted in the theorem.

References*

1*. Vinogradov, I.M.: Elements of number theory. Izdat. "Nauka", Moscow, 1965
2*. Vinogradov, I.M.: Some theorems in analytic number theory, Dokl. Akad. Nauk SSSR (1934, no. 4) 185–187
3*. Vinogradov, I.M.: On the upper bound for $G(n)$ in Waring's problem, Izv. Akad. Nauk SSSR (ser. 7, cl. fiz. mat.) (1934, no. 10) 1455–1469
4*. Vinogradov, I.M.: On Weyl sums. Mat. Sbornik *42* (1935) 521–530
5. Vinogradov, I.M.: Some theorems concerning the theory of primes. Mat. Sbornik *2* (*44*) no. 2 (1937) 179–195 (in English)
6*. Vinogradov, I.M.: Some general theorems relating to the theory of primes. Trudy Tbilisskogo Mat. Inst. *3* (1937) 1–37
7*. Vinogradov, I.M.: Certain general lemmas and their application in estimation trigonometrical sums. Mat. Sbornik *3* (*45*) (1938) 435–471 (English summary)
8*. Vinogradov, I.M.: On the estimates of trigonometric sums. Dokl. Akad. Nauk SSSR *34* no. 7 (1942) 199–200
9*. Vinogradov, I.M.: An improvement of the estimates of trigonometric sums. Izv. Akad. Nauk SSSR Ser. Mat. *6* (1942) 33–40
10*. Vinogradov, I.M.: The method of trigonometric sums in number theory. Trudy Mat. Inst. Steklov *23* (1947) 1–109 (Part of [12])
11*. Vinogradov, I.M.: General theorems on the upper bound for the modulus of trigonometrical sums. Izv. Akad. Nauk SSSR Ser. Mat. *15* (1951) 109–130
12*. Vinogradov, I.M.: Selected works, Izdat. Akad. Nauk SSSR, Moscow, 1952
13*. Vinogradov, I.M.: A new estimate for $\zeta(1+it)$. Izv. Akad. Nauk SSSR Ser. Mat. *22* (1958) 161–164
14*. Vinogradov, I.M.: Certain problems in analytical number theory Trudy Tret'ego Vsesoyuznogo mat. s'ezda, vol. 3 (1958)
15*. Vinogradov, I.M.: On the distribution of fractional parts of the values of a polynomial. Izv. Akad. Nauk SSSR Ser. Mat. *25* (1961) 749–754
16*. Hua, L.-K.: Additive theory of prime numbers. Trudy Mat. Inst. Steklov *22* (1947) 1–179 (revised English translation: see [16a])
16a. Hua, L.K.: Additive theory of prime numbers. Translations of mathematical monographs *13*, American Mathematical Society, Providence, Rhode Island, 1965 (revised English translation of [16])
17. Hua, L.K.: Die Abschätzung von Exponentialsummen und ihre Anwendung in der Zahlentheorie. In: Enzyklopädie der mathematischen Wissenschaften mit Einschluß ihrer Anwendungen. Band I, 2, Heft 13, Teil 1. B.G. Teubner Verlagsgesellschaft, Leipzig, 1959 (in German)

* All asterisked items were originally published in Russian.

Section III

Special Variants of the Method of Trigonometric Sums

Moscow, Nauka (1976) 120 p.

Contents

Notation

It is assumed that the reader is acquainted with my book[1] and the notation used therein. Besides, we shall also use the following notation:

C is a constant,
c is a positive constant number,
ε is an arbitrarily small positive constant less than unity, and
θ is a number such that $|\theta| \leqq I$.

For positive B, the symbolic inequality $A \ll B$ denotes that, for some c, we have $|A| \leqq cB$. For positive A and B, the symbolic inequality $B \gg A$ also means the same.

For a real h, the symbol (h) stands for the positive difference between h and the nearest integer, i.e. $\min(\{h\}, 1 - \{h\})$.

The symbol $\sum_a$ stands for the sum taken over the values of a specified under the symbol.

The punctuation mark ';' between two formulas stands for the word 'where'.

Introduction

The new method which I discovered in 1934 in the analytical theory of numbers was rapidly improved and its field of application widened. In the simple variants of this method, a key role is played by the estimates of sums of the type:

$$\sum_{\Omega}\sum \xi(u)\,\eta(v)\,\Phi(u,v), \tag{1}$$

where $\Phi(u,v) = e^{2\pi i f(u,v)}$ and $f(u,v)$ is a real-valued function (certain other types of functions can also be used as $\Phi(u,v)$). Double summation is taken over the integral points in a given domain Ω.

In 1934 in this way I made an important step forward in the Waring problem, by obtaining $G(n) = O(n \ln n)$ in place of the previous result $G(n) = O(n2^n)$. We know that it is impossible to obtain an order better than $G(n) = O(n)$. In the same year I was able to succeed in making considerable advance in another question as well, i.e. the problem of the number of integral points in a three-dimensional domain; in particular, as regards the asymptotic expression $T = V + R$ for the number of integral points inside a sphere $x^2 + y^2 + z^2 \leqq a^2$ in terms of its volume V. Instead of the old result $R = O(a^{3/2})$, I derived a new expression $R = O(a^{1.4+\varepsilon})$, which I improved to $R = O(a^{4/3}(\ln a)^6)$ in 1963.

Experience in building the scheme for the solution of the Waring problem helped me in finding a scheme for solving the problem of approximating any $\delta\,(0 \leqq \delta < 1)$ to a sufficient degree of accuracy by means of the fractional parts of the values of a polynomial. This scheme, in turn, suggested to me the use of an almost similar procedure to estimate Weyl sums, i.e. sums of the type:

$$T(\alpha_n, \ldots, \alpha_1) = \sum_{0<x\leqq P} e^{2\pi i(\alpha_n x^n + \ldots + \alpha_1 x)}. \tag{2}$$

My first new estimates of the Weyl sums were published in 1935. In 1938 I replaced the complicated estimate of the power function

$$|T(\alpha_n, \ldots, \alpha_1)|^{2b}$$

which lies at the base of these new estimates of the Weyl sums, by a quite simple estimate of the integral

$$\int_0^1 \ldots \int_0^1 |T(\alpha_n, \ldots, \alpha_1)|^{2b}\, d\alpha_n \cdots d\alpha_1.$$

The theorem which gives the estimate of this integral is known as the 'mean value theorem'. Both the new estimates of the Weyl sums and the mean value theorem were repeatedly improved in the subsequent years (they were radically refined in the monograph[2]).

In 1937 I found a new method for obtaining non-trivial estimates of sums of the type:

$$\sum_{p\leqq N} \Phi(p), \tag{3}$$

using the fact that these sums can be formed only by addition and subtraction of a relatively small number (as compared with N) of double sums of the type (1) and sums of the type:

$$\sum_{m'<m\leqq m''} \Phi(dm). \tag{4}$$

The sum (3) can be formed from simpler sums with the help of the following identity (or its generalizations):

$$\Phi(1) + \sum_{H<y\leqq N} \Phi(y) = \sum\sum_{dm\leqq N} \mu(d)\,\Phi(dm), \tag{5}$$

where H is any number such that $1 < H \leqq \sqrt{N}$, y runs through the numbers which are not divisible by primes not greater than H, d runs through the products of primes (including the null product, unity) not greater than H, and m runs through natural integers.

Identity (5) is known since long. Its simplest proof is based on the sieve of Eratosthenes. A corrollary of this identity is the famous Euler identity (p runs through all primes):

$$\sum_{m=1}^{\infty} \frac{1}{m^s} = \prod \frac{1}{1-\frac{1}{p^s}} = \frac{1}{\sum\limits_{d=1}^{\infty} \frac{\mu(d)}{d^s}}; \qquad s = \sigma + it, \quad \sigma > 1$$

(obtained by substituting $\Phi(m) = m^{-s}$ and then taking the limit) which lies, besides its generalizations to L-series, at the base of the modern theory of zeta functions and L-series. Special modifications of (5) underlie the Brun method.

My 1937 method opened broad vistas to the solution of a wide class of problems in number theory that defied solution by other methods.

Using this method, in 1937 I found a non-trivial estimate of the sum:

$$\sum_{p\leqq N} e^{2\pi i\alpha p},$$

with the help of which I derived an asymptotic expression for the number of representations of an odd number N in the form of the sum of three odd primes (from which the Goldbach ternary representation follows as a particular case for all sufficiently large N).

In the same year I also found a non-trivial estimate for the sum:

$$T'(\alpha_n, \ldots, \alpha_1) = \sum_{p\leqq N} e^{2\pi i(\alpha_n p^n + \ldots + \alpha_1 p)}, \tag{6}$$

which is similar to the Weyl sum, except that here summation is taken only over primes. The asymptotic formula in the Waring problem in primes follows as a corollary from this estimate.

Furthermore, non-trivial estimates for various other types of sums over primes (the sum $\sum_{p\leqq N} \chi(p+k)$, in particular) were found with the help of this method. Various other applications of these estimates were also investigated.

A need arose for a systematic presentation of the results obtained by my method. Thus, the monograph [2] appeared, and its aim, as mentioned in its Introduction, was to give a profound insight into my 1934 method as applied to a few selected important problems in the analytical theory of numbers. The problems chosen for this purpose were those in the presentation of which special variants of my method based on the mean value theorem are essential. This made the monograph, as regards presentation, uniform.

In this monograph we shall examine those problems in which simpler variants of my method that are not based on the mean value theorem play the decisive part. It was found convenient to arrange the monograph into six chapters in accordance with the various types of these variants.

Chapter 1 deals with the number of integral points in a three-dimensional domain.

Chapter 2 gives the estimate of $G(n)$ in the Waring problem.

Chapter 3 is devoted to the question of approximating a given δ $(0 \leqq \delta < 1)$ by the fractional parts of the values of a polynomial.

Chapter 4 gives the estimates for sums of the type:

$$\sum_{p \leqq N} e^{2\pi i \alpha p}, \qquad \sum_{k=1}^{K} \left| \sum_{p \leqq N} e^{2\pi i k \alpha p} \right|, \qquad \sum_{p \leqq N} e^{2\pi i k f \sqrt{p}}, \qquad \sum_{p \leqq N} \chi(p+k),$$

and some of their applications.

Chapter 5 contains the derivation of the asymptotic formula of Hardy and Littlewood in the Goldbach problem.

Chapter 6 deals with an elementary variant of my method.

Detailed information concerning each chapter is given at the beginning of the chapter.

Chapter 1. On the Number of Integral Points in a Three-Dimensional Domain

From the problem on the mean value of arithmetic functions formulated by Gauss, a problem arose in the first half of the 19th century of finding an asymptotic expression of the sum:

$$F(I) + \ldots + F(N),$$

where $F(n)$ is some arithmetic function, and N may grow unboundedly. This problem, in turn, gave rise to another problem, to which a great number of cases of the former problem are reduced, viz. that of finding an asymptotic expression for the number of integral points in certain domains of two, three or more dimensions.

Thus, Gauss himself studied the number T of integral points in a circle $x^2 + y^2 \leqq N$, and derived the following formula for T:

$$T = \pi N + R; \quad R = O(\sqrt{N}). \tag{1}$$

Dirichlet studied the number T' of integral points in an rectangular hyperbola $x > 0$, $y > 0$, $xy \leqq N$, and gave the following formula for T':

$$T' = N(\ln N + 2E - 1) + R; \quad R = O(\sqrt{N}), \tag{2}$$

where E is the Euler constant.

Only in the beginning of the 20th Century were more exact results obtained with the help of an elementary method elaborated by Voronoi. In 1903 he[3] obtained the estimate $R = O(N^{1/3} \ln N)$ for the remainder term in (2), while in 1906 Sierpinsky gave $R = O(N^{1/3})$ for the remainder term in (1). Subsequently, the present author developed his elementary method (1917), with the help of which, imposing only one restriction

$$\frac{1}{A} \leqq f''(x) \leqq \frac{k}{A} \tag{3}$$

(k is a constant and A may grow unboundedly) on the function $f(x)$, he derived the formula

$$\sum_{q < x \leqq r} \{f(x)\} = \frac{1}{2}(r - q) + R; \quad R = O\left(\left(\frac{r-q}{A} + 1\right)(A \ln A)^{2/3}\right), \tag{4}$$

which is useful in finding the number of integral points in a wide class of two-dimensional domains. Rectangular hyperbolic and circular domains mentioned in the previous paragraph fall under this class as particular cases (of course, with somewhat worsened estimates of R). Later, Yarnik showed that the expression (4) for a function satisfying the condition (3) cannot be improved any further to an accuracy up to the logarithmic factor.

Analytical methods were also developed side by side with elementary techniques. Some information about them can be found in Hua's monograph. In this chapter an important part is played by the method based on the expansion of $\{f(x)\}$ (or functions similar to it) in the form of a Fourier series and on the estimate of a sum of the type:

$$S_m = \sum_{q<x\leqq r} e^{2\pi i m f(x)},$$

where m is an integer, or on the approximation of this sum by a shorter sum. The idea underlying this method is obvious from Voronoi's paper[5]; Hardy and Littlewood[6] studied a particular case of approximating S_m by a shorter sum and I have applied this method in a rough form to particular two and three dimensional domains[7].

The method illustrating the application of a lemma on approximating S_m by a shorter sum to two-dimensional domains (that is sufficiently exact for this purpose) is given in paper[8] by Corput. Using the rectangular hyperbola as an example, he has shown that the application of the main method along with the estimate of shorter sums by the Weyl method gives $R = O(N^{0.33})$ in place of $R = O(N^{1/3} \ln N)$. Subsequently, the estimates of R for a rectangular hyperbola as well as for a circle were refined. Best results were obtained in the case of a circle. For example, Hua[9] found that $R = O(N^{13/40})$. Hardy and Littlewood have demonstrated that it is impossible to obtain an estimate better than $R = O(N^{1/4} (\ln N)^2)$.

As applied to three-dimensional domains, the main method built on a scheme which is an improvement of the procedure used in my paper[7] and has greater flexibility. With the help of this method we can study several cases with sufficient accuracy [10]; thus, we obtain $R = O(N^{3/4})$ instead of the trivial order $R = O(N)$ in the asymptotic expression

$$T = \tfrac{4}{3}\pi N^{3/2} + R,$$

for the number T of integral points in a sphere $x^2 + y^2 + z^2 \leqq N$. As is known, it is not possible to obtain a result better than $R = O(N^{1/2} \ln N)$.

In 1934 I found that a combination of the above method and a simple variant of the new method I discovered then yields far more accurate results[11]. In the subsequent years these primary results were improved. Thus, in 1963 I obtained[12] $R = O(N^{2/3} (\ln N)^6)$ for the case of a sphere. Derivation of this estimate is the aim of this chapter.

The lemmas used in the proof, if they appear either in[1] or in [2], will be stated without proof. The lemma on the approximation of S_m by a shorter sum is proved in an exact formulation necessary for the derivation (with additional restrictions imposed on $f(x)$).

The chapter also briefly outlines the simplest variant of the main form of the method (only the expansion of $\psi(f(x))$ as a Fourier series and a simple estimate of the sum S_m are used) as applied to two-dimensional domains. This is illustrated in the derivation of my formula (4) with an improved remainder term.

Regions of dimensions greater than three are not considered in this chapter. For the sake of completeness, however, I shall say a few words about them. Two classes of these domains have been investigated in detail.

Class I includes multi-dimensional rational ellipsoids. Here, if the volume of the ellipsoid is taken as the leading term in the asymptotic expression for the number of integral points in the domain, then for the remainder term R we obtain an unimprovable order $R = O(N^{n/2-1})$ for $n > 4$, or an almost unimprovable order $R = O(N \ln N)$ for $n = 4$ (the speculated final estimate being $R = O(N)$) with the help of a modified additive method of Hardy and Littlewood (Landau, Walfisz).

Class II includes multi-dimensional hyperboloids $x_1 \ldots x_n < N, x_1 > 0, \ldots, x_n > 0$. For this case, drawing heavily from the theory of the zeta function, Hardy and Littlewood[13] have obtained the estimate $R = O(N^{1-3/(n+2)})$ for the remainder term R in the asymptotic expression for the number of integral points. A.A. Karatsuba[14], essentially using my method as well, has derived an entirely new estimate $R = O(N^{1-c/n^{2/3}})$. The speculated final order is believed to be $R = O(N^{(n-1/2n)+\varepsilon})$.

Lemma 1. In the interval $q < x \leqq r$, let the second derivative $f''(x)$ be continuous and satisfy the conditions:

$$\frac{1}{A} \leqq |f''(x)| \leqq \frac{k}{A},$$

where k is a constant and $A \geqq 5k$. Further, let $\varphi(x) \ll H$ and be a monotonic function. Then the sum

$$S = \sum_{q<x\leqq r} \varphi(x)\, e^{2\pi i f(x)}$$

is bounded by the inequality:

$$S \ll H\left(\frac{r-q}{\sqrt{A}} + \sqrt{A} + \ln U\right); \qquad U = \max(r-q, A).$$

Proof. We shall confine ourselves only to the case $\varphi(x) = 1$ (the general case can be derived trivially from this particular case by means of the Abel transformation). Without loss of generality, suppose that $r - q > 5k\sqrt{A}, f''(x) > 0$ and, in the interval $q \leqq x \leqq r$, let

$$f'(x) \ll \frac{U}{A}$$

(otherwise, without altering the value of the sum S, we can substitute $f(x) - x[f'(r)]$ for $f(x)$).

Let $m = [4U^3 + 1]$, Q and R be the least and the greatest integers satisfying the condition $q \leqq Q - 0.5$, $R + 0.5 \leqq r$. Assuming

$$W_M = \int_{-0.5}^{0.5} \frac{\sin(2m+1)\pi x}{\sin \pi x} e^{2\pi i f(M+x)}\, dx,$$

we find that

$$\sum_{M=Q}^{R} W_M = \sum_{n=-m}^{m} I_n; \quad I_n = \int_{Q-0.5}^{R+0.5} e^{2\pi i \eta_n(x)}\, dx, \qquad (5)$$

$$\eta_n(x) = -nx + f(x).$$

Further, since

$$\int_{-0.5}^{0.5} \frac{\sin(2m+1)\pi x}{\sin \pi x}\, dx = 1,$$

we obtain

$$W_M = e^{2\pi i f(M)} + V_M;$$

$$V_M = \int_{-0.5}^{0.5} \frac{\sin(2m+1)\pi x}{\sin \pi x} \left(e^{2\pi i f(M+x)} - e^{2\pi i f(M)}\right) dx.$$

Hence, integrating by parts, we obtain

$$V_M = \int_{-0.5}^{0.5} \frac{\cos(2m+1)\pi x}{(2m+1)\pi} Y_x\,dx;$$

$$Y_x = \frac{e^{2\pi i f(M+x)}\,2\pi i f'(M+x)}{\sin \pi x} - \frac{(e^{2\pi i f(M+x)} - e^{2\pi i f(M)})\,\pi\cos\pi x}{\sin^2 \pi x}, \tag{6}$$

$$Y_x \ll U^2, \quad V_M \ll U^{-1}, \quad W_M = e^{2\pi i f(M)} + O(U^{-1}).$$

Now, consider the integral

$$I_\alpha^\beta = \int_\alpha^\beta e^{2\pi i \eta_n(x)}\,dx.$$

First, let $0 < \eta_n'(\alpha)$. Using the substitution $\eta_n(x) = u$, we get

$$I_\alpha^\beta = \int_{f'(\alpha)}^{f'(\beta)} \frac{e^{2\pi i u}}{\eta_n'(x)}\,du,$$

where $\eta_n'(x)$ should be regarded a function of u. Applying the usual technique, we shall express the real and imaginary parts of the integral I_α^β in the form of an alternating series. We find that each integral is

$$\ll \frac{1}{\eta_n'(\alpha)}.$$

At the same time, we have

$$I_\alpha^\beta \ll \frac{1}{\eta_n'(\alpha)}.$$

Besides, if $0 \leqq \eta_n'(\alpha)$, we have the following estimate

$$I_\alpha^\beta \ll \sqrt{A}.$$

Indeed, for $\beta - \alpha \leqq \sqrt{A}$, the estimate is trivial, while for $\beta - \alpha > \sqrt{A}$, it follows from $I_\alpha^\beta = I_\alpha^{\alpha+\sqrt{A}} + I_{\alpha+\sqrt{A}}^\beta$.

Now, let $\eta_n'(\beta) < 0$. Using the substitution $x = -x_1$, reduce the integral I_α^β to the form considered above. Thus, we find

$$I_\alpha^\beta \ll \frac{1}{-\eta_n'(\beta)} = \frac{1}{|\eta_n'(\beta)|}.$$

Besides, for $\eta_n'(\beta) < 0$, we have the estimate

$$I_\alpha^\beta \ll \sqrt{A}.$$

Indeed, for $\beta - \alpha \leqq \sqrt{A}$, the estimate is trivial, while for $\beta - \alpha > \sqrt{A}$, it follows from the equality $I_\alpha^\beta = I_\alpha^{\beta-\sqrt{A}} + I_{\beta-\sqrt{A}}^\beta$.

Finally, assume that $\eta_n'(\alpha) < 0 < \eta_n'(\beta)$. Then, there exists an x_n such that $\eta_n'(x_n) = 0$. Hence, we have

$$I_\alpha^\beta = I_\alpha^{x_n} + I_{x_n}^\beta,$$

from which, by what has been proved above, we obtain

$$I_\alpha^\beta \ll \sqrt{A}\,.$$

Let Q' be the largest number not greater than $Q - 0.5$ for which $f'(Q')$ is integral, and R' be the least number not less than $R + 0.5$ for which $f'(R')$ is integral. We have

$$\sum_{n=-m}^{m} I_n = T' + T'' + T''' ; \qquad T' = \sum_{n=f'(Q')-1}^{-m} I_n ,$$

$$T'' = \sum_{n=f'(R')+1}^{m} I_n , \qquad T''' = \sum_{n=f'(Q')}^{f'(R')} I_n ,$$

Hence, applying the estimates found for I_n, we obtain

$$T' \ll \sum_{s=1}^{m+f'(Q')} \frac{1}{s} \ll \ln U, \qquad T'' \ll \sum_{s=1}^{m-f'(R')} \frac{1}{s} \ll \ln U,$$

$$T''' \ll (f'(R') - f'(Q) + 1)\sqrt{A} \ll \left(\frac{\beta-\alpha}{\sqrt{A}} + \sqrt{A}\right).$$

For $H = 1$, now the lemma readily follows from (5) and (6).

Lemma 1a. (A weaker form of Lemma 3 in Chapter 2 of monograph[2].) In the interval $q \leqq x \leqq r$ let the function $f(x)$ be twice differentiable and let $f''(x)$ be of constant sign and let $|f''(x)|$ be not greater than a certain constant, and $f'(x)$ be of constant sign and for some positive δ less than 0.5 let $f'(x)$ satisfy the condition:

$$\delta \leqq |f'(x)| < 0.5\,.$$

Then

$$\sum_{q<x\leqq r} e^{2\pi i f(x)} \ll \frac{1}{\delta} + \ln(r-q) + 1\,.$$

Proof. Under the conditions stated in our lemma, repeat the arguments used in the proof of Lemma 1. From (5) and (6), we easily find that

$$\sum_{q<x\leqq r} e^{2\pi i f(x)} = \sum_{n=-m}^{m} I_n + O(1);$$

$$I_n = \int_{q_0}^{r_0} e^{2\pi i \eta_n(x)}\,dx; \qquad \eta_n(x) = -nx + f(x),$$

where, for $n = 0$, put $q_0 = q$, $r_0 = r$, and for other values of n, take $q_0 = Q - 0.5$ and $r_0 = R + 0.5$. Thus, we find that $I_0 \ll 1/\delta$ and (since $-0.5 < f'(x) < 0.5$) the sum $\sum I_n$ taken over the remaining values of n is $\ll \sum_{n=1}^{m} \frac{1}{n-0.5} \ll \ln(r-q) + 1$. This completes the proof of the lemma.

Lemma 2. Let ϱ be a positive integer, α and β be real numbers, $0 < \varDelta < 0.1$, $\varDelta \leqq \beta - \alpha \leqq 1 - \varDelta$. Then there exists a periodic function $\psi(x)$ of period 1 having the following properties:

1. $\psi(x) = 1$ in the interval $\alpha + 0.5\varDelta \leqq x \leqq \beta - 0.5\varDelta$,
2. $0 \leqq \psi(x) \leqq 1$ in the intervals $-0.5\varDelta \leqq x \leqq \alpha + 0.5\varDelta$ and $\beta - 0.5\varDelta \leqq x \leqq \beta + 0.5\varDelta$;

3. $\psi(x)=0$ in the interval $\beta+0.5\Delta \leqq x \leqq 1+\alpha-0.5\Delta$; and
4. $\psi(x)$ can be expanded as a Fourier series of the type

$$\psi(x)=\beta-\alpha+\sum_{m=1}^{\infty}(g_m e^{2\pi imx}+h_m e^{-2\pi imx}),$$

where g_m and h_m depend only on m, α, β, Δ; and

$$|g_m| \leqq \kappa_m, \quad |h_m| \leqq \kappa_m; \quad \kappa_m=\begin{cases} \dfrac{1}{\pi m}, & \text{if } m \leqq \Delta^{-1}, \\ \dfrac{1}{\pi m}\left(\dfrac{\varrho}{\pi m \Delta}\right)^{\varrho}, & \text{if } m > \Delta^{-1}. \end{cases}$$

Proof. This lemma is a modification of Lemma 2 proved in Chapter 2 of the monograph[2].

Lemma 3. Let $\delta_1, \delta_2, \ldots, \delta_Q$ be numbers such that $0 \leqq \delta_s \leqq 1$; for given ϱ, Δ and some R satisfying the condition $R > \Delta Q$, let the function $\psi(x)$ obey the conditions of Lemma 2. For the sum

$$U(\alpha, \beta)=\sum_{s=1}^{Q} \psi(\delta_s)$$

let

$$U(\alpha, \beta)-(\beta-\alpha) Q \ll R.$$

Then

(a) for any σ, such that $0<\sigma \leqq 1$, the number A_σ of the values of δ_s satisfying the inequality $\delta_s<\sigma$ is given by the expression:

$$A_\sigma=\sigma Q+R_\sigma; \quad R_\sigma \ll R,$$

and (b) we have

$$\sum_{s=1}^{Q} \delta_s-\frac{1}{2} Q \ll R.$$

Proof. For $0<\beta-\alpha \leqq 1$, let $D(\alpha, \beta)$ denote the number of values of δ_s satisfying the condition: $\alpha \leqq \delta_s \leqq \beta \pmod 1$.

When $2\Delta<\beta-\alpha \leqq 1-2\Delta$, from the obvious inequality

$$U(\alpha+0.5\Delta, \beta-0.5\Delta) \leqq D(\alpha, \beta) \leqq U(\alpha-0.5\Delta, \beta+0.5\Delta)$$

and from the conditions of the lemma, it follows that

$$D(\alpha, \beta)-(\beta-\alpha) Q \ll R, \tag{7}$$

which is readily extended to the remaining cases as well: to the case where $0<\beta-\alpha<2\Delta$ with the aid of the identity:

$$D(\alpha, \beta)=D(\alpha, \alpha+1-2\Delta)-D(\beta, \alpha+1-2\Delta),$$

and to the case where $1-2\Delta<\beta-\alpha \leqq 1$ with the aid of the identity:

$$D(\alpha, \beta)=D(\alpha, \alpha+0.5)+D(\alpha+0.5, \beta).$$

In (7) putting $\alpha=0$ and $\beta=\sigma$, we find that the assertion (a) is true.

To prove the assertion (b), first suppose that $RQ^{-1} < 0.1$ (otherwise, the assertion is trivial). Assuming that $n = [QR^{-1}]$, $v = 1/n$, by assertion (*a*), we find

$$\begin{aligned}
&A_v = vQ + R_v, && A_v = vQ + R_v,\\
&A_{2v} = 2vQ + R_{2v}, && A_{2v} - A_v = vQ + R_{2v} - R_v,\\
&A_{3v} = 3vQ + R_{3v}, && A_{3v} - A_{2v} = vQ + R_{3v} - R_{2v},\\
&\dots\dots\dots && \dots\dots\dots\dots\dots\dots\\
&A_{nv} = nvQ + R_{nv}, && A_{nv} - A_{(n-1)v} = vQ + R_{nv} - R_{(n-1)v}.
\end{aligned}$$

The sum of the products of the right-hand sides of the equalities in the second column by $0, v, 2v, \dots, (n-1)v$, respectively, gives the lower bound for the sum $\sum_\delta \delta_s$. The sum of the products of these right-hand sides by $v, 2v, 3v, \dots, nv$, gives the upper bound for the same sum. Both the upper and the lower bound can be reduced to the form $0.5Q + O(R)$. This completes the proof of assertion (b).

Lemma 4. In the interval $q \leqq x \leqq r$, let the function $f(x)$ be twice continuously differentiable, and

$$\frac{1}{A} \leqq f''(x) \leqq \frac{k}{A},$$

where k is a constant, and $A \geqq 5k$. Then

$$\sum_{q<x\leqq r} \{f(x)\} = \frac{1}{2}(r-q) + O(R); \quad R = (r-q)A^{-1/3} + A^{2/3}.$$

Proof. Without loss of generality, we shall assume that q and r are integers. Using the terms $\{f(x)\}$ of the sum on the left-hand side of the inequality being proved as the numbers δ_s, and then putting $\varrho = 1$, $\Delta = A^{-1/3}$, we shall apply the assertion (b) of Lemma 3. Our lemma is proved, if we can show that

$$\sum_{q<x\leqq r} \psi(f(x)) - \frac{1}{2}(r-q) \ll R. \tag{8}$$

By Lemma 2, we have

$$\sum_{q<x\leqq r} \psi(f(x)) = \sum_{m=1}^{\infty} (g_m U_m + h_m \overline{U}_m); \quad U_m = \sum_{q<x\leqq r} e^{2\pi i m f(x)},$$

$$g_m \ll \kappa_m, \quad h_m \ll \kappa_m; \quad \kappa_m = \begin{cases} \dfrac{1}{m}, & \text{if } m \leqq A^{1/3},\\[2mm] \dfrac{A^{1/3}}{m^2} & \text{if } m > A^{1/3}.\end{cases}$$

Since the condition

$$\frac{m}{A} \leqq mf''(x) \leqq \frac{mk}{A},$$

is satisfied in the interval $q \leqq x \leqq r$, by virtue of Lemma 1, for $m \leqq A^{2/3}$, we have

$$U_m \ll (r-q)\sqrt{\frac{m}{A}} + \sqrt{\frac{A}{m}}.$$

Therefore, we find

$$\sum_{q<x\leqq r}\psi(f(x))-\frac{1}{2}(r-q)\ll\sum_{0<m\leqq A^{1/3}}\frac{1}{m}\left((r-q)\sqrt{\frac{m}{A}}+\sqrt{\frac{A}{m}}\right)$$

$$+\sum_{A^{1/3}<m\leqq A^{2/3}}\frac{A^{1/3}}{m^2}\left((r-q)\sqrt{\frac{m}{A}}+\sqrt{\frac{A}{m}}\right)+\sum_{A^{2/3}<m}\frac{A^{1/3}}{m^2}(r-q),$$

Hence we find that the inequality (8) is valid. This completes the proof of the lemma.

Lemma 5 (Sonin formula). In the interval $Q\leqq x\leqq R$, suppose the function $f(x)$ has a continuous second derivative. Let

$$\varrho(x)=0.5-\{x\},\quad \sigma(x)=\int_0^x\varrho(z)\,dz,$$

then

$$\sum_{Q<x\leqq R}f(x)=\int_Q^R f(x)\,dx+\varrho(R)f(R)-\varrho(Q)f(Q)$$

$$-\sigma(R)f'(R)+\sigma(Q)f'(Q)+\int_Q^R\sigma(x)f''(x)\,dx.$$

Proof. See the solution of Question N8, Chapter 2 in the book[1].

Theorem 1. The number T of integral points in the circle $x^2+y^2\leqq a^2$ is

$$T=\pi a^2+O(a^{2/3}).$$

Proof. We have

$$T=1+4[a]+8\sum_{0<x\leqq\frac{a}{\sqrt{2}}}[\sqrt{a^2-x^2}]-4\left[\frac{a}{\sqrt{2}}\right]^2. \tag{9}$$

Assuming $f(x)=\sqrt{a^2-x^2}$, in the interval $0<x\leqq a/\sqrt{2}$, we find

$$f'(x)=-\frac{x}{\sqrt{a^2-x^2}},\quad f''(x)=\frac{-a^2}{(a^2-x^2)^{3/2}},\quad \frac{1}{a}\leqq|f''(x)|\leqq\frac{\sqrt{8}}{a}.$$

Further, by Lemma 4, we obtain

$$\sum_{0<x\leqq\frac{a}{\sqrt{2}}}\{\sqrt{a^2-x^2}\}=\frac{a}{2\sqrt{2}}+O(a^{2/3}).$$

And, by virtue of Lemma 5, we find

$$\sum_{0<x\leqq\frac{a}{\sqrt{2}}}\sqrt{a^2-x^2}=\frac{\pi a^2}{8}+\frac{a^2}{4}+\left(\frac{1}{2}-\left\{\frac{a}{\sqrt{2}}\right\}\right)\frac{a}{\sqrt{2}}-\frac{a}{2}+O(1).$$

In addition, we have

$$\left[\frac{a}{\sqrt{2}}\right]^2=\frac{a^2}{2}-2\frac{a}{\sqrt{2}}\left\{\frac{a}{\sqrt{2}}\right\}+O(1).$$

From these facts, and from (9) we find that the assertions of the theorem are true.

Lemma 6. Let H, U, A, q, and r be real numbers satisfying the condition:

$$H>0, \quad U \gg A \gg 1, \quad 0<r-q \leqq U.$$

Further, let $f(x)$ and $\varphi(x)$ be real-valued algebraic functions whose exponents are bounded, and in the interval $q \leqq x \leqq r$ let

$$A^{-1} \ll f''(x) \ll A^{-1}, \quad f'''(x) \ll \frac{1}{AU},$$

$$\varphi(x) \ll H, \quad \varphi'(x) \ll HU^{-1}, \quad \varphi''(x) \ll HU^{-2}.$$

Then

$$\sum_{q<x\leqq r} \varphi(x)\, e^{2\pi i f(x)} = \sum_{f'(q)\leqq n \leqq f'(r)} b_n Z_n + O(HT_q + HT_r + H\ln(U+1)),$$

where, defining x_n by the equality $f'(x_n)=n$, we obtain

$$Z_n = \frac{1+i}{\sqrt{2}} \frac{\varphi(x_n)}{\sqrt{f''(x_n)}} e^{2\pi i(-nx_n+f(x_n))},$$

Here, $b_n=1$, if n is different from $f'(q)$ and $f'(r)$, and $b_n=0.5$, if n is equal to one of these numbers. Then, we always have $Z_n \ll H\sqrt{A}$. Finally, $T_q=0$ for integral values of $f'(q)$, and is equal to

$$\min\left(\frac{1}{(f'(q))}, \sqrt{A}\right)$$

in other cases; $T_r=0$ for integral values of $f'(r)$ and is equal to

$$\min\left(\frac{1}{(f'(r))}, \sqrt{A}\right)$$

in other cases.

Proof. Assume that $r-q$ is greater than a certain sufficiently large constant exceeding three; otherwise the lemma is trivial. Besides, in the interval $q \leqq x \leqq r$, assume that the condition:

$$0 \leqq f'(x) \quad \text{and} \quad f'(x) \ll UA^{-1}$$

is satisfied; otherwise, without altering the value of the sums on the right-hand and left-hand sides of the equality being proved, we can replace $f(x)$ by a function $f(x)+gx$, choosing an integer g such that the condition $f'(r)-f'(q)<kUA^{-1}$ is satisfied.

Let $m=[4U^3+1]$, Q and R be the least and the greatest integers satisfying the conditions $q \leqq Q-0.5$, $R+0.5 \leqq r$. Assuming

$$W_M = \int_{-0.5}^{0.5} \frac{\sin(2m+1)\pi x}{\sin \pi x} \varphi(M+x)\, e^{2\pi i f(M+x)}\, dx,$$

we find

$$\sum_{M=Q}^{R} W_M = \sum_{n=-m}^{m} \int_{Q-0}^{R+0.5} \varphi(x)\, e^{2\pi i \eta_n(x)}\, dx; \quad \eta_n(x) = -nx+f(x).$$

Further, since

$$\int_{-0.5}^{0.5} \frac{\sin(2m+1)\pi x}{\sin \pi x}\, dx = 1,$$

we obtain

$$W_M = \varphi(M)\, e^{2\pi i f(M)} + V_m;$$

$$V_m = \int_{-0.5}^{0.5} \frac{\sin(2m+1)\pi x}{\sin \pi x} \left(\varphi(M+x)\, e^{2\pi i f(M+x)} - \varphi(M)\, e^{2\pi i f(M)}\right) dx,$$

from which, integrating by parts, we obtain

$$V_m = \int_{-0.5}^{0.5} \frac{\cos(2m+1)\pi x}{(2m+1)\pi} Y_x\, dx;$$

$$Y_x = \frac{e^{2\pi i f(M+x)}\left(\varphi(M+x)\, 2\pi i f'(M+x) + \varphi'(M+x)\right)}{\sin \pi x}$$

$$- \frac{\left(\varphi(M+x)\, e^{2\pi i f(M+x)} - \varphi(M)\, e^{2\pi i f(M)}\right) \pi \cos \pi x}{\sin^2 \pi x}.$$

Hence, we find

$$Y_x \ll HU^2, \quad V_m \ll HU^{-1}, \quad W_M = \varphi(M)\, e^{2\pi i f(M)} + O(HU^{-1}),$$

$$\sum_{M=Q}^{R} \varphi(M)\, e^{2\pi i f(M)} = \sum_{n=-m}^{m} \int_{Q-0.5}^{R+0.5} \varphi(x)\, e^{2\pi i \eta_n(x)}\, dx + O(H).$$

From the last formula we easily obtain the following more convenient expression:

$$\sum_{q<M\leqq r} \varphi(M)\, e^{2\pi i f(M)} = \sum_{n=-m}^{m} \int_{q_0}^{r_0} \varphi(x)\, e^{2\pi i \eta_n(x)}\, dx + O(H), \tag{10}$$

where q_0 and r_0 are defined as follows: let n_q be the value of n nearest to $f'(q)$ such that $n_q \geqq f'(q)$ and n_r be the value nearest to $f'(r)$ such that $n_r \leqq f'(r)$. Then, if $n_q \leqq n \leqq n_r$, assume that $q_0 = q$ and $r_0 = r$. In other cases, take that $q_0 = Q - 0.5$ and $r_0 = R + 0.5$.

If $\eta_n'(\alpha) > 0$, consider the integral

$$I_\alpha^\beta = \int_\alpha^\beta \varphi(x)\, e^{2\pi i \eta_n(x)}\, dx.$$

Using the substitution $\eta_n(x) = u$, reduce this integral to the form:

$$I_\alpha^\beta = \int_{\eta_n(\alpha)}^{\eta_n(\beta)} \Phi(u)\, e^{2\pi i u}\, du; \quad \Phi(u) = \frac{\varphi(u)}{\eta_n'(x)},$$

where $\eta_n'(x)$ is regarded a function of u. Since $\Phi(u)$ is an algebraic function of u of finite degree, the integral I_α^β can be divided into finite number of subintegrals, to each of which we can apply the usual technique. Thus, we obtain the estimate

$$|I_\alpha^\beta| \ll \frac{H}{\eta_n'(\alpha)}.$$

In addition to this estimate, if $\eta_n'(\alpha) \geqq 0$, we obtain the following estimate:

$$I_\alpha^\beta \ll H\sqrt{A}.$$

Indeed, for $\beta - \alpha \leqq \sqrt{A}$, the last estimate is trivial, and for $\beta - \alpha > \sqrt{A}$, this estimate follows from the equality:

$$I_\alpha^\beta = I_\alpha^{\alpha+\sqrt{A}} + I_{\alpha+\sqrt{A}}^\beta.$$

Thus, if $\eta_n'(\alpha) > 0$, we have

$$I_\alpha^\beta \ll H \min\left(\frac{1}{\eta_n'(\alpha)}, \sqrt{A}\right).$$

If $\eta_n'(\beta) < 0$, arguing in a similar manner, we obtain

$$I_\alpha^\beta \ll H \min\left(\frac{1}{-\eta_n'(\beta)}, \sqrt{A}\right).$$

Now we can estimate the sum T of the values of $I_{q0}^{r_0}$ satisfying the condition $\eta_n'(q_0) > 0\,(n < f'(q_0))$, as well as the sum T' of the values of $I_{q0}^{r_0}$ satisfying the condition $\eta_n'(r_0) < 0\,(f'(r_0) < n)$. We find

$$T \ll \sum_{n=n_{q-1}}^{-m} \min\left(\frac{H}{\eta_n'(q)}, H\sqrt{A}\right) \ll \min\left(\frac{H}{n_q - f'(q)}, H\sqrt{A}\right) + H\ln U. \tag{11}$$

By means of similar arguments, we derive

$$T' \ll \min\cdot\left(\frac{H}{-n_r + f'(r)}, H\sqrt{A}\right) + H\ln U. \tag{12}$$

Now we have only to consider the integrals I_q^r satisfying the condition $f'(q) \leqq n \leqq f'(r)$. Defining x_n by the equality $f'(x_n) = n$ $(\eta_n'(x_n) = 0)$, we shall represent the integral I_q^r as the sum:

$$I_q^r = I_q^{x_n} + I_{x_n}^r. \tag{13}$$

For $r > x_n$, consider the integral $I_{x_n}^r$. Putting $x = x_n + z$, $r = x_n + \delta$, $\lambda(z) = \eta_n(x_n + z) - \eta_n(x_n)$, we obtain

$$I_{x_n}^r = e^{2\pi i \eta_n(x_n)} J_0^\delta; \quad J_0^\delta = \int_0^\delta \varphi(x_n + z)\, e^{2\pi i \lambda(z)}\, dz.$$

Further, using the substitution $\lambda(z) = u$, reduce the integral J_0^δ to the form:

$$J_0^\delta = \int_0^{\lambda(\delta)} \varphi(x_n + z) \frac{e^{2\pi i u}}{\lambda'(z)}\, du,$$

where $\dfrac{\varphi(x_n + z)}{\lambda'(z)}$ is regarded a function of u. On comparing the integral J_0^δ with the integral

$$J' = \int_0^{\lambda(\delta)} \varphi(x_n) \frac{e^{2\pi i u}}{\lambda'(z)}\, du,$$

we find

$$J_0^\delta - J' = \int_0^{\lambda(\delta)} \Phi(u)\, e^{2\pi i u}\, du; \quad \Phi(u) = \frac{\varphi(x_n + z) - \varphi(x_n)}{\lambda'(z)}.$$

Since $\Phi(u)$ is an algebraic function of u of finite degree, the interval $0 \leqq u \leqq \lambda(\delta)$ can be divided into a finite number of intervals, in each of which the function $\Phi(u)$ is monotonic and of constant sign. Denoting by K the greatest value of the modulus of $\Phi(u)$ in the interval $0 \leqq u \leqq \lambda(\delta)$, and using the usual technique, we get

$$J_0^\delta - J' \ll K.$$

But, as K is the greatest value of the modulus of $\Phi(u)$ regarded as a function of z in the interval $0 \leqq z \leqq \delta$, from $\varphi(\gamma+z) - \varphi(\gamma) \ll HU^{-1}z$, $\lambda'(z) \gg zA^{-1}$, we obtain $K \ll HAU^{-1}$. Consequently,

$$J_0^\delta - J' \ll HAU^{-1}.$$

Now comparing the integral J' with the integral

$$J'' = \int_0^{\lambda(\delta)} \varphi(x_n) \sqrt{\frac{A_0}{2u}}\, e^{2\pi i u}\, du; \quad A_0 = (\lambda''(x_n))^{-1}.$$

we obtain

$$J' - J'' = \varphi(x_n) \int_0^{\lambda(\delta)} F(u)\, e^{2\pi i u}\, du; \quad F(u) = \frac{1}{\lambda'(z)} - \sqrt{\frac{A_0}{2u}}.$$

Hence we find

$$J' - J'' \ll HL,$$

where L is the greatest value of the modulus of $F(u)$ regarded as a function of z in the interval $0 \leqq z \leqq \delta$. But, as

$$|F(u)| < \left| \frac{\dfrac{1}{(\lambda'(z))^2} - \dfrac{A_0}{2\lambda(z)}}{\dfrac{1}{\lambda'(z)}} \right| = \left| \frac{2\lambda(z) - A_0(\lambda'(z))^2}{2\lambda(z)\,\lambda'(z)} \right|,$$

$$\lambda(z) = \frac{z^2}{2A_0} + O\left(\frac{z^3}{AU}\right), \quad \lambda'(z) = \frac{z}{A_0} + O\left(\frac{z^2}{AU}\right),$$

$$\lambda(z) \gg \frac{z^2}{A}, \qquad \lambda'(z) \gg \frac{z}{A},$$

we have

$$J' - J'' \ll HAU^{-1},$$

$$I'_{x_n} = \varphi(x_n)\, e^{2\pi i \eta_n(x_n)} \int_0^{\lambda(\delta)} \sqrt{\frac{A_0}{2u}}\, e^{2\pi i u}\, du + O(HAU)^{-1}. \tag{14}$$

Further, putting $u = z^2/2A_0$, we obtain

$$\int_0^\infty \sqrt{\frac{A_0}{2u}}\, e^{2\pi i u}\, du = \int_0^\infty e^{2\pi i \frac{z^2}{2A_0}}\, dz = \frac{1+i}{2}\sqrt{\frac{A_0}{2}}. \tag{15}$$

Moreover, assuming

$$D_n = \left| \int_{\lambda(\delta)}^\infty \sqrt{\frac{A_0}{2u}}\, e^{2\pi i u}\, du \right|$$

and then applying the technique already used, we obtain

$$D_n \leqq \sqrt{\frac{A_0}{2\lambda(\delta)}};$$

hence, as $\lambda(\delta) \geqq \delta^2/2A$, $\lambda'(\delta) \geqq k\delta/A$, we find

$$D_n \leqq \frac{1}{\lambda'(\delta)}.$$

If $\lambda'(\delta) < A^{-0.5}$, dividing the integration range into two intervals $\lambda(\delta) \leqq u \leqq \lambda(\delta)+1$ and $\lambda(\delta)+1 < u < \infty$, we obtain

$$D_n \ll \sqrt{A}.$$

Consequently,

$$D_n \ll \min\left(\frac{1}{\lambda'(\delta)}, \sqrt{A}\right) \ll \min\left(\frac{1}{f'(r)-f'(x_n)}, \sqrt{A}\right) \ll \min\left(\frac{1}{\eta_n'(r)}, \sqrt{A}\right).$$

From (14), (15) and the last inequality, we obtain

$$I^r_{x_n} = \frac{1+i}{2\sqrt{2}}\,\varphi(x_n)\, e^{2\pi i\eta_n(x_n)}\,\frac{1}{\sqrt{f''(x_n)}} + O\left(H\min\left(\frac{1}{\eta_n'(r)}, \sqrt{A}\right)\right),$$

where, when $n_r = f'(r)$, the first term on the right vanishes. Hence, we find

$$\sum_{n_q \leqq n \leqq n_r} I^r_{x^n} = \sum_{n_q \leqq n \leqq n_r} \frac{1}{2} Z_n + O\left(H\min\left(\frac{1}{\eta_n'(r)}, \sqrt{A}\right) + H\ln U\right), \tag{16}$$

where, when $n = n_r = f'(r)$, the term $\frac{1}{2}Z_n$ vanishes. Similarly, (using the substitution $x = -x_1$) we obtain

$$\sum_{n_q \leqq n \leqq n_r} I^{x_n}_q = \sum_{n_q \leqq n \leqq n_r} \frac{1}{2} Z_n + O\left(H\min\left(\frac{1}{\eta_n(q)}, \sqrt{A}\right) + H\ln U\right), \tag{17}$$

where, when $n = n_q = f'(q)$, the term $Z_n/2$ vanishes. From (10), (11), (12), (13), (16) and (17), we find that our lemma is true.

Theorem 2. The number T of integral points in a sphere $x^2+y^2+z^2 \leqq a^2$ is

$$T = \tfrac{4}{3}\pi a^3 + O(R); \quad R = a^{4/3}(\ln a)^6.$$

Proof. Divide the sphere into 48 equal domains with the help of the planes $x=0$, $y=0$, $z=0$, $x=y$, $y=z$, and $z=x$. Let a representative of these domains be the region Ω bounded by the inequalities:

$$0 \leqq y \leqq \frac{a}{\sqrt{3}}, \quad y \leqq x \leqq \sqrt{\frac{a^2-y^2}{2}}, \quad x \leqq z \leqq \sqrt{a^2-y^2-x^2}.$$

Let Π denote the projection of this region on the XOY plane, a two-dimensional region bounded by the inequalities:

$$0 \leqq y \leqq \frac{a}{\sqrt{3}}, \quad y \leqq x \leqq \sqrt{\frac{a^2-y^2}{2}},$$

To each integral point in the domain Π, put into correspondence the length Z of that part of the perpendicular drawn from this point to Π which is in Ω and a definite number γ. This number γ is equal to 1 for the internal points of the domain, is zero for the origin of coordinates and for the points on the line $x = \sqrt{\frac{a^2 - y^2}{2}}$, and is equal to 0.5 for the remaining points on the contour of the domain. We readily find that the sum $G = \sum_{\Pi} \gamma Z$ taken over all integral points in the domain Ω is equal to the volume V_Ω of the domain Ω with an error $\ll a$.

For a given z, the Diophantine equation $X^2 + Y^2 = a^2 - Z^2$ has $\ll a$ solutions. Therefore the number of integral points on the sphere is $\ll a^{1+\varepsilon}$. Now consider the sum $T_\Omega = \sum_{\Omega} \mu$ taken over the integral points in the domain Ω, where μ is equal to 1 for the internal points of the domain, is zero for the points on the edges and on the sphere, and is equal to 0.5 for the remaining points on the boundary of the domain. We easily find that T_Ω is equal to

$$\sum_{\Pi} \gamma(Z - \{Z\} + 0.5) = G - \sum_{\Pi} \gamma(\{Z\} - 0.5)$$

with an error $\ll R$, and, consequently, is equal to $V_\Omega - \sum_{\Pi} \gamma(\{Z\} - 0.5)$ with an error $\ll R$. On the other hand, on summing the values of T_Ω for all 48 regions of Ω, we find that it is equal to T with an error $\ll R$.

Consequently, for each region of the domain Ω, if we can show that

$$\sum_{\Pi} \gamma(\{Z\} - 0.5) \ll R,$$

then Theorem 2 follows immediately. Since $\{Z\} = \{z\} = \sqrt{a^2 - y^2 - x^2}$, the above inequality can be replaced by an equivalent inequality:

$$\sum_{(\Pi_0)} (\{\sqrt{a^2 - y^2 - x^2}\} - 0.5) \ll R, \tag{18}$$

in which summation is taken over the integral points in a domain Π_0 close to Π bounded by the inequalities:

$$0 < y \leqq \frac{a}{\sqrt{3}}, \quad y < x \leqq \sqrt{\frac{a^2 - y^2}{2}}.$$

Let Q denote the number of integral points in Π_0.

We shall apply the assertion (b) of Lemma 3 to prove the inequality (18). Taking the fraction $\{\sqrt{a^2 - y^2 - x^2}\}$ on the left-hand side of (18) as the number δ_s, consider the function $\psi(\delta_s)$ for $\varrho = 6$ and $\Delta = a^{-2/3}$. Inequality (18) is proved, if we can show that

$$\sum_{(\Pi_0)} \psi(\sqrt{a^2 - y^2 - x^2}) - \tfrac{1}{2}(\beta - \alpha)\, Q \ll R.$$

Applying the property (4) of Lemma 2, and assuming that

$$W_m = \sum_{(\Pi_0)} e^{-2\pi i m \sqrt{a^2 - y^2 - x^2}},$$

$$V = \sum_{m=1}^{\infty} g_m \overline{W}_m, \quad V' = \sum_{m=1}^{\infty} h_m W_m,$$

by virtue of the inequalities,

$$g_m \ll \kappa_m, \quad h_m \ll \kappa_m; \quad \kappa_m = \begin{cases} \dfrac{1}{m}, & \text{if} \quad m \leqq \Delta^{-1}, \\ \dfrac{1}{m^7 \Delta^6}, & \text{if} \quad m > \Delta^{-1}, \end{cases}$$

we find that it suffices to prove

$$V \ll R, \quad V' \ll R.$$

We shall only prove the second inequality. The first inequality can be established in a similar manner.

Consider W_m under the condition $m \leqq a^{0.8}$. We find

$$W_m = \sum_{0 < y \leqq \frac{a}{\sqrt{3}}} L_y; \qquad L_y = \sum_{y < x \leqq \sqrt{\frac{a^2-y^2}{2}}} e^{-2\pi i m \sqrt{a^2-y^2-x^2}}.$$

On applying Lemma 6 to the sum L_y, we obtain

$$\varphi(x) = 1, \qquad f(x) = -m\sqrt{a^2-y^2-x^2}, \qquad f'(x) = \frac{mx}{\sqrt{a^2-y^2-x^2}},$$

$$f''(x) = \frac{m(a^2-y^2)}{(a^2-y^2-x^2)^{3/3}}, \qquad f'''(x) = \frac{3m(a^2-y^2)\,x}{(a^2-y^2-x^2)^{5/2}}.$$

Putting $H = 1$, $U = a$, $A = a/m$, we find that the conditions of Lemma 6 are fulfilled. After certain simple calculations, we obtain

$$L_y = \frac{1+i}{\sqrt{2}} \sum_{\frac{my}{\sqrt{a^2-2y^2}} \leqq u \leqq m} \frac{m(a^2-y^2)^{1/4} e^{-2\pi i \sqrt{(a^2-y^2)(m^2+u^2)}}}{(m^2+u^2)^{3/4}}$$

$$- \frac{1+i}{2\sqrt{2}} \frac{m(a^2-y^2)^{1/4} e^{-2\pi i \sqrt{(a^2-y^2)\,2m^2}}}{(2m^2)^{3/4}} + O(T + \ln a);$$

$$T = \min\left(\frac{1}{(\Phi(y))}, \sqrt{\frac{a}{m}}\right); \qquad \Phi(y) = \frac{my}{\sqrt{a^2-2y^2}}.$$

Now we shall estimate the sum:

$$S = \sum_{0 < y \leqq \frac{a}{\sqrt{3}}} (T + \ln a).$$

All the values of y can be divided into $\ll m$ sequences, such that, for the numbers belonging to the same sequence, for some k, we have

$$k < \Phi(y) \leqq k+1.$$

The sum of the values of T corresponding to the numbers belonging to a given sequence is

$$\ll \sqrt{\frac{a}{m}} + \sum_{0 < s \ll \frac{a}{m}} \frac{a}{ms} \ll \frac{a}{m} \ln a.$$

Therefore, we have

$$S \ll a \ln a.$$

Further, we shall estimate the sum

$$S' = \sum_{0<y\leqq\frac{a}{\sqrt{3}}} \frac{m(a^2-y^2)^{1/4} e^{-2\pi i\sqrt{(a^2-y^2)2m^2}}}{(2m^2)^{3/4}}.$$

Assuming that

$$H \doteq \sqrt{\frac{a}{m}}, \quad U = a, \quad A = \frac{a}{m},$$

we find that the conditions of Lemma 1 are fulfilled for this sum. Therefore

$$S' = \sqrt{\frac{a}{m}}\left(\frac{\frac{a}{m}}{\sqrt{\frac{a}{m}}} + \sqrt{\frac{a}{m}}\right) \ll a.$$

As a result of all these facts, we have

$$W_m = \sum_{0<y\leqq\frac{a}{\sqrt{3}}} \frac{1+i}{\sqrt{2}} \sum_{\frac{my}{\sqrt{a^2-2y^2}}\leqq u\leqq m} \frac{m(a^2-y^2)^{1/4} e^{-2\pi i\sqrt{(a^2-y^2)(m^2+u^2)}}}{(m^2+u^2)^{3/4}} + O(a\ln a).$$

Now, changing the order of summation, we obtain

$$W_m = \frac{1+i}{\sqrt{2}} \sum_{0<u\leqq m} W_{m,u} + O(a\ln a);$$

$$W_{m,u} = \sum_{0<y\leqq\frac{au}{\sqrt{m^2+2u^2}}} \frac{m(a^2-y^2)^{1/4} e^{-2\pi i\sqrt{(a^2-y^2)(m^2+u^2)}}}{(m^2+u^2)^{3/4}}.$$

Now, putting

$$f(y) = -\sqrt{(a^2-y^2)(m^2+u^2)}, \quad H = \sqrt{\frac{a}{m}},$$

$$U = a, \quad A = \frac{a}{m},$$

we find that the conditions of Lemma 6 are fulfilled for the sum $W_{m,u}$. After certain simple calculations, we obtain

$$W_{m,u} = \frac{1+i}{\sqrt{2}} \sum_{0<v\leqq u} \frac{ame^{-2\pi i a\sqrt{m^2+u^2+v^2}}}{m^2+u^2+v^2} + O\left(\sqrt{\frac{a}{m}}\ln a\right).$$

Hence, we find

$$W_m = W'_m + O(a\ln a);$$

$$W'_m = \sum_{0<u\leqq m} \sum_{0<v\leqq u} \frac{iame^{-2\pi i a\sqrt{m^2+u^2+v^2}}}{m^2+u^2+v^2}.$$

In particular, we have $W'_m \ll am$, $W_m \ll am + a\ln a$.

Now we shall proceed to derive an inequality for V'. Let $M_0 = [a^{1/3}(\ln a)^2]$ and k_1 be the least integer satisfying the condition $M_0 \cdot 2^{k_1} > a^{11/15}$. We find

$$\sum_{0<m\leqq M_0} h_m W_m \ll a^{4/3}(\ln a)^2, \qquad \sum_{M_0\cdot 2^{k_1}<m\leqq a^{7/9}} h_m W_m \ll a^{4/3},$$

$$\sum_{m>a^{7/9}} h_m W_m \ll a^{4/3}, \qquad \sum_{M_0<m\leqq M_0\cdot 2^{k_1}} |h_m|\, a\ln a \ll a(\ln a)^2.$$

(In deriving the first and second inequalities, take $W_m \ll am + a\ln a$, while in deriving the third, $W_m \ll a^2$).

Therefore

$$V' = V_0 + O(a^{4/3}(\ln a)^2); \quad V_0 = \sum_{M_0<m\leqq M_0\cdot 2^{k_1}} h_m W'_m.$$

The sum V_0 can be divided into k_1 subsums of the type:

$$V_M = \sum_{M<m\leqq 2M} h_m W'_m.$$

Putting $\lambda_r = \frac{r}{4M}$, $\mu_s = \frac{s}{4M}$, we easily find that

$$V_M = \frac{1}{16M^2} \sum_{0<r\leqq 4M} \sum_{0<s\leqq 4M} \sum_{0\leqq k\leqq 2M} \sum_{0\leqq l\leqq 2M} \sum_{M<m\leqq 2M} h_m$$

$$\times \sum_{0<u\leqq 2M} \sum_{0<v\leqq 2M} \frac{iame^{-2\pi ia\sqrt{m^2+u^2+v^2}}}{m^2+u^2+v^2} \times e^{2\pi i\lambda_r(m-u-k)+2\pi i\mu_s(u-v-l)}.$$

Let $\xi(z)$ denote the number of solutions of the equation $m^2+n^2 = z$, and let $V_{M,r,s}$ stand for that part of V_M which corresponds to a given pair of values of r and s. Thus, we find

$$\sum_k e^{-2\pi i\lambda_r k} \ll \min\left(M, \frac{1}{(\lambda_r)}\right), \quad \sum_l e^{-2\pi i\mu_s l} \ll \min\left(M, \frac{1}{(\mu_s)}\right),$$

$$V_{M,r,s} \ll \frac{a\kappa_M}{M} \min\left(M, \frac{1}{(\lambda_r)}\right) \min\left(M, \frac{1}{(\mu_s)}\right) \times \sum_z \xi(z) \left|\sum_v \frac{e^{-2\pi i(\mu_s v + a\sqrt{z+v^2})}}{z+v^2}\right|.$$

Hence, summing over all values of r and s, and in the sum over x replacing μ_s by the value of μ at which the modulus of this sum is maximum, we obtain

$$V_M \ll a\kappa_M(\ln a)^2 \sum_z \xi(z) \left|\sum_v \frac{e^{-2\pi i(\mu v + a\sqrt{z+v^2})}}{z+v^2}\right|,$$

Hence, by virtue of the inequality

$$\sum_z (\xi(z))^2 \ll M^2(\ln a)^3,$$

we find

$$V_M^2 \ll a^2\kappa_M^2 M^4(\ln a)^7 E;$$

$$E = \sum_{0<v\leqq 2M} \sum_{0<v_1\leqq 2M} \sum_{M^2<z\leqq 8M^2} \frac{e^{-2\pi i(\mu(v-v_1)+a(\sqrt{z+v^2}-\sqrt{z+v_1^2}))}}{(z+v^2)(z+v_1^2)}.$$

That part of the sum E which corresponds to $v = v_1$ is obviously $\ll M^{-1}$. Divide the remaining part of the sum into two subsums: the subsum E_1 for which $v < v_1$ and the subsum E_2 for which $v > v_1$. Thus, we find $|E_1| = |E_2|$. Therefore,

$$E \ll M^{-1} + |E_1|.$$

Further we find

$$12M^2 E_1 = \sum_{0 \leqq j < 12M^2} \sum_{v} \sum_{v_1} \sum_{M^2 < z \leqq 8M^2} \sum_{M^2 < g \leqq 12M^2} e^{2\pi i \frac{j(g-z-v^2)}{12M^2}}$$

$$\times \frac{e^{2\pi i(\mu(v_1-v)+a\sqrt{g+v_1^2-v^2}-a\sqrt{g})}}{(g+v_1^2-v^2)g} = 12M^2 \sum_j E_j;$$

$$E_j \ll \frac{1}{M^2} \min\left(M^2, \frac{1}{(\gamma_j)}\right) U_f; \quad \gamma_j = \frac{j}{12M^2},$$

$$U_j = \sum_{\substack{v \\ v<v_1}} \sum_{v_1} \left| \sum_{M^2 < g \leqq 12M^2} \frac{e^{2\pi i(\gamma_j g + a\sqrt{g+v_1^2-v^2} - a\sqrt{g})}}{(g+v_1^2-v^2)g} \right| \ll \sum_{0 < t \leqq 4M^2} \eta(t)\,|R_t|,$$

where $\eta(t)$ is the number of solutions of the equation $v_1^2 - v^2 = t$, and

$$R_t = \sum_{M^2 < g \leqq 12M^2} \frac{e^{2\pi i(\gamma_j g + a\sqrt{g+t} - a\sqrt{g})}}{(g+t)g}.$$

Consider numbers of the type $\tau = 4M^3 2^{-k}$, where k is a natural integer. Let τ' be such a number with $k = k_1$, which is the least τ satisfying the condition $\tau' \geqq a^{2/3}$, and let τ'' be such a number with $k = k_0$, which is the least τ satisfying the condition $\tau'' \gg a^{1/9} M^{5/3}$. That part of the sum U_j which corresponds to the case $t \leqq \tau'$ is obviously

$$\ll \sum_{0 < t \leqq \tau'} \frac{\eta(t)}{M^2} \ll a^{2/3} M^{-2} \ln a.$$

Divide the remaining part into $k_1 \ll \ln a$ subsums of the type:

$$S_\tau = \sum_{\tau < t \leqq 2\tau} \eta(t)\,|R_t|.$$

The sum R_t is of the type stated in Lemmas 1 and 6. Now, putting

$$f(g) = \gamma_j g + a(\sqrt{g+t} - \sqrt{g}), \quad \varphi(g) = \frac{1}{(g+t)g},$$

we obtain

$$f'(g) = \gamma_j + \frac{a}{2}((g+t)^{-1/2} - g^{-1/2}).$$

$$f''(g) = \frac{a}{4}(g^{-3/2} - (g+t)^{-3/2}),$$

$$U = M^2, \qquad \frac{M^5}{at} \ll A \ll \frac{M^5}{at}, \qquad f'''(g) \ll \frac{1}{AU}, \qquad H \ll M^{-4}.$$

Applying Lemma 1, when $\tau' \leqq \tau < \tau''$, we obtain

$$R_t \ll a^{1/2} t^{1/2} M^{-9/2} + a^{-1/2} t^{-1/2} M^{-3/2},$$
$$S_\tau \ll (a^{1/2} \tau^{3/2} M^{-9/2} + a^{-1/2} \tau^{1/2} M^{-3/2}) \ln a.$$

Hence, we find

$$\sum_{\tau' \leqq \tau < \tau''} S_t \ll a^{2/3} M^{-2} (\ln a)^2.$$

Finally, consider the case where $\tau'' \leqq \tau \leqq 2M^2$. Apply Lemma 6 to the sum R_t. Defining $g_{w,t}$ by the equality $f'(g_{w,t}) = w$, and putting

$$f'(M^2) = w_1(t), \quad f'(12M^2) = w_2(t),$$

$$\Phi(w, t) = \frac{2}{g_{w,t}(g_{w,t} + t)\sqrt{a(g_{w,t}^{-3/2} - (g_{w,t} + t)^{-3/2})}},$$

$$\Psi(w, t) = \gamma_j g_{w,t} + a(\sqrt{g_{w,t} + t} - \sqrt{g_{w,t}}) - w g_{w,t},$$

we obtain

$$R_t = \frac{1+i}{\sqrt{2}} \sum_{w_1(t) < w \leqq w_2(t)} \Phi(w, t) e^{2\pi i \Psi(w,t)} + O(a^{-1/2} t^{-1/2} M^{-3/2}).$$

Hence, we find

$$S_\tau \ll G + a^{-1/2} \tau^{1/2} M^{-3/2} \ln a;$$

$$G = \sum_{\tau < t \leqq 2\tau} \eta(t) \left| \sum_{w_1(t) < w \leqq w_2(t)} \Phi(w, t) e^{2\pi i \Psi(w,t)} \right|.$$

Here, it should be noted that

$$\sum_{\tau'' \leqq \tau \leqq 2M^2} a^{-1/2} \tau^{1/2} M^{-3/2} \ll a^{-1/2} M^{-1/2} \ll a^{2/3} M^{-2}.$$

Now we shall estimate the sum G. Evidently, $w_1 = w_1(t)$ and $w_2 = w_2(t)$ are decreasing functions of t. Therefore, their inverse functions $t'(w_1)$ and $t''(w_2)$ are also decreasing functions (of w_1 and w_2, respectively). Under these conditions, w can only take those values which satisfy the condition $w_1(2\tau) < w \leqq w_2(\tau)$, while t, for a given w, only those values which satisfy the condition:

$$\max(\tau, t'(w)) < t \leqq \min(2\tau, t''(w)).$$

After squaring and then changing the order of summation, we find

$$G^2 \ll \tau (\ln a)^3 \sum_{w_1(2\tau) < w' \leqq w_2(\tau)} \sum_{w_1(2\tau) < w \leqq w_2(\tau)} B_{w',w};$$

$$B_{w',w} = \sum_{t' < t \leqq t''} \Phi_0(t) e^{2\pi i \Psi_0(t)};$$

$$\Phi_0(t) = \Phi(w', t) \Phi(w, t), \qquad \Psi_0(t) = \Psi(w', t) - \Psi(w, t);$$

$$t' = \max(\tau, t'(w'), \quad t'(w)), \qquad t'' = \min(2\tau, t''(w'), t''(w)).$$

Further, for the sum $B_{w',w}$, we obtain

$$\Phi_0(t) \ll H,\ t''-t' \ll U;\quad H = a^{-1}\tau^{-1}M^{-3},\ U = \tau.$$

Since $w_2(\tau) - w_1(2\tau) \ll a\tau M^{-3}$, the sum of all $|B_{w',w}|$ satisfying the condition $w' = w$ is

$$\ll a\tau M^{-3}HU \ll \tau M^{-6}.$$

Now consider the sum $B_{w',w}$ for $w' > w$ (evidently, $B_{w,w'} = \bar{B}_{w',w}$). Using the equality

$$\gamma + \frac{a}{2}\left((g_{w,t}+t)^{-1/2} - g_{w,t}^{-1/2}\right) = w,$$

and differentiating partially with respect to t, we find

$$\frac{\partial(g_{w,t}+t)}{\partial t} = \frac{g_{w,t}^{-3/2}}{g_{w,t}^{-3/2} - (g_{w,t}+t)^{-3/2}}.$$

Then, using the same equality and differentiating partially with respect to w, we find

$$\frac{\partial g_{w,t}}{\partial w} = \frac{4}{a}\,\frac{1}{g_{w,t}^{-3/2} - (g_{w,t}+t)^{-3/2}}$$

and also

$$\frac{\partial \Psi(w,t)}{\partial t} = \frac{a}{2}(g_{w,t}+t)^{-1/2}.$$

Hence, we obtain

$$\frac{\partial^2 \Psi(w,t)}{\partial t^2} = -\frac{a}{4}(g_{w,t}+t)^{-3/2}\frac{\partial(g_{w,t}+t)}{\partial t} = -\frac{a}{4}\,\frac{1}{(g_{w,t}+t)^{3/2} - g_{w,t}^{3/2}}.$$

But, as

$$\frac{\partial}{\partial w}\,\frac{\partial^2 \Psi(w,t)}{\partial t^2} = \frac{3}{8}\,a\,\frac{(g_{w,t}+t)^{1/2} - g_{w,t}^{1/2}}{((g_{w,t}+t)^{3/2} - g_{w,t}^{3/2})^2}\,\frac{\partial g_{w,t}}{\partial w} = \Omega(w,t);$$

$$\Omega(w,t) = \frac{3}{2}\,\frac{(g_{w,t}+t)^{1/2} - g_{w,t}^{1/2}}{(g_{w,t}^{-3/2} - (g_{w,t}+t)^{-3/2})\,((g_{w,t}+t)^{3/2} - g_{w,t}^{3/2})^2},$$

applying the Lagrange formula, for some w''' $(w < w''' < w')$, we have

$$\frac{\partial^2 \Psi_0(t)}{\partial t^2} = \frac{\partial^2 \Psi(w',t)}{\partial t^2} - \frac{\partial^2 \Psi(w,t)}{\partial t^2} = (w'-w)\,\Omega(w''',t).$$

Hence, putting $w' - w = \sigma$, $A = \tau^2 2M^{-2}\sigma^{-1}$, we obtain

$$\frac{1}{A} \ll \frac{\partial^2 \Psi(t)}{\partial t^2} \ll \frac{1}{A}.$$

Now applying Lemma 1, we find (here $H = a^{-1}\tau^{-1} - M^{-3}$, $U = \tau$) we get

$$B_{w',w} \ll a^{-1}\tau^{-1}M^{-2}\sigma^{1/2};$$

hence, we find that the sum of all $|B_{w',w}|$ for $w' > w$ is

$$\ll (w_2(\tau) - w_1(2\tau)) \sum_{0<\sigma<w_2(\tau)-w_1(2\tau)} a^{-1}\tau^{-1}M^{-2}\sigma^{1/2} \ll a^{3/2}\tau^{3/2}M^{-19/2}.$$

Consequently,

$$G^2 \ll \tau (\ln a)^3 (\tau M^{-6} + a^{3/2}\tau^{3/2} M^{-19/2}),$$

$$S_\tau \ll (\tau M^{-3} + a^{3/4}\tau^{5/4} M^{-19/4}) (\ln a)^{3/2} + a^{-1/2}\tau^{1/2} M^{-3/2},$$

$$U_j \ll a^{2/3} M^{-2} \ln a + (M^{-1} + a^{3/4} M^{-9/4}) (\ln a)^{3/2} + a^{-1/2} M^{-1/2},$$

$$E \ll M^{-1} + |E_1| \ll (a^{2/3} M^{-2} + M^{-1} + a^{3/4} M^{-9/4} + a^{-1/2} M^{-1/2}) (\ln a)^{5/2}$$

$$\ll (a^{2/3} M^{-2} + M^{-1}) (\ln a)^{5/2},$$

$$V_M^2 \ll a^2 \kappa_M^2 M^4 (a^{2/3} M^{-2} + M^{-1}) (\ln a)^{19/2},$$

$$V_M \ll a \kappa_M M^2 (a^{1/3} M^{-1} + M^{-1/2}) (\ln a)^{19/4}.$$

Therefore, for $M \leqq a^{2/3}$, we have

$$V_M \ll a^{4/3} (\ln a)^{19/4},$$

and for $M > a^{2/3}$,

$$V_M \ll \frac{a^5}{M^{11/2}} (\ln a)^{19/4}.$$

Hence, we find

$$V_0 \ll a^{4/3} (\ln a)^{23/4}, \quad V' \ll a^{4/3} (\ln a)^{23/4},$$

which completes the proof of the theorem.

Chapter 2. The Estimate of $G(n)$ in the Waring Problem

In this chapter we shall apply my method to derive an estimate of the function $G(n)$ introduced by Hardy and Littlewood to denote the least natural number r to which some positive number N_0 can be put into correspondence, so that the equation

$$N = x_1^n + \ldots + x_r^n$$

is solvable in positive integers $x_1, \ldots, x_r$ for any integer N not less than N_0.

In 1919 Hardy and Littlewood found the following upper bound for $G(n)$:

$$G(n) \leqq n2^{n-2}h; \quad \lim_{n\to\infty} h = 1,$$

which increases with increasing n as a quantity of the order of $n2^n$.

In 1934, using my method, I found an entirely new upper bound[15]

$$G(n) < n(6\ln n + 10).$$

This bound increases with n as a quantity of the order of $n \ln n$ and, by virtue of the obvious inequality $G(n) > n$, it cannot be improved essentially by any other bound of lesser order. Thereafter, I could only diminish the coefficient of $\ln n$ in brackets, and obtained

$$G(n) < n(3\ln n + 11) \text{ [16]},$$

$$G(n) < n(2\ln n + 4\ln\ln n + 2\ln\ln\ln n + 13) \text{ [17]}.$$

Derivation of the second estimate given above is the aim of this chapter. In the derivation we shall assume that $n > 170000$, otherwise the second estimate will always be worse than the first.

Lemma 1. Let k_0 be an integer, $k_0 \geqq 6$, $k = 2k_0$, $r = 2r_0$, $r_0 = [k^2(2\ln k + \ln\ln k + 2.6)]$, Y_0 be a positive integer, $y_1, \ldots, y_r$ run, independently of each other, through $1, \ldots, Y_0$, and finally, let $\eta_s = y_1^s + \ldots + y_{r_0}^s - y_{r_0+1}^s - \ldots - y_r^s$. Then, the number T of solutions of the system

$$\eta_k = l_k, \ldots, \eta_1 = l_1$$

is

$$T \ll Y_0^{r - \frac{k(k+1)}{2}}.$$

Proof. This lemma is a trivial modification of Lemma 4 in Chapter 4 of the monograph[2].

Lemma 2. In the interval $M \leqq x \leqq M'$, let $f(x)$ be a real-valued differentiable function, and inside this interval let $f'(x)$ be monotone and of constant sign. For a constant $\delta\,(0 < \delta < 1)$, let $|f'(x)| \leqq \delta$. Then

$$\sum_{M < x \leqq M'} e^{2\pi i f(x)} = \int_M^{M'} e^{2\pi i f(x)}\,dx + O\left(3 + \frac{2\delta}{1-\delta}\right).$$

Proof. This lemma is Lemma 3 given in Chapter 2 of the monograph[2].

Lemma 3. Let $P > 1$, z be a real number, and

$$I = \int_0^P e^{2\pi i z x^n}\, dx.$$

Then

$$|I| \leqq Z; \quad Z = \begin{cases} P, & \text{if } |z| \leqq P^{-n}, \\ \sqrt{2}\,|z|^{-\nu}, & \text{if } |z| > P^{-n}. \end{cases}$$

Proof. The case where $|z| \leqq P^{-n}$ is obvious. Therefore, we shall only consider the case where $|z| > P^{-n}$. Moreover, without loss of generality, we can assume that $z > 0$. Making the substitution $zx^n = u$, we get

$$I = U + iV; \quad U = \int_0^\sigma \psi(u) \cos 2\pi u\, du,$$

$$V = \int_0^\sigma \psi(u) \sin 2\pi u\, du, \quad \sigma = zP^n, \quad \psi(u) = \nu u^{\nu-1} z^{-\nu},$$

where $\psi(u)$ is a decreasing function of u. Consider the integral U. Using numbers of the type $0.25 + 0.5l$, where l is an integer, divide the integration range $0 \leqq u \leqq \sigma$ into subintervals. Accordingly, the integral U is represented as an alternating series. Hence, it follows that

$$|U| \leqq \int_0^1 \psi(u)\, du = z^{-\nu}.$$

Arguing in a similar manner, we obtain

$$|V| \leqq z^{-\nu}.$$

Therefore

$$|I| \leqq \sqrt{z^{-2\nu} + z^{-2\nu}} = \sqrt{2}\, z^{-\nu}.$$

Lemma 4. Let

$$S_{a,q} = \sum_{x=0}^{q-1} e^{2\pi i \frac{a}{q} x^n}, \quad q > 0, \quad (a, q) = 1.$$

Then

$$|S_{a,q}| \ll q^{1-\nu}.$$

Proof. This lemma is the lemma stated in Question 11b, Chapter 6 in the book [1].

Lemma 5. Let N_0 be a positive integer and let $S_{a,q}$ be as defined in Lemma 4. Then, assuming

$$A(q) = \sum_{\substack{(a,q)=1 \\ 0 \leqq a < q}} \left(\frac{S_{a,q}}{q}\right)^{4n} e^{-2\pi i \frac{a}{q} N_0}, \quad \mathfrak{S}(N_0) = \sum_{q=1}^{\infty} A(q),$$

we obtain

$$\mathfrak{S} \gg 1.$$

Proof. This lemma is Lemma 11, Chapter 6, in the monograph[2].

Lemma 6. Let r be a positive integer, $N_1 > 0$, $K_r(N_1)$ be the number of solutions in positive integers $x_1, \ldots, x_r$ of the inequality:

$$x_1^n + \ldots + x_r^n \leqq N_1 .$$

Then

$$K_r(N_1) = T_r N_1^{rv} - \theta r N_1^{rv-v}; \quad T_r = \frac{(\Gamma(1+v))^r}{\Gamma(rv+1)}, \quad \theta > 0 .$$

Proof. This lemma is Lemma 2 in Chapter 6 of the monograph[2].

Special notation. Let N_0 be a positive integer such that $0.5 \leqq N_0 \leqq N$, and $P = [N^v + 1]$,

$$\tau = 2nP^{n-1}, \quad X_0 = [\sqrt{P}], \quad \tau_0 = X_0^{n-0.5}, \quad \gamma = 0.25\,(1 - 0.5v),$$

$$L_\alpha = \sum_{x=1}^{P} e^{2\pi i \alpha x^n}.$$

We shall express any α in the interval $-\tau^{-1} \leqq \alpha \leqq 1 - \tau^{-1}$ as

$$\alpha = \frac{a}{q} + z, \qquad (a, q) = 1, \qquad 0 \leqq a < q, \qquad 0 < q \leqq \tau, \ |z| < \frac{1}{q\tau}, \tag{1}$$

which is always possible. For a given fraction a/q, any interval where $q \leqq P^\gamma$ shall be called a *main interval.* Obviously, main intervals do not overlap. The intervals which remain after the main intervals have been chosen are called *auxiliary intervals*.

Lemma 7. Let N_1 be an integer, $N_0 - P^{n-v} \leqq N_1 \leqq N_0$,

$$I'(N_1) = \int_0^1 L_\alpha^{4n} e^{-2\pi i \alpha N_1} \, d\alpha$$

and let $I_0'(N_1)$ be that part of the integral $I'(N_1)$ which is taken over main intervals. Then

$$I_0'(N_0) \gg N^3 .$$

Proof. Let $M = [0.5P^{n-v} + 1]$, $N_0 - 2M \leqq N_1 \leqq N_0$. Let $H_{a,q}(N_1)$ be that of the integral $I'(N_1)$ which is taken over the main interval of the fraction a/q. Let α be some number in this interval. Transforming the sum L by means of the substitution $x = qt + s$, where s runs through $0, 1, \ldots, q-1$, and t runs, for a given s, through the integers in the interval $-sq^{-1} < t \leqq (P-s)\,q^{-1}$, we obtain

$$L_\alpha = \sum_s \sum_t e^{2\pi i \left(\frac{as^n}{q} + z(qt+s)^n\right)} = \sum_s e^{2\pi i \frac{as^n}{q}} D(z);$$

$$D(z) = \sum_t e^{2\pi i z (qt+s)^n}.$$

But in the interval specified for t, we have

$$\left| \frac{d}{dt} z\,(qt+s)^n \right| \leqq \frac{1}{2}.$$

Therefore, by Lemma 2, we obtain

$$D(z) = \int_{-sq^{-1}}^{(P-s)q^{-1}} e^{2\pi i z (qt+s)^n}\, dt + 5\theta = \frac{1}{q}\psi + 5\theta;$$

$$\psi = \int_0^P e^{2\pi i z x^n}\, dx, \quad L_\alpha = \psi \frac{S_{a,q}}{q} + 5\theta' q;$$

but, by Lemmas 3 and 4, we get

$$\psi \ll Z, \quad \frac{S_{a,q}}{q} \ll q^{-\nu}, \quad \psi \frac{S_{a,q}}{q} \ll Zq^{-\nu};$$

$$Z = \begin{cases} P, & \text{if } |z| \leq P^{-n}, \\ |z|^{-\nu}, & \text{if } |z| > P^{-n}. \end{cases}$$

Obviously, $Zq^{-\nu} > q$; therefore

$$L_\alpha^{4n} e^{-2\pi i \alpha N_1} = \psi^{4n} \left(\frac{S_{a,q}}{q}\right)^{4n} e^{-2\pi i \frac{a}{q} N_1 - 2\pi i z N_0} + O(f);$$

$$f = Z^{4n-1} q^{-3+\nu} + Z^{4n} q^{-4} z P^{n-\nu}.$$

Hence, we obtain

$$H_{a,q}(N_1) = \left(\frac{S_{a,q}}{q}\right)^{4n} e^{-2\pi i \frac{a}{q} N_1} R_q(N_0) + F;$$

$$R_q(N_0) = \int_{-(q\tau)^{-1}}^{(q\tau)^{-1}} \psi^{4n} e^{-2\pi i z N_0}\, dz, \qquad F \ll \int_0^{(q\tau)^{-1}} f\, dz \ll P^{3n-\nu} q^{-3+\nu}.$$

Further, we easily find

$$\int_{-\infty}^{-(q\tau)^{-1}} \psi^{4n} e^{-2\pi i z N_0}\, dz + \int_{(q\tau)^{-1}}^{\infty} \psi^{4n} e^{-2\pi i z N_0}\, dz \ll P^{3n-3} q^3.$$

Consequently,

$$H_{a,q}(N_1) = \left(\frac{S_{a,q}}{q}\right)^{4n} e^{-2\pi i \frac{a}{q} N_1} R(N_0) + O(P^{3n-\nu} q^{-3+\nu});$$

$$R(N_0) = \int_{-\infty}^{\infty} \psi^{4n} e^{-2\pi i z N_0}\, dz.$$

But, we have

$$I'(N_1) = \int_{-\tau^{-1}}^{\tau^{-1}} L_\alpha^{4n} e^{-2\pi i \alpha N_1}\, d\alpha + \int_{\tau^{-1}}^{1-\tau^{-1}} L_\alpha^{4n} e^{-2\pi i \alpha N_1}\, d\alpha,$$

where the left-hand side is the number of representations of N_1 of the form $x_1^n + \ldots + x_{4n}^n$ ($x_1, \ldots, x_{4n}$ are positive integers), while the first term on the right-hand side is equal to

$$H_{0,1}(N_1) = R(N_0) + O(P^{3n-\nu}).$$

Therefore

$$\sum_{N'=1}^{M} \sum_{N''=1}^{M} I'(N_0 - N' - N'') = M^2 R(N_0) + G,$$

where

$$G \ll M^2 P^{3n-\nu} + P^{4n} \int_{\tau^{-1}}^{1-\tau^{-1}} \frac{d\alpha}{4(\alpha)^2} \ll M^2 P^{3n-\nu}.$$

But, by Lemma 6, the left-hand side is equal to $4T_{4n} N^3 M^2 + O(N^{3-\nu^2} M^2)$. Hence, it follows that

$$R(N_0) = 4T_{4n} N_0^3 + O(P^{3n-\nu}),$$

$$H_{a,q}(N_0) = 4T_{4n} N_0^3 \left(\frac{S_{a,q}}{q}\right)^{4n} e^{-2\pi i \frac{a}{q} N_0} + O(P^{3n-\nu} q^{-3+\nu}).$$

Summing this expression over all a's corresponding to a given q, and then summing the resulting expression over all $q = 1, \ldots, [P^\gamma]$, we obtain

$$I'_0(N_0) = 4T_{4n} N_0^3 \sum_{0<q \leqq P^\gamma} A(q) + O(P^{3n-\nu}).$$

But $A(a) \ll q^{-3}$. Therefore

$$\sum_{q > P^\gamma} A(q) \ll P^{-2\gamma},$$

$$I'_0(N_0) = 4T_{4n} N_0^3\, \mathfrak{S}(N_0) + O(P^{3n-\nu}).$$

Consequently,

$$I'_0(N_0) \gg N^3$$

which completes the proof of the lemma.

Theorem. For an integer $n > 170{,}000$, we have

$$G(n) < n(2\ln n + 4\ln\ln n + 2\ln\ln\ln n + 13).$$

Proof. Let

$$k_0 = [0.5 \ln n], \quad k = 2k_0, \quad r' = [k^2(\ln k + 0.5\ln\ln k + 1.3)],$$

$$r_0 = 2r', \quad r = 2r_0, \quad \delta = \frac{n - k_0 - 0.5}{n - 0.5}, \quad k_2 = 3n.$$

Assuming that

$$X_j = [X_0^{\delta^j}], \quad Y_j = [\sqrt{X_j}], \quad j = 0, \ldots, k_2 - 1,$$

consider the variables A'_j and A''_j which run, independently of each other, through

$$(X_j + y_{j,1})^n + \ldots + (X_j + y_{j,r'})^n, \qquad y_{j,t} = 1, \ldots, Y_f.$$

Further, consider the variables $A_{j,1}$ and $A_{j,2}$ which run, independently of each other, through the values of $A'_j - A''_j$. Finally, consider the variable A_j which runs through the values of $A_{j,1} - A_{j,2}$. We readily find that all the values of $A_{j,1}$ (and $A_{j,2}$) obey the inequalities:

$$-r_0 n X_j^{n-1} Y_j < A_{j,1} < r_0 n X_j^{n-1} Y_j,$$

and all the values of A_j obey the inequalities:

$$-rnX_j^{n-1} Y_j < A_j < rnX_j^{n-1} Y_j.$$

Now consider the variables W_1 and W_2 which run, independently of each other, through

$$A_{0.1} + \ldots + A_{k_2-1,1},$$

and the variable W which runs through the values of $W_1 - W_2$.

Evidently, for some c_2, all the values of W_1 (and of W_2) obey the inequalities:

$$-c_2 X_0^{n-0.5} < W_1 < c_2 X_0^{n-0.5}.$$

For every x in the interval $-c_2 X_0^{n-0.5} < x < c_2 X_0^{n-0.5}$, let $\xi(x)$ denote the number of solutions of the equation $W_1 = x$, then the sum

$$\sum (\xi(x))^2$$

taken over all x in this interval is equal to the number of solutions of the equation $W = 0$. Now we shall estimate this number. The variable W can be represented in the following form:

$$W = A_0 + \ldots + A_{k_2-1}, \tag{2}$$

where, obviously,

$$A_j = X_j^{n-1} U_j^{(1)} + \ldots + X_j^{n-k} U_j^{(k)} + \ldots + X_j U_j^{(n-1)} + U_j^{(n)};$$

$$U_j^{(s)} = \binom{n}{s} \eta_j^{(s)}; \quad \eta_j^{(s)} = y_{j,1}^s + \ldots + y_{j,r_0}^s - y_{j,r_0+1}^s - \ldots - y_{j,r}^s$$

(each $y_{j,t}$ runs through $1, \ldots, Y_j$). Under these conditions, $U_j^{(s)}$ obeys the inequalities $-r_0 n^s Y_j^s < U_j^{(s)} < r_0 n^s Y_j^s$. By Lemma 1, the number of solutions of the system $U_j^{(1)} = l_j^{(1)}, \ldots, U_j^{(k)} = l_j^{(k)}$ is

$$\ll Y_j^{r - \frac{k(k+1)}{2}}.$$

Consider the sum

$$B_j = X_j^{n-1} l_j^{(1)} + \ldots + X_j^{n-k} l_j^{(k)},$$

where $l_j^{(s)}$ runs through the integers in the interval $-r_0 n^s Y_j^s < l_j^{(s)} < r_0 n^s Y_j^s$.

It is easily seen that an interval of length X_j^{n-k} contains

$$\ll \frac{Y_j^2}{X_j} \frac{Y_j^3}{X_j} \cdots \frac{Y_j^k}{X_j} = Y_j^{\frac{k(k+1)}{2} - 1} X_j^{-k+1}$$

values of B_j. Therefore, an interval of length $X_j^{n-k_0-0.5}$ contains

$$\ll X_j^{k_0-0.5} Y_j^{\frac{k(k+1)}{2} - 1} X_j^{-k+1} Y_j^{r - \frac{k(k+1)}{2}} \ll Y_j^r X_j^{-k_0}$$

values of A_j. Hence, from (2) it follows that an interval of length $X_{k_2-1}^{n-k_0-0.5}$ contains

$$\ll D; \quad D = (Y_0 \ldots Y_{k_2-1})^r (X_0 \ldots X_{k_2-1})^{-k_0} \ll (Y_0 \ldots Y_{k_2-1})^r X_0^{-(n-0.5)(1-\delta^{k_2})}$$

values of W. Consequently, the number of solutions of the equation $W = 0$ is $\ll D$.

Let $R = [X_0^{1-0.5\nu}]$ and let v run through the primes in the interval $R < v \leqq 2R$ (for some c_3 and c_4, we know that the number R_0 of primes is equal to

$$c_3 \frac{R}{\ln R} < R_0 < c_4 \frac{R}{\ln R}.$$

Further, let w run through the numbers

$$(X_0+y_0)^n+\ldots+(X_{k_2-1}+y_{k_2-1})^n;\quad y_j=1,\ldots,Y_j.$$

Consider the sum

$$Q_\alpha=\sum_v\sum_w e^{2\pi i\alpha v^n w}.$$

We find

$$Q_\alpha^{r_0}\ll R_0^{r_0-1}V;\quad V=\sum_v\left|\sum_w e^{2\pi i\alpha v^n w}\right|^{r_0},$$

where V can be expressed as

$$V=\sum_v\sum_W e^{2\pi i\alpha v^n W}=\sum_v\sum_x \xi(x)\,e^{2\pi i\alpha v^n x}. \tag{3}$$

Representing α as $\alpha=a/q+z$, $(a,q)=1$, $0\leqq a<q$, $0<q\leqq\tau_0$, $|z|\leqq 1/q\tau_0$, we shall estimate the sum Q_α for the α's belonging to the auxiliary intervals. Here, without loss of generality, we may assume that $z\geqq 0$.

First consider the case where $q\geqq R$. We find

$$V\ll\sum_x\left|\sum_{y=0}^{q-1}\xi(x)\,\eta(y)\,e^{2\pi i x\frac{ay+\psi(y)}{q}}\right|,$$

where $0\leqq\psi(y)<1$ and $\eta(y)$ is the number of solutions of the congruence $v^n\equiv y \pmod q$. Here we have

$$\sum(\xi(x))^2\ll D,\quad \eta(y)\ll q^\varepsilon.$$

Further, we find (y_1 runs, independently of y, through the same values as y):

$$V^2\ll D\sum_x\sum_y\sum_{y_1}\eta(y)\,\eta(y_1)\,e^{2\pi i x\Phi(y,y_1)};$$

$$\Phi\,\Phi y,y_1)=\frac{a(y-y_1)+\psi(y)-\psi(y_1)}{q},$$

$$V^2\ll Dq^{2\varepsilon}\sum_{y_1}\sum_y\min\left(X_0^{n-0.5},\frac{1}{(\Phi(y,y_1))}\right).$$

For a given y_1, let ϱ denote the least residue of $d(y-y_1)$ modulo q. Then

$$(\Phi(y,y_1))=\left(\frac{\varrho+\sigma(\varrho)}{q}\right);\quad |\sigma(\varrho)|<1.$$

Therefore, rearranging the terms of the sum over y so that the values of $|\varrho|$ form a non-decreasing sequence, we find that the value of this sum does diminish if not more than the first three terms are replaced by $X_0^{n-0.5}$, not more than the next two terms by $q/1$, not more than the next two terms by $q/2$, and so on. Thus we obtain

$$V^2\ll Dq^{2\varepsilon}\sum_{y_1}\left(X_0^{n-0.5}+\sum_{s=1}^{q}\frac{q}{s}\right)\ll Dq^{3\varepsilon}R_0X_0^{n-0.5}$$

$$\ll(Y_0\ldots Y_{k_2-1})^r R_0^2 X_0^{-(1-0.5\nu)+(n-0.5)\delta^{k_2}+\varepsilon_1}, \tag{4}$$

$$Q_\alpha\ll Q'P^{-\sigma};\quad \sigma=\frac{1-0.5\nu-(n-0.5)\delta^{k_2}+\varepsilon'}{2r_0},$$

where $Q'=(Y_0\ldots Y_{k_2-1})R_0$ is the number of terms in the sum Q_α.

Now consider the case where $\sqrt{R} < q \leqq R$. Here, after slight modification, the proof used above can be applied to demonstrate this case, too. In (4), take $\frac{R^2}{q^2} q^{\varepsilon}$ and q for $q^{3\varepsilon}$ and R_0, respectively. Thus, we obtain

$$V^2 \ll D \frac{R^2}{q} q^{\varepsilon} X_0^{n-0.5},$$

Accordingly,

$$\sigma_1 = \frac{0.5(1-0.5\nu)-(n-0.5)\delta^{k_2}+\varepsilon'}{2r_0}.$$

Finally, consider the case where $q < \sqrt{R}$. Here, we have

$$\frac{1}{q\tau} < z < \frac{1}{q\tau_0}.$$

That part of the sum (3) which corresponds, for a given s taken from the sequence $0, \ldots, q-1$, to the values of v of the type $qt+s$ (t is an integer) can be expressed as

$$V_s = \sum_x \sum_t \xi(x)\,\eta(t)\,e^{2\pi i x \Phi(t)}; \quad \Phi(t) = \frac{as^n}{q} + z(qt+s)^n,$$

where t runs through the integers satisfying the condition $R < (gt+s) \leqq 2R$, while $\eta(t)$ is equal to 1 or 0 depending on whether $gt+s$ is a prime or not. Hence, we obtain

$$V_s^2 \ll X' \sum_{t_1} F;$$

$$F = \sum_t \min\left(X_0^{n-0.5}, \frac{1}{(\Phi(t)-\Phi(t_1))}\right); \quad X' = \sum_x (\xi(x))^2.$$

But, on replacing any value of t by the same value plus one (if there exists such a value), we find that the value of $\Phi(t)$ increases by a number of the type

$$nz(qt+q\theta+s)^{n-1}q,$$

which lies between the limits qA^{-1} and $q\cdot 2^{n-1}A^{-1}$; $A^{-1} = nzR^{n-1}$. These limits in turn lie between the bounds $1/R$ and $1/(X_0^{n-0.5-0.5\nu})$. Therefore, after rearranging the terms in the sum F so that the values of $(\Phi(t)-\Phi(t_1))$ form a non-decreasing sequence, we find that this sum does not diminish if not more than the first three terms are replaced by $X_0^{n-0.5}$, not more than the next two terms by $A/1$, not more than the next two terms by $A/2$, and so on. Thus, we obtain

$$F \ll X_0^{n-0.5} + \sum_{s=1}^{R} \frac{A}{s} \ll X_0^{n-0.5},$$

$$V^2 \ll X' R_0 X_0^{n-0.5}, \quad Q_\alpha \ll Q' P^{-\sigma_1}.$$

Thus, we always have

$$Q_\alpha \ll Q' P^{-\sigma_1}.$$

Now, let $P_0 = [0.25P]$, $P_1 = [0.5P_0^{1-\nu}]$, $\ldots$, $P_{k_3-1} = [0.5P_{k_3-2}^{1-\nu}]$, $k_3 = [n(\ln n + 2\ln\ln n + \ln\ln\ln n + 3)]$ and let ξ_s run through $P_s^n, (P_s+1)^n, \ldots, (2P_s-1)^n$. Now let us form the numbers

$$u = \xi_0 + \xi_1 + \ldots + \xi_{k_3-1}.$$

These numbers are not equal to each other. They satisfy the condition:

$$P_0^n < u < (2P_0)^n;$$

the total number U of these numbers is equal to $P_0 P_1 \ldots P_{k_3-1}$. It satisfies the condition:

$$U \gg P^{n-n(1-\nu)^{k_3}}.$$

Consider the integral:

$$I(N) = \int_{-\tau_0^{-1}}^{-\tau_0^{-1}+1} L_\alpha^{4n} Q_\alpha S_\alpha^2 e^{-2\pi i \alpha N} d\alpha;$$

$$L_\alpha = \sum_{x=1}^{P} e^{2\pi i \alpha x^n}, \quad S_\alpha = \sum_u e^{2\pi i \alpha u}.$$

In accordance with the division of its integration range into main and auxiliary intervals, the integral $I(N)$ can be expressed as the sum of two subintegrals:

$$I(N) = I_0(N) + I_1(N).$$

First, we shall estimate $I_1(N)$. For some c_5, we find

$$I_1(N) \ll P^{4n} Q' U^2 P^{-n-\sigma_1+n(1-\nu)^{k_3}} \ll P^{4n} Q' U^2 P^{-n} P^{-c_5}.$$

Now, we shall estimate $I_0(N)$. Evidently, (u_1 runs through the same values as u), we have

$$I_0(N) = \sum_v \sum_w \sum_u \sum_{u_1} I_0'(N - vw - u - u_1),$$

where $0.5N < N - vw - u - u_1 < N$. Therefore, applying Lemma 7, we find

$$I_0(N) \gg P^{3n} Q' U^2 = P^{4n} Q' U^2 P^{-n},$$

from which it follows that $I(N) \gg P^{4n} Q' U^2 P^{-n} > 0$. Consequently, N can represented as the sum of

$$4n + k_2 + 2k_3 \leq n(2\ln n + 4\ln\ln n + 2\ln\ln\ln n + 13)$$

terms of the type x^n, where x is positive. This completes the proof of our theorem.

Chapter 3. Approximation by Means of the Fractional Parts of the Values of a Polynomial

In this chapter we shall apply my method to approximate a given proper fraction by means of the fractional parts of the values of a polynomial [18]. The theorem derived in this chapter is remarkable in that with a small number g of nonzero terms of a polynomial we can obtain far more accurate results than those that can be derived from Theorem 1 of Chapter 5 in the monograph [2].

Special Notation. Let $f(x) = a_h x^h + \ldots + a_n x^n$ $(n > 4)$ be a real polynomial with g nonzero terms whose exponents are positive and arranged in increasing order of magnitude, and the sum of the exponents $h + \ldots + n$ is equal to D. Let f be one of the exponents, and

$$\gamma = \frac{1}{g}, \quad \varphi = \frac{1}{f}, \quad \nu = \frac{1}{n}, \quad \varrho = \frac{\gamma\varphi\nu}{4} \frac{\ln D}{(\ln D + 1)\ln(D\ln D + D)}.$$

Let a_f be represented as

$$a_f = \frac{a}{q} + \frac{\theta}{q^2}, \quad (a, q) = 1, \quad q \geqq c_0,$$

where c_0 is sufficiently large. Then, we shall assume that

$$b = \left[\frac{\ln(D\ln D + D)}{-\ln(1-\nu)} + 1\right],$$

$$p = \left[\left(\frac{q}{2}\right)^{\varphi}\right], \quad p_t = p^{(1-\nu)^{t-1}}; \quad t = 1, \ldots, b, \quad \sigma = (1-\nu)^b.$$

Finally, let

$$\Delta = q^{-\varrho}, \quad M = q^{\varrho(1+\varepsilon_1)}, \quad \beta = [(b+1)\, g^{\varepsilon_1^{-1}} + 1],$$

where ε_1 is sufficiently small.

Having chosen an increasing sequence of positive constants $c_1, \ldots, c_g$ less than unity with a nonzero determinant

$$\begin{vmatrix} c_1^{h-1} & \ldots & c_g^{h-1} \\ \cdot & \cdot\;\cdot\;\cdot & \cdot \\ c_1^{n-1} & \ldots & c_g^{n-1} \end{vmatrix}$$

and taking

$$X_{t,s} = [p_t c_5], \quad \xi_t = [p_t^{1-\varepsilon}]; \quad t = 1, \ldots, b, \quad s = 1, \ldots, g,$$

we shall use the symbol $x_{t,s}$ to denote a variable that runs through the integers in the interval

$$X_{t,s} - \xi_t < x_{t,s} \leqq X_{t,s} + \xi_t$$

and $G_{t,s}$ to denote the number $2\xi_t$ of numbers $x_{t,s}$ and G to denote the number of systems $(x_{1,1}, \dots, x_{1,g}, \dots, x_{b,1}, \dots, x_{b,g})$. Evidently,

$$G = G_{1,1} \dots G_{1,g} \dots G_{b,1} \dots G_{b,g} \gg (p_1 \dots p_b)^{g(1-\varepsilon)}.$$

For each $t = 1, \dots, b$, taking some definite integers $m_{t,1}, \dots, m_{t,g}$ satisfying the conditon $0 < m_{t,s} \leqq M$, we shall assume that

$$U_{t,r} = m_{t,1} x_{t,1}^r + \dots + m_{t,g} x_{t,g}^r,$$
$$U_r = U_{1,r} + \dots + U_{b,r}; \quad r = h, \dots, n.$$

1. Number of Solutions of Some Systems of Equations. After choosing certain intervals of lengths

$$Mp_t^{h-1}, \dots, Mp_t^{n-1},$$

we shall estimate the number Φ_t' of systems $(x_{t,1}, \dots, x_{t,g})$ for which $U_{t,h}, \dots, U_{t,n}$ lie in these intervals, respectively. Let $(x_1, \dots, x_g)$, $(x_1 + \zeta_1, \dots, x_g + \zeta_g)$ be two such systems. Then

$$m_{t,1}((x_1+\zeta_1)^h - x_1^h) + \dots + m_{t,g}((x_g+\zeta_g)^h - x_g^h) = \theta_h M p_t^{h-1}$$
$$\dots\dots\dots\dots\dots\dots\dots\dots\dots\dots$$
$$m_{t,1}((x_1+\zeta_1)^n - x_1^n) + \dots + m_{t,g}((x_g+\zeta_g)^n - x_g^n) = \theta_n M p_t^{n-1},$$

which reduces to the form

$$m_{t,1} c_1^{h-1} \beta_{h,1} \zeta_1 + \dots + m_{t,g} c_g^{h-1} \beta_{h,g} \zeta_g = \frac{\theta_h}{h} M$$
$$\dots\dots\dots\dots\dots\dots\dots\dots$$
$$m_{t,1} c_1^{n-1} \beta_{n,1} \zeta_1 + \dots + m_{t,g} c_g^{n-1} \beta_{n,g} \zeta_g = \frac{\theta_n}{n} M,$$

where, for sufficiently large $c_0 = c_0(\varepsilon, n)$, all $\beta_{r,s}$ will be so close to unity that the determinant

$$\begin{vmatrix} c_1^{h-1}\beta_{h,1}, \dots, c_g^{h-1}\beta_{h,g} \\ \dots\dots\dots\dots \\ c_1^{n-1}\beta_{n,1}, \dots, c_g^{n-1}\beta_{n,g} \end{vmatrix}$$

is $\gg 1$. Hence, we easily find that

$$m_{t,s}\zeta_s \ll M, \quad \zeta_s \ll \frac{M}{m_{t,s}}; \quad s = 1, \dots, g.$$

Therefore

$$\Phi_t' \ll \frac{M^g}{m_{t,1} \dots m_{t,g}}.$$

Further, after choosing certain intervals of lengths

$$Mp_t^{h(1-\nu)}, \dots, Mp_t^{n(1-\nu)},$$

we shall estimate the number Φ_t of systems $(x_{t,1}, \ldots, x_{t,g})$ for which $U_{t,h}, \ldots, U_{t,n}$ lie in these intervals, respectively. Since

$$\frac{Mp_t^{h(1-\nu)}}{Mp_t^{h-1}} \cdots \frac{Mp_t^{n(1-\nu)}}{Mp_t^{n-1}} = p_t^{-\nu D+g},$$

we have

$$\Phi_t \ll \frac{M^g p_t^{-\nu D+g}}{m_{t,1} \ldots m_{t,g}}.$$

For the numbers $U_{t,r}$, we have also the inequality

$$U_{t,r} \ll Mp_t^r.$$

Let $\psi(z_h, \ldots, z_n)$ denote the number of solutions of the system

$$U_h = z, \ldots, U_n = z_n.$$

For the systems

$$\begin{gathered}(z_h, \ldots, z_n)\\(U_{1,h}, \ldots, U_{1,n})\\\ldots\ldots\ldots\ldots\\(U_{t-1,h}, \ldots, U_{t-1,n})\end{gathered}$$

the numbers $(U_{t,h}, \ldots, U_{t,n})$ will occur in definite intervals of lengths

$$\ll Mp_t^{h(1-\nu)}, \ldots, \ll Mp_t^{n(1-\nu)}$$

respectively, and, consequently, the number of different such systems is $\ll \Phi_t$. Therefore

$$\psi(z_h, \ldots, z_n) \ll \Phi;$$

$$\Phi = \Phi_1 \ldots \Phi_b = \frac{M^{gb}(p_1 \ldots p_b)^{g-\nu D}}{m_{1,1} \ldots m_{1,g} \ldots m_{b,1} \ldots m_{b,g}}. \tag{1}$$

The number U_r is also such that

$$U_r \ll Mp^r. \tag{2}$$

2. Estimate of the Main Sum. We shall now estimate the sum

$$T = \sum_{y=1}^{p} \left| \sum_0 e^{2\pi i(\alpha_h y^h U_h + \ldots + \alpha_n y^n U_n)} \right|,$$

where $\sum_0$ denotes summation taken over all G systems $(x_{1,1}, \ldots, x_{1,g}, \ldots, x_{b,1}, \ldots, x_{b,g})$, i.e. over all G systems $(U_h, \ldots, U_n)$. We find

$$T^2 \ll p \sum_{y=1}^{p} \sum_0 \sum_0 e^{2\pi i(\alpha_h y^h(U_h' - U_h) + \ldots + \alpha_n y^n(U_n' - U_n))},$$

where the systems $(U_h', \ldots, U_n')$ are the same as the systems $(U_h, \ldots, U_n)$.

Let $\psi(u_h, \ldots, u_n)$ be the number of solutions of the system

$$U_h' - U_h = u_h, \ldots, U_n' - U_n = u_n. \tag{3}$$

By virtue of (1), we obviously have

$$\Psi(u_h, \ldots, u_n) \ll G\Phi.$$

According to (2), there exist $c^{(h)}, \ldots, c^{(n)}$ such that the system (3) is possible only when $u_h, \ldots, u_n$ lie in the intervals

$$-c^{(h)}Mp^h \leqq u_h \leqq c^{(h)}Mp^h, \ \ldots, \ -c^{(n)}Mp^n \leqq u_n \leqq c^{(n)}Mp^n,$$

respectively. Making $u_h, \ldots, u_n$ run through the integers in these intervals, we find

$$\sum_{u_h} \cdots \sum_{u_n} (\Psi(u_h, \ldots, u_n))^2 \ll G\Phi \sum_{u_h} \cdots \sum_{u_n} \Psi(u_h, \ldots, u_n) = G^3\Phi.$$

Hence it follows that

$$|T|^2 \leqq p \sum_{u_h} \cdots \sum_{u_n} \Psi(u_h, \ldots, u_n) \sum_{y=1}^{p} e^{2\pi i(\alpha_h y^h u_h + \ldots + \alpha_n y^n u_n)},$$

$$T^4 \ll \frac{p^2 G^3 \Phi}{Mp^f} \sum_{u_h} \cdots \sum_{u_f'} \sum_{u_f} \cdots \sum_{u_n} \left| \sum_{y=1}^{p} e^{2\pi i(\alpha_h y^h u_h + \ldots + \alpha_f y^f (u_f' + u_f) + \ldots + \alpha_n y^n u_n)} \right|^2,$$

where u_f' is an additional variable which runs through the integers in the interval

$$-2c^{(j)}Mp^j \leqq u_f' \leqq 2c^{(j)}Mp^j.$$

Hence we find

$$T^4 \ll p^2 G^3 \Phi M^{-1} p^{-f} M^{g-1} p^{D-f} \sum_{u_f'} \sum_{u_f} \sum_{y_1=1}^{p} \sum_{y=1}^{p} e^{2\pi i \alpha_f (y_1^f - y^f)(u_f' + u_f)}$$

$$\ll G^3 \Phi M^{g-2} p^{D-2f+2} \sum_{y_1}^{p} \sum_{y=1}^{p} \min\left(M^2 p^{2f}, \frac{1}{4(\alpha_f(y_1^f - y^f))^2}\right).$$

For a given y_1, the difference $t = y_1^f - y^j$ runs through certain integers in the interval $-0.5q + 1 \leqq t \leqq 0.5q - 1$. Denoting by r the least residue of $at \pmod q$, we shall express $(\alpha_f t)$ as

$$(\alpha_f t) = \left(\frac{r + 0.5\theta_r}{q}\right).$$

Here, for $|r| > 0$, $(\alpha_f t)$ will always be greater than $(|r| - 0.5)/q$. Therefore, substituting $M^2 P^{2t}$ and $g/(|r| - 0.5)$ for the summand in the double sum when $t = 0$ and $|r| > 0$, respectively, we obtain

$$\sum_{y_1=1}^{p} \sum_{y=1}^{p} \min\left(M^2 p^{2f}, \frac{1}{4(\alpha_f(y_1^f - y^f))^2}\right) \ll pM^2 p^{2f},$$

$$T^4 \ll G^3 \Phi M^g p^{D+3},$$

$$\frac{T}{pG} \ll (G^{-1}\Phi M^g p^{D-1})^{1/4} \ll \left(\frac{M^{g(b+1)}(p_1 \ldots p_b)^{g\varepsilon - \nu D} p^{D-1}}{m_{1,1} \ldots m_{1,g} \ldots m_{b,1} \ldots m_{b,g}}\right)^{1/4},$$

$$T \ll pG\left(\frac{q^{\varrho g(b+1) - \varphi + \varphi\sigma D + \varepsilon'}}{m_{1,1} \ldots m_{1,g} \ldots m_{b,1} \ldots m_{b,g}}\right)^{1/4}.$$

Theorem. There exists a $c = c(n)$ such that for any real A, the system of inequalities

$$|f(z) - v - A| < cq^{-\varrho}, \quad 0 < z < p^2$$

is satisfied by integers z and v.

Proof. By Lemma 2 of Chapter 1, there exists a function $\lambda(z)$ with period 1 having the following properties:

1. $0 \leqq \lambda(z) \leqq 1$ in the interval $A - \Delta \leqq z \leqq A + \Delta$,
2. $\lambda(z) = 0$ in the interval $A + \Delta \leqq z \leqq 1 + A - \Delta$,
3. $\lambda(z)$ can be expanded as a Fourier series of the type:

$$\lambda(z) = \Delta + \lambda_0(z); \quad \lambda_0(z) = \sum_{m=1}^{\infty} (B_m e^{2\pi imz} + B'_m e^{-2\pi imz}),$$

where B_m and B'_m depend only on m, A, Δ, and satisfy the inequalities $B_m \ll F(m)$, $B'_m \ll F(m)$, where

$$F(m) = \begin{cases} \Delta, & \text{if } m \leqq \dfrac{1}{\Delta}, \\ \dfrac{1}{\Delta^\beta m^{\beta+1}}, & \text{if } m > \dfrac{1}{\Delta}. \end{cases}$$

Consider the sum

$$H = \sum_{y=1}^{p} \left| \sum_{x_{1,1}} \lambda_0(f(yx_{1,1})) \times \ldots \right.$$
$$\left. \times \sum_{x_{1,g}} \lambda_0(f(yx_{1,g})) \ldots \sum_{x_{b,1}} \lambda_0(f(yx_{b,1})) \ldots \sum_{x_{b,g}} \lambda_0(f(yx_{b,g})) \right|.$$

Here we have

$$\sum_{x_{t,s}} \lambda_0(f(yx_{t,s})) \ll \sum_{m_{t,s}=1}^{\infty} F(m_{t,s}) \left| \sum_{x_{t,s}} e^{2\pi i m_{t,s} f(yx_{t,s})} \right|.$$

Therefore

$$H \ll \sum_{m_{1,1}=1}^{\infty} F(m_{1,1}) \ldots \sum_{m_{1,g}=1}^{\infty} F(m_{1,g}) \ldots \sum_{m_{b,1}=1}^{\infty} F(m_{b,1}) \times \ldots \times \sum_{m_{b,g}=1}^{\infty} F(m_{b,g}) \, |T|, \tag{4}$$

where T is the sum considered at the beginning of this paragraph.

Since

$$\sum_{m=1}^{\infty} F(m) \ll 1, \quad \sum_{m \leqq M} \frac{F(m)}{m^{1/4}} \ll \Delta M^{3/4}$$

and

$$\sum_{m > M} F m) \ll \frac{1}{\Delta^\beta M^\beta} \ll q^{-\varrho g(b+1)}$$

the sum of the terms on the right-hand side of (4), provided at least one $m_{t,s}$ exceeds M, is

$$\ll pGq^{-\varrho g(b+1)} = pG\Delta^{gb} q^{-\varrho g},$$

while the sum of the remaining terms is

$$\ll pGq^{\frac{\varrho g(b+1)}{4}-\frac{\varphi}{4}+\frac{\varphi\sigma D}{4}+\frac{3\varrho gb}{4}+\varepsilon\Delta gb} \ll pG\Delta^{gb} q^{\varrho g\left(b+\frac{1}{4}\right)-\frac{\varphi}{4}+\frac{\varphi\sigma D}{4}+\varepsilon'},$$

where the exponent of q is also negative. Therefore, for certain c_1 and c_2, we have

$$H < pGc_1^{gb}\Delta^{gb}q^{-c_2 gb}.$$

Hence it follows that for some y and certain t and s, the inequality

$$\frac{\left|\sum_{x_{t,s}} \lambda_0(f(yx_{t,s}))\right|}{G_{t,s}} < c_1\Delta q^{-c_2} < \Delta,$$

is satisfied, where c_0 is sufficiently large, while for some particular value of $x_{t,s}$, the inequality

$$\lambda(f(yx_{t,s})) > 0$$

is also satisfied. This completes the proof of our theorem.

Example. Let α be real and

$$f(x) = \alpha x^n + x\sqrt{2}.$$

Expanding $\sqrt{2}$ as a continued fraction, consider that pair of adjacent fractions, whose denominators q and q' satisfy the condition $q \leqq \sqrt{\tau} < q'$. Since the partial quotients are bounded, we have $q' \ll q$. Therefore, applying our theorem, we find that two integers z and v indeed satisfy the system:

$$|f(z) - v - A| < c\tau^{-\varrho}, \quad 0 < z < \tau;$$

$$\varrho = \frac{\ln(n+1)}{32n(\ln(n+1)+1)\ln((n+1)\ln(n+1)+(n+1))}.$$

Chapter 4. Estimates of the Simplest Sums over Prime Numbers

In this chapter, using the sums of the type (1) mentioned in the Introduction, we shall derive the estimates of several simple sums over prime numbers. The procedure used here in the derivation is a modification of the scheme which I used first in 1937 to evaluate the sum

$$\sum_{p \leqq N} e^{2\pi i k \alpha p}$$

in a paper dealing with the asymptotic expression in the Goldbach ternary representation problem. We may note that a more sophisticated variant of this method to evaluate a sum of the type

$$\sum_{p \leqq N} e^{2\pi i f(p)}; \quad f(p) = \alpha_n p^n + \ldots + \alpha_1 p$$

is given in detail in Chapter 7 of the monograph[2].

Because of the diversity of the sums being studied, I feel it useful to divide the chapter into three sections.

Section I includes the general theorems concerning the estimates of the sums

$$\sum_{p \leqq N} e^{2\pi i \alpha p}, \quad \sum_{k=1}^{K} \left| \sum_{p \leqq N} {}^{2\pi i k \alpha p} \right|$$

and some of the applications of these estimates.

Section II is devoted to the estimate of the sum[21]

$$\sum_{p \leqq N} e^{-2\pi i k f \sqrt{p}}$$

and its application in deriving the law of distribution of the values of the fraction:

$$\{f\sqrt{p}\}; \quad p \leqq N.$$

Finally, Section III deals with the estimate of the sum[22]:

$$\sum_{p \leqq N} \chi(p+k).$$

It should be noted that my method, in combination with the results obtained by the Hasse-Weyl method, also gives a non-trivial estimate of this sum under more stringent conditions[23, 24].

Section I. Estimates of the Sums $\sum_{p \leqq N} e^{2\pi i \alpha p}$ and $\sum_{k=1}^{K} \left| \sum_{p \leqq N} e^{2\pi i k \alpha p} \right|$

Lemma 1. For a real α satisfying the condition

$$\alpha = \frac{a}{q} + \frac{\theta}{q^2}, \quad (a, q) = 1, \quad 1 < q < N,$$

a) the sum

$$V_g = \sum_{z=g}^{g+q'} \min\left(U, \frac{1}{(\alpha z)}\right); \quad q' < q,\ U > 0,$$

obeys the inequality

$$V_g \ll U + \ln q;$$

and

b) the sum

$$V = \sum_{0<z\leqq 0.5q} \frac{1}{(\alpha z)}$$

obeys the inequality

$$V \ll q \ln q.$$

Proof. Without loss of generality, we can assume that $q \geqq 6$.

a) Using the substitution $z = g + t$, and expressing $q\alpha g$ as $B + \theta'$, where B is an integer and $\theta' < 0$, when $\theta \geqq 0$, and $\theta' \geqq 0$, when $\theta < 0$, we obtain

$$(\alpha z) = \left(\frac{B + \theta' + at + \theta t q^{-1}}{q}\right) = \left(\frac{\varrho + \theta_\varrho}{q}\right),$$

where ϱ is the least non-negative residue of $B + a \cdot t$ modulo q. The sum V_g takes the form:

$$V_g = \sum_{\varrho} \min\left(U, \frac{1}{\left(\dfrac{\varrho + \theta_\varrho}{q}\right)}\right).$$

Now substitute U for the terms with $\varrho = q - 1$, $\varrho = 0$, $\varrho = 1$ and substitute q/s for the terms with $\varrho = q - 1 - s$ and $\varrho = s + 1$, when $0 < s < 0.5q$ (terms with lesser s are not considered). Thus, we get

$$V_g \ll U + q \ln q.$$

b) We have

$$(\alpha z) = \left(\frac{az + \theta q^{-1} z}{q}\right) = \left(\frac{\varrho + 0.5\theta_\varrho}{q}\right)$$

where ϱ is the least non-negative residue of az modulo q. The sum V takes the form:

$$V = \sum_{\varrho} \frac{1}{\left(\dfrac{\varrho + 0.5\theta_\varrho}{q}\right)}.$$

When $0 < s \leqq 0.5q$, each term with $\varrho = q - s$ and $\varrho = s$ has been replaced by $1/(s - 0.5q)$ which is not less than or equal to the term. Thus, we obtain

$$V \ll q \ln q.$$

This completes the proof of our lemma.

Lemma 2. Let u and v run through increasing sequences of natural numbers, and let

$$1 < U < N; \quad 2U \leqq U' < 4U,$$
$$\alpha = \frac{a}{q} + \frac{\theta}{q^2}, \quad (a, q) = 1, \quad 1 < q < N,$$
$$S = \sum_{U < u \leqq U'} \sum_{uv \leqq N} e^{2\pi i \alpha uv}.$$

Then

$$S \ll N \sqrt{\frac{1}{q} + \frac{q \ln q}{N} + \frac{\ln q}{U} + \frac{U}{N}}.$$

Proof. We find that

$$S^2 \ll U \sum_{U < t \leqq U'} \sum_{tv \leqq N} \sum_{tv_1 \leqq N} e^{2\pi i \alpha t (v - v_1)},$$

where t runs through all the natural numbers within the limits shown. Changing the order of summation (first, for a given v and v_1, summing over t), we obtain

$$S^2 \ll U \sum_{v \leqq N/U} \sum_{v_1 \leqq N/U} \min\left(U, \frac{1}{(\alpha (v - v_1))}\right).$$

That part of the expression on the right-hand side which corresponds to a given v_1, by Lemma 1a, is

$$\ll U\left(\frac{N}{Uq} + 1\right)(U + q \ln q).$$

Therefore

$$S^2 \ll N\left(\frac{N}{Uq} + 1\right)(U + q \ln q) \ll N^2 \left(\frac{1}{q} + \frac{q \ln q}{N} + \frac{\ln q}{U} + \frac{U}{N}\right);$$

hence, we find that our lemma is true.

Theorem 1. Let $H = e^{0.5\sqrt{r}}$; $r = \ln N$,

$$\alpha = \frac{a}{q} + \frac{\theta}{q^2}, \quad (a, q) = 1, \quad 1 < q < N, \quad S = \sum_{p \leqq N} e^{2\pi i \alpha p}.$$

Then

$$S \ll N r \ln r \left(\sqrt{\frac{1}{q} + \frac{q \ln q}{N}} + H^{-1} \sqrt{\ln q}\right).$$

Proof. Let P be the product of all primes not exceeding H^2. Let d run through the divisors of P, m run through natural numbers, and y through the products of different primes exceeding H^2. We find that

$$\sum_{y \leqq N} e^{2\pi i \alpha y} + O(NH^{-2}) = \sum_{d \leqq N} \mu(d) S_d; \qquad (1)$$
$$S_d = \sum_{0 < m \leqq Nd^{-1}} e^{2\pi i \alpha d m}.$$

Express the right-hand side of (1) in the form $T'+T''$, where T' contains the terms satisfying the condition $d > NH^{-2}$, while T'' contains the terms satisfying the condition $d \leqq NH^{-2}$. First we shall evaluate T'. Let κ be the number of prime divisors of d contained in T'. Then, we have

$$H^{2\kappa} > NH^{-2}, \quad \kappa > \sqrt{r} - 1, \quad \tau(d) < 2^{\sqrt{r}-1}.$$

Therefore, by virtue of the inequalities

$$|S_d| \leqq \frac{N}{d}, \qquad \sum_{0<d\leqq N} \frac{\tau(d)}{d} \ll r^2$$

(inequality (1) given in Chapter 1 of the monograph[2]), we get

$$T' \ll \sum_{NH^{-2}<d\leqq N} \frac{N}{d}\,\frac{\tau(d)}{2^{\sqrt{r}-1}} \ll \frac{Nr^2}{2^{\sqrt{r}-1}} \ll NH^{-1.3}.$$

Now we shall estimate T''. Let s_0 be the largest integer s that satisfies the condition $(s_0-0.5)\,q < NH^{-2}$. We find

$$S_d \ll \min\left(\frac{N}{d}, \frac{1}{(\alpha d)}\right), \qquad T'' \ll \sum_{d\leqq NH^{-2}} \min\left(\frac{N}{d}, \frac{1}{(\alpha d)}\right).$$

Expanding the right-hand side as follows

$$\sum_{0<d\leqq 0.5q} + \sum_{0.5q<d\leqq 1.5q} + \ldots + \sum_{(s-0.5)q<d\leqq(s+0.5)q} + \ldots + \sum_{(s_0-0.5)q<d\leqq NH^{-2}}$$

and then applying the estimate (b) of Lemma 1 to the first term, and the estimate (a) to the rest of the terms, we obtain

$$T'' \ll q\ln q + \sum_{0<s\leqq s_0} \left(\frac{N}{qs} + q\ln q\right)$$

$$\ll q\ln q + \frac{Nr}{q} + NH^{-2}\ln q \ll Nr\left(\frac{q}{N} + \frac{1}{q} + H^{-2}\right).$$

Finally, consider the left-hand side of (1). Denoting by y_k the product of y's which have exactly k prime factors, consider the sum

$$T_k = \sum_{y_k\leqq N} e^{2\pi i\alpha y_k}.$$

Let $k>1$. Compare the sum T_k with the sum

$$T_k' = \sum_{H^2<y_1\leqq NH^{-2(k-1)}} \ \sum_{y_1 y_{k-1}\leqq N} e^{2\pi i\alpha y_1 y_{k-1}}.$$

The number of terms in the sum T_k' containing $y_1 y_{k-1}$ that are divisible by the squares of numbers other than unity, is evidently $\ll NH^{-2}$. The remaining terms are the same as those in the sum T_k, but each term of the sum T_k is repeated k times in the sum T_k'. Therefore

$$T_k = \frac{1}{k}\,T_k' + O\left(\frac{1}{k}\,NH^{-2}\right).$$

Now we shall evaluate T_k'. Divide the interval $H^2 < y_1 \leqq NH^{-2(k-1)}$ into $\ll r$ subintervals of the type:

$$Y_1 < y_1 \leqq Y_1'; \quad 2Y_1 \leqq Y_1' < 4Y_1.$$

By Lemma 2, that part of the sum T_k' which corresponds to a subinterval of this type is

$$\ll N\sqrt{\frac{1}{q}+\frac{q\ln q}{N}+\frac{\ln q}{Y_1}+\frac{Y_1}{N}} \ll N\left(\sqrt{\frac{1}{q}+\frac{q\ln q}{N}}+H^{-1}\sqrt{\ln q}\right).$$

On multiplying this expression by r, we obtain the bound for T_k'. Consequently, we have

$$T_k \ll \frac{1}{k}Nr\left(\sqrt{\frac{1}{q}+\frac{q\ln q}{N}}+H^{-1}\sqrt{\ln q}\right).$$

On summing this inequality over all $\ll r$ values of k, and since $T_1 = S + O(H)$, we find that the left-hand side of (1) is equal to

$$S+R; \quad R \ll Nr\ln r\left(\sqrt{\frac{1}{q}+\frac{q\ln q}{N}}+H^{-1}\sqrt{\ln q}\right).$$

Summing up all these facts, we find that our theorem is true.

Lemma 3. Let $\varepsilon \leqq 0.01$, $b_0 = e^{r^{1-\varepsilon}}$, $b = e^{r^{1-2\varepsilon}}$, $0 < q < e^{\sqrt{r}}$, $(l, q) = 1$, $0 \leqq l < q$, $U \geqq 0$, $W \geqq b_0$, F be the product of primes not greater than b that do not divide q. Let d run through the products of different primes not exceeding b_0 that do not divide q. Then,

a) the number T of numbers of the type $qx + l$ coprime to F in the interval

$$U < qx + l \leqq U + W,$$

satisfies the condition

$$T \ll \frac{Wr^{2\varepsilon}}{r\varphi(q)};$$

b) the following inequality holds:

$$\sum_d \frac{\mu(d)}{d} \ll \frac{r^{2\varepsilon}}{r};$$

c) the following inequality holds:

$$\sum_{d > N^{0.8}} \frac{1}{d} \ll e^{-r^{\varepsilon}}.$$

Proof. Assertion (a) follows from Lemma 9 proved in Chapter 7 of the monograph[2], while assertion (b) follows from the following estimate:

$$\sum_d \frac{\mu(d)}{d} = \prod_{p \leqq b_0}\left(1-\frac{1}{p}\right)\prod_{p \backslash q}\left(1-\frac{1}{p}\right)^{-1} \ll \frac{1}{\ln b_0}\,\frac{q}{\varphi(q)} \ll \frac{r^{2\varepsilon}}{r}.$$

We shall prove assertion (c). For $d > N^{0.8}$, we have $b_0^{\Omega(d)} > N^{0.8}$, and $\Omega(d) > 0.8r^{\varepsilon}$. Let $p_1, \ldots, p_{\sigma_0}$ be primes not exceeding b_0 that do not divide q. We find that

$$\sum_{\substack{d>N^{0.8}\\ \Omega(d)=s}} \frac{1}{d} < \frac{\left(\frac{1}{p_1}+\ldots+\frac{1}{p_{\sigma_0}}\right)^s}{s!} \ll \left(\frac{e(c+\ln r)}{s}\right)^s \ll e^{-1.25s},$$

$$\sum_{d>N^{0.8}} \frac{1}{d} \ll \sum_{s>0.8r^{\varepsilon}} e^{-1.25s} < e^{-r^{\varepsilon}}.$$

Theorem 2. Let $\varepsilon < 0.01$,

$$\alpha = \frac{a}{q} + z, \quad \delta = |zN|, \quad (a,q)=1,$$
$$0 \leqq a < q, \quad \delta \leqq e^{0.25r^{\varepsilon}}, \quad q \leqq e^{0.25r^{\varepsilon}}.$$

Then the sum

$$S = \sum_{p \leqq N} e^{2\pi i \alpha p}$$

is bounded by the inequality

$$S \ll N r^{-1+\varepsilon_1} q^{-0.5},$$

or, if $\delta \geqq 1$, by the inequality

$$S \ll N r^{-1+\varepsilon_1} q^{-0.5} \delta^{-0.5}.$$

Proof. Let $b_0 = e^{r^{1-\varepsilon}}$, and let F_0 be the product of primes which are not greater than b_0 and do not divide q. Let d run through the divisors of F_0, m through the natural numbers that are coprime to q, and, finally, y through the natural numbers coprime to $F_0 q$. We find

$$\sum_{y \leqq N} e^{2\pi i \alpha y} = \sum_{d \leqq N} \mu(d) S_d; \quad S_d = \sum_{m \leqq Nd^{-1}} e^{2\pi i \alpha d m}. \tag{2}$$

Express the right-hand side of (2) as $U' + U''$, where U' includes those terms for which $d > N^{0.8}$, while U'' contains those terms for which $d \leqq N^{0.8}$.

First we shall evaluate U'. Applying assertion (c) of Lemma 3, we obtain

$$U' \ll N e^{-r^{\varepsilon}}.$$

Now we shall estimate U''. Taking $m = qs + l$, where l runs through numbers such that $(l,q)=1$ $(0 \leqq l < q)$, and s runs, for a given l, through numbers such that $0 < qs + l \leqq Nd^{-1}$, we get (Lemma 3 of Chapter 2 in the monograph[2]):

$$S_d = \sum_l e^{2\pi i \frac{a}{q} dl} \sum_{-lq^{-1} < s \leqq Nd^{-1}q^{-1} - lq^{-1}} e^{2\pi i z d q s} + O((\delta+1)q)$$
$$= \frac{\mu(q)}{q} \frac{N}{d} I + O((\delta+1)q);$$
$$I = \int_0^1 e^{2\pi i z N v}\, dv, \quad I \ll \min(1, \delta^{-1}).$$

Hence, we obtain assertions (b) and (c) of Lemma 3:

$$U'' = \frac{\mu(q)}{q} N I' \sum_{d \leqq N^{0.8}} \frac{\mu(d)}{d} + O(N^{0.8}(\delta+1)q) \ll N \frac{r^{2\varepsilon}}{rq} \min(1, \delta^{-1}).$$

Finally, consider the left-hand side of (2). Denoting by y_k the product which has exactly k prime factors, consider the sum

$$T_k = \sum_{y_k \leqq N} e^{2\pi i \alpha y_k}.$$

For $k > 1$, compare this sum with the sum

$$T_k' = \sum_{b_0 < y_1 \leqq N b_0^{-(k-1)}} \sum_{y_1 y_{k-1} \leqq N} e^{2\pi i \alpha y_1 y_{k-1}}.$$

The number of terms in the sum T_k' containing $y_1 y_{k-1}$ that are divisible by the squares of numbers other than unity is evidently $\ll N b_0^{-1}$. The remaining terms are the same as those in the sum T_k, but each term of the sum T_k is repeated k times in the sum T_k'. Therefore

$$T_k = \frac{1}{k}\,(T_k' + O(N b_0^{-1})).$$

Now we shall evaluate the sum T'. Divide the interval $b_0 < y_1 \leqq N b_0^{-(k-1)}$ into $\ll r$ subintervals of the type:

$$X < y' \leqq X'; \quad 2X \leqq X' < 4X,$$

We shall use the notation:

$$T_k'(X) = \sum_{X < y_1 < X'} \sum_{y_1 y_{k-1} \leqq N} e^{2\pi i \alpha y_1 y_{k-1}}.$$

Putting $b = e^{r^{1-2\varepsilon}}$, divide the first coordinate quadrant into squares:

$$\begin{aligned} &b\xi < y_1 \leqq b(\xi+1), \\ &b\eta < y_{k-1} \leqq b(\eta+1) \end{aligned} \tag{3}$$

where ξ and η are integers (y_1 is the abscissa and y_{k-1} is the ordinate). In place of the exponential function

$$e^{2\pi i \alpha y_1 y_{k-1}} = e^{2\pi i \left(\frac{a}{q} y_1 y_{k-1} + z y_1 y_{k-1}\right)},$$

we shall now put a new exponential function

$$e^{2\pi i \left(\frac{a}{q} lh + z b^2 \xi\eta\right)}$$

into correspondence with each integral point in the square (3), where l and h are the least non-negative residues of y_1 and y_{k-1} modulo q, respectively. The new exponent differs from the previous one in that it contains terms which are $\ll \delta b^2 b_0^{-1}$. Consequently, the new sum $T_k''(X)$, into which the sum $T_k'(X)$ is transformed, differs from the latter by a term which is

$$\ll N\delta b^2 b_0^{-1} \ll \frac{N}{r^2}\, e^{-0.9 r^{1-\varepsilon}}.$$

Replace the sum $T_k''(X)$ by an approximate sum $T_k'''(X)$ containing terms of the same type as in the sum $T_k''(X)$, but with summation taken over all integral points (y_1, y_{k-1}) belonging to those squares (3) containing a point $(b\xi, b\eta)$ belonging to the range of summation of $T_k''(X)$. Evidently, the sum $T_k'''(X)$ differs from the sum $T_k''(X)$ by a term which is

$$\ll b\left(X+\frac{N}{X}\right) \ll Nbb_0^{-1} \ll \frac{N}{r^2} e^{-0.9r^{1-\varepsilon}}.$$

Now consider that part $t(\xi,\eta)$ of the sum $T_k'''(X)$ which corresponds to the square (3). Let $L_\xi(t)$ be the number of values of y_1 of the type $qx+l$ that satisfy the first inequality in (3), and $H_\eta(h)$ be the number of values of y_{k-1} of the type $qy+h$ that satisfy the second inequality in (3). Then, by assertion (a) of Lemma 3, we have ($L_\xi(l)=0$ for $(l,q)>1$ and $H_\eta(h)=0$ for $(h,q)>1$)

$$L_\xi(l) \ll K, \quad H_\eta(h) \ll K, \quad K=\frac{br^{4\varepsilon}}{r\varphi(q)}.$$

Therefore

$$t(\xi,\eta)=\sum_{l=0}^{q-1} L_\xi(l) \sum_{h=0}^{q-1} H_\eta(h)\, e^{2\pi i \frac{a}{q} lh}\, e^{2\pi i z b^2 \xi\eta}.$$

The sum $T_k'''(X)$ takes the form:

$$T_k'''(X)=\sum_{X_1<\xi\leqq X_1'} \sum_{\xi\eta\leqq N_1} t(\xi,\eta); \; X_1=\frac{X}{b},\; X_1'=\frac{X'}{b},\; N_1=\frac{N}{b^2}.$$

Hence we find

$$T_k'''(X) \ll \sum_{X_1<\xi\leqq X_1'} \sum_{l=0}^{q-1} K \left| \sum_{\eta\leqq N_1\xi^{-1}} e^{2\pi i z b^2 \xi\eta} \sum_{h=0}^{q-1} H_\eta(h)\, e^{2\pi i \frac{a}{q} lh} \right|,$$

$$|T_k'''(X)|^2 \ll X_1 q K^2 \sum_\xi \sum_l \sum_\eta \sum_{\eta_1} e^{2\pi i z b^2 \xi(\eta-\eta_1)}$$

$$\times \sum_h \sum_{h_1} H_\eta(h) H_{\eta_1}(h_1)\, e^{2\pi i \frac{a}{q} l(h-h_1)},$$

where η_1 runs through the same values as η, and h_1 through the same values as h. Therefore, summing over l (the values of other variables being fixed), we obtain

$$|T_k'''|^2 \ll X_1 q^2 K^2 \sum_\xi \sum_\eta \sum_{\eta_1} e^{2\pi i z b^2 \xi(\eta-\eta_1)} \sum_{h=0}^{q-1} H_\eta(h) H_{\eta_1}(h).$$

Now change the order of summation, bearing in mind that η and η_1 satisfy the conditions $0<\eta\leqq N_1X^{-1}$, $0<\eta_1\leqq N_1X^{-1}$, while ξ, for given η and η_1, runs through the natural numbers which satisfy the condition:

$$X_1<\xi\leqq X_0; \quad X_0=\min\left(X_1', \frac{N_1}{\eta}, \frac{N_1}{\eta_1}\right).$$

Then, summing over ξ (the values of other variables being fixed) and after certain simple calculations, we obtain

$$|T_k'''(X)|^2 \ll N_1 q^3 K^4 \sum_{0\leqq s\leqq N_1X_1^{-1}} \min\left(X_1, \frac{1}{(zb^2 s)}\right).$$

First consider the case where $\delta \leqq 1$. Substituting X_1 for each term on the right-hand side of the sum, we obtain

$$|T_k'''(X)|^2 \ll N_1^2 q^3 K^4, \quad T_k'''(X) \ll N \frac{r^{\varepsilon'}}{r^2 q^{0.5}}, \quad T_k' \ll N \frac{r^{\varepsilon'}}{r q^{0.5}}.$$

Now consider the case where $\delta > 1$. Putting

$$s_0 = \frac{N_1}{X_1 \delta},$$

we find

$$\sum_{0 \leqq s \leqq N_1 X_1^{-1}} \min\left(X_1, \frac{1}{(zb^2 s)}\right) \ll \sum_{0 \leqq s \leqq s_0} X_1 + \sum_{s_0 < s \leqq N_1 X_1^{-1}} \frac{N_1}{\delta s} \ll N_1 \frac{\ln(e\delta)}{\delta},$$

$$T_k' \ll N \frac{r^{\varepsilon''}}{r q^{0.5}} \delta^{-0.5}.$$

From the estimate found for T_k' $(k > 1)$ and from the equality

$$S = T_1 + O(r^{1-\varepsilon})$$

it follows that the left-hand side of (2) differs from S by a term which is

$$\ll N r^{-1+\varepsilon_1} q^{-0.5} \min(1, \delta^{-0.5}).$$

Hence, from the estimates found for U' and U'', we find that our theorem is true.

Lemma 4. Let $\varepsilon_0 < 0.001$, $N^{0.2} \leqq H \leqq N^{0.5}$, F be the product of primes $p \leqq H$. Then, assuming that

$$D = r^{\frac{\ln r - 1}{\ln(1+\varepsilon_0)}}; \quad r = \ln N,$$

the divisors d of F that do not exceed N can be distributed among $< D$ sets having the following properties:

a) The numbers d belonging to the same set have the same number β of prime factors and, consequently, have the same $\mu(d) = (-1)^\beta$;

b) One of the sets, called the simplest set, consists of only one number, i.e. $d = 1$. For this set, putting $\varphi = 1$, we obtain $\varphi = d = 1$. To each of the remaining sets, there corresponds a definite φ such that all the numbers in this set satisfy the condition:

$$\varphi < d \leqq \varphi^{1+\varepsilon_0}.$$

c) For any $U (0 \leqq U < \varphi)$, there exist two such sets of numbers d' and d'' (the set of numbers d'' may also be the simplest set) with corresponding φ' and φ'' satisfying the condition $\varphi' \varphi'' = \varphi$, $U \leqq \varphi' < UH$, that for some natural number B, each number d in the chosen set is repeated B times, if from all products of $d' d''$ we choose only those which satisfy the condition $(d', d'') = 1$.

Proof. Distribute all prime divisors of F over $(\tau + 1)$ intervals of the type

$$e^{(1+\varepsilon_0)^l} < p \leqq e^{(l+\varepsilon_0)^{l+1}},$$

which are obtained by making t (the ordinal number of the interval) run through $0, 1, \ldots, \tau$, where τ stands for the largest integer satisfying the condition:

$$e^{(1+\varepsilon_0)^\tau} < H.$$

From this condition, it readily follows that the number $(\tau+1)$ of all intervals satisfies the inequality:

$$\tau + 1 < \frac{\ln r - \ln 2}{\ln(1+\varepsilon_0)} + 1 < \frac{\ln r - 1}{\ln(1+\varepsilon_0)}.$$

We shall put each divisor $d\,(1 < d \leqq N)$ of F into correspondence with a non-decreasing sequence $l_0, l_1, \ldots, l_\tau$, where l_t stands for the number of prime factors of d that lie in the interval with the ordinal number t. The set of values of d associated with the same sequence is precisely the set that is stated in our lemma.

Since each d is the product of not more than r primes, we have $l_t \leqq r$ for each $t = 0, 1, \ldots, \tau$. Consequently, the number of different sets does not exceed

$$r^{\tau+1} < D.$$

The property (a) follows from the definition.

Now take some set other than the simplest set. Let $d = p_1 \ldots p_\beta$ be the ordinal number of this set, its factors being arranged in increasing order of magnitude. Let φ_s be the left-hand limit of the interval which bounds p_s. Then the right-hand limit can be expressed in the form $\varphi_s^{1+\varepsilon_0}$. Therefore, for any $s = 1, \ldots, \beta$, we have

$$\varphi_s < p_s \leqq \varphi_s^{1+\varepsilon_0}.$$

Hence, putting $\varphi = \varphi_1 \ldots \varphi_\beta$, we obtain

$$\varphi < d \leqq \varphi^{1+\varepsilon_0}.$$

Let U be a number such that $0 < U < \varphi$. Denoting by λ the least number which satisfies the inequality $U < \varphi_1 \ldots \varphi_\lambda$, consider two sets of numbers: the set of numbers d' which run through the products of the type $p_1 \ldots p_\lambda$ and the set of numbers d'' which run through the products of the type $p_{\lambda+1} \ldots p_\beta$. Then the number $\varphi' = \varphi_1 \ldots \varphi_\lambda$ will correspond to the set of d', while the number $\varphi'' = \varphi_{\lambda+1} \ldots \varphi_\beta$ to the set of d''. Now we have

$$\varphi'\varphi'' = \varphi, \quad U < \varphi' \leqq UH.$$

The equality $d = d'd''$ is possible only when $(d', d'') = 1$ and in this case it has

$$B = \binom{k_1 + k_2}{k_1}$$

solutions, where k_1 is the number of factors in the product φ' which are equal to φ_λ, and k_2 is the number of factors in the product φ'' which are equal to φ_λ.

Lemma 5. For $x \leqq N$, let the function $\Phi(x)$ obey the condition $|\Phi(x)| \leqq \Phi_0$. Let p run through primes, and Q denote the product of primes p; $N^{0.2} < p \leqq N$, and finally let

$$S = \sum_{p \leqq N} \Phi(p), \quad W_s = \sum_{\substack{y_1 \backslash Q \\ y_1 \ldots y_s \leqq N}} \ldots \sum_{y_s \backslash Q} \Phi(y_1 \ldots y_s).$$

Then, for some constants $\lambda_1, \lambda_2, \lambda_3, \lambda_4$,

$$S = \lambda_1 W_1 + \lambda_2 W_2 + \lambda_3 W_3 + \lambda_4 W_4 + O(N^{0.8}\Phi_0).$$

Proof. Let

$$S_h = \sum_{z_h \leqq N} \Phi(z_h),$$

where z_h runs through the divisors of Q which have exactly h different prime factors. Among the W_s products $y_1 \ldots y_s$, the number of products which are divisible by the sequares of numbers other than unity is $\ll N^{0.8}$, and the number of products equal to a given z_h is s_h. Therefore

$$W_s = sS_1 + \ldots + s^4 S_4 + O(N^{0.8}\Phi_0).$$

Putting $s = 1, 2, 3, 4$, we obtain a system of four equations with four unknowns S_1, S_2, S_3 and S_4 with determinants consisting of their non-zero coefficients. Hence (as $S = S_1 + O(N^{0.2}\Phi_0)$), we find that our lemma is true.

Lemma 6. For $x < N$, let the function $\Phi(x)$ be such that $|\Phi(x)| \leqq \Phi_0$. Let p run through primes, and let Q denote the product of primes p; $N^{0.2} < p \leqq N$, and finally for a natural number $s \leqq 4$, we have

$$W_s = \sum_{\substack{y_1 \backslash Q \\ y_1 \ldots y_s \leqq N}} \cdots \sum_{y_s \backslash Q} \Phi(y_1 \ldots y_s).$$

Then

$$W_s \ll \sum |T| + N^{0.8+\varepsilon}\Phi_0,$$

where summation is taken over $\ll N^{\varepsilon}$ terms of two types.

A term $|T|$ of the first type satisfies the inequality:

$$|T| \leqq \sum_{\delta} |T_\delta|,$$

where δ runs through an increasing sequence of natural numbers. If the sequence converges to a unique limit $\delta = 1$, the term is called the simplest:

$$T_\delta = \sum_{\substack{X \leqq \delta x \ll X^{1+\varepsilon_0} \\ \delta^2 xy \leqq N}} \sum_{Y \leqq \delta y \ll Y^{1+\varepsilon_0}} \Phi(\delta^2 xy);$$

$$N^{0.4} \leqq X \ll N^{0.6}, \quad N^{0.8} \leqq XY \leqq N, \quad X \geqq Y.$$

Here x and y run through a non-decreasing sequence of natural numbers satisfying the following condition: for a given x_0, the equation $x = x_0$ has $\ll N^{\varepsilon}$ solutions; and for a given y_0, the equation $y = y_0$ has $\ll N^{\varepsilon}$ solutions.

A term $|T|$ of the second type is given by the following equation:

$$T = \sum_{\substack{X \leqq x \ll X^{1+\varepsilon_0} \\ xm \leqq N}} \sum_{M \leqq m < M'} \Phi(xm), \quad X < N^{0.8}, \quad M > N^{0.2},$$

where x runs through a non-decreasing sequence of natural numbers satisfying the following condition: for a given x_0, the equation $x = x_0$ has $\ll N^{\varepsilon}$ solutions, and m runs through the sequence of natural numbers.

Besides this 'basic' division of the terms $|T|$ into two types, sometimes we shall also use a 'special' division of the terms into two types; the first special type differs from the first basic type in the following condition:

$$N^{1/3} < X \ll N^{2/3} \quad \text{in place of} \quad N^{0.4} < X \ll N^{0.6},$$

and the second special type differs from the second basic type in the following condition:

$$X < N^{1/3}, \quad M > N^{2/3} \qquad \text{in place of} \qquad X < N^{0.8}, \quad M > N^{0.2}.$$

Proof. Let P be the product of primes $p \leqq N^{0.2}$. Then the following identity

$$W_s = \sum_{\substack{d_1 \ m_1 \\ d_1 m_1 \dots d_s m_s \leqq N}} \sum \dots \sum_{d_s} \sum_{m_s} \mu(d_1) \dots \mu(d_s)\, \Phi(d_1 m_1 \dots d_s m_s),$$

holds true, where d_j runs through the divisors of P, and each m_j runs through natural numbers. Indeed, put the terms on the right-hand side of the identity being proved into correspondence with the integral points in an s-dimensional hyperboloid $h_1 > 0, \dots, h_s > 0, h_1 \dots h_s \leqq N$ by associating those terms which satisfy the condition $d_1 m_1 = h_1, \dots, d_s m_s = h_s$ with the point $(h_1, \dots, h_s)$. The sum of these terms is equal to the product

$$\sum_{d_1 \backslash (h_1, P)} \mu(d_1) \dots \sum_{d_s \backslash (h_s, P)} \mu(d_s)\, \Phi(h_1 \dots h_s),$$

which takes the form of the sum W_s, if and only if all $h_1, \dots, h_s$ are coprime to P; in the contrary case, the sum is equal to zero.

Without loss of generality, we shall confine ourselves only to the case of $s = 4$. By Lemma 4, for each $j = 1, 2, 3$, and 4, the values of d are distributed among $< D$ sets ($H = N^{0.2}$). And the values of m_j are distributed among $\ll r$ intervals of the type:

$$M_j \leqq m_j < M_j'; \quad 2M_j \leqq M_j' < 4M_j.$$

Let

$$T = \sum_{\substack{d_1 \\ d_1 d_2 d_3 d_4 m_1 m_2 m_3 m_4 \leqq N}} \sum_{d_2} \sum_{d_3} \sum_{d_4} \sum_{m_1} \sum_{m_2} \sum_{m_3} \sum_{m_4} \Phi(d_1 d_2 d_3 d_4 m_1 m_2 m_3 m_4),$$

where summation is taken over four sets of the values of d_1, d_2, d_3 and d_4 bounded by the inequalities

$$\varphi^{(1)} \leqq d_1 < F^{(1)}, \quad \varphi^{(2)} \leqq d_2 < F^{(2)}, \quad \varphi^{(3)} \leqq d_3 < F^{(3)},$$
$$\varphi^{(4)} \leqq d_4 < F^{(4)}; \quad F^{(j)} = (\varphi^{(j)})^{1+\varepsilon_0},$$

and over four sets of the values of m_1, m_2, m_3 and m_4 bounded by the inequalities:

$$M_1 \leqq m_1 < M_1', \quad M_2 \leqq m_2 < M_2',$$
$$M_3 \leqq m_3 < M_3', \quad M_4 \leqq m_4 < M_4';$$

we find

$$|W_4| \leqq \sum |T|,$$

where summation is taken over all $\ll (Dr)^4 \ll N^\varepsilon$ terms of the sum T.

First consider a 'trivial' sum T satisfying the condition:

$$\varphi^{(1)} \varphi^{(2)} \varphi^{(3)} \varphi^{(4)} M_1 M_2 M_3 M_4 \leqq N^{0.8}.$$

For this sum, obviously, we have $|T| \ll N^{0.8+\varepsilon} \Phi_0$. Further, consider the sums T satisfying the condition:

$$\varphi^{(1)} \varphi^{(2)} \varphi^{(3)} \varphi^{(4)} > N^{0.4}.$$

Let t be the least number satisfying the condition $\varphi^{(1)} \dots \varphi^{(t)} > N^{0.4}$, and let

$$\varphi^{(1)} \dots \varphi^{(t-1)} N^{\gamma} = N^{0.4}$$

(when $t = 1$, the coefficient of N^{γ} is taken to be equal to 1). We find that $\varphi^{(t)} > N^{\gamma}$. Therefore (by Lemma 4), there exist a natural number B and two sets of numbers d' and d'' with the respective φ' and φ'' satisfying the condition

$$\varphi' \varphi'' = \varphi^{(t)}, \quad N^{\gamma} < \varphi' \leqq N^{\gamma + 0.2},$$

such that all the values of d_t, each being taken B times, are obtained, if we choose from among all the products $d'd''$ only those that satisfy the condition $(d', d'') = 1$. Assuming

$$u = d_1 \dots d_{t-1}, \qquad v = d_{t+1} \dots d_4 m_1 m_2 m_3 m_4,$$
$$U = \varphi^{(1)} \dots \varphi^{(t-1)}, \quad V = \varphi^{(t+1)} \dots \varphi^{(4)} M_1 M_2 M_3 M_4,$$

we obtain

$$N^{0.4} < U\varphi' < N^{0.6},$$
$$BT = \sum_{U\varphi' \leqq ud' < (U\varphi')^{1+\varepsilon_0}} \sum \; \sum_{\substack{\varphi'' V \leqq d''v \lessdot (\varphi'' V)^{1+\varepsilon_0} \\ ud'd''v \leqq N \\ (d', d'') = 1}} \sum \Phi(ud'd''v).$$

Hence we find

$$BT = \sum_{\delta} \mu(\delta) T_{\delta};$$

$$T_{\delta} = \sum_{U\varphi' \leqq \delta ud_0' < (U\varphi')^{1+\varepsilon_0}} \sum \; \sum_{\substack{\varphi'' V \leqq \delta d_0'' v \leqq (\varphi'' V)^{1+\varepsilon_0} \\ \delta^2 ud_0' d_0'' v \leqq N}} \sum \Phi(\delta^2 ud_0' d_0'' v),$$

where δ runs through natural numbers, d_0' and d_0'' run, for a given δ, through the quotients obtained on dividing d' and d'' (that are multiples of δ) by δ. Putting $U\varphi' = X$, $ud_0' = x$, $d_0''v = y$, we find that T is a term of the first type.

Now consider the sum T satisfying the condition:

$$\varphi^{(1)} \varphi^{(2)} \varphi^{(3)} \varphi^{(4)} < N^{0.4}.$$

Without loss of generality, we can assume that M_1, M_2, M_3 and M_4 are arranged in a non-decreasing order of magnitude.

First, let $M_4 < N^{0.2}$. Let t be the least number satisfying the condition:

$$\varphi^{(1)} \varphi^{(2)} \varphi^{(3)} \varphi^{(4)} M_1 \dots M_t > N^{0.4}.$$

Putting

$$X = \varphi^{(1)} \varphi^{(2)} \varphi^{(3)} \varphi^{(4)} M_1 \dots M_t, \quad Y = M_{t+1} \dots M_4,$$
$$x = d_1 d_2 d_3 d_4 m_1 \dots m_t, \quad y = m_{t+1} \dots m_4,$$

we obtain

$$N^{0.4} < X < N^{0.6},$$
$$T = \sum_{X \leqq x \lessdot X^{1+\varepsilon_0}} \; \sum_{\substack{Y \leqq y \lessdot Y^{1+\varepsilon_0} \\ xy \leqq N}} \Phi(x, y).$$

Hence we find that T is a sum of the first type (for $\delta=1$).

Finally, let $M_4 > N^{0.2}$. Then, putting

$$X=\varphi^{(1)}\varphi^{(2)}\varphi^{(3)}\varphi^{(4)}M_1M_2M_3, \qquad M=M_4,$$
$$x=d_1d_2d_3d_4m_1m_2m_3, \qquad m=m_4,$$

we obtain

$$T=\sum_{X\leqq x\ll X^{1+\varepsilon_0}}\ \sum_{\substack{M\leqq m<M'\\ xm\leqq N}}\Phi(xm); \quad X<N^{0.8}, \quad M>N^{0.2}.$$

Hence we find that T is a sum of the second type.

Thus, we have demonstrated the lemma for the 'basic' division of the terms $|T|$ into two types. Now we shall take up the 'special' division. Obviously (since $N^{1/3}<N^{2/5}<N^{3/5}<N^{2/3}$), the first type of basic division is also a first type of special division. Therefore we shall now redistribute only the terms of the second type. Thus, let

$$\varphi^{(1)}\varphi^{(2)}\varphi^{(3)}\varphi^{(4)}<N^{0.4}, \quad M>N^{0.2}.$$

If $M_1>N^{2/3}$, then $(M=M_4)\ |T|$ is a term of the second special type. If, however, $N^{1/3}\leqq M_4<N^{2/3}$, then putting $X=M$, we find that $|T|$ is a term of the first special type. Let $M_4<N^{1/3}$. If $M_3\geqq N^{0.2}$, then putting $X=M_3M_4$, we find that $|T|$ is a term of the first special type. If, however, $M_3<N^{0.2}$, then taking the least number of the type $\varphi^{(1)}\varphi^{(2)}\varphi^{(3)}\varphi^{(4)}M_1\ldots M_j$ greater than $N^{0.4}$ as the number X, we find that $|T|$ is a term of the first special type. This completes the proof of our lemma.

Generalization. Lemma 6 remains valid, even if the summation variables x, y, δ, and m run through only those values which are coprime to a certain given natural number q not exceeding N.

Proof. This generalization can be proved in the same way as Lemma 6, assuming that d_j, m_j, h_j run through only those integers that are coprime to the given natural number.

Lemma 7. Let K be an integer, $K\leqq N$, $\Phi(z)=e^{2\pi i\alpha kz}$, let α be a real number,

$$\alpha=\frac{a}{q}+\frac{\theta}{q^2}, \quad (a,q)=1, \quad 0<q<N,$$

and let $|T|$ be a term of the first type stated in Lemma 6. Then, assuming

$$S_K=\sum_{k=1}^{K}|T|, \quad \Delta=\sqrt{\frac{1}{q}+\frac{q}{N}+N^{-0.2}},$$

we obtain

$$S_K\ll KN^{1+\varepsilon}\Delta.$$

Proof. We have

$$|T|\leqq\sum_{\delta}|T_\delta|;$$

$$T_\delta=\sum_{X\leqq\delta x\ll X^{1+\varepsilon_0}}\ \sum_{\delta^2xy\leqq N}e^{2\pi i\alpha k\delta^2xy}; \quad N^{0.4}\leqq X\ll N^{0.6}.$$

Making ξ run through consecutive natural numbers, we obtain

$$T_\delta \ll N^\varepsilon \sum_{X \leqq \delta\xi \ll X^{1+\varepsilon_0}} \left| \sum_{y \leqq \frac{N}{\delta^2 \xi}} e^{2\pi i \alpha k \delta^2 \xi y} \right|,$$

$$T_\delta^2 \ll N^{\varepsilon'} \frac{X}{\delta} \sum_{X \leqq \delta\xi \ll X^{1+\varepsilon_0}} \sum_{y_1 \leqq \frac{N}{\delta^2 \xi}} \sum_{y \leqq \frac{N}{\delta^2 \xi}} e^{2\pi i \alpha k \delta^2 \xi (y_1 - y)},$$

where y_1 runs through the same values as y. Changing the order of summation (first summing over ξ for given y_1 and y), we obtain

$$T_\delta^2 \ll N^{\varepsilon''} \frac{X}{\delta} \sum_{y_1 \leqq \frac{N}{\delta X}} \sum_{y \leqq \frac{N}{\delta X}} \min\left(\frac{X}{\delta}, \frac{1}{(\alpha k \delta^2 (y_1 - y))}\right)$$

$$\ll \frac{N^{1+\varepsilon'''}}{\delta^2} \sum_{0 \leqq \eta \leqq \frac{N}{\delta X}} \min\left(\frac{X}{\delta}, \frac{1}{(\alpha k \delta^2 \eta)}\right),$$

where η runs through consecutive natural numbers.

Without loss of generality, we assume that $\Delta^{-1} \geqq c_0$, where c_0 is a sufficiently large number greater than unity. We have

$$|T| \leqq T' + T''; \quad T' = \sum_{\delta \leqq \Delta^{-1}} |T_\delta|, \quad T'' = \sum_{\Delta^{-1} < \delta \leqq \sqrt{N}} |T_\delta|,$$

$$T'' \ll \sum_{\Delta^{-1} < \delta \leqq \sqrt{N}} \frac{N^{1+2\varepsilon}}{\delta^2} \ll N^{1+2\varepsilon} \Delta, \quad \sum_{k=1}^{K} T'' \ll K N^{1+2\varepsilon} \Delta.$$

Divide the interval $0 < \delta \leqq \Delta^{-1}$ into $\ll r$ subintervals of the type

$$D \leqq \delta < D'; \quad 2D \leqq D' < 4D,$$

and put

$$T(D) = \sum_{D \leqq \delta < D'} |T_\delta|.$$

We find

$$(T(D))^2 \ll D \sum_{D \leqq \delta < D'} |T_\delta|^2$$

$$\ll \frac{N^{1+\varepsilon^{(4)}}}{D} \sum_{D \leqq \delta < D'} \sum_{0 \leqq \eta \leqq \frac{N}{DX}} \min\left(\frac{X}{D}, \frac{1}{(\alpha k \delta^2 \eta)}\right)$$

$$\leqq \frac{N^{1+\varepsilon^{(5)}}}{D} \left(\sum_{0 < u < \frac{16DN}{X}} \min\left(\frac{X}{D}, \frac{1}{(\alpha k u)}\right) + X \right),$$

$$\left(\sum_{k=1}^{K} T(D)\right)^2 \ll K \sum_{k=1}^{K} (T(D))^2$$

$$\ll K \frac{N^{1+\varepsilon^{(5)}}}{D} \sum_{k=1}^{K} \left(\sum_{0 < u < \frac{16DN}{X}} \min\left(\frac{X}{D}, \frac{1}{(\alpha k u)}\right) + X \right)$$

$$\ll K \frac{N^{1+\varepsilon^{(6)}}}{D} \left(\sum_{0 < v < \frac{16KDN}{X}} \min\left(\frac{X}{D}, \frac{1}{(\alpha v)}\right) + KX \right).$$

Hence, applying Lemma 1a, we obtain

$$\left(\sum_{k=1}^{K} T(D)\right)^2 \ll K\frac{N^{1+\varepsilon^{(7)}}}{D}\left(\left(\frac{KDN}{Xq}+1\right)\left(\frac{X}{D}+q\right)+KX\right)$$

$$\ll K^2N^{2+\varepsilon^{(7)}}\left(\left(\frac{1}{Xq}+\frac{1}{KN}\right)\left(\frac{X}{D}+q\right)+\frac{X}{N}\right) \ll K^2N^{2+\varepsilon^{(7)}}\Delta^2,$$

$$\sum_{k=1}^{K} T(D) \ll KN^{1+0.5\varepsilon^{(7)}}\Delta, \quad \sum_{k=1}^{K} T' \ll KN^{1+\varepsilon^{(7)}}\Delta,$$

$$S_K \ll KN^{1+\varepsilon_1}\Delta.$$

This completes the proof of our lemma.

Lemma 8. Let K be an integer, $K \leqq N$, $\Phi(z) = e^{2\pi i \alpha k z}$, let α be a real number,

$$\alpha = \frac{a}{q}+\frac{\theta}{q^2}, \quad (a, q) = 1, \quad 0 < q < N,$$

and let T be a term of the second type stated in Lemma 6. Then, assuming

$$S_K = \sum_{k=1}^{K} |T|, \quad \Delta_1 = \frac{1}{q}+\frac{q}{N}+N^{-0.2},$$

we obtain

$$S_K \ll KN^{1+\varepsilon_1}\Delta_1.$$

Proof. Without loss of generality, we can take that $\Delta_1^{-1} \geqq c_1$, where c_1 is a sufficiently large number greater than unity. We have (take $x \leqq cX^{1+\varepsilon_0}$ in place of $x \ll X^{1+\varepsilon_0}$)

$$S_K = \sum_{k=1}^{K}\left|\sum_{\substack{X \leqq x \leqq cX^{1+\varepsilon_0}\\ xm \leqq N}} \sum_{M \leqq m < M'} e^{2\pi i \alpha k x m}\right|;$$

$$X < N^{0.8}, \quad M > N^{0.2};$$

For given k and x, summing over m, we find

$$S_K \ll N^{\varepsilon'} \sum_{0 < z \leqq KcX^{1+\varepsilon_0}} \min\left(M, \frac{1}{(\alpha z)}\right).$$

If $KcX^{1+\varepsilon_0} \leqq 0.5q$, by Lemma 1b we obtain

$$S_K \ll N^{\varepsilon'} q \ln q \ll N^{1+\varepsilon''}\frac{q}{N}.$$

And if $KcX^{1+\varepsilon_0} > 0.5q$, by Lemma 1*a* we obtain

$$S_K \ll N^{\varepsilon'}\frac{KcX^{1+\varepsilon_0}}{q}(M+q\ln q)$$

$$\ll N^{\varepsilon''}KXM\left(\frac{1}{q}+\frac{1}{M}\right) \ll KN^{1+\varepsilon''}\left(\frac{1}{q}+N^{-0.2}\right).$$

Thus, the lemma is true in both cases.

Theorem 3. Let K be an integer, $K \leqq N$, and α a real number,

$$\alpha = \frac{a}{q}+\frac{\theta}{q^2}, \quad (a, q) = 1, \quad 0 < q < N.$$

Then, assuming

$$V_K = \sum_{k=1}^{K} \left| \sum_{p \leqq N} e^{2\pi i\alpha kp} \right|,$$

we find

$$V_K \ll KN^{1+\varepsilon}\left(\sqrt{\frac{1}{q}+\frac{q}{N}+N^{-0.2}}\right).$$

Proof. The theorem follows from Lemmas 5, 6, 7 and 8.

Theorem 4. Let α be a real number,

$$\alpha = \frac{a}{q} + \frac{\theta}{q^2}, \quad (a, q) = 1, \quad 0 < q < N.$$

Then, for any σ: $0 < \sigma \leqq 1$, the number A_σ of the values of $\{\alpha p\}$; $p \leqq N$ satisfying the condition $\{\alpha p\} < \sigma$ is given by the expression:

$$A_\sigma = \sigma\pi(N) + R_\sigma; \quad R_\sigma \ll N^{1+\varepsilon_1}\Delta, \quad \Delta = \sqrt{\frac{1}{q}+\frac{q}{N}+N^{-0.2}}.$$

Proof. Taking $\Delta < 0.1$ (which does not violate the generality) and using the notation of Lemma 3 of Chapter 1, put $\varrho = 1$ and take the sequence of fractions $\{\alpha p\}$ where $p \leqq N$ (consequently, $Q = \pi(N)$) as the sequence $\delta_1, \ldots, \delta_Q$. Apply assertion (a) of Lemma 3 of Chapter 1. Then, for the sum

$$U(\alpha, \beta) = \sum_{p \leqq N} \psi(\alpha p),$$

we obtain

$$U(\alpha, \beta) - (\beta - \alpha)\pi(N) \ll \sum_{m=1}^{\infty} \kappa_m |T_m|; \quad T_m = \sum_{p \leqq N} e^{2\pi i m\alpha p};$$

$$\kappa_m = \begin{cases} \dfrac{1}{m}, & \text{if } m \leqq \dfrac{1}{\Delta}, \\[2ex] \dfrac{1}{\Delta m^2}, & \text{if } m > \dfrac{1}{\Delta}. \end{cases}$$

Distribute the numbers m over intervals of the type:

$$K \leqq m < 2K; \quad K = 2^n \quad (n = 0, 1, 2, \ldots),$$

Assuming

$$W_K = \sum_{m=K}^{2K-1} \kappa_m |T_m|,$$

we find (by Theorem 1), that for $K \leqq 1/\Delta$,

$$W_K \ll \frac{1}{K} KN^{1+\varepsilon}\Delta = N^{1+\varepsilon}\Delta;$$

and for $1/\Delta < K \leqq \frac{1}{2}N$,

$$W_K \ll \frac{1}{\Delta K^2} KN^{1+\varepsilon}\Delta = \frac{1}{\Delta K} N^{1+\varepsilon}\Delta.$$

Moreover, we trivially find that for $\frac{1}{2}N < K$,

$$W_K \ll \frac{1}{\Delta K} N = \frac{N}{\Delta^2 NK} N\Delta < \frac{N}{K} N\Delta .$$

And we obtain

$$U(\alpha, \beta) - (\beta - \alpha)\,\pi(N) \ll \sum_K W_K \ll N^{1+\varepsilon_1}\Delta ,$$

which, by virtue of assertion 1a, completes the proof of our theorem.

Theorem 5. Let q be an integer, $0 < q < N$, and let $\varrho(p)$ denote the remainder which p leaves on dividing by q. Then, for any integer q_1 $(0 < q_1 \leqq q)$, the number B_{q_1} of the values of $\varrho(p)$, $p \leqq N$ such that $\varrho(p) < q_1$ is

$$B_{q_1} = \frac{q_1}{q}\,\pi(N) + O\left(N^{1+\varepsilon_1}\left(\sqrt{\frac{1}{q} + \frac{q}{N}} + N^{-0.2}\right)\right).$$

Proof. This theorem follows as a corollary from Theorem 4 for $a = 1, \theta = 0, \sigma = q_1/q$.

Section II. Estimates of the Sum $\sum_{p \leqq N} e^{2\pi i k f \sqrt{p}}$

Lemma 9. Let $|T|$ be a term of the first special type stated in Lemma 6 and $\Phi(z) = e^{2\pi i k f \sqrt{z}}$, where k is a natural number, and f is a positive constant. Then

$$|T| \ll N^{1+\varepsilon_1 - 1/8} k^{1/4}.$$

Proof. We shall take $\delta x \leqq c_1 X^{1+\varepsilon_0}$ and $\delta y \leqq c_2 Y^{1+\varepsilon_0}$ in place of $\delta x \ll X^{1+\varepsilon_0}$ and $\delta y \ll Y^{1+\varepsilon_0}$, respectively. We find that

$$|T| \ll \sum_\delta |T_\delta|; \quad T_\delta = \sum_{X \leqq \delta x \leqq c_1 X^{1+\varepsilon_0}} \sum_{\substack{Y \leqq \delta y \leqq c_2 Y^{1+\varepsilon_0} \\ \delta^2 xy \leqq N}} e^{2\pi i k f \delta \sqrt{xy}};$$

$$X \geqq Y, \quad N^{0.8} < XY \leqq N.$$

Making ξ run through consecutive natural numbers, we obtain

$$T_\delta^2 \ll N^\varepsilon \frac{X}{\delta} \sum_{X \leqq \delta\xi \leqq c_1 X^{1+\varepsilon_0}} \sum_{\substack{Y \leqq \delta y_1 \leqq c_2 Y^{1+\varepsilon_0} \delta^2 \xi y_1 \leqq N \\ \delta^2 \xi y_1 \leqq N}} \sum_{\substack{Y \leqq \delta y \leqq c_2 Y^{1+\varepsilon_0} \\ \delta^2 \xi y \leqq N}} e^{2\pi k f \delta \sqrt{\xi}(\sqrt{y_1} - \sqrt{y})},$$

where y_1 runs through the same values as y. Changing the order of summation, i.e. for given y_1 and y, first summing over ξ within the limits

$$\frac{X}{\delta} \leqq \xi \leqq \min\left(c_1 \frac{X^{1+\varepsilon_0}}{\delta}, \frac{N}{\delta^2 y_1}, \frac{N}{\delta^2 y}\right);$$

$$y_1 \leqq Y_0, \quad y \leqq Y_0; \quad Y_0 = \min\left(c_2 \frac{Y^{1+\varepsilon_0}}{\delta}, \frac{N}{\delta X}\right),$$

by Lemma 1 of Chapter 1, we obtain

$$T_\delta^2 \ll N^\varepsilon \frac{X}{\delta} \sum_{0 \le y_1 \le Y_0} \sum_{0<y\le Y_0}$$

$$\cdot \min\left(\frac{X}{\delta}, \frac{X}{\delta}\left(\frac{k^{1/2}\delta^{1/2}|\sqrt{y_1}-\sqrt{y}|^{1/2}}{\left(\frac{X}{\delta}\right)^{3/4}} + \frac{1}{\left(\frac{X}{\delta}\right)^{1/4} k^{1/2}\delta^{1/2}|\sqrt{y_1}-\sqrt{y}|^{1/2}}\right)\right),$$

Hence, by virtue of the fact that

$$|\sqrt{y_1}-\sqrt{y}| = \frac{s}{\sqrt{y_1}+\sqrt{y}} < \frac{s}{\sqrt{y_1}}; \quad s = |y_1 - y|,$$

we obtain

$$T_\delta^2 \ll N^\varepsilon \frac{X^2}{\delta^2} \sum_{0<y_1\le Y_0} \sum_{0\le s\le Y_0} \min\left(1, \frac{k^{1/2}\delta^{1/2}s^{1/2}}{\left(\frac{X}{\delta}\right)^{3/4} y_1^{1/4}} + \frac{Y_0^{1/4}}{\left(\frac{X}{\delta}\right)^{1/4}\delta^{1/2}s^{1/2}}\right)$$

$$\ll N^{\varepsilon'} \frac{X^2}{\delta^2}\left(Y + \frac{k^{1/2}Y^{9/4}}{X^{3/4}} + \frac{Y^{7/4}}{X^{1/4}}\right)$$

$$\ll N^{\varepsilon'} \frac{X^2Y^2}{\delta^2}\left(\frac{1}{Y} + \frac{k^{1/2}Y^{1/4}}{X^{3/4}} + \frac{1}{X^{1/4}Y^{1/4}}\right) \ll N^{2+\varepsilon'-1/4}k^{1/2},$$

$$T_\delta \ll \frac{N^{1+0.5\varepsilon'-1/8}}{\delta}k^{1/4}, \quad T \ll N^{1+\varepsilon_1-1/8}k^{1/4}.$$

This completes the proof of our lemma.

Lemma 10. Let $|T|$ be a term of the second special type stated in Lemma 6, and $\Phi(z) = e^{2\pi ikf\sqrt{z}}$, where k is a natural number and f is a positive constant. Then

$$|T| \ll N^{1+\varepsilon_2-1/4}k^{1/2}.$$

Proof. We find

$$|T| \ll \sum_{X\le x \ll X^{1+\varepsilon_0}} \left|\sum_{\substack{M\le m\le M' \\ xm\le N}} e^{2\pi ikf\sqrt{xm}}\right|.$$

Hence, by Lemma 1 of Chapter 1, we obtain

$$|T| \ll N^{\varepsilon_1} \sum_x M\left(\frac{k^{1/2}x^{1/4}}{M^{3/4}} + \frac{1}{M^{1/4}x^{1/4}}\right) \ll N^{\varepsilon_2} XM\left(\frac{k^{1/2}X^{1/4}}{M^{3/4}} + \frac{1}{X^{1/4}M^{1/4}}\right)$$

$$\ll N^{1+\varepsilon_2-1/4}k^{1/2}.$$

This completes the proof of our lemma.

Theorem 6. Let k be a natural number, $k \le N^{0.1}$, and f a positive constant, and

$$S_k = \sum_{p\le N} e^{2\pi ikf\sqrt{p}}.$$

Then

$$S_k \ll N^{1+\varepsilon-1/8}k^{1/4}.$$

Proof. The theorem follows directly from Lemmas 6, 9 and 10.

Theorem 7. Let f be a positive constant. Then, for any σ such that $0<\sigma\leqq 1$, the number A_σ of the values of $\{f\sqrt{p}\}$; $p\leqq N$, obeying the condition $\{f\sqrt{p}\}<\sigma$ is

$$A_\sigma=\sigma\pi(N)+R_\sigma;\quad R_\sigma\ll N^{1+\varepsilon-0.1}.$$

Proof. We shall use the notation of Lemma 3, Chapter 1. Taking any arbitrarily large natural number l, which is a multiple of 10, put $\varepsilon_{(0)}=1/l$, $\Delta=N^{0.1+\varepsilon_{(0)}}$ and $\varrho=0.1\,l$. Taking the fractions $\{f\sqrt{p}\}$; $p\leqq N$ (consequently, $Q=\pi(N)$), as the numbers $\delta_1,\dots,\delta_Q$, we shall apply the assertion (a) of Lemma 3 of Chapter 1. Thus, for the sum

$$U(\alpha,\beta)=\sum_{p\leqq N}\psi(f\sqrt{p}),$$

we obtain the inequality

$$U(\alpha,\beta)-(\beta-\alpha)\,\pi(N)\ll\sum_{k=1}^{\infty}H(k)\,|S_k|;\quad S_k=\sum_{p\leqq N}e^{2\pi ikf\sqrt{p}};$$

$$H(k)=\begin{cases}\dfrac{1}{k}, & \text{if } k\leqq\Delta^{-1},\\[2ex] \dfrac{1}{\Delta^{\varrho}k^{\varrho+1}}, & \text{if } k>\Delta^{-1},\end{cases}$$

$$S_k\ll\begin{cases}N^{1+\varepsilon-1/8}k^{1/4}\\ N\ \text{always}\end{cases},\quad \text{if } k\leqq N^{0.1}.$$

Hence we find

$$\sum U(\alpha,\beta)-(\beta-\alpha)\,\pi(N)\ll N^{1+\varepsilon_1-0.1},$$

and consequently, by the assertion (a) of Lemma 3 of Chapter 1, we find that our theorem is true.

Example. Taking $f=\frac{1}{2}$, $\sigma=\frac{1}{2}$, we find that the number $D(N)$ of primes satisfying the condition

$$p\leqq N,\quad 0\leqq\{\tfrac{1}{2}\sqrt{p}\}<\tfrac{1}{2}.$$

In other words, the number $D(N)$ of primes which do not exceed N and lie in intervals of the type $(2n)^2\leqq p<(2n+1)^2$ (where n is an integer) is

$$D(N)=\tfrac{1}{2}\pi(N)+O(N^{1+\varepsilon-0.1}).$$

Indeed, how close this expression is to reality is evident from the following table which lists the values of N of the type $(2m)^2-m$.

Values of N of the type $(2m)^2 - m$:

N	3	14	33	60	95	138	189	248	315	390
$2D(N)$	0	4	10	18	26	36	46	54	68	80
$\pi(N)$	2	6	11	17	24	38	42	58	65	77
N	473	564	663	770	885	1008	1139	1278	1425	1580
$2D(N)$	94	108	126	140	158	174	192	210	228	252
$\pi(N)$	91	103	121	136	153	168	189	206	224	249
N	1743	1914	2093	2280	2475	2678	2889	3108	3335	3570
$2D(N)$	276	294	316	336	366	388	420	446	470	496
$\pi(N)$	271	293	316	338	366	388	418	442	470	499
N	3813	4064	4323	4590	4865	5148	5439	5738	6045	6360
$2D(N)$	528	562	590	620	654	696	726	760	804	840
$\pi(N)$	529	560	590	620	651	686	718	754	788	828
N	6683	7014	7353	7700	8055	8418	8789	9168	9565	9950
$2D(N)$	866	904	938	980	1016	1056	1094	1132	1180	1222
$\pi(N)$	861	902	937	972	1007	1046	1090	1131	1178	1221

Section III. Estimate of the Sum $\sum_{p \leqq N} \chi(p+k)$

Lemma 11. Let q be an odd prime, $(k, q) = 1$, $\chi(a)$ be the non-trivial character modulo q, $\xi(x)$ and $\eta(y)$ be non-negative,

$$S = \sum_{x=0}^{q-1} \sum_{y=0}^{q-1} \xi(x)\, \eta(y)\, \chi(xy+k);$$

$$\sum_{x=0}^{q-1} (\xi(x))^2 \leqq X_0, \quad \sum_{y=0}^{q-1} (\eta(y))^2 \leqq Y_0.$$

Then

$$|S| \leqq \sqrt{2X_0 Y_0 q}.$$

Proof. We find

$$|S|^2 \leqq X_0 \sum_{x=0}^{q-1} \sum_{y_1=0}^{q-1} \sum_{y=0}^{q-1} \eta(y_1)\, \eta(y)\, \chi\left(\frac{xy_1+k}{xy+k}\right)$$

(terms with $xy_1 + k$ or $xy + k$ multiples of q are zero). That part of the right-hand side which corresponds to $y_1 = y$ does not exceed $X_0 Y_0 q$. That part which corresponds to a pair of unequal y_1 and y (but one of them being equal to zero) is numerically $\leqq X_0 \eta(y) \eta(y_1)$. And that part which corresponds to a pair of unequal and non-zero y and y_1 is

$$X_0 \eta(y_1) \eta(y) \chi\left(\frac{y_1}{y}\right) \sum_{x=0}^{q-1} \chi\left(1 + \frac{k(yy_1^{-1} - 1)}{xy + k}\right),$$

which is numerically $\leqq X_0 \eta(y_1) \eta(y)$. From these facts and from

$$\sum_{x=0}^{q-1} \sum_{y=0}^{q-1} \eta(y_1) \eta(y) \leqq qY_0$$

we find that our lemma is true.

Lemma 12. Let q be an odd prime, $(k, q) = 1$, $\chi(a)$ be the non-trivial character modulo q, $\xi(x)$ and $\eta(y)$ be non-negative. Let M and N be non-negative integers, X and Y be positive integers,

$$S = \sum_{x=M+1}^{M+X} \sum_{y=N+1}^{N+Y} \xi(x) \eta(y) \chi(xy+k);$$

$$\xi(x) \ll \alpha, \quad \eta(y) \ll \beta.$$

Then

$$S \ll \alpha\beta XYF; \quad F = \sqrt{\frac{1}{X} + \frac{1}{Y} + \frac{1}{q} + \frac{q}{XY}}.$$

Proof. We shall apply Lemma 11. If $X_0 \leqq q$, substitute $a^2 X$ for X_0, and if $X > q$, substitute $(\alpha X/q)^2 q$ for X_0. If $Y \leqq q$, substitute $\beta^2 Y$ for Y_0, and if $Y > q$, substitute $(\beta Y/q)^2 q$ for Y_0. Thus, we find that our lemma is true.

Lemma 13. Let q be an odd prime, $(k, q) = 1$, $\chi(a)$ be the non-trivial character modulo q,

$$T \ll \sum_{\delta} |T_\delta|,$$

where δ runs through a certain sequence of natural numbers, and T is a sum of the type

$$T_\delta = \sum_{X/\delta < x \leqq (X/\delta) g} \sum_{\substack{Y/\delta < y \leqq (Y/\delta) h \\ xy \leqq N/\delta^2}} \chi(\delta^2 xy + k); \quad g = c' X^{\varepsilon_0}, \quad h = c'' Y^{\varepsilon_0},$$

where X and Y are numbers such that $N^{1/3} < X < N^{2/3}$, $XY > N^{2/3}$; and for a given x_0, the equation $x = x_0$ has $\ll N^{\varepsilon'}$ solutions, and like-wise, for a given y_0, the equation $y = y_0$ has $\ll N^{\varepsilon'}$ solutions. Then

$$T \ll N^{1+\varepsilon_1} \Delta; \quad \Delta = \sqrt{\frac{1}{q} + \frac{q}{N}} + N^{-1/6}.$$

Proof. Without loss of generality, we can confine ourselves to the case $\Delta^{-1} \geqq c_0$, where c_0 is a sufficiently large constant greater than unity.

Similarly, we shall also take that $Y > N^{1/6}$ (otherwise, $XY \leqq N^{5/6} \ll N\Delta$).

We can restrict ourselves only to those values of $\delta < \Delta^{-1}$, because

$$\sum_{\delta \geqq \Delta^{-1}} |T_\delta| \ll \sum_{\delta \geqq \Delta^{-1}} \frac{rN^{1+2\varepsilon}}{\delta^2} \ll N^{1+\varepsilon'}\Delta .$$

Since $\delta^2 < q$, and, consequently, δ^2 is not divisible by q, the terms $\chi(\delta^2 xy + k)$ can be replaced by $\chi(xy+k')$; $\delta^2 k' \equiv k \pmod q$. Assuming

$$\frac{X}{\delta} = X_\delta, \quad \frac{Y}{\delta} = Y_\delta, \quad \frac{N}{\delta^2} = N_{\delta^2},$$

we obtain

$$|T_\delta| = |T'_\delta|; \quad T'_\delta = \sum_{\substack{X_\delta \leqq x \leqq X_\delta g \\ xy \leqq N_{\delta^2}}} \sum_{Y_\delta \leqq y \leqq Y_\delta h} \chi(xy+k').$$

Dividing the interval $X_\delta \leqq x \leqq X_\delta g$ into $\ll r$ intervals of the type:

$$X'_\delta \leqq x < X''_\delta, \quad 2X'_\delta \leqq X''_\delta < 4X'_\delta,$$

put

$$T''_\delta = \sum_{\substack{X'_\delta \leqq x < X''_\delta \\ xy \leqq N_{\delta^2}}} \sum_{Y'_\delta \leqq y < Y''_\delta} \chi(xy+k').$$

Change the range of summation of T''_δ by the domain Ω bounded by the inequalities

$$X'_\delta \leqq x < X''_\delta, \quad xy \leqq N_{\delta^2},$$

assuming that the equation $y = y_0$ has no solutions at the points where the condition $Y'_\delta \leqq y < Y''_\delta$ is not satisfied. Divide the domain Ω into two subdomains: Ω_1 bounded by the inequalities

$$X'_\delta \leqq x < X''_\delta, \quad 0 < y \leqq \frac{N_{\delta^2}}{X''_\delta},$$

and Ω_2 bounded by the inequalities

$$X'_\delta \leqq x < X''_\delta, \quad \frac{N_{\delta^2}}{X''_\delta} \leqq y < \frac{N_{\delta^2}}{x}.$$

Accordingly, the sum T_δ can be expressed as the sum of two terms:

$$T''_\delta = T''_{\delta,\Omega_1} + T''_{\delta,\Omega_2}.$$

First we shall estimate T''_{δ,Ω_1}. In Lemma 12, on substituting $X''_\delta - X'_\delta$ and N_{δ^2}/X''_δ for X and Y, and $N^{\varepsilon'}$ for α and β, we get

$$T''_{\delta,\Omega_1} \ll N_{\delta^2} N^{2\varepsilon'} F_\delta; \quad F_\delta = \sqrt{\frac{1}{X'_\delta} + \frac{X'_\delta}{N_{\delta^2}} + \frac{1}{q} + \frac{q}{N_{\delta^2}}}.$$

The problem now is to find an estimate for the sum T''_{δ,Ω_2} with an order of accuracy similar to that of T''_{δ,Ω_1}. Therefore, we shall assume that F_δ is less than a sufficiently small positive constant less than unity; otherwise the estimate of similar order of accuracy will be trivial. Let $s = s_0$ be the largest integer that satisfies the condition: $2^s \leqq F_\delta^{-1}$. From the domain Ω_2, choose one 'first', $2 = 2^1$ 'second', $4 = 2^2$ 'third', ...,

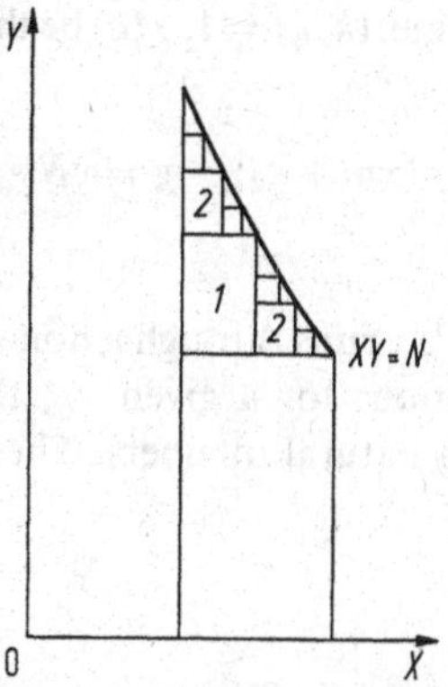

and finally 2^{s_0-1} 's_0'-th regions as shown in the figure. Here the s-th region is a right-angled triangle with the base $(X''_\delta - X'_\delta)/2^s$ and a height whose order is exactly equal to $N_{\delta^2}/X'_\delta 2^s$. By Lemma 12, that part of the sum T''_{δ,Ω_2} which corresponds to one of the (s-th) region is

$$\ll \frac{N_{\delta^2}}{2^{2s}} N^{2\varepsilon'} \sqrt{\frac{2^s}{X'_\delta} + \frac{X'_\delta 2^s}{N_{\delta^2}} + \frac{1}{q} + \frac{q 2^{2s}}{N_{\delta^2}}} \ll \frac{N_{\delta^2}}{2^s} N^{2\varepsilon'} F_\delta .$$

Consequently, that part of the sum T''_{δ,Ω_2} which corresponds to all s-th regions is $\ll N_{\delta_2} N^{2\varepsilon'} F_\delta$. Since that part of the sum T''_{δ,Ω_2} which corresponds to the remaining part of the domain Ω_2, is

$$\ll \frac{X'_\delta}{2^{s_0}} \frac{N_{\delta^2}}{X'_\delta} N^{2\varepsilon'} \ll N_{\delta^2} N^{2\varepsilon'} F_\delta ,$$

we obtain the estimate

$$T''_{\delta,\Omega_2} \ll N_{\delta^2} N^{2\varepsilon'} F_\delta .$$

This estimate, in combination with the estimate found above for T''_{δ,Ω_1}, gives

$$T''_\delta \ll N_{\delta^2} N^{2\varepsilon'} F_\delta .$$

Hence, from $X'_\delta < X_\delta g \ll X_s N^{\varepsilon_0}$, we find

$$T''_\delta \ll N_{\delta^2} N^{2\varepsilon' + 0.5\varepsilon_0} \sqrt{\frac{1}{X_\delta} + \frac{X_\delta}{N_{\delta^2}} + \frac{1}{q} + \frac{q}{N_{\delta^2}}} .$$

Since the sum T'_δ consists of $\ll r$ subsums of the type T_δ, substituting the above expressions for X_δ and N_{δ^2}, we obtain

$$T_\delta \ll \frac{N^{1+2\varepsilon'+0.5\varepsilon_0} r}{\delta} \sqrt{\frac{1}{X} + \frac{X}{N} + \frac{1}{q} + \frac{q}{N}} .$$

Summing this inequality over δ, and since $N^{1/3} < X < N^{2/3}$, we get the estimate of T stated in the lemma.

Lemma 14. Let q be an odd prime, $(k, q) = 1$, $\chi(a)$ be the non-trivial character modulo q and

$$T = \sum_{X < x \leqq Xg} \sum_{\substack{M \leqq m < M' \\ xm \leqq N}} \chi(xm + k); \quad g = cN^{\varepsilon_0}; \quad M > N^{2/3},$$

where $1 \leqq X < N^{1/3}$, $M > N^{2/3}$, x runs through a non-decreasing sequence of integers satisfying the following condition: for a given x_0, the equation $x = x_0$ has $\ll N^{\varepsilon'}$ solutions, and m runs through natural numbers. Then

$$T \ll N^{1+\varepsilon_1} \sqrt{\frac{1}{q} + \frac{q}{N}}.$$

Proof. Let T_x be the sum of the terms in T with given x. For an x not divisible by q, by a well known theorem on the sum of the values of a character (see, for instance, Question 12a, Chapter 6 in [1]), we obtain

$$T_x \ll N^{\varepsilon'} r \sqrt{q}.$$

For an x which is a multiple of q, we trivially find

$$T_x \ll N^{\varepsilon'} M.$$

Therefore

$$T \ll N^{\varepsilon'} \frac{N}{M} \left(r\sqrt{q} + \frac{M}{q} \right) \ll N^{1+\varepsilon_1} \sqrt{\frac{q}{N} + \frac{1}{q}}.$$

Theorem 8. Let q be an odd prime, $(k, q) = 1$, $\chi(a)$ be the non-trivial character modulo q, and

$$T = \sum_{p \leqq N} \chi(p + k).$$

Then

$$T \ll N^{1+\varepsilon_2} \left(\sqrt{\frac{1}{q} + \frac{q}{N}} + N^{-1/6} \right).$$

Proof. This theorem follows as a corrolary from Lemmas 6 (special division), 13 and 14.

Corollary. Let q be an odd prime, $(k, q) = 1$. The number of quadratic residues (non-residues) modulo q of the type $p + k$, $p \leqq N$ is equal to

$$0.5\pi(N) + R(N),$$

where

$$R(N) \ll N^{1+\varepsilon_2} \left(\sqrt{\frac{1}{q} + \frac{q}{N}} + N^{-1/6} \right).$$

Chapter 5. The Asymptotic Formula in the Goldbach Ternary Representation Problem

The history of the Goldbach ternary representation problem, which arose in the correspondence between Goldbach and Euler in 1742, is briefly outlined in the Introduction of the monograph[2]. This problem is a hypothesis that any even number not less than six can be expressed as the sum of two odd primes (Goldbach's binary representation) and any odd number not less than nine can be expressed as the sum of three odd primes (Goldbach's ternary representation). Hardy and Littlewood were the first to study this question in a far broader formulation than as a mere problem of the existence of representations: they investigated an asymptotic expression for the number of representations, and gave a rough derivation for this expression in the Goldbach ternary problem, in particular.

The method which I discovered in 1937 for estimating sums over primes[19, 20] permits, in the first instance, the evaluation of an estimate for the simplest of such sums, i.e. a sum of the type:

$$\sum_{p \leqq N} e^{2\pi i \alpha p}.$$

This estimate in combination with the previously known theorems concerning the distribution of primes in arithmetic progressions with a common difference not greater than a certain slowly growing function $\psi(N)$ (the Page theorem or the Walfisz theorem based on Siegel's lemma that gives a far more accurate order of the remainder term) paved the way for establishing unconditionally the asymptotic formula of Hardy and Littlewood in the Goldbach ternary representation problem.

In the proof given here, in addition to my estimates (Theorems 1 and 2 of Chapter 4), I shall only use the simplest variant of the Page theorem; moreover, the proof is so constructed that this theorem is applied as sparingly as possible. By complicating the theorem and widening its field of application, we can obtain an essentially more accurate remainder term[26].

Lemma 1 (Page). For a given ε_0, let there be given arbitrarily large c_1 and c. Then, the number $\pi(N, q, l)$ of primes not exceeding N occurring in the arithmetic progression

$$qx + l; \quad 0 < q \leqq r^c, \quad (q, l) = 1, \quad 0 \leqq l < q; \quad r = \ln N,$$

is

$$\pi(N, q, l) = \frac{1}{q_1} \int_2^N \frac{dx}{\ln x} + H; \quad q_1 = \varphi(q),$$

where for all q, except possibly for a sequence of exclusive multiples of some $q = q_0$ satisfying the condition

$$q \geqq r^{2-\varepsilon_0},$$

we have

$$H \ll \frac{N r^{-c}}{q_1 r}.$$

Proof. This lemma follows as a corollary from the well-known Page theorem[3].

Lemma 2. Let $N \geqq 2$, $\ln N = r$, z be a real number, and

$$I(z) = \int_2^N \frac{e^{2\pi izx}}{r}\, dx, \quad J(z) = \int_2^N \frac{e^{2\pi izx}}{\ln x}\, dx.$$

Then

$$I(z) \ll Z, \quad J(z) \ll Z;$$

$$Z = \begin{cases} Nr^{-1}, & \text{if } |z| \leqq N^{-1}, \\ |z|^{-1} r^{-1}, & \text{if } N^{-1} < |z| \leqq N^{-0.5}. \end{cases}$$

Proof. For the integral $I(z)$, both the cases follow trivially. For the integral $J(z)$, the first case is trivial, we shall therefore only consider the second case. Moreover, without loss of generality, we can assume that $z > 0$.

Applying the transformation used in the proof of Lemma 3 of Chapter 2, we find (using the substitution $2zx = u$) that

$$J(z) = U + iV; \quad U = \int_{\sigma_0}^{\sigma} \psi(u) \cos \pi u\, du, \quad V = \int_{\sigma_0}^{\sigma} \psi(u) \sin \pi u\, du;$$

$$\sigma_0 = 4z, \quad \sigma = 2Nz, \quad \psi(u) = \frac{1}{2z \ln \dfrac{u}{2z}},$$

$$|U| \leqq \int_{\sigma_0}^{\sigma_0+1} \psi(u)\, du = \int_2^{2+(2z)^{-1}} \frac{dx}{\ln x} \ll z^{-1} r^{-1}, \quad |V| \ll z^{-1} r^{-1}.$$

Hence, we find that the lemma is true for the integral $J(z)$ as well.

Lemma 3. Let $\tau = Nr^{-c}$, where $c \geqq 4$, and let

$$R = \int_{-\tau^{-1}}^{\tau^{-1}} (J(z))^3 e^{-2\pi izN}\, dz,$$

where $J(z)$ is as in Lemma 2. Then

$$R = \frac{N^2}{2r^3} + O\left(\frac{N^2}{r^4}\right).$$

Proof. Using the notation of Lemma 2, we shall compare the integral R with the integral

$$R_0 = \int_{-0.5}^{0.5} (I(z))^3 e^{-2\pi iz}\, dz.$$

We find

$$R - R_0 = \int_{-\tau^{-1}}^{\tau^{-1}} ((J(z))^3 - (I(z))^3) e^{-2\pi izN}\, dz + \left(\int_{-\tau^{-1}}^{\tau^{-1}} (I(z))^3 e^{-2\pi izN}\, dz - R_0 \right).$$

The first term on the right-hand side, by virtue of Lemma 2 and the inequality

$$|J(z) - I(z)| < \int_2^N \left(\frac{1}{\ln x} - \frac{1}{r} \right) dx \ll \frac{N}{r^2},$$

is

$$\ll \int_{-\tau^{-1}}^{\tau^{-1}} Z^2 \frac{N}{r^2}\,dx \ll \int_0^{N^{-1}} \frac{N^3}{r^4}\,dz + \int_{N^{-1}}^{r^{-1}} \frac{N}{r^4 z^2}\,dz \ll \frac{N^2}{r^4}.$$

While the second term, by virtue of Lemma 2, is

$$\ll \int_{\tau^{-1}}^{0.5} Z^3\,dz \ll \int_{\tau^{-1}}^{0.5} \frac{dz}{z^3 r^3} \ll \frac{N^2}{r^{11}}.$$

Consequently, we have $R - R_0 \ll N^2 r^{-4}$. Further, assuming that

$$R' = \int_{-0.5}^{0.5} (S(z))^3 e^{-2\pi i z N}\,dz; \quad S(z) = \sum_{x=3}^{N} \frac{e^{2\pi i z x}}{r},$$

by virtue of Lemma 2 of Chapter 2, we obtain $I(z) - S(Z) \ll r^{-1}$; $Z \gg r^{-1}$.

$$R_0 - R' \ll \int_0^{0.5} Z^2 r^{-1}\,dz = \int_0^{N^{-1}} \frac{N^2}{r^2}\,dz + \int_{N^{-1}}^{0.5} \frac{dz}{r^3 z^2} \ll \frac{N}{r^3}.$$

Therefore, $R - R' \ll N^2 r^{-4}$. We easily find that $r^3 R'$ gives the number of representations of N in the form:

$$N = x_1 + x_2 + x_3$$

with integers x_1, x_2, x_3 exceeding 2. For every $x_1 = 3, 4, \ldots, N-6$, the equality $x_2 + x_3 = N - x_1$ occurs $N - x_1 - 5$ times; consequently,

$$r^3 R' = \sum_{x_1=3}^{N-6} (N - x_1 - 5) = \frac{(N-7)(N-8)}{2} = \frac{N^2}{2} + O(N),$$

Hence, we find that our lemma is true.

Theorem. The number $I(N)$ of representations of an odd positive number N in the form

$$N = p_1 + p_2 + p_3$$

with odd primes p_1, p_2, p_3 is given by the expression:

$$I(N) = \frac{N^2}{2r^3} S(N) + O\left(\frac{N^2}{r^{3.5-\varepsilon}}\right);$$

$$S(N) = \prod_p{}' \left(1 + \frac{1}{(p-1)^2}\right) \prod{}'' \left(1 - \frac{1}{p^2 - 3p - 1}\right),$$

where the product $\prod'$ is taken over all primes, and the product $\prod''$ only over the prime divisors of N. $S(N) > 1$.

Corollary (Goldbach ternary representation). There exists a c_0 such that any odd N not less than c_0 is the sum of three odd primes.

Proof. Putting $\tau = N r^{-7}$, we obtain

$$I(N) = \int_{-\tau^{-1}}^{-\tau^{-1}+1} S_\alpha^3 e^{-2\pi i \alpha N}\,d\alpha; \quad S_\alpha = \sum_{2 < p \leq N} e^{2\pi i \alpha p}.$$

Divide the range of integration of $I(N)$ into intervals of two classes: Class I and Class II. Intervals which contain the values of α of the type

$$\alpha = \frac{a}{q} + z; \quad (a, q) = 1, \quad -\tau^{-1} < z \leqq \tau^{-1}, \quad 0 < q \leqq r^3.$$

are said to be intervals of Class I. Evidently (for sufficiently large c_0), the intervals of Class I do not overlap each other. The intervals which remain after the intervals of Class I have been chosen are said to be intervals of Class II. Any α belonging to an interval of Class II can be represented as

$$\alpha = \frac{a}{q} + z; \quad (a, q) = 1, \quad -\frac{1}{q\tau} \leqq z \leqq \frac{1}{q\tau}, \quad r^3 < q \leqq \tau.$$

In accordance with the division of the integration range, the integral $I(N)$ can be expressed as the sum of two subintegrals:

$$I(N) = I_1(N) + I_2(N).$$

1. Estimate of $I_2(N)$: By Theorem 1 of Chapter 4, for $q \geqq r^7$, we have

$$S_\alpha \ll Nr \ln r \left(\sqrt{\frac{1}{q} + \frac{qr}{N}} + re^{-0.5\sqrt{r}} \right) \ll Nr^{-2.5+\varepsilon_1}.$$

And, by Theorem 2 of Chapter 4, for $r^3 < q \leqq r^7$, we have

$$S_\alpha \ll Nr^{-2.5+\varepsilon_1}.$$

Therefore

$$I_2(N) \ll Nr^{-2.5+\varepsilon_1} \int_0^1 |S_\alpha|^2 \, d\alpha$$

$$= Nr^{-2.5+\varepsilon} \int_0^1 \sum_{2<p'\leqq N} \sum_{2<p\leqq N} e^{2\pi i\alpha(p'-p)} \, d\alpha \ll N^2 r^{-3.5+\varepsilon_1}.$$

2. Intervals of Class I corresponding to the values of q are not exclusive: Let $I_{a,q}$ be that part of the integral $I_1(N)$ taken over an interval of Class I which contains the fraction a/q whose denominator q is not exclusive. Taking some $\alpha = a/q + z$ in this interval, we shall divide the sum S_α into $[r^{12}]$ subsums of the type:

$$S_{\alpha, N_1} = \sum_{N_1 - A < p \leqq N_1} e^{2\pi i \left(\frac{a}{q} + z\right) p}; \quad A = N[r^{12}]^{-1}.$$

For the terms in $S_{\alpha, N_1'}$ we have $|zN_1 - zp| \leqq zA$. The number of such terms is $\ll Ar^{-1}$. Therefore

$$S_{\alpha, N_1} - e^{2\pi i z N_1} \sum_{N_1 - A < p \leqq N_1} e^{2\pi i \frac{a}{q} p} \ll zA^2 r^{-1} \ll Ar^{-6}.$$

But (by Lemma 1, for $c = 17$), for a given $l (0 \leqq l < q, (l, q) = 1)$, the number of primes of the types $qx + l$ in the interval $N_1 - A < p \leqq N_1$ is

$$\frac{1}{q_1} \int_{N_1 - A}^{N_1} \frac{dx}{\ln x} + O\left(\frac{Nr^{-17}}{q_1 r}\right).$$

Therefore

$$S_{\alpha, N_1} = \sum_l e^{2\pi i \frac{a}{q} l} \frac{1}{q_1} \int_{N_1 - A}^{N_1} \frac{e^{2\pi z N_1}}{\ln x} dx + O(Ar^{-6}).$$

Further, (using the notation of Lemma 2), we find

$$\sum_l e^{2\pi i \frac{a}{q} l} = \mu(q), \quad |zN_1 - zx| \leqq zA, \quad \int_{N_1 - A}^{N_1} \frac{zA}{\ln x} dx \ll Ar^{-6},$$

$$S_{\alpha, N_1} = \frac{\mu(q)}{q_1} \int_{N_1 - A}^{N_1} \frac{e^{2\pi i z x}}{\ln x} dx + O(Ar^{-6}),$$

$$S_\alpha = \frac{\mu(q)}{q_1} J(z) + O(Nr^{-6}),$$

$$\frac{J(z)}{q_1} \ll \frac{Z}{q_1}, \quad S_\alpha^3 - \frac{\mu(q)}{q_1^3} (J(z))^3 \ll \frac{Z^2}{q_1^2} Nr^{-6} + N^3 r^{-18},$$

$$I_{a,q} - \int_{-\tau^{-1}}^{\tau^{-1}} \frac{\mu(q)}{q_1^3} (J(z))^3 e^{-2\pi i \left(\frac{a}{q} + z\right) N} dz$$

$$\ll \int_0^{\tau^{-1}} \left(\frac{Z^2}{q_1^2} Nr^{-6} + \frac{N^3}{q_1^2} r^{-12} \right) dz \ll N^2 r^{-8} q_1^{-2}.$$

Making a run through a reduced system of residues modulo q, we obtain

$$\sum_a I_{a,q} = G(q)\, R + O(Nr^{-5} q_1^{-1}); \quad G(q) = \frac{\mu(q)}{q_1^3} \sum_a e^{2\pi i \frac{a}{q} N},$$

hence, by virtue of Lemma 3 and the inequality $G(q) \leqq q_1^{-2}$, we get

$$\sum_a I_{a,q} = \frac{N^2}{2r^3} G(q) + O\left(\frac{N^2}{r^4 q_1}\right).$$

3. Main intervals corresponding to exclusive values of q: Let q take an exclusive value. Then $q = q_0 k$, where k is an integer satisfying the condition $0 < k \leqq r^{1+\varepsilon_0}$ (since $q \leqq r^3$ and $q_0 \geqq r^{2-\varepsilon_0}$). Assuming

$$\alpha = \frac{a}{q} + z, \quad -\tau \leqq z \leqq \tau, \quad \delta = |z| N,$$

$$S(q, \delta) = Nr^{-1+\varepsilon} q^{-0.5}, \qquad \text{if} \quad \delta \leqq 1,$$

$$S(q, \delta) = Nr^{-1+\varepsilon} q^{-0.5} \delta^{0.5}, \quad \text{if} \quad \delta \geqq 1,$$

by Theorem 2 of Chapter 4, we obtain $S_\alpha \ll S(q, \delta)$. At the same time we have

$$I_{a,q} \ll \int_0^{\tau^{-1}} (S(q, \delta))^3 dz \ll N \int_0^{r^3} (S(q, \delta))^3 d\delta \ll N^2 r^{-3+3\varepsilon} q^{-1.5}.$$

Making a run through a reduced system of residues modulo q, we obtain

$$\sum_a I_{a,q} \ll N^2 r^{-3+3\varepsilon} q^{-0.5}$$

or, for the sake of uniformity (since the order of the first term is less than the order of the second term), we write

$$\sum_a I_{a,q} = \frac{N^2}{2r^3} G(q) + O(N^2 r^{-3+3\varepsilon} r^{-0.5}).$$

4. Preliminary expression for $I(N)$: We find, from paragraphs 1, 2, 3 that

$$I_1(N) - \sum_{q \leqq r^3} \frac{N^2}{2r^3} G(q) \ll \sum_{q \leqq r^3} \frac{N^2}{r^4 q_1^2} + \sum_{k < r^{1+\varepsilon_0}} \frac{N^2 r^{-3+3\varepsilon}}{q_0^{0.5} k^{0.5}} \ll \frac{N^2 r^{\varepsilon_4}}{r^{3.5}},$$

$$\sum_{q > r^3} \frac{N^2}{2r^3} G(q) \ll \sum_{q > r^3} \frac{N^2}{r^3 q_1^2} \ll \frac{N^2}{r^4},$$

$$I(N) = \frac{N^2}{2r^3} S(N) + O\left(\frac{N^2 r^{\varepsilon_0}}{r^{3.5}}\right);$$

$$S(N) = \sum_{q=1}^{\infty} G(q).$$

5. Transformation and study of $S(N)$: Evidently, $G(q)$ can be different from zero only when the canonical expansion of q is of the type: $q = p_1 \ldots p_k$ (for $q = 1$, in particular). In this case, we have

$$G(p_1) \ldots G(p_k) = G(p_1 \ldots p_k).$$

Indeed, for $k = 2$, the validity of this identity follows from the equality:

$$G(p_1)\, G(p_2) = \frac{1}{\varphi(p_1 p_2)} \sum_{0 < a_1 < p_1} \sum_{0 < a_2 < p_2} e^{2\pi i \frac{a_1 p_2 + a_2 p_1}{p_1 p_2} N},$$

where $a_1 p_2 + a_2 p_1$ runs through a reduced system of residues modulo $p_1 p_2$. This identity can be extended trivially to the case where $k > 2$.

For $x > 2$, we have

$$\prod_{p \leqq x} (1 + G(p)) = \sum_{q \leqq x} G(q) + \sum_{q > x}{}' G(q),$$

where $\sum'$ is taken over those values of g which are not divisible by primes greater than x. Since $S(N)$ converges absolutely as x increases to infinity, the first term on the right-hand side tends to $S(N)$, while the second term tends to zero. Therefore, making p run through all primes, we obtain

$$S(N) = \prod_p (1 + G(p)).$$

Further, we easily find that

$G(p) = 1/(p-1)^3$, if N is not divisible by P, and

$G(p) = -1/(p-1)^2$, if N is divisible by P.

Therefore

$$S(N) = \prod{}' \left(1 + \frac{1}{(p-1)^3}\right) \prod{}'' \left(1 - \frac{1}{(p-1)^2}\right),$$

where $\prod'$ is taken over the values of p that do not divide N, whereas $\prod''$ is the product over the values of p that divide N.

The first product on the right-hand side is greater than 2. Since N is odd, and $p = 2$ does not occur in the second product, and p is always less than or equal to $p_1 - 1$, where p_1 is the nearest prime greater than p, the second product is

$$> \prod_p \left(1 - \frac{1}{p^2}\right) = \frac{6}{\pi^2} > 0.6 .$$

Hence we find

$$S(N) > 1 .$$

We can also write the expression for $S(N)$ as follows:

$$S(N) = \prod_p \left(1 + \frac{1}{(p-1)^3}\right) \prod{}'' \left(\frac{1 - \dfrac{1}{(p-1)^2}}{1 + \dfrac{1}{(p-1)^3}}\right)$$

$$= \prod_p \left(1 + \frac{1}{(p-1)^3}\right) \prod{}'' \left(1 - \frac{1}{p^2 - 3p + 3}\right).$$

Thus, the theorem has been proved.

Chapter 6. An Elementary Variant of the Method of Trigonometric Sums

My method of trigonometric sums admits elementary modification (made without the help of the techniques of analysis). For example, of great importance is the version in which Lemmas 3 and 4 (or their corollaries) of this chapter play the key part. The method which I used in 1925 to prove the theorem on the distribution of residues and non-residues of power n modulo p can be reduced to this version. In 1953 I also applied this variant in combination with Lemma 8 (which reduces sums over primes to well known sums) to derive the theorem on the distribution of primes with respect to a given modulus.

The aim of this chapter is to illustrate the application of this variant to prove a somewhat simplified version of my theorem published in 1925 and to prove the theorem published in 1953.

Special Notation. For $0 < \sigma \leqq 1$, let $\psi(x)$ denote the function defined as follows:

$$\psi(x) = 1 - \sigma, \quad \text{if} \quad \{x\} < \sigma,$$
$$\psi(x) = -\sigma, \quad \text{if} \quad \{x\} \geqq \sigma.$$

Lemma 1. Let $\tau \geqq 1$. Then, every real number α can be represented in the following form:

$$\alpha = \frac{A}{Q} + \frac{\theta}{Q\tau}; \quad (A, Q) = 1, \quad 0 < Q \leqq \tau, \quad |\theta| < 1.$$

Proof. See the solution to Question 4 in Chapter 1 of book[1].

Lemma 2. Let $\tau \geqq 1$, α and β be real numbers, d an integer, and

$$\alpha = \frac{A}{Q} + \frac{\theta}{Q\tau}; \quad (A, Q) = 1, \quad 0 < Q \leqq \tau, \quad |\theta| < 1.$$

Then the sum

$$T = \sum_{x=d}^{d+Q-1} \{\alpha x + \beta\}$$

is bounded by the equality $T = \frac{1}{2}Q + \theta'$; $|\theta'| < 1$.

Proof. For $Q \leqq 2$, the lemma is trivial. Therefore, assume that $Q > 2$. Putting $x = y + d$, we obtain

$$T = \sum_{y=0}^{Q-1} \left\{\frac{Ay + f(y)}{Q}\right\}; \quad f(y) = (\alpha d + \beta)\, Q + \frac{\theta y}{\tau}.$$

Expressing the least value of $f(y)$ as $k + x$, where k and x are the integral and fractional parts, we find that the greatest value of $f(y)$ is of the form: $k + x + H$;

$H = |\theta|(Q-1)/\tau$ (consequently, $0 \leqq H < 1$). If r is the least non-negative residue of $Ay + k \pmod{Q}$, then

$$T = \sum_{r=0}^{Q-1} \left\{\frac{r+\lambda(r)}{Q}\right\}; \quad \lambda(r) = \kappa + \frac{|\theta| y}{\tau}, \quad \kappa \leqq \lambda(r) \leqq \kappa + H.$$

Further, putting

$$\left\{\frac{r+\lambda(r)}{Q}\right\} = \frac{r+\lambda(r)}{Q} - g_r, \tag{1}$$

we find that $g_r = 0$ for all r, except possibly for $r = Q-1$, for which g_r may be equal to unity, if, in addition, $x + H \geqq 1$. Summing the Q equalities which follow from (1) for $r = 0, 1, \ldots, (Q-1)$, we find

$$T = \tfrac{1}{2}Q - \tfrac{1}{2} + \kappa + \tfrac{1}{2}H - g_{Q-1},$$

Hence we find that our lemma is valid both when $x + H < 1$ and when $x + H \geqq 1$.

Lemma 3. Let $\tau \geqq 1$, α and β be real numbers, d be an integer, and

$$\alpha = \frac{A}{Q} + \frac{\theta}{Q\tau}; \quad (A, Q) = 1, \quad 0 < Q \leqq \tau, \quad |\theta| < 1.$$

Then the sum

$$S = \sum_{x=d}^{d+Q-1} \psi(\alpha x + \beta)$$

is bounded by the inequality:

$$|S| < 2.$$

Proof. Let T_β denote the sum T stated in Lemma 2. Since

$$\{\alpha x + \beta - \sigma\} - \{\alpha x + \beta\} = 1 - \sigma, \quad \text{if} \quad \{\alpha x + \beta\} < \sigma,$$
$$\{\alpha x + \beta - \sigma\} - \{\alpha x + \beta\} = -\sigma, \quad \text{if} \quad \{\alpha x + \beta\} \geqq \sigma,$$

we have

$$T_{\beta-\sigma} - T_\beta = S.$$

Hence, by Lemma 2, we find that our lemma is true.

Corollary. Let d be an integer, $(A, q) = 1$,

$$S = \sum_{\substack{x=d \\ (x,q)=1}}^{d+q-1} \psi\left(\frac{ax}{q}\right).$$

Then

$$|S| \leqq 2\tau(q).$$

Proof. We find that

$$S = \sum_{\substack{0 \leqq r < q \\ (r,q)=1}} \psi\left(\frac{r}{q}\right) = \sum_{\delta \backslash q} \mu(\delta) \sum_{r'=0}^{q\delta^{-1}-1} \psi\left(\frac{r'}{q\delta^{-1}}\right); \quad |S| \leqq \sum_{\delta \backslash q} 2 \leqq 2\tau(q).$$

Lemma 4. Let q be a natural number, h an integer, and let x run through X_0 consecutive numbers in the sequence $1, \ldots, q$,

$$S_y = \sum_x \psi\left(\frac{xy+h}{q}\right).$$

Then

$$\sum_{\substack{0<y\leq q\\(y,q)=1}} |S_y| < 4q(\ln q)^2.$$

Proof. For $q \leq 60$, the lemma is trivial. Therefore, take $q > 60$. For $X_0 = q$, the lemma follows from Lemma 3. Therefore, assume that $X_0 < q$.

We shall estimate S_y. For this purpose, express y/q as

$$\frac{y}{q} = \frac{A_0}{Q_0} + \frac{\theta_0}{Q_0 X_0}, \quad (A_0, Q_0) = 1, \quad 0 < Q_0 \leqq X_0, \quad |\theta_0| < 1.$$

Let X_1 be remainder obtained on dividing X_0 by Q_0. If $X_1 > 0$, express y/q as

$$\frac{y}{q} = \frac{A_1}{Q_1} + \frac{\theta_1}{Q_1 X_1}, \quad (A_1, Q_1) = 1, \quad 0 < Q_1 \leqq X_1, \quad |\theta_1| < 1,$$

and so on till we arrive at some $X_{n+1} = 0$. Now applying Lemma 3, we find that

$$|S_y| < 2\left[\frac{X_0}{Q_0}\right] + 2\left[\frac{X_1}{Q_1}\right] + \ldots + 2\left[\frac{X_n}{Q_n}\right],$$

where the sum on the right-hand side (as well as the number n of its terms) is completely determined by the value of y. Hence we find that

$$\sum_y |S_y| \leqq \sum_y \left(2\frac{X_0}{Q_0} + 2\frac{X_1}{Q_1} + \ldots + 2\frac{X_n}{Q_n}\right).$$

To each term of the type $2X/Q$ within brackets, there corresponds a definite system of conditions of the type:

$$\frac{y}{q} = \frac{A}{Q} + \frac{\theta}{QX}, \quad (A, Q) = 1, \quad 0 < Q \leqq X, \quad |\theta| < 1. \tag{2}$$

Let B_Q be the sum of those terms to which a given value of Q corresponds. Let $(q, Q) = \delta$, $q = q_1\delta$, $Q = Q_1\delta$. The first condition in (2) implies that a t exists such that

$$Q_1 y \equiv t \pmod{q_1}, \quad 0 < |t| < \frac{q_1}{X},$$

and one value of t may correspond to not more than δ different values of y. Furthermore,

$$0 < |t| < \frac{q_1}{Q_1\delta}, \quad X < \frac{q_1}{|t|}, \quad 2\frac{X}{Q} < 2\frac{q_1}{|t|Q_1\delta}.$$

Hence, we easily obtain

$$B_Q < \sum_{Q_1 < q_1} \sum_{0<|t|<q_1/Q_1\delta} 2\frac{q_1}{\delta Q_1 |t|} \leqq 4 \sum_{Q<q} \sum_{0<t<q/Q} \frac{q}{Qt} < 4q(\ln q)^2.$$

Now we shall prove a simplified version of my theorem[27] published in 1925.

Lemma 5. Let p be an odd prime, $p-1=nf, n$ and f are such integers that $1<n<p-1$. Let h be an integer satisfying the condition $1<h<p-1$ and $\sigma=h/(p-1)$. Let y run through the numbers of Class i (i.e. such numbers that $\operatorname{ind} y \equiv i \pmod{n}$) and let c_i be the total number of numbers of Class i (i $= 1, \ldots, h$). Then, assuming

$$U = \sum_{0<y<p} \sum_{x=1}^{h} \psi\left(\frac{yx}{p}\right),$$

we obtain

$$U = c_0 c_i + c_1 c_{1+i} + \ldots + c_{n-1} c_{n-1+i} - \frac{fh^2}{p-1}.$$

Proof. The validity of the theorem follows from the fact that among the terms $\psi(yx/p)$ which correspond to an x belonging to the s-th class, c_{s+i} terms are equal to $1-\sigma$, while the rest are equal to $-\sigma$.

Corollary. Let

$$c_s = \frac{h}{n} + \delta_s.$$

Then

$$U = \delta_0 \delta_i + \delta_1 \delta_{1+i} + \ldots + \delta_{n-1} \delta_{n-1+i}.$$

In particular, for $i=0$,

$$U = \delta_0^2 + \delta_1^2 + \ldots + \delta_{n-1}^2. \tag{3}$$

Proof is trivial.

Theorem. We have

$$\delta_0^2 + \delta_1^2 + \ldots + \delta_{n-1}^2 < 4p(\ln p)^2.$$

In particular, for each s,

$$|\delta_s| < 2\sqrt{p}\ln p.$$

Proof. The theorem follows from Eq. (3) and Lemma 4.

Lemma 6. Let $q>2$, $(a,q)=1$, let x run through X different numbers in the sequence $0, \ldots, q-1$, each being repeated $\leqq \alpha$ times, and let y run through Y different numbers of the same sequence that are coprime to q, each being repeated $\leqq \beta$ times. Then, the sum

$$S = \sum\sum \psi\left(\frac{axy}{q}\right)$$

is bounded by the inequality:

$$|S| < 2\alpha\beta\sqrt{XYq}\ln q.$$

Proof. Without loss of generality, we can take it that $q > 60$. We find (y_1 runs through the same values as y)

$$S^2 \leqq X\alpha^2 \sum_{\xi=0}^{q-1} \sum_{y} \sum_{y_1} \psi\left(\frac{a\xi y}{q}\right) \psi\left(\frac{a\xi y_1}{q}\right) = X\alpha^2 \sum_{y} S_y;$$

$$S_y = \sum_{u=0}^{q-1} \sum_{v} \psi\left(\frac{u}{q}\right) \psi\left(\frac{uv}{q}\right),$$

where v runs through Y numbers from the sequence $0, \ldots, q-1$ that are coprime to q, each being repeated $\leqq \beta$ times. Dividing the sum S_y into two parts such that one part contains the terms satisfying the condition $u < \sigma q$, while the other part contains the terms satisfying the condition $u \geqq \sigma q$, we shall apply Lemma 1 to each of these parts. Thus, we obtain

$$|S_y| < 4\beta q (\ln q)^2\, 2(1-\sigma)\sigma < 2\beta q (\ln q)^2,$$

$$S^2 < 2X\alpha^2 Y\beta^2 q (\ln q)^2, \quad |S| < \alpha\beta\sqrt{2XYq}\, \ln q.$$

The lemma has been demonstrated.

Lemma 7. Let $q > 2$. $(a, q) = 1$, let x run through X consecutive integers, each being repeated $\leqq \alpha$ times, and y through Y consecutive integers coprime to q, each being repeated $\leqq \beta$ times. Then, the sum

$$S = \sum_{x} \sum_{y} \psi\left(\frac{axy}{q}\right)$$

is bounded by the inequality:

$$S \ll XY\alpha\beta (\ln q) \sqrt{\frac{1}{X} + \frac{1}{Y} + \frac{1}{q} + \frac{q}{XY}}.$$

Proof. The sum S can be reduced to the form stated in Lemma 6 on replacing x and y by x_0 and y_0 that are congruent modulo q to x and y, from the sequence $0, \ldots, q-1$. Accordingly, the numbers X, α, Y, β are to be replaced by $X_0, \alpha_0, Y_0, \beta_0$, respectively. If $X \leqq q$, evidently, we can take $X_0 = X$, $\alpha_0 = \alpha$; and if $X > q$, we can put $X_0 = q$, $\alpha_0 = 2\alpha X q^{-1}$. If $Y \leqq q$, we can take $Y_0 = Y$, $\beta_0 = \beta$; and if $Y > q$, we can take $Y_0 = q$, $\beta_0 = 2\beta Y q^{-1}$. Thus, we always obtain

$$|S| < 2\alpha_0\beta_0\sqrt{X_0 Y_0 q}\, \ln q.$$

Here

$$\alpha_0\sqrt{X_0} = \begin{cases} \alpha X \sqrt{\dfrac{1}{X}} & \text{for } X \leqq q, \\[2ex] 2\alpha X \sqrt{\dfrac{1}{q}} & \text{for } X > q, \end{cases}$$

$$\beta_0\sqrt{Y_0} = \begin{cases} \beta Y \sqrt{\dfrac{1}{Y}} & \text{for } Y \leqq q, \\[2ex] 2\alpha Y \sqrt{\dfrac{1}{q}} & \text{for } Y > q. \end{cases}$$

Hence, we readily find that our lemma is true for each of these four possible cases.

Lemma 8a. Let

$$1 \leqq U < N, \quad 1 < \Delta \leqq U, \quad U + \Delta \leqq N,$$

$$S = \sum_x \sum_y \psi\left(\frac{axy}{q}\right),$$

where x and y run through integers belonging to two non-decreasing sequences of natural numbers satisfying the condition: for a given x_0, the equation $x = x_0$ has $\ll N^{\varepsilon}$ solutions, and for a given y_0, the equation $y = y_0$ has $\ll N^{\varepsilon}$ solutions, where y runs through the values coprime to q and summation is taken over the domain:

$$U < x \leqq U + \Delta, \quad \frac{N}{U+\Delta} < y \leqq \frac{N}{x}. \tag{4}$$

Then

$$S \ll \frac{N^{1+\varepsilon_1}\Delta^2}{U^2} F, \quad F = \sqrt{\frac{1}{\Delta} + \frac{U^2}{N\Delta} + \frac{1}{q} + \frac{qU^2}{N\Delta^2}}.$$

Proof. Let $F \ll 1$ (otherwise, the lemma is trivial). Let r_0 be the largest integer such that $2^{r_0} < F^{-1}$. From the domain (4), choose the 'first', 'second', ..., 'r_0-th' regions as shown in the figure. The r-th region is a right-angled triangle with a base $\Delta/2^r$ and a height of exact order $N\Delta/U^2 2^r$. The height and the base of the triangle are not included in the domain. The total number of r-th regions is equal to 2^{r-1}. By Lemma 7, that part of the sum S which corresponds to one of the r-th regions is

$$\ll \frac{\Delta}{2^r} \frac{N^{1+\varepsilon'}\Delta}{U^2 2^r} \sqrt{\frac{2^r}{\Delta} + \frac{U^2 2^r}{N\Delta} + \frac{1}{q} + \frac{qU^2 2^r}{N\Delta^2}} \ll \frac{N^{1+\varepsilon'}\Delta^2}{U^2 2^r} F.$$

That part of the sum S which corresponds to an un-numbered region is

$$\ll \frac{N^{1+\varepsilon'}\Delta}{U^2} \frac{\Delta}{2^{r_0}} \ll \frac{N^{1+\varepsilon'}\Delta^2}{U^2} F.$$

Therefore

$$S \ll \frac{N^{1+\varepsilon'}\Delta^2}{U^2}\left(\sum_{r=1}^{r_0} \frac{2^{r-1}}{2^r \ln q} + 1\right) F \ll \frac{N^{1+\varepsilon_1}\Delta^2}{U^2} F.$$

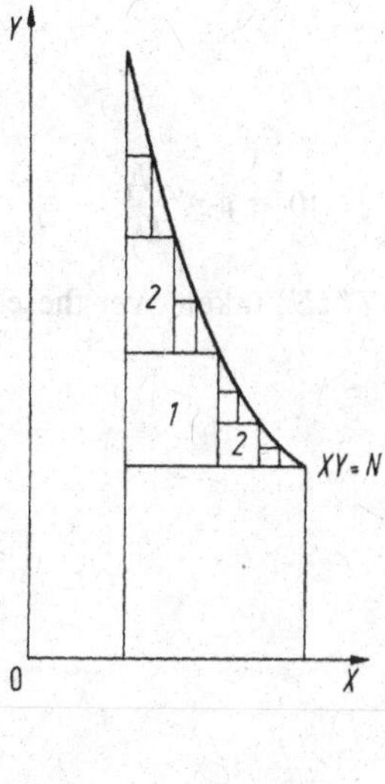

Lemma 8b. Let the conditions be as stipulated in Lemma 8a, but let summation be taken over the domain:

$$U < x \leqq U + \Delta, \quad Y_0 < y \leqq Y_0 + Y;$$

$$\Delta \ll U, \quad 0 \leqq Y_0 < Y_0 + Y \leqq \frac{N}{U + \Delta}.$$

Then

$$S \ll \Delta Y N^{\varepsilon_1} F; \quad F = \sqrt{\frac{1}{\Delta} + \frac{1}{Y} + \frac{1}{q} + \frac{q}{\Delta Y}}.$$

Proof. This lemma follows directly from Lemma 7.

Lemma 8. Let $|T|$ be a term of the first special type stated in Lemma 6 of Chapter 4 and $\Phi(z) = \psi(az/q)$. Then

$$T \ll N^{1+\varepsilon_2} F_2; \quad F_2 = \sqrt{\frac{1}{q} + \frac{q}{N}} + N^{-1/6}.$$

Proof. We have

$$T \ll \sum_{\delta} |T_\delta|; \quad T_\delta = \sum_{\frac{X}{\delta} < x \ll \frac{X^{1+\varepsilon_0}}{\delta}} \; \sum_{xy \leqq \frac{N}{\delta^2}} \psi\left(\frac{a\delta^2 xy}{q}\right); \quad N^{1/3} < x \leqq N^{2/3}.$$

Without loss of generality, we can take that $F_2 \ll 1$. Since

$$T_\delta \ll \frac{N^{1+\varepsilon_2}}{\delta^2}, \quad \sum_{\delta > F_2^{-1}} \frac{N^{1+\varepsilon_2}}{\delta^2} \ll N^{1+\varepsilon_2} F_2,$$

we shall only consider the case where $\delta \leqq F_2^{-1}$.

Divide the interval $X/\delta < x \ll X^{1+\varepsilon_2}/\delta$ into $\ll \ln N$ subintervals of the type

$$U < x \leqq U'; \quad 2U \leqq U' < 4U.$$

Accordingly, the sum T_δ is divided into the same number of subsums of the type

$$T_\delta(U) = \sum_{U < x \leqq U'} \; \sum_{xy \leqq N_1} \psi\left(\frac{a\delta^2 xy}{q}\right); \quad N_1 = \frac{N}{\delta^2}.$$

Dividing the region of summation of T_δ into two subregions:

$$U < x \leqq U', \quad \frac{N_1}{U'} < y \ll \frac{N_1}{x} \quad \text{and} \quad U < x \leqq U', \quad 0 < y \leqq \frac{N_1}{U'},$$

we shall apply Lemmas 8a and 8b to the subsums $T'_\delta(U)$ and $T''_\delta(U)$ taken over these subregions, respectively. We thus obtain

$$\left(\Delta = U' - U, \; Y = \frac{N_1}{U'}\right)$$

$$T'_\delta(U) \ll N_1^{1+\varepsilon'} F', \quad T''_\delta(U) \ll N_1^{1+\varepsilon'} F';$$

$$F' = \sqrt{\frac{1}{U} + \frac{X\delta}{N_1} + \frac{1}{q} + \frac{q}{N_1}}.$$

Hence, it follows that

$$T_\delta(U) \ll N_1^{1+\varepsilon'} F',$$

$$T_\delta \ll N_1^{1+\varepsilon''} F' \ll N^{1+\varepsilon''} \sqrt{\frac{\delta}{X} + \frac{X\delta}{N} + \frac{1}{q} + \frac{q\delta^2}{N}}$$

$$\ll \frac{N^{1+\varepsilon''}}{\delta} \sqrt{\frac{1}{X} + \frac{X}{N} + \frac{1}{q} + \frac{q}{N}} \ll \frac{N^{1+\varepsilon''}}{\delta} \left(\sqrt{\frac{1}{q} + \frac{q}{N}} + N^{-1/6} \right),$$

$$\sum_{\delta \leqq F_2^{-1}} |T_\delta| \ll N^{1+\varepsilon_2} F_2 .$$

The lemma has been proved.

Lemma 9. Let $|T|$ be a term of the second special type stated in Lemma 6 of Chapter 6, and $\Phi(z) = \psi(az/q)$. Then

$$T \ll N^{1+\varepsilon_1} F; \quad F = \sqrt{\frac{1}{q} + \frac{q}{N}}.$$

Proof. We have

$$T = \sum_{\substack{X < x \ll X^{1+\varepsilon_0} \\ xm \leqq N}} \sum_{M < m \leqq M'} \psi\left(\frac{axm}{q}\right).$$

Divide the interval $X < x \ll X^{1+\varepsilon_0}$ into $\ll \ln N$ subintervals of the type

$$U < x \leqq U'; \quad 2U \leqq U' < 4U,$$

Accordingly, the sum T is divided into the same number of subsums of the type

$$T(U) = \sum_{\substack{U < x \leqq U' \\ xm \leqq N}} \sum_{M < m \leqq M'} \psi\left(\frac{axm}{q}\right).$$

For the sake of clarity, we shall divide the sum $T(U)$ into two terms: the sum $T'(U)$ taken over the domain

$$U < x \leqq U', \quad M < m \leqq M_1; \quad M_1 = \min\left(M', \frac{N}{U'}\right),$$

and the sum $T''(U)$ taken over the domain

$$U < x \leqq U', \quad M_2 < m \leqq M_3;$$

$$M_2 = \max\left(M, \frac{N}{U}\right), \quad M_3 = \min\left(M', \frac{N}{x}\right).$$

First, we shall estimate the sum $T'(U)$. Divide this sum into a subsum $T_1'(U)$ taken over the domain

$$U < x \leqq U'; \quad M < m \leqq M + \left[\frac{M_1 - M}{q}\right],$$

and a second subsum $T_2'(U)$ taken over the domain

$$U < x \leqq U', \quad M + \left[\frac{M_1 - M}{q}\right] < m \leqq M_1 .$$

The subsum $T_1'(U)$ is non-zero only when $M_1 - M \geqq q$. In this case, that part of this sum which corresponds to a given x can be subdivided into $[(M_1 - M)/q]$ subsums, each of which, by the corollary of Lemma 3, is $\ll \tau(q)$. Consequently,

$$T_1'(U) \ll XN^{\varepsilon'} \frac{M}{q} \tau(q) \ll N^{1+\varepsilon'} F.$$

We shall estimate $T_2'(U)$ by means of Lemma 7. Putting

$$Y = M_1 - M - \left[\frac{M_1 - M}{q}\right]$$

(consequently, $Y > q$), we obtain

$$T_2'(U) \ll UY \sqrt{\frac{1}{U} + \frac{1}{Y} + \frac{1}{q} + \frac{q}{UY}}\, N^{\varepsilon''}$$

$$\ll XY \sqrt{\frac{1}{q} + \frac{q}{XY}}\, N^{\varepsilon''} \ll N^{1+\varepsilon'} F.$$

Therefore

$$T'(U) \ll N^{1+\varepsilon'} F.$$

Now we shall estimate a sum of the type:

$$T_0(U) = \sum_{U < x \leqq U'} \sum_{\substack{m > M_0 \\ xm \leqq N}} \psi\left(\frac{axm}{q}\right); \quad M_0 > \left[\frac{N}{U'}\right],$$

and show that $T_0(U) \ll N^{1+\varepsilon'} F$.

First, let $U \leqq N/q$. Then, for a given x, the sum over m is divided into $[(Nx^{-1} - M_0)/q]$ complete sums (i.e. with a summation range equal to an interval of length q) and possibly one incomplete sum (with a summation range of a length less than q). Applying the corollary of Lemma 3 to each of the complete sums, and then estimating the incomplete sum trivially, we obtain

$$T_0'(U) \ll N^{\varepsilon'} \left(\frac{N}{q} \tau(q) + Uq\right) \ll N^{1+\varepsilon''} \left(\frac{1}{q} + \sqrt{\frac{q}{N}}\right) \ll N^{1+\varepsilon''} F.$$

Now let $U > \sqrt{N/q}$. From the interval $M_0 < m \leqq N/x$, choose $[(Nx^{-1} - M_0)/q]$ complete intervals. The sum $\sum_m \psi(axm/q)$ taken over each of these intervals is of an order not greater than $\tau(q)$. The sum of all these sums is $\ll N\tau(q)/xq$, and

$$\sum_{U < x \leqq U'} \frac{N}{xq} \tau(q) \ll N^{1+\varepsilon''} F.$$

The remaining part T_0' of the sum T_0 is taken over the region consisting of curvilinear triangles (if $N/U - M_0 < q$, this region consists of only one triangle, i.e. the domain of summation of the sum T). Let

$$s_0 = \left[\frac{NU^{-1} - M_0}{q}\right].$$

Then, to each $s = 0, \ldots, s_0$, there corresponds a definite curvilinear triangle bounded by the curve $m = N/x$, the line $m = M_0 + sq$, and the line $x = N/(M_0 + (s+1)q)$ if $s < s_0$, or the line $x = U$, if $s = s_0$.

The base Δ of the triangle satisfies the inequality:

$$\Delta \leqq \frac{N}{M_0 + sq} - \frac{N}{M_0 + (s+1)q} \ll \frac{Nq}{M_0^2} \ll \frac{qU^2}{N}.$$

Therefore, by Lemma 8a, that part $T'_{0,s}$ of the sum T'_0 which is taken over the s-th triangle is

$$\ll \frac{N^{1+\varepsilon}\Delta^2}{U^2}\sqrt{\frac{1}{\Delta} + \frac{U^2}{N\Delta} + \frac{1}{q} + \frac{qU^2}{N\Delta^2}}$$

$$\ll \frac{N^{1+\varepsilon}\Delta}{U}\left(\frac{\sqrt{\Delta}}{U} + \sqrt{\frac{\Delta}{N}} + \sqrt{\frac{1}{q} + \frac{q}{N}}\right).$$

And since

$$\sqrt{\frac{\Delta}{N}} \ll \frac{\sqrt{\Delta}}{U} \ll \sqrt{\frac{q}{N}},$$

we have

$$T'_{0,s} \ll \frac{N^{1+\varepsilon}\Delta}{U}\sqrt{\frac{1}{q} + \frac{q}{N}}, \quad T'_0 \ll N^{1+\varepsilon}\sqrt{\frac{1}{q} + \frac{q}{N}}.$$

Hence, using the facts already proved above, we find

$$T_0(U) \ll N^{1+\varepsilon} F.$$

But as the sum $T''(U)$ can be expressed as the difference of two sums of the type $T_0(U)$, we have

$$T''(U) \ll N^{1+\varepsilon_1} F.$$

From the inequalities established for $T'(U)$ and $T''(U)$, it follows that

$$T(U) \ll N^{1+\varepsilon_1} F,$$

hence, finally, we obtain

$$T \ll N^{1+\varepsilon_2} F.$$

which completes the proof of our lemma.

Theorem. Let p run through primes, and

$$S = \sum_{p \leqq N} \psi\left(\frac{ap}{q}\right).$$

Then

$$S \ll N^{1+\varepsilon} F_3; \quad F_3 = \sqrt{\frac{1}{q} + \frac{q}{N}} + N^{-1/6}.$$

Proof. This theorem follows from the generalization of Lemma 5 (Chapter 4) and also from Lemmas 8 and 9 of this chapter.

References*

1*. Vinogradov, I.M.: Fundamentals of Number Theory. Nauka, 1971
2*. Vinogradov, I.M.: The Method of Trigonometric Sums in Number Theory. 1971
3. Voronoi, G.: Sur un problème du calcul des fonctions asymptotiques. J. für Math., *126*, 241–282, 1903 (in French)
4*. Vinogradov, I.M.: A new method of deriving asymptotic expressions of arithmetic functions. Izv. Rossiiskoi Akademii Nauk, ser. 6, *II*, 1347–1378, 1917
5*. Voronoi, G.: On a transcendental function and its applications in summation of certain sequences. Annales scientifiques de l'Ecole normale superieure (3), *XXI*, 207–267, 1904
6. Hardy, G.H. and J.E. Littlewood: The trigonometrical series associated with the elliptic ϑ function. Acta Math., *37*, 193–239, 1914 (in English)
7*. Vinogradov, I.M.: On the mean value of the number of classes of proper primitive forms of a negative discriminant. Soobshcheniya Khar'k. Matem. obshestva, 1918 (preprints dated 1917 came out of press the same year).
8. van der Corput, J.G.: Verschärfung der Abschätzungen beim Teileproblem. Math. Ann., *87*, 39–65, 1922 (in German)
9. Hua, L.K.: The lattice points in a circle. Quart. J. Math., Oxford, *13*, 18–29, 1942 (in English)
10*. Vinogradov, I.M.: On the distribution of fractional parts of the values of a function of two variables. Izv. Leningradskogo Politekh. Instituta, *30*, 31–52, 1927
11*. Vinogradov, I.M.: The number of integral points in a sphere. Trudy Matem. Instituta im. V.A. Steklova, AN SSSR, *9*, 17–38, 1935
12*. Vinogradov, I.M.: The number of integral points in a sphere. Izv. AN SSSR, *27*, 957–968, 1963
13. Hardy, G.H., and J.E. Littlewood: The approximate functional equation in the theory of the zeta-function with applications to the divisor problems of Dirichlet and Piltz. Proc. London Math. Soc. (2) *21*, 39–74, 1922 (in English)
14*. Karatsuba, A.A.: Estimates of trigonometric sums obtained by the Vinogradov method and their applications. Trudy ordena Lenina Matematicheskogo Instituta im. V.A. Steklova, *CXII*, Nauka, 1971, 241–255
15*. Vinogradov, I.M.: A new estimate of $G(n)$ in Waring's problem. Doklady AN SSSR, *5*, 249–253, 1934
16*. Vinogradov, I.M.: The method of trigonometric sums in number theory. Trudy Matem. Instituta im. V.A. Steklova, *23*, 1–109, 1947
17*. Vinogradov, I.M.: On the upper bound of $G(n)$. Izv. AN SSSR, *29*, 637–642, 1959
18*. Vinogradov, I.M.: New results on the distribution of fractional parts of a polynomial. Doklady AN SSSR, *2*, 355–357, 1936

* All asterisked items were originally published in Russian.

19*. Vinogradov, I.M.: Representation of an odd number as the sum of three primes. Doklady AN SSSR, *15*, 291–294, 1937
20*. Vinogradov, I.M.: Estimates of certain simple trigonometric sums with prime numbers. Izv. AN SSSR, *3*, 391–398, 1938
21*. Vinogradov, I.M.: Certain general properties of distribution of prime numbers. Matem. Sbornik, *7*, 365–372, 1940
22*. Vinogradov, I.M.: An improvement of the method for estimating sums over primes. Izv. AN SSSR, *7*, 17–34, 1943
23*. Vinogradov, I.M.: Estimate of a sum taken over primes in an arithmetic progression. Izv. AN SSSR, *30*, 481–496, 1966
24*. Karatsuba, A.A.: Sums of characters with primes. Izv. AN SSSR, *34*, 299–321, 1970
25. Page, A.: On the number of primes in arithmetic progression. Proc. London Math. Soc., (2) *39*, 116–141, 1935 (in English)
26*. Mardzhanishvili, K.K.: On the proof of the Goldbach-Vinogradov theorem. Doklady AN SSSR, *30*, 681–684, 1941
27*. Vinogradov, I.M.: An elementary proof for a general assertion in analytical number theory. Izv. Leningradskogo Politekh. Instituta, *29*, 3–12, 1925
28*. Vinogradov, I.M.: An elementary proof of a theorem in the theory of primes. Izv. AN SSSR, *17*, 3–12, 1953

Ivan Matveevič Vinogradov: A Brief Outline of His Life and Works

By K.K. Mardzhanishvili *

Ivan Matveevič Vinogradov was born on September 14, 1891, in the village of Milolyub in the Vilikoluki district of the Pskov province. His father was a village priest and his mother, a school teacher. He completed his secondary education in the city of Velikiye Luki. Then he entered the Faculty of Mathematics and Physics at St. Petersburg University. While a student of mathematics, his main interest lay in the field of number theory.

Ivan Vinogradov graduated from university in 1914, and in recognition of his work on the distribution of quadratic residues and nonresidues, he was permitted to stay on for a post-graduate course. At the recommendation of V.A. Steklov, he was awarded a scholarship in 1915.

While studying for a master's degree, which he obtained successfully, Vinogradov began to solve the most difficult problems in the theory of numbers. Even in 1914–1918, he wrote papers that are in no way inferior to the works of the then eminent specialists in number theory. Unfortunately, far from all his early works were published at that time. Moreover, communication between Soviet and foreign scholars was interrupted in the early post-revolution years: Soviet readers could not receive foreign journals, while the advances made by Soviet mathematicians were unknown abroad. The works of Weyl on the estimates of trigonometric sums, the works of Hardy and Littlewood on the Waring problem and many other important advances were not known in the Soviet Union, whereas Vinogradov's early works remained unknown abroad. But, some of his papers reached the West in a roundabout way: this was the case with paper [2] in the present collection of Vinogradov's works which was published by the Kharkov Mathematical Society in 1918. Preprints were produced in 1917 and it was later found that it had been sent by Ya.V. Uspensky to Professor E. Landau in Göttingen, as well as elsewhere abroad. In 1918–1920 Ivan Vinogradov worked at Perm State University, first as senior lecturer and then as professor. He had returned to Petrograd by the end of 1920 as he was appointed professor at the Polytechnical Institute and somewhat later professor at Petrograd University. At the Polytechnical Institute he gave an original course in higher mathematics and at Petrograd University, a course in number theory, which formed the basis of this widely known textbook *Elements of Number Theory*. This book, translated in many countries, is concise, yet it acquaints the reader with the elements of number theory beginning from the basic fundamentals, the definition of divisibility of an integer by another

* Born 26th August 1903, died 13th February 1981

integer; it also gives exercises (and a key to them) that gradually take the reader up to the most complicated current problems in the field.

Side by side with his lecturing activities, Ivan Vinogradov was also engaged in intensive research work. In particular, he developed methods yielding a solution to new additive problems in number theory, giving estimates of trigonometric sums more general than those of Weyl and which are useful in solving certain other prolems.

Gradually Vinogradov's works began to be known abroad. Thus, in 1927 a chapter entitled *Vinogradov's Method* appeared in the wellknown *Lectures on Number Theory* by Landau. The results obtained by Vinogradov were evaluated highly.

Ivan Vinogradov was elected an academician of the U.S.S.R. Academy of Sciences in January 1929; from that moment on he began his activities as a science organizer, which he performed alongside his research activities. Jointly with S.I. Vavilov, he elaborated a plan to completely reorganize the Physics and Mathematics Institute of the U.S.S.R. Academy of Sciences; thereafter he became the director of the Mathematical Division of the Institute. In 1934, the Physics and Mathematics Institute was split into two independent institutions: the Steklov Institute of Mathematics and the Lebedev Institute of Physics. Vinogradov was appointed the director of the Institute of Mathematics, which post he holds to this day.

The first scientific work by Vinogradov (begun in 1915) deals with the distribution of quadratic residues and non-residues in a small (as compared with the modulus) interval of natural numbers. The problem consists in the following. With respect to a given prime, called the modulus, all the natural numbers that are not multiples of the modulus are divided into two classes: Class I contains those numbers that, when divided by the modulus, leave the same remainder as the squares of integers, whereas Class II includes all other natural numbers. The numbers belonging to Class I are called quadratic residues, while those belonging to Class II, are quadratic non-residues. The classical Euler-Gauss theorem asserts that quadratic residues and non-residues occur in equal numbers in a complete system of residues. Vinogradov has shown that an asymptotically equal number of quadratic residues and non-residues are contained in any interval of natural numbers of a length slightly greater than the square root of the modulus (greater than, say, the square root of the modulus multiplied by the square of the logarithm of the modulus). Thus, Vinogradov's well known hypothesis came into being on the least quadratic nonresidue, which so far remains unproved (the least quadratic non-residue does not exceed a fixed power, as small as we please, of the modulus). The estimate of the Legendre symbol, i.e. the inequality

$$\left|\sum_{n=1}^{N}\left(\frac{n}{p}\right)\right| < \sqrt{p}\ln p,$$

where (n/p) is the Legendre symbol, derived in this paper, and its generalizations to sums of characters are known as "Vinogradov's estimate for the sum of characters", and are so far used in different areas of number theory (it is for this paper that Vinogradov was able to stay on as a post-graduate student at university).

The next series of Vinogradov's works relates to the asymptotic behaviour of the number of integral points in plane domains. This problem dates back to Gauss and Dirichlet who studied the number of lattice points in a circle (Gauss) and under a hyperbola (Dirichlet). They derived asymptotic expressions for the number of integral points in the form of a sum of two terms: a principal term and a remainder which does not exceed the square root of the principal term in order of magnitude. Thus, the question arose of how to find the correct order of the remainder term in these problems (called "the Gauss problem of lattice points in a circle" and "the Dirichlet divisor problem"). In 1903, G. Voronoi worked out a method of proving that the remainder term in Dirichlet's asymptotic expression does not exceed the cube root of the principal term in order of magnitude (Serpinsky applied the Voronoi method to the Gauss problem and obtained a similar result).

Ivan Vinogradov investigated a much more general problem, viz. that of finding an asymptotic expression for the number of integral points in an arbitrary two-dimensional domain (1917). He developed a new arithmetic method to prove the theorem on the number of integral points in plane domains formed from a finite number of curvilinear trapezia: $x = a$, $x = b$, $y = 0$ and $y = f(x)$, where $f(x)$ obeys certain conditions regarding the order of growth of $f''(x)$ in the interval $a \leqslant x \leqslant b$ (circles and parabolas, in particular, are examples of this type of domain). This theorem is formulated as follows: Let the function $f(x)$ be twice continuously differentiable, and, in the interval $[a, b]$, let

$$\frac{1}{A} \ll f''(x) \ll \frac{1}{A}, \quad A \gg 1 .$$

Then the number of integral points in a plane domain bounded by the lines $x = a$, $x = b$, $y = 0$ and the curve $y = f(x)$ differs from the area of this domain by a quantity not greater than R, where

$$R = O\left[\left(\frac{(b-a)}{A} + 1\right)(A \ln A)^{2/3}\right].$$

Later, V. Yarnik demonstrated that, for the above domains, the remainder in this formula cannot be further improved to an accuracy better than the logarithmic factor.

From 1924 onwards Ivan Vinogradov undertook the study of additive problems in number theory. The first main result that he obtained in 1927 is a fresh proof of the Waring problem. This proof was instrumental in creating a powerful tool of the modern theory of numbers, the method of trigonometric sums. In 1770, Waring put forward a hypothesis generalizing the Lagrange theorem on the representation of a natural number in the form of a sum of the squares of four integers. This hypothesis reads as follows: For any fixed integer $n \geqq 3$, every natural number can be represented as the sum of the n-th powers of a fixed number of natural numbers. The general solution to the Waring problem was given by Hilbert in 1909. The Hilbert method lacked generality, and was complicated and thus did not find further application. In 1922, Hardy and Littlewood found a new method (the so-called circle method of Hardy, Littlewood and Ramanujan) of solving a wider class of additive problems, with the help of which they provided a new proof of the Waring problem, in particular. It is indeed an extension of the method of the generating functions used by Euler to solve linear additive problems and is essentially based on the theory of functions of a complex variable (infinite series in a complex domain are taken as the generating functions and the Cauchy integral formula is applied). In 1926, Ivan Vinogradov applied finite trigonometric sums to prove the Waring problem: this not only resulted in the discovery of a new but sufficiently simple proof, but also in opening up a broad vista for solving complicated new problems in number theory (say, the representation of natural numbers as the sum of the values of a positive integral-valued polynomial). Integers occur periodically among real numbers. The simplest trigonometric functions, sine and cosine, being periodic functions of period 2π, provide a means of analytically separating integers from the set of real numbers. Trigonometric sums are finite sums of sines and cosines, whose arguments are real integral functions. With the help of the integrals of such sums, we can write in a simple form expressions which give the number of solutions of any equation in integers. For instance, from a simple equality

$$\int_0^1 e^{2\pi i\alpha n}\,d\alpha = \begin{cases} 1, & \text{if } n = 0, \\ 0, & \text{if } n \neq 0 \text{ is an integer}, \end{cases}$$

it follows that the number of representations of an integer N in the form of a sum of the type

$$N = x_1^n + x_2^n + \ldots + x_k^n,$$

where $x_1, x_2, \ldots, x_k$ are natural numbers, is equal to the integral

$$J = \int_0^1 S^k(\alpha)\, e^{-2\pi i\alpha N}\, d\alpha, \quad S(\alpha) = \sum_{1 \leqq x \leqq N^{1/n}} e^{2\pi i\alpha x^n}.$$

The sums $S(\alpha)$ (as well as their generalizations) are called trigonometric sums.

The problem regarding the distribution of the fractional parts of the values of a function of an integral argument is also formulated in a simple fashion in terms of trigonometric sums: for this purpose, it suffices to expand the characteristic function of the interval as a Fourier series. The problem of the number of integral points in two- and three-dimensional domains can also be reduced to trigonometric sums: for this purpose, it is sufficient to expand the "fractional part of x" as a Fourier series. Thus, a wide class of diverse problems in number theory can be formulated uniformly in terms of trigonometric sums.

The general procedure by which these problems in number theory are investigated with the help of the Vinogradov method of trigonometric sums is as follows: the number of solutions of an equation, or the number of fractional parts of a function that lie in a given interval, or the number of integral points in a given domain is expressed as an integral of trigonometric sums or in the form of a series whose coefficients are trigonometric sums. The exact expression is represented in the form of a sum of two terms: a principal term and an auxiliary term (for example, in the case of the Fourier series of the characteristic function of an interval, the principal term is derived from the zero-th coefficient term in the Fourier expansion). The principal term approaches the leading term in the asymptotic expression, while the auxiliary term is the remainder. In such additive problems as those of Waring, Goldbach and others, the principal term is investigated with the help of a method similar to the circle method of Hardy, Littlewood and Ramanujan (at present, this method is known as "the circle method of Hardy, Littlewood and Ramanujan in the form of Vinogradov's trigonometric sums"). In most other problems (the distribution of fractional parts, integral points in domains, etc.) the principal term is derived trivially. Now the problem is that of estimating the remainder term: if we could succeed in demonstrating that the remainder is a quantity of an order less than that of the principal term, then we would prove the validity of the asymptotic formula. The central problem in estimating the remainder lies in obtaining estimates for trigonometric sums with the best possible accuracy.

Using the Weyl method, Vinogradov estimated the trigonometric sums encountered in the proof of the Waring problem (1926). While continuing the study of the Waring problem, in 1934 he discovered a new method for estimating trigonometric sums which is incomparably more accurate than the Weyl method. Applying this new method, he obtained much better results in the problem of the distribution of the fractional parts of the values of a polynomial, in the Waring problem itself, in approximating a real number by the fractional parts of a polynomial and in many other problems. Besides these questions, the Vinogradov method has found fruitful application in the theory of zeta functions (N.G. Chudakov), in the Hilbert-Kamke problem (K.K. Mardzhanishvili) and in various other mixed additive problems.

First we shall dwell on the results obtained by Ivan Vinogradov in the Waring problem. In studying the Waring problem, Hardy and Littlewood introduced the function $G(n)$, the minimum number of terms sufficient to represent any large number in the form of a sum of the n-th powers of integers. They found an upper bound of the order of $n2^n$ for $G(n)$. They also derived an asymptotic expression for the number of representations of a natural number in the form of a sum of the n-th powers of integers, the number of terms being of the order of $n2^n$.

Using his 1924 method, Vinogradov derived for $G(n)$ an upper bound of the order of $n \ln n$. This bound, as $G(n)$ is obviously greater than n, cannot be further improved.

Slightly later, he showed that the asymptotic formula established by Hardy and Littlewood in the Waring problem holds valid for a number of terms of the order of $n^2 \ln n$. In this period Vinogradov essentially strengthed some of the previously known results such as the estimate of the Weyl sums and, as a consequence, the distribution of the fractional parts of a polynomial, approximation of a given fraction by the fractional parts of a polynomial and many other questions. For instance, in estimating trigonometric sums of the type:

$$S = S(\alpha_n, \ldots, \alpha_1) = \sum_{x=1}^{P} e^{2\pi i f(x)}, \quad f(x) = \alpha_n x^n + \ldots + \alpha_1 x,$$

which he called Weyl sums, he obtained the following result: if $\alpha_r = \frac{a}{q} + \frac{\theta}{q^2}$, for some r, where $2 \leqq r \leqq n$ and $P \leqq q \leqq p^{r-1}$, then

$$S \ll P^{1-\varrho}, \quad \varrho = \frac{1}{5n^2 \ln n}.$$

The estimate known till then and derived by the Weyl method was a much poorer one with $\varrho = 2^{-n+1}$.

Vinogradov's mean value theorem forms the basis of his method of estimating the Weyl sums. The trigonometric sums $S = S(\alpha_n, \ldots, \alpha_1)$, as functions of $\alpha_n, \ldots, \alpha_1$, are periodic in each argument with the same period equal to 1. The integral

$$J_b = \int_0^1 \ldots \int_0^1 |S(\alpha_n, \ldots, \alpha_1)|^{2b} \, d\alpha_n \ldots d\alpha_1,$$

known today as the Vinogradov integral, is the mean value of the $2b$-th power of the absolute value of the sum S. The mean value theorem is formulated as follows: Let n be an integer greater than 2, $\nu = n^{-1}$, $f(x) = \alpha_n x^n + \ldots + \alpha_1 x$, $P \geqq 1$, l be a positive integer, $b \geqq nl$,

$$J_b = J_b(P) = \int_0^1 \ldots \int_0^1 \left| \sum_{0<x\leqq P} e^{2\pi i f(x)} \right|^{2b} d\alpha_n \ldots d\alpha_1 .$$

Then

$$J_b = J_b(P) < D_l P^{2b - \Delta(l)},$$

where

$$D_l = (nl)^{3nl} (2n)^{3n(n+1)l/2}, \quad \Delta(l) = \frac{n(n+1)}{2} (1 - (1 - \nu)^l).$$

For $b \geqq b_0 = [n^2 (2 \ln n + \ln \ln n + 2.6)]$, from this theorem we obtain the following unimprovable estimate for J_b:

$$J_b \ll P^{2b - \frac{n^2+n}{2}},$$

which plays the key role in estimating Weyl sums, in the Hilbert-Kamke problem, in the derivation of asymptotic expressions in the Waring problem and in several other problems of number theory. The mean value theorem in itself allows the Weyl sums to be estimated even when the degree n of the polynomial may grow as the principal parameter P increases. Such estimates are used in the problem of distribution of primes, in the multidimensional Dirichlet divisor problem and in many other questions.

A direct corollary of the Vinogradov mean value theorem is the following assertion: Let k be a constant $\geqq 8$; from the n-dimensional domain Π_n of volume 1 of the points $(\alpha_n, \ldots, \alpha_1)$, where $0 \leqq \alpha_n < 1, \ldots, 0 \leqq \alpha_1 < 1$, it is possible to choose such a small region of volume V bounded by the inequality:

$$V < c(n,k) P^{\frac{-n(n+1)}{2} g_k}, \quad g_k = 1 - \frac{4 \ln k - 1}{k}$$

($c(n,k)$ is a constant that depends solely on n and k), that for the points in the remaining part of the domain Π_n, we have

$$\left| \sum_{x \leqq P} e^{2\pi i (\alpha_n x^n + \ldots + \alpha_1 x)} \right| \leqq P^{1 - 1/k}.$$

Under these conditions, we can easily prove that it is not possible to improve the bound of V by another bound $V < c'(n,k) P^{-n(n+1)g'/2}$ of the same type, where $g' = g(n,k) > 1$. Thus, the exact upper bound holds valid for all Weyl sums, except in a negligibly small number of cases.

J. van der Corput, N.G. Chudakov, Hua Loo-Keng, Yu.V. Linnik, K.K. Mardzhanishvili, A.A. Karatsuba, and many others have taken an active part in the development and application of Vinogradov's method.

In 1937, Ivan Vinogradov elaborated a new method of estimating trigonometric sums over primes, i.e. trigonometric sums in which summation is taken over prime numbers.

Using the estimate obtained in this method and the method described above for solving additive problems, he derived an asymptotic expression for the number of representations of an odd number in the form of the sum of three primes. From this expression, it follows that any sufficiently large odd number is the sum of three primes. Thus, the Goldbach ternary representation problem, which defied solution for almost two centuries, was solved (this problem arose in the letters between Goldbach and Euler). This eminent theorem of Vinogradov on three primes is formulated as follows: The number $I(N)$ of representations of an odd number N in the form of a sum of three primes is given by the following asymptotic expression:

$$I(N) = \frac{N^2}{2(\ln N)^3} \cdot S(N) + O\left(\frac{N^2}{(\ln N)^{3.5-\varepsilon}}\right),$$

$$S(N) = \prod_p \left(1 + \frac{1}{(p-1)^3}\right) \prod_p{}' \left(1 - \frac{1}{p^2 - 3p + 3}\right) > 1,$$

where the product $\prod_p$ is taken over all primes and the product $\prod_p{}'$, only over prime divisors of N.

The Vinogradov method of estimating trigonometric sums over primes opened up a broad vista for solving several new problems, for which there was no approach to the solution till then (one such problem is the Waring problem in primes).

In subsequent years Ivan Vinogradov made several improvements in his method of trigonometric sums. Thus, in 1953 he obtained an estimate for the sum of non-trivial characters in the sequence of shifted primes by the method of trigonometric sums. This estimate cannot be derived even from any hypothesis on the distribution of primes, from the Riemann extended hypothesis, in particular. This theorem is formulated as follows: Let χ be a non-trivial character modulo Q, where Q is a prime, $(k, Q) = 1$, then

$$\sum_{p \leqq N} \chi(p+k) \ll N^{1+\varepsilon}(Q^{1/4} N^{-1/3} + N^{-1/10}).$$

From this estimate we obtain the following arithmetic result: the number of quadratic residues and non-residues modulo Q of the type $p + k$, where $p \leqq x$, is equal to

$$\frac{\pi(x)}{2}(1 + o(1)),$$

provided $x > Q^{0.75+\varepsilon}$. The extended Riemann hypothesis on the zeroes of L-series gives this result only for $x > Q^{1+\varepsilon}$.

In 1957, using his method of estimating Weyl's sums, Vinogradov found a new bound for the zeroes of the Riemann zeta function, and hence, as a consequence, a new remainder term in the asymptotic expression for the distribution of primes not exceeding a given bound:

$$\pi(x) = \int_2^x \frac{du}{\ln u} + O(xe^{-c(\ln x)^{0.6}\lambda(x)}), \quad \lambda(x) = (\ln \ln x)^{-0.2}.$$

In 1963, Vinogradov obtained a radically new remainder term in the asymptotic expression for the number of integral points in a sphere. From 1958 to 1971 he was engaged in proving several new theorems concerning the estimates of Weyl sums which can be used in judging the magnitude of their moduli for any possible values of the coefficients of the polynomial in the exponent (these results are largely presented in his monograph *The Method of Trigonometric Sums in Number Theory* (1971) and in its second edition (1980)).

His monograph *Special Variants of the Method of Trigonometric Sums* was published in 1976. It presents the main results that were not included in his first monograph. It traces the history of the development of the Vinogradov method of trigonometric sums: solutions of particular problems in number theory gave rise to new ideas and fresh problems which, in turn, while being a tool in solving difficult and general problems, were instrumental in creating the powerful technique of the analytical theory of numbers. In addition, this monograph gives a detailed account of Vinogradov's works relating to the problem of integral points in two and three dimensional domains; it gives proof of the theorem that the number T of integral points in a sphere $x^2 + y^2 + z^2 \leqq a^2$ is

$$T = \tfrac{4}{3}\pi a^3 + O(a^{4/3}(\ln a)^6)$$

and also of the theorem on the estimate of the function $G(n)$:

$$G(n) < n(2\ln n + 4\ln\ln n + 2\ln\ln\ln n + 13);$$

as well as of the theorem on the estimates of the simplest sums over primes, in particular, of the sums:

$$\sum_{p \leqq N} e^{2\pi i k f\sqrt{p}} \quad \text{and} \quad \sum_{p \leqq N} \chi(p+k).$$

Applications of these estimates in arithmetical problems are illustrated; an elementary method is given to estimate trigonometric sums over primes.

Vinogradov's method finds important applications in various fields of mathematics: in mathematical analysis, in approximate calculations, in the theory of probability and in discrete mathematics.

A Chronological List of I. M. Vinogradov's Works*

1917 * A new method of deriving asymptotic expressions for arithmetic functions. Izv. Ross. Akad. Nauk, Ser. 6, 16 (1917) 1347–1378 (*see pp. 3–27 within these selecta*)

1918 * On the mean value of the number of classes of proper primitive forms of negative discriminant. Soobshch. Khar'k. Matem. Ob-va, *16* no. 1–2 (1918) 10–38 (*see pp. 28–52 within these selecta*)

* On an asymptotic equality in the theory of quadratic forms. Zh. Fiz. Mat. Obshchestva pri Perm. Universitete, *1* (1918) 18–28 (abstract in French)

On the distribution of power residues and non-residues. Zh. Fiz. Mat. Obshchestva pri Perm. Universitet, *1* (1918) 94–98 (*see pp. 53–56 within these selecta*)

1919 * On the distribution of quadratic residues and non-residues. Zh. Fiz. Mat. Obshchestva pri Perm. Universitete, *2* (1919) 1–16 (abstract in French)

1921 * On the asymptotic equalities in number theory. Izv. Ross. Akad. Nauk, *15* (1921) 158–160

1922–1924 Sur un théorème général de Waring. Mat. Sb., *31* (1922–24) 490–507 (abstract in Russian)

1925 * An elementary proof of a general theorem in analytic number theory. Izv. Akad. Nauk SSSR, Ser. 6, *19* no. 16–17 (1925) 785–796 (abstract in French) (*see pp. 57–64 within these selecta*)

* An elementary proof for a general assertion in the analytic number theory. Izv. Leningrad. Polytekh. Instituta, *29* (1925) 3–12 (abstract in French)

1926 * On a bound for the least n^{th} power non-residue. Izv. Akad. Nauk SSSR, Ser. 6, *20* no. 1–2 (1926) 47–58 (*see pp. 69–77 within these selecta*)

* On the fractional parts of a polynomial of integral argument. Izv. Akad. Nauk SSSR, *20* no. 9 (1926) 585–600

* On the distribution of indices. Dokl. Akad. Nauk SSSR, Ser. A, (1926) 73–76 (*see pp. 65–68 within these selecta*)

* On the distribution of fractional parts of values of a function of one variable. Zh. Leningrad. Fiz. Mat. Obshchestva, *1* no. 1 (1926) 56–65 (abstract in French) (*see pp. 78–84 within these selecta*)

On a general theorem concerning the distribution of power residues and non-residues. Bull. Am. Math. Soc., *32* no. 6 (1926) 596

1927 * An analytical proof of the theorem on the distribution of fractional parts of a polynomial of integral argument. Izv. Akad. Nauk SSSR, *21* no. 7–8 (1927) 567–578

* On the distribution of fractional parts of values of a function of two variables. Izv. Leningrad. Polytekh. Instituta, *30* (1927) 31–52 (abstract in French) (*see pp. 85–100 within these selecta*)

Démonstration élémentaire d'un théorème de Gauss. Zhur. Leningrad. Fiz. Mat. Obshchestva, *1* no. 2 (1927) 187–193 (abstract in Russian).

---

* All asterisked items are originally published in Russian.

On a general theorem concerning the distribution of power residues and non-residues. Trans. Am. Math. Soc., *29* no. 1 (1927) 209–217

On the bound for the least n^{th} power non-residue. Ibid., 218–226

1928 * On Waring's theorem. Izv. Akad. Nauk SSSR, Otd. Fiz.-Mat. Nauk, no. 4 (1928) 393–400 (*see pp. 101–106 within these selecta*)

* On the representation of a number by means of a polynomial of several variables. Ibid, 401–414

1929 * On a class of systems of Diophantine equations. Izv. Akad. Nauk SSSR, Otd. Fiz.-Mat. Nauk, no. 4 (1929) 355–376

* Mathematics and the needs of national economy. Torg.-prom. gaz., *13/1* no. 11 (1929)

* On the question of organization of election of academicians. Izvestiya, *30/1* no. 24 (1929)

1930 * On the least primitive root. Dokl. Akad. Nauk SSSR, Ser. A, no. 1 (1930) 7–11

1931 * Struggle at the mathematics front (from the declaration of mathematicians of a materialist viewpoint). Stud. Pravda, *9/IV* no. 16 (1931) (in collaboration with others)

1932 * Elements of Higher Mathematics, Vol. I, Analytical Geometry. Leningrad, KUBUCH Pub., (1932), 248 p.

* On the number of integral points in a circle. Izv. Akad. Nauk SSSR, Otd. Mat. Estestv. Nauk, no. 3 (1932) 313–336

1933 * Elements of Higher Mathematics, Vol. II, Differential Calculus. Leningrad, KUBUCH Pub., (1933) 176 p.

* On problems in analytic number theory. In: "Trudy Noyabr'skoi Yubileinoi Sessii Akad. Nauk SSSR". Leningrad. Izd. Akad. Nauk SSSR, (1933) 1–11

* On a trigonometric sum and its application in number theory. Dokl. Akad. Nauk SSSR, no. 5 (1933) 195–204. (abstract in English).

* On some trigonometric sums and their applications. Dokl. Akad. Nauk SSSR, no. 6 (1933) 249–255 (abstract in English)

* Application of finite trigonometric sums to distributions of the fractional parts of a polynomial of integral argument. Tr. Inst. Fiz.-Mat., *4* (1933) 5–8

1934 * On the upper bound for $G(n)$ in Waring's problem. Izv. Akad. Nauk SSSR, Otd. Mat. Estestv. Nauk, no. 10 (1934) 1455–1469 (abstract in English) (*see pp. 110–123 within these selecta*)

* New applications of trigonometric sums. Dokl. Akad. Nauk SSSR, *2* no. 1 (1934) 10–14 (abstract in English)

* New asymptotic expressions. Dokl. Akad. Nauk SSSR, *2* no. 2 (1934) 49–51 (abstract in English).

* Trigonometric sums of a compound modulus. Dokl. Akad. Nauk SSSR, *2* no. 5 (1934) 225–229 (abstract in English)

* New theorems concerning the distribution of quadratic residues. Dokl. Akad. Nauk SSSR, *2* no. 6 (1934) 289–290 (abstract in French)

* New theorems concerning the distribution of primitive roots. Dokl. Akad. Nauk SSSR, *2* no. 7 (1934) 366–369 (abstract in English)

* A new solution to Waring's problem. Dokl. Akad. Nauk SSSR, *3* no. 6 (1934) 337–341 (abstract in English)

* Some new problems in number theory. Dokl. Akad. Nauk SSSR, *4* no. 1 (1934) 1–6 (in Russian and in English)

* Some theorems in analytic number theory. Dokl. Akad. Nauk SSSR, *5* no. 4 (1934) 185–187 (abstract in English)

* A new estimate for $G(n)$ in Waring's problem. Dokl. Akad. Nauk SSSR, *5* no. 5–6 (1934) 249–253 (in English and in Russian) (*see pp. 107–109 within these selecta*)

* Certain theorems concerning the distribution of indices and primitive roots. Tr. Inst. Fiz.-Mat., *5* (1934) 87–93

Sur quelques nouveaux résultats en théorie analytique des nombres. C. R. Acad. Sci. Paris, Ser. I, *199* no. 3 (1934) 174–175

* On the congress of Soviet mathematicians. Leningrad, Pravda, *28/IV* no. 150 (1934) (jointly with B.I. Segal)

1935 * Opening address delivered at the II All-Union Mathematical Congress. In: * Proceedings of the II All-Union Mathematical Congress. Leningrad, June 24–30, 1934, Vol. I, Plenary sessions and Review reports. Leningrad-Moscow, Izd. Akad. Nauk SSSR, (1935) 19–20

* On the approximation by means of rational fractions with a denominator of exact power. Dokl. Akad. Nauk SSSR, *7* no. 1 (1935) 1–5 (abstract in French)

* On some rational approximations. Dokl. Akad. Nauk SSSR, *8* no. 1 (1935) 3–6

* On some rational approximations. C. R. Acad. Sci. URSS, *8* no. 1 (1935) 3–6

* On the fractional terms of polynomials and other functions. Dokl. Akad. Nauk SSSR, *8* no. 3 (1935) 99–100

On fractional terms of polynomials and of other functions. C. R. Acad. Sci. URSS, *8* no. 3 (1935) 99–100

* New estimates for Weyl sums. Dokl. Akad. Nauk SSSR, *8* no. 5 (1935) 195–198 (*see pp. 124–128 within these selecta*)

Nouvelles évaluations des sommes de Weyl. C. R. Acad. Sci. URSS, *8* no. 5 (1935) 195–198

* A new method of proving Waring's theorem. Tr. Mat. Inst., *9* (1935) 5–15

* The number of integral points in a sphere. Ibid., 17–18

On approximation to zero with the help of numbers of certain general form. Mat. Sb., *42* no. 2 (1935) 149–156 (abstract in Russian)

On Weyl sums. Mat. Sb., *42* no. 5 (1935) 521–530 (abstract in Russian)

An asymptotic formula for the number of representations in Waring's problem. Ibid., 531–534 (abstract in Russian)

Une nouvelle variante de la démonstration du théorème de Waring. C. R. Acad. Sci. Paris, Ser. I, *200* no. 3 (1935) 182–184

Sur les sommes de M. H. Weyl. C. R. Acad. Sci. Paris, Ser. I, *201* no. 13 (1935) 514–516

On Waring's problem. Ann. Math., *36* no. 2 (1935) 395–405

1936 * Elements of Number Theory (College textbook). Moscow-Leningrad, Ob "edinennoe Nauchno-Tekhnicheskoe Izdatel'stvo, (1936) 96 p.

* A new improvement of the estimation of trigonometric sums. Dokl. Akad. Nauk SSSR, *1* no. 10 (1936) 195–196

A new improvement of the estimation of trigonometric sums. C. R. Acad. Sci. URSS, *1* no. 10 (1936) 199–200

* New results concerning the distribution of the fractional parts of a polynomial. Dokl. Akad. Nauk SSSR, *11* no. 9 (1936) 355–357

New results concerning the distribution of fractional parts of a polynomial. C. R. Acad. Sci. URSS, *11* no. 9 (1936) 361–364

On the number of fractional parts of a polynomial lying in a given interval. Mat. Sb. Nov. ser., *1* no. 1 (1936) 3–8 (abstract in Russian)

A new method of resolving of certain general questions of number theory. Ibid., 9–20 (abstract in Russian)

Approximation by means of the fractional parts of a polynomial. Ibid., 21–27 (abstract in Russian)

On asymptotic formula in Waring's problem. Mat. Sb., Nov. Ser., *1* no. 2 (1936) 169–174 (abstract in Russian)

A new method of estimation of trigonometric sums. Ibid., 175–188 (abstract in Russian)

Supplement to the paper "On the number of fractional parts of a polynomial lying in a given interval". Mat. Sb., Nov. Ser., *1* no. 3 (1936) 405–408 (abstract in Russian)

Sur les nouveaux résultats de la théorie analytique des nombres. C. R. Acad. Sci. Paris, Ser. I, *202* no. 3 (1936) 179–180

---

* From 1933–1948 the Doklady were published both in Russian and in a foreign language edition entitled Comptes Rendus de l'Académie de l'URSS.

Sur quelques inégalités nouvelles de la théorie des nombres. C. R. Acad. Sci. Paris, Ser. I, *202* no. 16 (1936) 1361–1362

Approximations with the help of certain fractions. Ann. Math., *37* no. 1 (1936) 101–106

On fractional parts of certain functions. Ann. Math., *37* no. 2 (1936) 448–455

1937 * A new method in analytic number theory. Leningrad-Moscow, Tr. Mat. Inst., *10* (1937) 122 p.

* Distribution of the fractional parts of the values of a polynomial when the argument runs through primes in an arithmetic progression. Izv. Akad. Nauk SSSR, Ser. Mat., no. 4 (1937) 505–514 (abstract in English)

* Representation of an odd number as the sum of three primes. Dokl. Akad. Nauk SSSR, *15* no. 6–7 (1937) 291–294 (*see pp. 129–132 within these selecta*)

Representation of an odd number as the sum of three primes. C. R. Acad. Sci. URSS, *15* no. 6–7 (1937) 291–294

* Some general theorems relating to the theory of primes. Trudy Tbilisskogo Mat. Inst. *3* (1937) 1–37

* Some new problems in the theory of primes. Dokl. Akad. Nauk SSSR, *16* no. 3 (1937) 139–141

Some new problems in the theory of primes. C. R. Acad. Sci. URSS, *16* no. 3 (1937) 131–132

* New estimates of trigonometric sums containing primes. Dokl. Akad. Nauk SSSR, *17* no. 4 (1937) 165–166

New estimates of trigonometric sums containing primes. C. R. Acad. Sci. URSS, *17* no. 4 (1937) 165–166

Some theorems concerning the theory of primes. Mat. Sb. Nov. ser., *2* no. 2 (1937) 179–195 (abstract in Russian)

A new estimate of a sum containing primes. Mat. Sb. Nov. Ser., *2* no. 5 (1937) 783–792 (abstract in English)

1938 * Elements of Number Theory (College textbook). 2-nd Edition, revised. Moscow-Leningrad, Ob "edinennoe Nauchno-Tekhnicheskoe Izdatel'stvo, (1938) 88 p.

* A new estimate of a trigonometric sum containing primes. Izv. Akad. Nauk SSSR, Ser. Mat., *2* no. 1 (1938) 3–14 (abstract in English)

* A refinement of the estimate of a trigonometric sum containing primes. Ibid., 15–24 (abstract in English)

* Estimates of certain simple trigonometric sums with prime numbers. Izv. AN SSSR, *3* (1938) 391–398

* Estimates of certain sums containing primes. Izv. Akad. Nauk SSSR, Ser. Math., *2* no. 4 (1938) 399–416 (abstract in English)

* Estimates of trigonometric sums. Izv. Akad. Nauk SSSR, Ser. Mat., *2* no. 5–6 (1938) 505–524 (abstract in English)

* Some new estimates in analytic number theory. Dokl. Akad. Nauk SSSR, *19* no. 5 (1938) 339–340

Some new estimations of analytic number theory. C. R. Acad. Sci. URSS, *19* no. 5 (1938) 339–340

* Sergei L'vovich Sobolev. Vestn. Akad. Nauk SSSR, no. 11–12 (1938) 35–37

* Distribution of quadratic residues and non-residues of the type $(p + k)$ with respect to prime modulus. Mat. Sb. Nov. Ser., *3* no. 2 (1938) 311–319 (abstract in English)

* Certain general lemmas and their application in the estimation of trigonometric sums. Mat. Sb. Nov. Ser., *3* no. 3 (1938) 435–471 (abstract in English)

* Certain general theorems relating to the theory of primes. Tr. Tbilis. Mat. Inst., *3* (1938) 1–67 (in Russian and in German)

* Two theorems in analytic number theory. Tr. Tbilis. Mat. Inst., *5* (1938) 153–180 (in Russian and in German)

1939 * Elementary estimates of a trigonometric sum with primes. Izv. Akad. Nauk SSSR, Ser. Mat., *3* no. 2 (1939) 111–122 (abstract in English)

* Estimates of certain simple trigonometric sums with prime numbers. Izv. Akad. Nauk SSSR, Ser. Mat., *3* no. 4 (1939) 371–398 (abstract in English) (*see pp. 133–159 within these selecta*)

* First meeting to review the research work carried out at the departments of mathematics and mechanics at higher educational institutions (held at the Division of Technical Sciences, The U.S.S.R. Academy of Sciences, April 15–17, 1938. Izv. Akad. Nauk SSSR, Otd. Tekh. Nauk, no. 1 (1939) 128–130 (in collaboration with S.L. Sobolev and V.K. Turkin)

* A new improvement of the method of estimating trigonometric sums with primes. Dokl. Akad. Nauk SSSR, *22* no. 2 (1939) 59–60

A new improvement of the method of estimation of trigonometric sums with primes. C. R. Acad. Sci. URSS, *22* no. 2 (1939) 59–60

* Simplest trigonometric sums with primes. Dokl. Akad. Nauk SSSR, *23* no. 7 (1939) 615–617

Simplest trigonometric sums with primes. C. R. Acad. Sci. URSS, *23* no. 7 (1939) 615–617

1940 * Elements of Number Theory (College textbook). 3-rd Edition, revised. Moscow-Leningrad, Gostekhizdat, (1940) 112 p.

* Distribution of primes in an arithmetic progression with respect to a given modulus. Izv. Akad. Nauk SSSR, Ser. Mat., *4* no. 1 (1940) 27–36 (abstract in English)

* Certain general properties of the distribution of primes. Mat. Sb. Nov. Ser., *7* no. 2 (1940) 365–372 (abstract in English)

1941 * Two theorems relating to the theory of distribution of prime numbers. Dokl. Akad. Nauk SSSR, *30* no. 4 (1941) 285–286

Two theorems relating to the theory of distribution of prime numbers. C. R. Acad. Sci. URSS, *30* no. 4 (1941) 287–288

* Some general properties of the distribution of products of prime numbers. Dokl. Akad. Nauk SSSR, *30* no. 8 (1941) 675–676

Some general properties of the distribution of products of prime numbers. C. R. Acad. Sci. URSS, *30* no. 8 (1941) 681–682

* Two theorems relating to the theory of distribution of primes. In: * "Research work done at the institutes belonging to the Physics and Mathematics Division of the USSR Academy of Sciences in 1940, Abstracts." Moscow-Leningrad, Izd. Akad. Nauk SSSR, (1941) 9–10

* A general property of the distribution of products of primes. Ibid., 10–11

* The award that is especially precious (on the confernment of the USSR State Prize). Pravda, *17/III* no. 75 (1941)

1942 * An improvement of the estimates of trigonometric sums. Izv. Akad. Nauk SSSR, Ser. Mat., *6* no. 1–2 (1942) 33–40 (abstract in English)

* On the estimation of trigonometric sums. Dokl. Akad. Nauk SSSR, *34* no. 7 (1942) 199–200

On the estimation of trigonometric sums. C. R. Acad. Sci. URSS, *34* no. 7 (1942) 182–183

* Improvement of some theorems in the theory of primes. Dokl. Akad. Nauk SSSR, *37* no. 4 (1942) 135–137

Improvement of some theorems in the theory of primes. C. R. Acad. Sci. URSS, *37* no. 4 (1942) 115–117

* Russian mathematics. Slavyane, no. 5–6 (1942) 74–75

1943 * An improvement of the method of estimating sums over primes. Izv. Akad. Nauk SSSR, Ser. Mat., *7* no. 1 (1943) 17–34 (abstract in English)

1944 * Elements of Number Theory (College textbook). 4-th edition, revised and enlarged. Moscow-Leningrad, Gostekhizdat, (1944) 142 p.

* General theorems of the estimations of trigonometric sums. Dokl. Akad. Nauk SSSR, *43* no. 2 (1944) 51–52

General theorems of the estimations of trigonometric sums. C. R. Acad. Sci. URSS, *43* no. 2 (1944) 47–48
* Chebyshev P.L. Complete Collection of Works. Vol. I: Number Theory. Ed. Moscow-Leningrad, Izd. Akad. Nauk SSSR, (1944) 342 (jointly with B.N. Delone)

1945 * Works of P.L. Chebyshev on number theory. In: *Scientific Heritage of P.L. Chebyshev, Part I: Mathematics, Moscow-Leningrad, Izd. Akad. Nauk SSSR (1945) 69–87 (jointly with B.N. Delone).
* Analytic number theory. Izv. Akad. Nauk SSSR, Ser. Mat., *9* no. 3 (1945) 159–168 (abstract in English)
* General theorems relating to the estimate of trigonometric sums. In: *Research work done during 1943–1944. The Physics and Mathematics Division of the U.S.S.R. Academy of Sciences. Abstracts. Moscow-Leningrad, Izd. Akad. Nauk SSSR, (1945) 64
* Works of Russian Mathematicians. Izvestiya, *12/VI* no. 136 (1945)

1946 * A general distribution law for the fractional parts of the values of a polynomial with the variable running over the primes. Doklady AN SSSR, 51 no. 7 (1946) 489–490
A general distribution law for the fractional parts of the values of a polynomial with the variable running over the primes. C. R. Acad. Sci. URSS, *51* no. 7 (1946) 491–492
* A general law of distribution of the fractional parts of the values of a polynomial when its argument runs over primes. In: *Research work done in 1945. The Physics and Mathematics Division of the U.S.S.R. Academy of Sciences, Abstracts. Moscow-Leningrad, Izd. Akad. Nauk SSSR, (1946) 51
* Universities and development of science. (Letter to the editor). Pravda, *6/III* no. 55 (1946) (jointly with I.S. Galkin)
* Ditto. Moscow University, 1946, 2/IV, N 17 (jointly with I.S. Galkin)
* Eminent Soviet Mathematicians (On the election of new academicians). Izvestiya, *28/XI* no. 279 (1946)
* Chebyshev P.L. Complete Collection of Works. Vol. I: Number Theory. (2-nd stereotype edition). Ed. Moscow-Leningrad. Izd. Akad. Nauk SSSR (1946) 342 p. (jointly with B.N. Delone).

1947 * The method of trigonometric sums in number theory. Moscow, 1947, 109 p., Trudy Matem. Instituta, vol. 23
* Additive problems in the theory of primes. In: *Jubilee Number, 30-th Anniversary of the Great October Socialist Revolution. Part I, Moscow-Leningrad, Izd. Akad. Nauk SSSR (1947) 65–79
* Analytic number theory. In: *Jubilee session of the U.S.S.R. Academy of Sciences, June 15–July 3, 1945, Part II, Moscow-Leningrad, Izd. Akad. Nauk SSSR (1947) 34–40 (220-th Anniversary of the U.S.S.R. Academy of Sciences)
* A general law of the theory of primes. Dokl. Akad. Nauk SSSR, *55* no. 6 (1947) 475–476
A general law of the theory of primes. C. R. Acad. Sci. URSS, *55* no. 6 (1947) 471–472
* Address delivered at the session of the USSR Academy of Sciences dedicated to the 30-th Anniversary of the Great October Socialist Revolution. Vestn. Akad. Nauk. SSSR, no. 11 (1947) 74–75 (under the Great October Banner)

1948 * Additive problems in the theory of primes. In: * "General Collection of the U.S.S.R. Academy of Sciences dedicated to the 30-th Anniversary of the Great October Socialist Revolution". Moscow-Leningrad, Izd. Akad. Nauk SSSR (1948) 458–464
* Estimate of trigonometric sums over primes. Izv. Akad. Nauk SSSR, Ser. Mat., *12* no. 3 (1948) 225–248
* The distribution of products of primes and the values of the Möbius function. Izv. Akad. Nauk SSSR, Ser. Mat., *12* no. 4 (1948) 341–350
El mundo de las matematicas. Cult. sov., *9* no. 50 (1948) 4–5
* Outstanding works of our mathematicians. Pravda, *1/VI* no 128 (1948)

1949 * Elements of Number Theory (College textbook). 5-th edition, revised. Moscow-Leningrad, (1949) 180 p.

* Refinement of the remainder term in an asymptotic expression. Izv. Akad. Nauk SSSR, Ser. Mat., *13* no. 2 (1949) 97–110

1950 Matemática sovietica. Bucuresti, Ed. Acad. RPR, (1950) 14 p. (In colaborare cu N.I. Mushelisvili)
* The upper bound of the modulus of trigonometric sums. Izv. Akad. Nauk SSSR, Ser. Mat., *14* no. 3 (1950) 199–214
* A new improvement of the method of estimating double sums. Dokl. Akad. Nauk SSSR, *73* no. 4 (1950) 635–638
* Sophia Kovalevskaya. (Centenary celebrations). Oktyabr', no. 1 (1950) 129–137
* Tables of the values of Bessel functions of imaginary argument. Moscow-Leningrad, Izd. Akad. Nauk SSSR, (1950) 404 p. Ed. (in collaboration with N.G. Chetaev)
* Preface. Ibid., III-IV (jointly with N.G. Chetaev)

1951 * General theorems on the upper bound of the modulus of trigonometric sums. Izv. Akad. Nauk SSSR, Ser. Mat., *15* no. 2 (1951) 109–130
* Arithmetic method as applied to the distribution of numbers with a given property of the index. Izv. Akad. Nauk SSSR, Ser. Mat., *15* no. 4 (1951) 297–308

1952 * Selected works. Moscow, Izd. Akad. Nauk SSSR, (1952) 436 p.
* Elements of Number Theory (College textbook). 6-th edition, corrected, Moscow-Leningrad, Gostekhizdat, (1952) 180 p.
* Elements of Number theory, Kiev, Rad. Shk., (1952) 152 p. (in Ukrainian)
A szàmelmélet alapjai. Budapest, Tankönyvkiadó, (1952) 140 p.
* A new approach to estimating the values of $\chi(p+k)$. Izv. Akad. Nauk SSSR, Ser. Mat., *16* no. 3 (1952) 197–210
* Mathematics in the Union of Soviet Socialist Republics. Priroda, no. 11 (1952) 62–63

1953 * Elements of Number Theory, Shanghai, (1953) 210 p. (in Chinese)
Zaklady theorie cisel. Praha, Gslke akad. ved, (1953) 173 p.
* An elementary proof of a theorem in the theory of primes. Izv. Akad. Nauk SSSR, Ser. Mat., *17* no. 1 (1953) 3–12
* An improved estimate of the sums of $\chi(p+k)$. Izv. Akad. Nauk SSSR, Ser. Mat., *17* no. 4 (1953) 285–290

1954 Elements of number theory. New York, Dover Publ., (1954) 227 p.
Elementy teorii liczb. Warszawa, Panstw. wyd. nauk., (1954) 207 p.
Method of trigonometric sums in the theory of numbers. London-New York, Inter. Sci. Publ., (1954) 180 p.
* Distribution of primes with a given value of Legendre symbol with respect to a prime modulus. Izv. Akad. Nauk SSSR, Ser. Mat., *18* no. 2 (1954) 105–112

1955 The method of trigonometric sums in the theory of numbers. Peking, (1955) 106 p. (Shusyue szin chzhan, vol. 1, no. 1) (in Chinese)
Elemente der Zahlentheorie. Berlin, VEB Dt. Wiss., (1955) 156 p. (Hochschulbücher für Mathematik. Bd. 22)
An introduction to the theory of numbers. London – New York, Pergamon Press, (1955) 155 p.
* An improvement of the asymptotic expressions for the number of integral points in a three-dimensional domain. Izv. Akad. Nauk SSSR, Ser. Mat., *19* no. 1 (1955) 3–10
* Address delivered at the meeting dedicated to the 100-th anniversary of the death of C.F. Gauss (abridged). Vestn. Akad. Nauk SSSR, no. 4 (1955) 105–107
* Chebyshev P.L. Selected Works. Ed., Moscow, Izd. Akad. Nauk SSSR, (1955) 926 p.

1956 Elemente der Zahlentheorie. München, Oldenbourg, 1956, 156 p.
* Problems in analytic number theory. In: * Proceedings of the III-rd All-Union Mathematical Congress. Moscow, June–July, 1956, vol. 2, Abstracts, Moscow, Izd. Akad. Nauk SSSR (1956) 5
* Speech delivered at the meeting of the All-Union Society for Cultural Relationship with Foreign Countries and of the Physics and Mathematics Division of the U.S.S.R. Academy of Sciences held on February 23, 1955 on the occasion of the 100-th

anniversary of C.F. Gauss. In: * Carl Friedrich Gauss. A collection of papers. Moscow, Izd. Akad. Nauk SSSR (1956) 7–10
* Particular cases of trigonometric sums. Izv. Akad. Nauk SSSR, Ser. Matl, *20* no. 3 (1956) 289–302
* A review of the All-Union Mathematical Congress. Vestn. Akad. Nauk SSSR, no. 9 (1956) 3–7
* Recent advances of Soviet mathematicians. On the eve of the III-rd All-Union Mathematical Congress. Priroda, no. 12 (1956) 68–69
* Achievements of Soviet mathematicians. Pravda, *5/IX* no. 249 (1956)

1957 * Trigonometric sums containing the values of a polynomial. Izv. Akad. Nauk SSSR, Ser. Mat., *21* no. 2 (1957) 145–170
* Living traditions of Russian science. On the development of Soviet mathematics. Pravda, *27/X* no. 300 (1957)

1958 * Certain problems in analytic number theory. In: * Proceedings of the III-rd All-Union Mathematical Congress, Moscow, June–July, 1956, Vol. 3, Reviews, Moscow, Izd. Akad. Nauk SSSR (1958) 3–13 (*see pp. 160–171 within these selecta*)
* The secrets of nature. In: *Nauka i Molodezh'. Moscow, Izd. Akad. Nauk SSSR (1958) 150
* An improvement of the method of estimating trigonometric sums. In: Abstracts, Short Communications. International Congress of Mathematicians in Edinburgh. Edinburgh University Press, (1958) 36 p.
* A particular case of the estimate of trigonometric sums over primes. Izv. Akad. Nauk SSSR, Ser. Mat., *22* no. 1 (1958) 3–14
* A new estimate of the function $\zeta(1+it)$. Izv. Akad. Nauk SSSR, Ser. Mat., *22* no. 2 (1958) 161–164
* On a multiple integral. Izv. Akad. Nauk SSSR, Ser. Mat. *22* no. 5 (1958) 577–584
* On the function $\zeta(s)$. Dokl. Akad. Nauk SSSR, *118* no. 4 (1958) 631

1959 * Elements of Number Theory. Tashkent, Uchitel', (1959) 158 p. (in Uzbek)
* Carl Friedrich Gauss. In: *C.F. Gauss. Works on the theory of numbers. Moscow, Izd. Akad. Nauk SSSR (1959) 875–878
* An estimate of a trigonometric sum over primes. Izv. Akad. Nauk SSSR, Ser. Mat., *23* no. 2 (1959) 157–164
On the upper bound of $G(n)$. Izv. Akad. Nauk SSSR, Ser. Mat., *23* no. 5 (1959) 637–642

1960 * Number of integral points in a given domain. Izv. Akad. Nauk SSSR, Ser. Mat., *24* no. 6 (1960) 777–786

1961 Elements of Number Theory. Tokio, (1961) 200 p. (in Japanese)
* On the distribution of the fractional parts of values of a polynomial. Izv. Akad. Nauk SSSR, Ser. Mat., *25* no. 6 (1961) 749–754 (*see pp. 172–177 within these selecta*)

1962 * The distribution of systems of the fractional parts of values of several polynomials. Izv. Akad. Nauk SSSR, Ser. Mat., *26* no. 6 (1962) 793–796
* A new page in the science of mathematics. Econ. Gaz., no. 3 (1962) 13 (jointly with M.M. Postnikov)

1963 * The number of integral points in a three-dimensional domain. Izv. Akad. Nauk SSSR, Ser. Mat., *27* no. 1 (1963) 3–8
* The number of integral points in a sphere. Izv. Akad. Nauk SSSR, Ser. Mat., 27 no. 5 (1963) 957–968

1964 * On estimates for trigonometric sums. Appendix 1. In: *Hua Loo-Keng: Method of trigonometric sums and its application in number theory. Moscow, Mir, (1965) 169–170

1965 * Elements of Number Theory (College textbook). 7-th edition, revised, Moscow, Nauka (1965) 172 p.
* Elements of Number Theory, Tashkent, Uchitel', (1965) 158 p. (in Uzbek)

* On estimates for trigonometric sums. Izv. Akad. Nauk SSSR, Ser. Mat., *29* no. 3 (1965) 493–504

1966 * Estimate of a sum taken over primes in an arithmetic progression. Izv. Akad. Nauk SSSR, Ser. Mat., *30* no. 3 (1966) 481–496

1967 * Discussion of the scientific activities of the Steklov Mathematics Institute (Report read at the meeting of the Presidium of the U.S.S.R. Academy of Sciences in 1967) Vestn. Akad. Nauk SSSR, no. 9 (1967) 9–11

* The state-of-the-art and main development trends of research in analytic number theory (Report read at the sessional meeting of the Division of Mathematics of the U.S.S.R. Academy of Sciences held in February 1966. Abridged). Uspekhi matem. nauk, *22* no. 1 (1967) 3–14 (in collaboration with A.G. Postnikov)

* Their vocation is mathematics. Sov. Rossiya, *10/VI* no. 133 (1967) (about Lenin Prize winners)

1968 A szàmelmélet alapjai, Budapest, Tankönyvkiadó. (1968) 1. 171 p.

* Recent advances in analytic number theory. In: „Trudy Mezhdunarodnogo kongressa matematikov (Moskva, 1966)". Moscow, Mir, (1968) 163–176 (jointly with A.G. Postnikov)

Recent developments in analytic number theory. Trans. Am. Math. Soc., Ser. 2, *70* (1968) 240–254

1969 * Reason should win a victory over madness. (Letter of Soviet scientists pleading for a ban on chemical and bacteriological weapons). Nauka i zhizn', no. 10 (1969) 25; Nov. vremya, no. 21 (1969) 5 (in collaboration with others)

1970 * Mathematics and scientific progress. In: *Lenin i sovremennaya nauka. Kn. 2", Moscow, Nauka, (1970) 77–80

1971 * The method of trigonometric sums in number theory. Moscow, Nauka, (1971) 160 p.

Fundamentos de la teoria de los números. Moscow, Mir, (1971) 207 p.

* Problems in analytic number theory. Report read at the annual meeting of the U.S.S.R. Academy of Sciences held in March 1971. Vestn. Akad. Nauk SSSR, no. 5 (1971) 52–56

1972 * Elements of Number Theory (College textbook). 8-th edition, corrected, Moscow, Nauka, (1972) 168 p.

* Preface in: *"Mezhdunarodnyi kongress matematikov v Nitstse, 1970. Doklady sovetskikh matematikov" (International Congress of Mathematicians, Nice. Reports of Soviet Mathematicians), Moscow, Nauka, (1972) 5–6

1974 * Talks with the correspondent of "Nedeli". Nedelya, no. 52 (1974) 4

1975 Trigonometric sums in number theory. Calcutta, Statist. Publ. Soc., (1975) 152 p.

1976 * Special Variants of the Method of Trigonometric Sums. Moscow, Nauka, (1976) 120 p. (*see pp. 299–383 within these selecta*)

1980 * The method of trigonometric sums in number theory. 2-nd edition, revised and enlarged, Moscow, Nauka, (1980) 144 p. (*see pp. 181–295 within these selecta*)

Printed in the United States
By Bookmasters